BIOTECHNOLOGY OF PENAEID SHRIMPS

(PERSPECTIVES ON PHYSIOLOGY OF GROWTH, REPRODUCTION AND DISEASE THERAPEUTICS)

The main objective of writing the present book is to collect comprehensive information on various aspects of physiology and biotechnology focusing mainly on reproduction, growth, disease control and therapeutics of penaeid shrimps. Generally shrimps are inhabiting in seas and come across different types of climatic conditions during their life cycle. Their life span is very short. In order to augment shrimp production in captive conditions one should understand thoroughly various aspects of their growth, reproduction, immune system and disease physiology and also interaction of hormones with various physiological processes. This kind of knowledge will definitely help in developing innovative technologies for increasing shrimp production under mariculture practices.The present book covers some fundamental aspects and few applied aspects of biotechnology concerning basic genomics and proteomics, reproduction, growth and disease control and also on therapeutics of shrimp. This information will be quite useful not only to the aqua-farmers/mariculture experts of the shrimp industry to augment quality shrimp production in captive condition but also to the faculties and students working in different organizations involved in teaching and research activities in shrimp biotechnology.

Dr A.D. Diwan is Former Assistant Director General (ADG) in the Fisheries Division of the Indian Council of Agricultural Research (ICAR), Ministry of Agriculture, Govt of India, New Delhi. Before assuming this position, he was Professor and Head, Fish Genetics & Biotechnology Division and also as a Dean (Faculty of Fishery Science) at the Central Institute of Fisheries Education (CIFE) a Deemed University, Mumbai. He started his career as Associate Professor in the Centre of Advanced Studies (CAS) in Mariculture, a ICAR/UNDP project of Government of India at the Central Marine Fisheries Research Institute (CMFRI), Kochi. He is specialized in animal physiology and neurosciences with vast background and experience in marine and oceanic fisheries and mariculture and published more than 130 research papers of both national and international repute. He has also edited and authored several books on physiology, biotechnology and fisheries.

BIOTECHNOLOGY OF PENAEID SHRIMPS

(PERSPECTIVES ON PHYSIOLOGY OF GROWTH, REPRODUCTION AND DISEASE THERAPEUTICS)

A.D. DIWAN

Institute of Biosciences and Technology,
Mahatma Gandhi Mission's University of Health Sciences,
AURANGABAD-431003, Maharashtra, India.

CRC Press
Taylor & Francis Group
Boca Raton London New York

CRC Press is an imprint of the
Taylor & Francis Group, an **informa** business

NPH

NARENDRA PUBLISHING HOUSE
DELHI-110006 (INDIA)

First published 2021
by CRC Press
2 Park Square, Milton Park, Abingdon, Oxon, OX14 4RN

and by CRC Press
6000 Broken Sound Parkway NW, Suite 300, Boca Raton, FL 33487-2742

CRC Press is an imprint of Informa UK Limited

British Library Cataloguing-in-Publication Data
A catalogue record for this book is available from the British Library

Library of Congress Cataloging-in-Publication Data
A catalog record has been requested

ISBN: 978-0-367-74073-3 (hbk)
ISBN: 978-1-003-15596-6 (ebk)

CONTENTS

M.S. SWAMINATHAN RESEARCH FOUNDATION

M.S. Swaminathan
Founder Chairman
Ex-Member of Parliament (Rajya Sabha)

FOREWORD

Shrimps are common to worldwide capture fisheries and to aquaculture. Demand for shrimp in the international markets, has increased greatly over time. The high demand has led to over-exploitation of the natural shrimp stocks in many parts of the world. A decline in the capture fisheries increased aquaculture production as an alternative source for sustaining and expanding the seafood industry. Due to overexploitation of the wild stock of marine shrimp and the decrease in total natural productivity, shrimp farming activity has developed rapidly in last two decades surpassing the annual global production of 700, 000 mt. In many South Asian countries the shrimp is considered as one of the important agricultural commodities and contributing its major share in agricultural economy. In the Indian sea food export trade, shrimp's share is nearly 65% of the total foreign currency earnings. In recent past the Department of Animal Husbandry, Dairy & Fisheries under the Ministry of Agriculture & Farmers Welfare and the ICAR are making their continuous efforts to enhance production of quality shrimp by all possible means, but unfortunately, at present there is no proper information on basic aspects of shrimp like its reproductive physiology, neuro-endocrinology, genomics and proteomics, diseases management and therapeutics. This kind of information is the pre-requisites for development of any biotechnological tools to augment productivity. The present book on "BIOTECHNOLOGY OF PENAEID SHRIMPS" covers some fundamental aspects and few applied aspects of biotechnology concerning genomics, proteomics, reproduction, growth and disease control and therapeutics of the penaeid shrimps.

I personally appreciate and congratulate Dr A D Diwan for taking efforts in writing such important reference book covering all these above mentioned aspects. I believe this kind of information will be quite useful not only to the aqua-farmers/mariculture experts to augment shrimp production in captive condition but also to the faculties and students working in different organizations involved in teaching and research activities.

M S Swaminathan

3rd Cross Road, Taramani Institutional Area, Chennai (Madras) - 600 113, India
Phone: +91-44-2254 2790, 2254 1698 Fax: +91-44-2254 1319
E-mail: founder@mssrf.res.in, swami@mssrf.res.in

PREFACE

The shrimp is considered as one of the important agricultural commodities and contributing its major share in our agricultural economy. In the sea food export of India, nearly 65% of the total foreign currency earning is contributed by frozen shrimps. The Ministry of Agriculture and Farmers Welfare and the Indian Council of Agricultural Research (ICAR) are making continuous efforts to enhance production of quality shrimp by all means. Unfortunately, at present there is no proper information on basic aspects of shrimp like its physiology, endocrinology, basics of genomics and proteomics and disease control and therapeutics. This kind of information is rather prerequisites for development of any biotechnological tool to augment production and productivity of the shrimp at farm level.

The main objective of writing present book is to collect comprehensive information on various aspects of physiology and biotechnology focusing mainly on reproduction, growth, disease control and therapeutics of penaeid shrimps. Generally shrimps are inhabiting in seas and come across different types of climatic conditions during their life cycle. Their life span is very short. In order to augment shrimp production in captive conditions one should understand thoroughly various aspects of their growth, reproduction, immune system and disease physiology and also interaction of hormones with various physiological processes. This kind of knowledge will definitely help in developing innovative technologies for increasing shrimp production under mariculture practices.

The present book covers some fundamental aspects and few applied aspects of biotechnology concerning basic genomics and proteomics, reproduction, growth and disease control and also on therapeutics of shrimp. This information will be quite useful not only to the aqua-farmers/mariculture experts of the shrimp industry to augment quality shrimp production in captive condition but also to the faculties and students working in different organizations involved in teaching and research activities in shrimp biotechnology.

In this endeavour, I wish to thank sincerely to the Science and Engineering Research Board (SERB) of Ministry of Science and Technology, Government of India, New Delhi for providing financial support for writing the present book under USERS scheme SB/UR/32/2013. I also wish to record my indebtedness and gratitude to shri Ankushrao Kadam, Secretary, Mahatma Gandhi Mission Trust, Aurangabad, Maharashtra for providing excellent facilities to undertake this scheme and for encouragement. I also wish to extend my sincere thanks to Dr Sanjay Harke, Director, MGM's Institute Biosciences &Technology for providing all necessary facilities during the tenure of the scheme. Last but not least I express my thanks to Ms Pooja Desai, M.Sc. (Biotech) and project technical assistant of the scheme who has supported me immensely not only in typing the manuscript but also going through all the chapters carefully and ensuring corrections wherever possible.

A.D. Diwan
New Delhi, India

SECTION-I

BASICS

CHAPTER 1

BASICS OF GENOMICS IN SHRIMP

1.1 INTRODUCTION

Aquaculture, is the fastest growing animal production industry, despite the fact that most species farmed have undergone very little genetic improvement. Increases in production within the industry have resulted from improvements in animal husbandry and nutrition, however, recently there has been an increased impetus towards the application of quantitative and molecular-based animal breeding methodologies to the improvement of Aquaculture, species. A whole genome sequence helps in the identification of candidate genes responsible for production and performance traits for a given species. Such genes can then be used for gene-assisted selection for superior broodstock and can lead to better management of shrimp stocks as well as a better understanding of pathogens and disease resistance. In general, a whole genome sequence can be viewed as the ultimate map of the genome, allowing geneticists to mine the genes and genetic variations associated with important production traits for the best utilization of genetics for Aquaculture. One of the important aspects of a whole genome sequence is to provide a map for spatial assignments of genetic variations referred to as single nucleotide polymorphisms (SNPs). SNPs are one of the fundamental reasons why one individual performs differently from another. The whole genome sequence also allows the identification of candidate genes responsible for production and performance traits. Once identified and validated, such genes can be used for gene assisted selection for superior broodstock. A sequenced genome can lead to better management of shrimp stocks as well as a better understanding of pathogens and disease resistance.

From modest beginnings more than four decades ago, shrimp harvested from commercial shrimp Aquaculture, operations now contributes more than 55% of the world's total supply of shrimp. The year 2007 marked the first time in history that the production of shrimp from farms exceeded the global capture of all marine shrimps. The global shrimp farming industry was valued at an estimated USD 16.7 billion in 2010 (FAO, 2010). The global shrimp farming industry's value has been growing at a rate of just under USD 1 billion per year for the trailing 10 year period of 2000-2010. In order of importance, the top five shrimp producing countries in the world as of 2010 are China, Thailand, Vietnam, Ecuador, and India. With production of 1.4 million tons per year in 2010, China is producing more than 2.5 times more farm-raised shrimp per year than its next closest competitor, Thailand. As a group, Asia dominates global shrimp farming production and is responsible for 88.80 % of the global production figure of 3.8 million tons in 2010 (FAO, 2010). Among 110 species of 12 genera belonging to the family Penaeidae (Flegel, 2007), *Penaeus, Litopenaeus, Metapenaeus*, and *Fenneropenaeus* are commonly cultured worldwide. As on today only two species, *Penaeus monodon* and *Litopenaeus vannamei* are the most widely reared species and contribute 87% to world production of *Aquaculture*, d shrimps (Andriantahina *et al.*, 2013). In the last more than forty years there is major improvement in the shrimp production system starting from larval rearing, nutrition, disease diagnostics and therapeutics, water quality management and genetic improvement of production traits. In the whole operation system reared strains of *P. monodon* in the beginning showed disease resistance, fast growth, short cultured period and low production cost. All these findings have latter resulted in selective breeding programs for improving genetic quality among the new traits of shrimps (Andriantahina *et al.*, 2013). Since then efforts are being made in traditional selective breeding, biotechnology and molecular genetics of penaeid shrimps. Andriantahina *et al.*, (2013) in their review paper provided useful information about cultured shrimp genetics and genomics including related research works, like domesticated stocks and genetically improved strains, genetic engineering, molecular and genomic shrimp resources, combination genetic improvement schemes, genotype-environment interactions etc. Although significant research has been conducted in Asia on the above mentioned aspects, the improvement of reared strains has been performed disproportionately by the institutions and industries (table 1). Research on shrimp has been focusing not only on developing disease resistant strains or shrimp production but also on breeding of some specific strains. Recent collaborative research between the Commonwealth Scientific and Industrial Research Organization (CSIRO) in Australia and commercial shrimp farming has resulted in successful captive breeding of *M. japonicus*. Coman *et al.*, (2007) reported that decrease in densities from three to one shrimp per square meter of second generation reared female *P. monodon* resulted in an improved growth of the individuals and reproductive performance of females. Likewise Sellars *et al.*, (2006) mentioned that brood stocks of *M. japonicus* reared in indoor tanks are less at risk of Mourilyan (MoV) infection and have greater capacity to tolerate infection compared to those reared in farm ponds. Now realising the importance of genomic research in shrimp culture, attempts are being made in several laboratories on developing molecular genetic resources including genetic maps, expressed sequence tags (ESTs), gene sequences, and few DNA libraries for shrimps. As far as inbreeding effects on maintenance of genetic quality is concerned there are few established reports where it has been shown that there is decrease in growth of the shrimp more than 30% or more and simultaneously decrease in survival rate and also reproductive performance. Reduction in genetic variation in cultured stocks of *F. chinensis* (Li *et al.*, 2006), *M. japonicus* (Luan *et al.*, 2006), *L. vannamei* (Freitas *et al.*, 2007) and *P. monodon* (Macbeth *et al.*, 2007) has been clearly correlated to the deleterious effects of inbreeding. The impact of genetic selection has grown significantly in recent past in shrimp culture and considerable work has been done on this aspect. Number of workers have reported that using the technique of genetic selection there is gain in growth of the shrimps to the tune of 4-10% per generation particularly in *L. stylirostris* (Goyard *et al.*, 2001), *F. chinensis* (Li *et al.*, 2005), *P. monodon* (Argue *et al.*, 2008) and *L. vannamei* (De Donato *et al.*, 2005, Andriantahina *et al.*, 2012). Considerable information is also available to correlate genetic selection and development of disease resistance strains (Andriantahina *et al.*, 2013). This kind of information is quite useful to produce the desired traits of shrimps and develop appropriate infrastructure and management strategies for their production in culture operations.

In tropical climates where most farmed shrimp is produced, it takes approximately three to six months to raise market-sized shrimp, with many farmers growing two to three crops per year. A steady stream of organic waste, chemicals and antibiotics from shrimp farms can pollute groundwater or coastal estuaries. Salt from the ponds can also seep into the groundwater and onto agricultural land. This has had lasting effects, changing the hydrology that provides the foundation of wetland ecosystems.

The introduction of pathogens can lead to major outbreaks of disease in shrimp with devastating consequences. When the shrimp become ill with some diseases, they swim on the surface rather than on the bottom of the production pond. Seagulls swoop down, consume the diseased shrimp, and then may subsequently defecate on a pond a few miles away, spreading the pathogen. When shrimp farms are shut down due to disease, there are socioeconomic impacts, including loss of employment.

Approximately 80 percent of farmed shrimp are raised from just two species *Penaeus vannamei* (Pacific white shrimp) and *Penaeus monodon* (giant tiger prawn). These monocultures are very susceptible to disease (Fig 1).

Table 1. Existing main farmed genetic resources of the main reared penaeid shrimp species

Species	Country of the core breeding population	Generations estabilished	Origin of stock	Proprietor	Sale	Source
Litopenaeus vannamei	Hawaii, USA	23	Mexico and Ecuador Mexico and Ecuador	Private	Internationals sales	1
	Forida, USA,	18+	Columbia, Costa Rica, Ecuador, Hawaii,	Private	Internationals sales	
	Colombia	15+	Panama, Peru, Salvador, and Venezuela	Private/ government	Local industry	2
	Mexico (several)	various upto 14+	Mexico, Ecuador, Venezuela, Colombia and Florida	Private	Local industry	3,4
	Venezuela	19 years	Mexico, Panama and Colombia	Private	Private use	5
	Brazil (several)	various upto 11	CostaRica, Ecuador, Mexico, Panama and Venezuela	Private	Local industry	6
	China (several)	Unknown	USA and South America	Private	Local industry	7
	Thailand (several)	Uknown	USA and South America		Local industry	7

[Table Contd.

Contd. Table

Species	Country of the core breeding population	Generations estabilished	Origin of stock	Proprietor	Sale	Source
Penaeus monodon	Hawaii, USA	5	Indo-Pacific	Private	Open sales	8
	Madagascar	20+	Southwest Indian Ocean	Private	No	
	Australia	11+	East and North coast of Australia	Private/ government	Local industry	9
	Thailand	8+	Thailand waters	Government	Local industry	10
	Thailand	8+	Thailand waters	Private	Local industry	10
Fennerpenaeus merguiensis	Thailand	17	Andaman sea	Government	Experimental local industry	10
Fennerpenaeus chinensis	China	15	China	Government	Local industry	11
Marsupenaeus japonicus	Australia	13+	Australia	Private	Local industry	12
Litopenaeus stylirostris	New Caledonia	38	Mexico and Panama	Private	Local industry	13
	Hawaii, USA	17+	Ecuador	Private	International sales	13
Fennerpenaeus indicus	Saudi Arabia	Not known	Saudi Waters	Private	No	7
	Iran	Not known	Persian Gulf	Cooperatve	Local industry	

(**Source:** Andriantahina *et al.*, 2013)

Fig. 1. Farmed penaeid shrimps

The tiger shrimp *P. monodon* is extensively cultured in tropical and sub-tropical waters. The brood stock is generally caught from wild and since the amount of brood stock could be depleting, many attempts have been made to domesticate *P. monodon* and to produce brood stock in a closed system.

Over the last decade there is a considerable growth of genomics in the shrimps, and the new tools of molecular biology are contributing to the growth in field at a rapid pace.

Though genomic sequencing programmes for several agricultural species have now been either completed or are in progress, genomic information on the shrimp is still limited and fragmented. Preliminary information on genetic maps for *P. japonicus*, *P. monodon and P. vannamei* is available at present; likewise limited expressed sequence tag (EST) databases have been developed, mainly in relation to diseases and reproduction. Some basic information is also available on likely roles of genes that are discovered in recent past. The technique of molecular markers could provide a better tool to assist the selection of individuals in shrimp breeding programmes. Microsatellites have become the markers of choice for a wide range of applications in genetic mapping and genome analysis (Beckmann and Soller, 1990). Microsatellites represent ideal molecular markers because they have multiple alleles that with highly polymorphic among individuals and microsatellite loci are highly abundant dispersed evenly throughout eukaryotic genomes. Microsatellites have been used to study several penaeid species such as *P. vannamei* (Wolfus *et al.*, 1997) *P. japonicus* (Moore *et al.*, 1999), *P. stylirostris* (Bierne *et al.*, 2000) and *P. monodon* (Tassanakajon *et al.*, 1998, Xu *et al.*, 1999,

Brooker *et al.*, 2000 and Pongsomboon *et al.*, 2000). Wuthisuthimethavee *et al.*, (2003) developed microsatellite markers in *P. monodon* which would be quite useful tool in breeding programs of this shrimp. Nine synthesized repeat sequences labelled with biotin were used and created an enriched microsatellite library for *P. monodon*. This technique was also used for *P. vannamei* shrimps collected from different locations and also wild females collected from various places. It was observed that all of the offspring shown to be genetic descendants of their presumed parents. All the families showed Mendelian inheritance. From the studies it was concluded that microsatellite markers are useful for informing animal breeders of genetic diversity levels in the population of interest before implementing a breeding program (1997).

Many shrimp industries move towards domestication and selective breeding techniques for their requirement of good quality brood stock. Therefore, it is imperative that to identify the pedigree of animals we require molecular tools and these tools will enable the performance of individuals within families, families and stocks as a whole. Many a times inbreeding leads to homozygosity, which in some species can lead to reduced growth, viability and reproductive performance (Goyard *et al.*, 2008). Inbreeding can also be linked to biochemical disorders and deformities from lethal and sub lethal recessive alleles (Dunham, 2004). Dixon *et al.*, (2008) reported shifts in genetic diversity during domestication of shrimp *P. monodon* which they have monitored using two multiplexed microsatellite system. This study indicated rapid loss of genetic diversity and reduction in effective population size that can occur during domestication when rigid genetic management of the stocks is not possible. This study also demonstrates the importance of value of using microsatellite marker system in shrimp breeding programmes to monitor genetic diversity and the effective population size.

So far most genetic improvement programs for shrimp have on growth and disease resistance improvement (Argue *et al.*, 2002, Gitterle *et al.*, 2005a, b, Goyard *et al.*, 2002, Hetzel *et al.*, 2000, Perez-Rostro and Ibarra, 2003a, b) and also on nutrition, environmental control, and management aspescts (for reviews see Harrison, 1990, Browdy, 1992, Wouters *et al.*, 2001 and Racotta *et al.*, 2003). However, very few studies have addressed the issue of developing genetic selection programs directed to increase reproductive performance simultaneously to other productive traits. Brady *et al.*, (2013) while working on candidates genes that are potentially associated with reproductive dysfunction of captive reared shrimp *P. monodon* found significant differences between wild caught and captive reared *P. monodon*. It is mentioned that reproductive performance of captive reared females is often characterised by longer latency period, lower egg production, egg hatch rates and post larval survivorship compared with their wild-caught counterparts. Therefore, details of understanding of the cellular and associated molecular events occurring during ovarian maturation could be detrimental for the success of reproduction in captive reared shrimps. Brady *et al.*, (2013) considered both gene expression and histological analysis of ovarian tissue in their studies and found that the transcript encoding vitellogenin- the major egg yolk protein precursor and lipid storage droplet protein which is involved in lipid accumulation have higher expression in wild caught shrimps when compared to captive reared shrimps. Efforts have been made here to summarise the current state of the shrimp genomics in the following prime areas of research.

1.2 AN EXPRESSED SEQUENCE TAG (EST) DATABASE AS AN APPROACH TO GENE DISCOVERY IN SHRIMP

Expressed Sequence Tag are short DNA sequences which provide information on genes that are expressed in given tissues, in given environments or developmental stage. EST sequencing can provide an efficient and rapid means for discovering novel genes, alleles and polymorphisms, thereby providing data on gene expression and regulation, and also for the development of genome maps. Collection of EST database has been done for several species of shrimps viz., *P. monodon, F. chinensis, M .japonicus, L setiferus,* and *L. vannamei* (Benzei, 2005) (table 2). Most of these analyses have been small scale sequencing efforts by individual workers studying mostly on shrimp disease and immunity (Robalino *et al.*, 2012). Information on ESTs from *L. Vannamei* and *L. setiferus* is the only available resource which is now available in public domain. It has been mentioned that when large sequencing efforts are conducted the rate of discovery of new genes rapidly decreases as sequencing progresses (Robalino *et al.*, 2012). There is considerable information available on small collection of ESTs relating to white spot diseases, immunity, respiration, endocrinology, digestion and reproduction in shrimp. It has been reported that work on ESTs from haemocyte and hepatopancreas derived from *P. vannamei* and *P. setiferus* have indicated that genes are spread in several functional groups (Gross *et al.*, 2001). However further it has been said that these data show that about 5o% or more of the unigenes obtained lack homologues in the Gene Bank databases and therefore have no known functions. Public access large insert libraries are available for *P. monodon* and *L. vannamei* with work in progress towards developing physical maps and genome sequences for these species (Saski *et al.*, 2009). In addition large number of molecular markers are expected to be explored out of these sequences, making them available for shrimp breeding programmes interested in using marker assisted selection. These facts indicate that further research work on shrimp genome is urgently required.

Table 2. Studies reporting characterization of EST in Penaeid shrimps.

Species	Tissues and conditions under study	#ESTs/unique sequences	Principal findings/ genes	Refrences
Pm	Testes	896/NA	Discovery of genes involved in reproductive maturation and sex determination.Description of testes specific genes.	(Lee latawit *et al.*, 2009)
Pm	Lymphoid organ from *vibrio harveyi* infected animals compared to non-infected animals	408 normal; 625 infected/ NA	Discovery of genes differential expressed in response to bacterial challenge. Discussion of the catheps infamily (L and B).p *p monodon*.biotec.or.th/home.jsp.	(Pongsomboon *et al.*, 2008b)
Pm	Vitellogenic ovaries	1051/559	Identification of sex-related genes. Further analysis of chromobox protein (CBX) which is preferentially expressed in ovaries.	(Preechaphol *et al.*, 2007)
Pm	Postlarvae infected with WSSv	6,671 normal and 7,298 infected/9622	Gene discovery in postlarvaem shrimp. Xbio .lifescience.ntu.edu.tw/pm.	(Leu *et al.*, 2007)
Pm	Eyestalk,hepatopancrease, haematopoietic tissue, haemocyte, lymphoid organ, and ovary from normal, heat stressed, WSSV, YHV, and *V. harveyi* infected shrimp	10100/4845	Large scale gene discovery, Go analysis Pmonodon.biotec.or.th	(Tassanakajon *et al.*, 2006)
Pm	Hemocytes from *Vibrio harveyi* infected animals compared to non- infected animals.	1062/NA	Immune gene discovery with a focus on antimicrobial peptides.	(Supungul *et al.*, 2004)

[Table Contd.

Contd. Table]

Species	Tissues and conditions under study	#ESTs/unique sequences	Principal findings/ genes	Refrences
Fc	Cephalothorax	10446/NA	Immune gene discovery. Immune genes including lectins, serine proteases, serpins, and lysozyme are discussed.	(Shen *et al.*, 2004)
Pm	Hemocytes from non-infected animals	615/Na	Identification of human genes expressed in normal hemocytes with focus on penaeidins, heat shock proteins and anti-LPs protein	(Supungul *et al.*, 2002)
Mj	Hemocytes from Wssv-infected animals compared to non-infected animals	635 normal 370 WSSV/NA	Immune gene discovery	(Rojtinnakorn *et al.* 2002)
Lv, Lse	Hemocytes and Hepatopancreas	2045/268	Immune gene discovery	(Gross *et al.*, 2001)
Pm	Cephalothorax, eyestalk, and pleopod tissue	151/NA	Gene discovery with a discussion of tissue specific genes.	(Lehnert *et al.*, 1999)
Lv	Hemocytes, gills, hepatopancreas, Lymphoid organ, eyestalk and ventral nerve cord.	13656/7466	Large-scale gene discovery with focus on genes from immune related tissues. www.marinegenomics.org	(O'Leary *et al.*, 2006)
Lv	Abdominal muscle	311/NA	Profile of gene expression in shrimp muscle tissue.	(Cesar *et al.* 2008)

(**Source:** Robalino *et al.*, 2012)

1.3 DIFFERENTIAL GENE EXPRESSION STUDIES AND GENE DISCOVERY

Differential gene expression studies helps in understanding gene function. The expression of a gene is either restricted to the cell or tissue where it is needed. In shrimp culture practices gene expression studies are mostly being carried out in response to different diseases and stimulators of the immune system. Once the genes that play a role in a subset of functions have been identified, these can then be selected for a smaller array required to the needs of particular project. Such studies are important in shrimp where many of the unigenes being discovered have no homologues in Gene Bank. Benzei (1998) in his report has mentioned that Marine Genomics Group in South Carolina has printed information about 3000 unigenes of *P. vannamei*. The Australian Institute of Marine Sciences has the data on microarray of *P. monodon* with 4032 PCR products pertaining to immune related gene expression in the face of the viral disease (De La Vega *et al.*, 2004). Using a microarray of the WSSV viral genome has established the expression of almost 90% of the viral genes in shrimp cells (Wang *et al.*, 2004). Some work has mentioned in depth studies particularly on molecular characterization and functional investigation of genes involved in ovarian function or disease response. In this regard Yamano *et al.*, (2004) and Khayat *et al.*, (2001) have mentioned the nature and dynamics of gene expression, and later functional translation of cortical rod proteins related thrombospondin in the kuruma shrimp *P. japonicus* and *P. semisulcatus* respectively. Leelatanawit *et al.*, (2004) described the isolation and characterization of genes differentially expressed in the ovaries and testes of *P. monodon*. Aoki *et al.*, (2004) characterized several genes earlier identified from EST analysis that are involved in disease response in *P. japonicus* such as prophenoloxidase, β1,3-glucan binding protein, α2-microglobulin, serine proteinase homologue, lysozyme, crustin-like peptide and transglutamase. Wang *et al.*, (2004) characterized the expression of white spot virus protein in shrimp cells, whereas Gueguen *et al.*, (2004) investigated the role of some of antimicrobial genes in *P. vannamei*. Robalino *et al.*, (2012) in their paper supplemented the information on shrimps that have been studied using differential expression cloning approaches include gametogenesis, abiotic stress responses, and immune responses (table 3). Further they mentioned that many of the sequences isolated from these studies do not allow identification of genes with predictable homology or function. Robalino *et al.*, (2012) also mentioned in their paper regarding differential gene expression studies in shrimp using microarray technology (table 4). Microarray with gene contents ranging from a few dozen to a few thousand genes have reported and used for expression studies in *L. vannamei*, *L. stylirostris*, *M. japonicus* and *F. chinensis*. Some of these studies are summarised in the table 4. Such investigations will help to understand

Table 3. Differential expression cloning studies in Penaeid shrimp.

Species	Tissues and conditions under study	Method	#unique sequences identifies	Principal findings	Rerferences
Ls	Hepatopancreas from shrimp experimentally infected with WSSV vs. uninfected shrimp.	DD	32	One of the earliest indications of the challenge of assigning function to novel sequences in shrimp.	(Astrofsky *et al.*, 2002)
Pm	Hepatopancreas from shrimp surviving a WSSV outbreak vs. uninfected shrimp.	DD	NA	PmAV, a C-type lectin with apparent antiviral activity in a heterologous (non shrimp) virus-host system.	(Luo *et al.*, 2003)
Mj	Hemocytes from shrimp stimulated by heat-killed microbes vs shrimp not stimulated.	SSH	77 with homology	Diverse functional groups of genes appear regulated, by heat-killed microbes, including protease inhibitors	(He *et al.*, 2004)
Mj	Hemocytes from shrimp surviving a WSSV outbreak vs. uninfected shrimp.	SSH	30	Diverse functional categories including chaperons, lectins, and protease inhibitors enriched in animals surviving an outbreak.	(He *et al.* 2005)
Mj	Hepatopancreas from shrimp surviving a WSSV outbreak VS. uninfected shrimp	SSH	31	Diverse functional categories including lec tins, glucan binding proteins, and small GTpase enriched in animals surviving an outbreak.	(Pan *et al.*, 2005)
Ls	Hematocytes from animals surviving a bacterial challenge compared to non-surviviong animals	SSH	184	Increased expression of several antimicrobials (e.g Penaedian, lysozyme, cryptdin-like) co-relating with survival to bacterial infection.	(de Lorgeril *et al.*, 2005)
Pm	Hemocytes from *Vibrio harveyi* infected animals compared to non-infected animals	DD	24 with homology	Induction of expected immune effectors (e.g lysozyme, ALF, transglutaminase), and of Argonaute, a gene of the RNAi pathway.	(Somboon-wiwat *et al.*, 2006)

[Table Contd.

Contd. Table]

Species	Tissues and conditions under study	Method	#unique sequences identifies	Principal findings	Rerferences
Lv	Hemocytes, gills, or hepatopancreas from shrimp induced with heat-inactivated microbes, dsRNA, or WSSY compared to mock treated animals. Same tissues from shrimp infected with WSSV at permissive and non-permissive temperatures.	SSH	3,231 from both SSH and standard cDNA libraries	A wide range of genes with potential roles in immunity, including antimicrobials, signaling factors, transcription factors, regulators of apoptosis, were suggested to be regulated by immune stimuli.	(Robalino *et al.*, 2007)
Lv	Hepatopancreas from uninfected members of a shrimp family selected based on its reduced susceptibility to WSSV, compared to a family with susceptibility	SSH	193, with 40 matches to known genes	Diverse functional categories of genes suggested to be enriched in WSSV-resistant family. These included lysozymes, cathepsins, lectins, and other potential antimicrobials.	(Zhao *et al.*, 2007)
Pm	Hemocytes from shrimp subjected to osmotic, hypoxic, or thermal stress, compared to non-stressed animals	SSH	176, with 58 matches to known genes	Some known immune factors and, strikingly, many retrotransposon-relted sequences are regulated during abiotic stress.	(de la Vega *et al.*, 2007a)
Me	Hepatopancreas from shrimp at different stages of ovarian development	DD	15 clones with homology	Vitellogenin and some functionally diverse enzymes regulated during ovarian maturation.	(Wong *et al.*, 2008)
Pm	Testes from broodstock vs juvenile shrimp	SSH	80 with homology	Discovery of a progestin receptor membrane component 1 gene.	(Leelatanawit *et al.*, 2008)

Ls- *Litopenaeus stylirostris*, Pm- *Penaeus monodon*, Mj- *Marsupenaeus japonicus*, Lv- *Litopenaeus vannamei*, Me- *Metapenaeus ensis*, DD- *differential display*, SSH- suppression subtractive hybridization, NA- not reported.

(**Source:** Robalino *et al.*, 2012)

Table. 4. Showing the summary of microarray studies in Penaeid Shrimps.

Species	**Estimated gene content (unique sequences)**	**Tissue and conditions studied**	**Principal findings**	**References**
Ls	84	Hepatopancreas from shrimp experimentally infected with WSSV vs. uninfected shrimp.	Several potential pattern recognition proteins, such as lectins and LPS/glucan binding protein, induced by WSSV	(Dhar *et al.*, 2003)
Fc	1,578 unique cDNAs plus 1,536 SSH clones	Whole cephalothorax from shrimp experimentally infected with WSSV vs uninfected shrimp, and from naturally infected shrimp vs. uninfected wild animals.	Diverse groups of genes regulayed during experimental and natural WSSV infection, including chaperones and genes involved in metabolism and cell structure	(Wang *et al.*, 2006)
Lv	2,469	Hepatopancreas from shrimp experimentally infected with WSSV vs. uninfected shrimp	Induction of some known antimicrobials and repression of oxidative stress genes and of the immune transcription factor STAT	(Robalino *et al.*, 2007)
Pm	NA	Hemocytes from shrimp exposed to either osmotic, hypoxic, or thermal stress	A complex response to abiotic stress, which included regulation of known immune factors (e.g crustin, lysozyme, transglutaminase), as well as changes in mRNAs corresponding to retrotransposons	(de la Vega *et al.*, 2007b)

[Table Contd.

Contd. Table]

Species	Estimated gene content (unique sequences)	Tissue and conditions studied	Principal findings	References
Pm	2,028	Hemocytes from YHv infected shrimp vs. mock infected animals	Known immune genes with complex patterns of temporal regulation. Cathepsin L highly induced in YHV infected hemocytes	(Pongsomboon *et al.*, 2008a)
Mj	2,306	Hemocytes from peptidoglycan stimulated shrimp vs. animals not stimulated	Known immune factors such as antimicrobial proteins response to peptidoglycan stimulation	(Fagutao *et al.*, 2008a)
FC	3,114	Hepatopancreas, Hemocytes, gills, and lymphoid organ from shrimp infected with WSSV vs. from animals stimulated with heat-killed Vibrio anguillarum vs mock stimulated	Very functionally diverse groups of genes responsive to either immune stimulus, but also an overlapping response was observed	(Wang *et al.*, 2008b)

Ls- *Litopenaeus stylirostris*, Fc- *Fenneropenaeus chinensis*, Lv- *Litopenaeus vannamei*,

Pm- *Penaeus monodon*, Mj- *Marsupenaeus japonicus*, NA- not reported.

(**Source:** Robalino *et al.*, 2012)

the immune pathways and improve our basic understanding of shrimp immunity. This will also help in finding out genes responsible for disease resistance while developing brood stock in captive condition and provide important avenues to the development of biotechnological tools for aquaculture.

Mitochondrial genome sequence and structure is widely used to provide information on phylogenetic relationships and on the genetic structure of populations and patterns of gene flow. This information may derive from studies of gene order, the sequences of individual genes, restriction fragment length polymorphism (RFLP) analysis of mtDNA, or the sequences of complete genomes. Though shrimp *P. monodon* is one of the most important aquaculture, species in the world, yet, there is no much information anything about their genomic structure. To understand the organization and evolution of the *P. monodon* genome, Huang *et al.*, (2011) worked on construction of a fosmid library consisting of 288,000 colonies which is equivalent to 5.3-fold coverage of the 2.17 Gb genome. In their findings they reported, approximately 11.1 Mb of fosmid end sequences (FESs) from 20,926 non-redundant reads representing 0.45% of the *P. monodon* genome. The redundancy of various repeat types in the *P. monodon* genome was observed to be highly repetitive nature. These results provided substantial improvement to our current knowledge not only for shrimp but also for marine crustaceans of large genome size. The presence of large quantities of microsatellite sequences seems to be a distinct characteristic of penaeid genomes, like those of *F. chinensis* (Gao and Kong, 2005) and *P. vannamei* (Meehan *et al.*, 2003). The mechanism that determines and maintains the abundance of tandem repeats is not well understood, but apparently reflects the response of the whole genome to overall selective and mutational pressures (Karlin, 1998). Similar results were obtained by Maneeruttanarungroj *et al.*, 2006 which revealed that 9.9% of the *P. monodon* ESTs (997/10,100) contained microsatellites. In addition, by reviewing the literature and by examining the *P. monodon* EST dataset in the *Penaeus* Genome Database, it has been reported that many shrimp genes/ESTs contain long stretches of microsatellites. As longer repeats generally have higher mutation rates, the abundance and long stretches of microsatellites are unusual, raising the possibility that they may have functional roles. Microsatellites have been hypothesized to be an important source of quantitative genetic variation and evolutionary adaptation (Kunzler *et al.*, 1995, Kashi *et al.*, 1997). The high mutational rate suggest that microsatellites can act like adjustable tuning knobs through which specific genes are able to rapidly adjust the norm of reaction in response to minor or major shifts in evolutionary demands (King *et al.*, 1997).

1.4 RNA INTERFERENCE BASED APPLICATIONS IN SHRIMP AQUACULTURE

The technique of RNA interference (RNAi)-based application is now being used in shrimp aquaculture, particularly to control the diseases that are frequently occurring leading to mass mortality and heavy losses. This is the technique of reverse genomics where sequence specific targeted genes are silenced by mediating RNA interference. In this technique the small RNAs are incorporated onto the RNA-induced silencing complex and related complexes of protein, which mediate targeted degradation and other silencing phenomena by means of complementary base-pairing. Because short interfering RNA can be supplied exogenously to trigger specific gene silencing, this technique has rapidly become the most widely used gene silencing tool. While explaining the mechanism of RNA interference machinery Robalino *et al.*, (2005) demonstrated in the shrimp *L. vannamei* that in vivo administration of dsRNA induced a down regulation of endogenous or viral gene expression in a sequence –specific manner. This finding has been supported later by the work done by Dechklar *et al.*, (2008) and Su *et al.*, (2008) on the identification of possible RNAi pathway components in the shrimp *P. monodon*. These results strongly suggest the existence of functional RNAi in the shrimp and pave the possible way of using reverse genomic in understanding gene function in these organisms. In shrimp, for inducement of targeted gene silencing, different methods are involved and the same is being carried out by in vivo injection, in vitro delivery to primary cell culture or by feeding bacteria carrying dsRNA in vivo. The EST mining and differential expression cloning has been applied in shrimp to gain insight into gene function. In recent past, it has been shown experimentally that functional genes are involved in moulting process, osmoregulation, reproduction and glucose metabolism in shrimp by using gene-specific dsRNA technology (Robalino *et al.*, 2012) (table 5). Using dsRNA technology some work has been carried out to relate immune system with gene function in the shrimp (de la Vega *et al.*, 2008, Shockey *et al.*, 2008, and Amparyup *et al.*, 2009) (table 6). These studies are of great importance in the field of biotechnology as they pave the way to develop relationship between functional genes and shrimps of commercial importance. To control the spread of viral disease new methods are being developed by exposing the shrimps to inactivated virus or viral proteins (Johnson *et al.*, 2008) (table7). In this particular method RNA interference machinery allows gene silencing in a highly sequence specific manner with no risk or side effects on target issues. This method of injection of viral gene specific dsRNA in to shrimp may prove to be powerful tool to inhibit viral replications and protect the shrimp from viral infections (Robalino *et al.*, 2005, Yodmuang *et al.*, 2006, Tirasophon *et al.*, 2007, Attasart *et al.*, 2010). Apiratikul *et al.*, (2014) while studying efficiency of dsRNA transfection of penaeid shrimp using cationic liposomes reported that the injection of a Yellow Head Virus (YHV)-Pro-dsRNA/cationic liposome complex markedly increase protection against YHV as compared with naked YHV-Pro-dsRNA in *L.vannamei*, suggesting that cationic liposome in association with dsRNA protect the shrimp against YHV (Fig 2). The development of genetic markers for breeding and reproductive performance in penaeid shrimps has been the area of significant interest for number of workers for several years. Different types of markers including microsatellites, Restricted Fragment Length Polymorphisms(RFLPs), Amplified Fragment Length Polymorphisms (AFLPs), and Randomly Amplified Polymorphic DNAs (RAPD) have been developed in several penaeid species (Xu *et al.*, 1999, Wilson *et al.*, 2002, Meehan *et al.*, 2003).

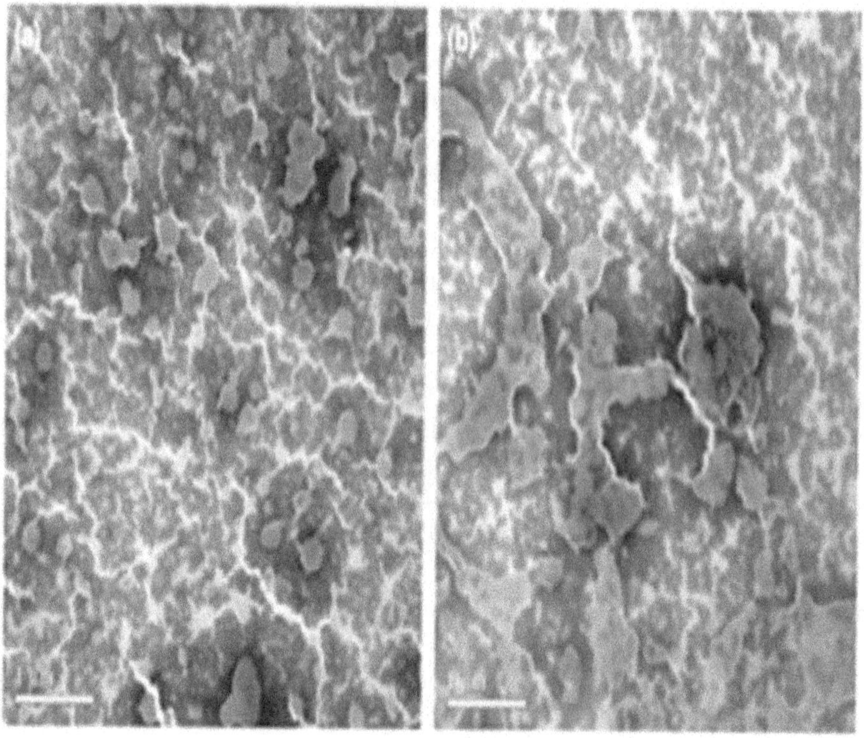

Fig. 2. Transmission electron microscopy (TEM) of cationic liposome and YHV-Pro dsRNA/ cationic liposome of gill cells. Either (a) cationic liposomes or (b) YHV-Pro dsRNA/cationic liposomes was stained with PTA and observed under a TEM. The bar represents 1 μm. (**Source:** Apiratikul *et al.*, (2014)

Table 5. Showing RNAi- based experiments in Penaeid shrimps.

Species	**Target gene**	**RNAi-based application**	**Phenotype**	**References**
Shrimp gene physiological function studies				
Pm	Argonaute (Pem-AGO)	dsRNA transfection into Oka cells	impaired RNAi ability	(Dechklar *et al.*, 2008)
Lv	putative farnesoic acid O-methyltransferase (LvFAMeT)	dsRNA injection into the 5th pereiopod	role in molting lethal phenotype induced	(Hui *et al.*, 2008)
Lsc	crustacean hyperglycemic hormone	dsRNA injection into abdominal body cavity	decrease in hemolymph glucose levels	(Lugo *et al.*, 2006)
Lv	putative ion transport peptide (Lv ITP)	dsRNA injection into the 5th pereiopod	role in osmoregulatory function lethal phenotype induced	(Tiu *et al.*, 2007)
Lv	hemocyanin	dsRNA/siRNA intramuscular injection	-reduction in hemocyanin mRNA levels after dsRNA injection -siRNA failed to induce genetic interference	(Robalino *et al.*, 2005)
Lv	CDP (CUB domain protein)	dsRNA intramuscular injection	reduction in CDP mRNA levels	(Robalino *et al.*, 2005)
Me	Molt-inhibiting hormone (MeMIH-B)	dsRNA injection into the pereiopod	gonad stimulatory function	(Tiu and Chan 2007)
Pm	Gonad-inhibiting hormone (Pem-GIH)	dsRNA-injerction into the pereiopod-incubation in explant culture	gonad-inhibitory function	(Treerattrakool *et al.*, 2008)
Pm	Vitellogenin receptor (VgR)	dsRNA intramuscular injection	role in the processing of vitellogenin	(Tiu *et al.*, 2008)

Pm- *Penaeus monodon*, Lv- *Litopenaeus vannamei*, Lsc- *Litopenaeus schmitti*, Me- *Metapenaeus ensis*, Mj- *Marsupenaeus japonicus*, Fc- *Fenneropenaeus chinensis* (**Source:** Robalino *et al.,*, 2012)

Table 6. Showing RNAi- based experiments in Penaeid shrimps.

Species	Target gene	RNAi-based application	Phenotype	References
Host- pathogen interaction studies				
Lv	Anti-lipopolysaccharide factor (LvALF)	dsRNA intramuscular injection	role in immune function against bacterial and fungal infections	(de la Vega *et al.*, 2008)
Lv	Crustin (LvABP1)	dsRNA intramuscular injection	role in anti-bacterial response	(Shockey *et al.*, 2008)
Pm	Prophenoloxidases (PmproPO1,2)	dsRNA intramuscular injection	role in anti-bacterial response	(Amparyup *et al.*, 2009)
Mj	Rab-GTPase (PjRab)	siRNA intramuscular injection	increased viral replication	(Wu *et al.*, 2008)
Pm	Small GTPase protein (PmRAb7)	dsRNA intramuscular injection	role in the endosomal trafficking pathway	(Ongvarrasopone *et al.*, 2008)
Mj	- Transglutaminase (TGase) - Clotting protein	dsRNA intramuscular injection	role in immune function against bacterial and fungal infections	(Maningas *et al.*, 2007)
Mj	Caspase (PjCaspase)	siRNA intramuscular injection	role in virus-induced apoptosis	(Wang *et al.*, 2008b)
Mj	β-integrin	siRNA intramuscular injection	cellular receptor for WSSV infection	(Li *et al.*, 2007)
Lv	Caspase -3 homologue (Cap-3)	dsRNA intramuscular injection	role in virus-induced apoptosis	(Rijiravanich *et al.*, 2008)
Pm	Dicer 1 (Pm Dcr1)	dsRNA intramuscular injection	increased susceptibility to viral infection	(Su *et al.*, 2008)
Pm	YHV binding protein (pmYRP65)	dsRNA transfection into Oka cells	inhibition of YHV infection	(Assavalapsakul *et al.*, 2006)

Pm- *Penaeus monodon*, Lv- *Litopenaeus vannamei*, Lsc- *Litopenaeus schmitti*, Me- *Metapenaeus ensis*, Mj- *Marsupenaeus japonicus*, Fc- *Fenneropenaeus chinensis* (**Source:** Robalino *et al.*, 2012)

Table 7. Showing RNAi- based experiments in Penaeid shrimps

Species	Target gene	RNAi-based application	Phenotype	References
RNAi-mediated antiviral silencing				
Fc	-VP28 (WSSV)-VP281 (WSSV)	ds RNA intramuscular junction	higher survival rates	(Kim *et al.*, 2007)
Pm	-Helicase coding gene (YHV) -Polymerase coding gene (YHV) -protease coding gene (YHV)	dsRNA transfection into Oka cells	inhibition of YHV replication	(Tirasophon *et al.*, 2005)
	-gp116 (YHV)-gp64 (YHV)	dsRNA intramuscular injection	inhibition of YHV multiplication	(Tirasophon
Pm	YHV-protease dsRNA intramuscular injection		in infected shrimp	*et al.*, 2007)
Pm	coding region of a protease gene (YHV)	dsRNA intramuscular injection	-inhibition of YHV replication -protection from YHV infection	(Yodmuang *et al.*, 2006)
Pm	- VP28 (WSSV) -VP15 (WSSV)	siRNA-transfection in insect cells -intramuscular injection	-silencing of homologous genes in a heterologous expression system -siRNA failed to induce sequence specific antiviral immunity	(Westenberg *et al.*, 2005)
Mj	Vp28 (WSSV)	siRNA intramuscular injection	Reduction in viral DNA production of infect infected animals	(Xu *et al.*, 2007)
Lv	Vp19 (WSSV)	dsRNA/siRNAs intramuscular injection	-higher survival rates after dsRNA injection -siRNA failed to induce antiviral immunity	(Robalino *et al.*, 2005)

[Table Contd.

Contd. Table]

Species	Target gene	RNAi-based application	Phenotype	References
Lv	-DNA polymerase (WSSV) -ribonucleotide reductase small subunit (WSSV) -thymidine kinase (WSSV) thym, idylate kinase (WSSV) -Vp24 (WSSV) -Vp28 (WSSV)	siRNA intramuscular injection	-inhibition of WSSV replication -suppression of selected WSSV gene expression -higher survival rates	(Wu *et al.*, 2007)
Lv	-ribonucleotide reductase small subunit (WSSV) -DNA polymerase DP (WSSV) -ORF WSV252 (WSSV) -Vp28 (WSV)	dsRNA intramuscular injection	protection from WSSV infection	(Robalino *et al.*, 2005)
Lv	predicted protease gene (TSV)	dsRNA intramuscular injection	protection from TSV infection	(Robalino *et al.*, 2005)
Pm	Vp28 (WSSV)	bacterially expressed dsRNA oral administration	protection from WSSV infection	(Sarathi *et al.*, 2008b)
Pm	Vp28 (WSSV)	bacterially expressed dsRNA intramuscular injection	protection from WSSV infection	(Sarathi *et al.*, 2008a)

Pm- *Penaeus monodon*, Lv- *Litopenaeus vannamei*, Lsc- *Litopenaeus schmitti*, Me- *Metapenaeus ensis*, Mj- *Marsupenaeus japonicus*, Fc- *Fenneropenaeus chinensis*.

(**Source:** Robalino *et al.*, 2012)

Lee *et al.*, (2015) while studying effective RNA-silencing strategy of myostatin controlling gene and its effects on growth in the shrimp *L. vannamei* found that the long dsRNA was most effective in decreasing the level of myostatin gene in both the heart and skeletal muscles. It was also reported that higher doses of dsRNA did not lead to greater decrease in myostatin transcripts. It is known that the major production sites for myostatin gene are the muscular tissues including the heart, thoracic muscle and claw muscle. Its expression is related to the moulting cycle (Covi *et al.*, 2008, Kim *et al.*, 2009) as skeletal muscle show a high degree of plasticity and its growth is linked to the moulting cycle (Mykles, 1997).

Neuroparsins are known to be involved in reproduction and it has been mentioned that in insects during early maturation of the female locusts, the transcript level of neuroparsin is found to be low but the level increased to 5 times during the onset of vitellogenesis (Yang *et al.*, 2014). In crustaceans, the presence of neuroparsin has been reported initially from differential expression studies and more recently from EST projects (Yang *et al.*, 2014). Yang *et al.*, (2014) in their studies carried out the work on RNAi silencing of the full length *Metapenaeus ensis* neuroparsin cDNA which resulted in inhibition of vitellogenesis in the ovaries during maturation process in the shrimp *M. ensis*. This study has provided the first evidence that neuroparsin is involved in the initial stage of ovary maturation in shrimp.

1.5 GENOMICS AND REPRODUCTION IN SHRIMP

It is important to note that penaeid shrimp larval production comes from a small proportion of females with multiple spawns which contribute to the production of the majority of nauplii. There is no much evidence for deterioration in the condition of the females and in offspring quality over consecutive spawning in a single generation. Multiple spawning behaviour is genetically determined and can be a target in the selection programme to improve the quality of traits in terms of number of spawns, fecundity, egg size, egg vitelline, egg acylglycerides, egg protein content, body weight, oocyte diameter and ovarian maturity in sub-adult females (Ibarra *et al.*, 2007). Ibarra *et al.*, (2007) in their review paper covered in-depth studies on the recent developments on the genetics of reproduction in shrimp and genes involved in multiple spawning capacities in shrimp.

Valosin containing protein (VCP) has been reported to be required in humans for the mitotic M-phase (Wojcik *et al.*, 2004). Sasagawa *et al.*, (2007) reported that this particular protein is required not only for progression of meiotic metaphase I but also for chromosome condensation at the diakinesis in meiotic prophase I. Witchulada *et al.*, (2014) done the work on characterization, expression and localization of valosin containing protein in the ovaries of shrimp *P. monodon*. They found that valosin-mRNA expression in the ovaries was greater than testes in both juveniles and brood stock. VCP further they have mentioned that was significantly up-regulated in the stages II and IV ovaries in the intact wild brood stock but it was not differentially expressed during ovarian development in the eye stalk ablated brood stock. However, when 5HT was administered exogenously, it resulted in an increase in valosin- mRNA in the ovaries of 18-month-old shrimp. Cellular localization also revealed the presence of VCP in the ooplasm of previtellogenic oocytes and subsequently in to the germinal vesicle of vitellogenic oocytes.

Chen *et al.*, (2014) while carrying out the studies on gene expression of vitellogenin-inhibiting hormone (VIH) in the shrimp *L. vannamei* mentioned that both the brain and eye stalk are the major sources for VIH mRNA expression. In vitro studies on primary culture of shrimp hepatopancreatic cells, it was reported that some endogenous inhibitory factors existed in *L. vannamei* that supressed hepatopancreatic vitellogenin gene expression. Further it is mentioned that purified recombinant VIH protein was effective in inhibiting viltellogenin mRNA expression in *in vitro* primary hepatopancreatic cells culture. Injection of recombinant VIH has reversed the ovarian growth which was induced by eyestalk ablation. These studies have provided new insight on VIH regulation of shrimp reproduction. Klinbunga *et al.*, (2015) studied expression levels of vitellogenin receptor (Vtgr) during ovarian development and association between its single nucleotide polymorphisms (SNPs) and reproduction-related parameters of the giant tiger shrimp *P. monodon*. It is reported that the expression level of Vitellogenin receptor in premature ovaries of juveniles was low compared to that of the matured shrimps. In the matured shrimps the level of Vtgr- mRNA has been reported to be significantly high, whereas in eyestalk ablated shrimps the same was comparably expressed at all the stages of ovarian development.

Uawisetwathana *et al.*, (2011) while reviewing the eyestalk ablation mechanism to induce ovarian maturation in *P. monodon* mentioned the importance DNA microarray studies and its approach to study responses from the organism to stimuli. In their studies they employed the cDNA microarray to reveal molecular mechanism of the eyestalk ablation effects by comparing gene expression levels of ovaries from non-ablated and ablated female brood stock over the course of seven days after the ablation. Based on their findings it is mentioned that there was significant dramatic increase in the levels known reproductive genes involved in vitellogenesis during the ablation. Besides these transcripts, the transcripts whose functions involved in electron transfer mechanism, immune responses and calcium signal transduction have reported to be significantly altered following the ablation. The authors further reported that such findings may lead to develop focused research investigation to pave the way for developing alternate method of inducement of maturation in place of eye stalk ablation.

Urtgam *et al.*, (2015) while studying correlation between gonad inhibiting hormone (GIH) and vitellogenin (Vg) during ovarian maturation in the domesticated shrimp *P. monodon* examined Vg and GIH mRNA expression and their physiological concentration at the protein level during ovarian maturation. It is reported that GIH m RNA expressed at the highest level in the eyestalk ganglia of the shrimp with immature ovary while the GIH peptide released actively in to the haemolymph. Further they observed that the release of GIH dropped drastically at stage I of the ovary, confirming to its negative regulatory function on Vg synthesis. Simultaneously Vg m RNA expression study confirmed that Vg gets synthesized in both the ovary and hepatopancreas. The expression Vg is found to be increased as ovarian maturation progressed. They also further observed that the Vg protein is actively released in the haemolymph sine the stage I of ovarian maturation which suggested that there is rapid release of Vg in to the haemolymph before deposition into oocytes. From these studies they speculated that incorporation of vitelline from follicle cells to the oocytes occurred more slowly in domesticated shrimp which may account slow maturation process in domesticated shrimp *P. monodon*.

1.6 GENETIC MAPS AND GENOMICS LIBRARIES IN SHRIMPS

Gene maps are fundamental to genomic work of any species as they represent a summary of genes and their relationship in the mapped populations. Considerable information is already available on low resolution maps for *P. japonicus*, (Li *et al.*, 2003), *P. monodon* (Wilson *et al.*, 2002) and *P. vannamei* (Perez *et al.*, 2004). Andriantahina *et al.*, (2013) while discussing molecular and genomic shrimp resources, mentioned in their paper the first order of genetic maps for *F. chinensis*, *L. vannamei*, *M. Japonicus* and *P. monodon* (table 8). Benzei (2005) in his paper has mentioned in detail about the main features of gene maps of various species of shrimps which is reflected in the table 9. The majority of the markers upon which these maps are based are Amplified Fragment Length Polymorphism (AFLPs). Only limited numbers of type 1 markers such as microsatellite or EST based markers have been mapped for *P. monodon* and *P. japonicus*. In recent years linkage maps based on microsatellite and AFLP markers have been published for shrimps like *L. vannamei* (Zhang *et al.*, 2007) and *P. monodon* (Maneeruttanarungroj *et al.*, 2006). Staelens *et al.*, (2008) has given more complete linkage map for *P. monodon* which has been constructed based on AFLP markers. A new linkage map for *L. vannamei* is currently under construction which incorporates a large number of SNP markers.

Table 8. Genomic resources available for the main reared penaeid shrimp species.

Species	**DNA sequences (including Est's)**	**Large insert DNA Libraries**	**Genetic maps**					
			Marker type Source (no. mappe)	**Marker seperation (cM)**	**Genome cover (%)**	**No. linkage groups**	**Expected linkage group no.**	**Source**
Litopenaeus vannamei	>160,000s	BAC library FOSMID library	AFLP SSR (30)	15-17	59-62	45	43	1
Penaeus monodon	10,000s	FOSMID library	AFLP (400-547)	4-5	92-100	42-45	44	3
Fennerpenaeus merguiensis	None	None	None	N/A	N/A	N/A	N/A	N/A
Fennerpena-euschinensis	10,000s	BAC (Small insert Size)	AFLP (194-197)	11-14	73-74	35-36	44	4
Marsupenaeus japonicus	1,000s	BAC (Small insert Size)	AFLP (139-245)	8	48-88	33-43	43	5
Litopenaeus stylirostris	1,000s	None	None	N/A	N/A	N/A	N/A	N/A
Fennerpenaeus indicus	None	None	None	N/A	N/A	N/A	N/A	N/A

(**Source:** Andriantahina *et al.,* 2013)

Table 9. Summary of published gene maps for marine shrimp.

Species Map type	*japonicus* **female**	**male**	*monodon* **combined**	*vannamei* **female**	**male**
Mapping population: No of 1 full sib families (no. of progeny (102) screened)					
No. of markers	150	251	116	249	228
No. of markers mapped	125	227	63	212	182
No. of linkage groups	31	43	19	51	47
Average no. (and range) of markers per linkage group	4.0(2-8)	5.3(2-10)	3.3	4.2(2-10)	3.9(2-11)
Average length (and range) of linkage groups (cM)	33.1(5.1-78.4)	41.4(0-102.7)	74.3	55.3(0-108)	45(2-159)
Length of genome mapped (cM)	1026	1780.8	1412	2771	2116
Average space (and range) between markers (cM)	10.9(-)	9.7(-)	22.4(-)	17.1(0-30)	15.6(2-30)
Actual haploid chromosome no	43	43	44	44	44
Estimate of average genome size	2300	2300	2000	4445	3583
Percent map coverage	44.6	77.4	70.6	62.3	59.0

(**Source:** Benzei 2005)

1.7 CHALLENGES OF GENOMICS RESEARCH IN SHRIMPS

In recent past considerable progress in the advancement of genomic research in penaeid shrimp has taken place. Information pertaining to gene expression studies in shrimp is now available in public databases and time has come that these genetic resources the researchers should make use of for identification of genetic markers, to identify disease resistance genes, and genes linked to the reproduction and other Aquaculture, relevant issues. The number of initiatives now is growing in mining the available sequence data to implement markers and generate increasingly more extensive linkage gene maps. The technologies are also in place to refine the selection of candidate relevant genes through the characterisation of two key aspects of the function of a gene i.e. its expression and its loss of function phenotype. Interestingly, it is now possible to block the expression of a gene in a targeted manner in vivo using RNA interference technique. It is also now possible to give complete protection to the shrimp in culture system from highly pathogenic viruses by delivering dsRNA molecules. The future development in this field is closely linked to the ability of the researchers who work in close association, sharing information and resources in benefit of the overall advancement of the science. All these efforts will lead to develop a full genome sequencing picture for penaeid shrimps in the near future.

References

Aoki T, Rattanachi A, Tomura N, Watanabe K, Itami T, Takahashi, Y, Ohira T, Hirono I. 2004. Molecular mechanisms of biodefense in Kuruma shrimp *Marsupenaeus japonicus*. *Mar. Biotechnol.* 6: S153-S157.

Andriantahina F, Liu XL, Huang H, Xiang JH, Yang CM, 2012. Comparison of reproductive performance and offspring quality ofdomesticated Pacific white shrimp, *Litopenaeus vannamei*. *Aquaculture*, , 324–325: 194–200.

Andriantahina Liu X, Feng T, Xiang J, 2013. Review article on 'Current Status of Genetics and Genomics of Reared Penaeid Shrimp: Information Relevant to Acccess and Benefit Sharing'. Springer Science+Business Media New York.

Andriantahina,F., Liu,X. and Huang, H, 2013. *Chin. J. Oceanol. Limn.*31: 534-541.

Argue BJ, Arce SM, Lotz JM, Moss SM, 2002. Selective breeding of Pacific white shrimp (*Litopenaeus vannamei*) for growth and resistance to Taura syndrome virus. *Aquaculture*, 204:447–460.

Apiratikul Nuttapon, Boon-ek Yingyongnarongkul, Wanchai Assavalapsakul, 2014. Highly efûcient double-stranded RNA transfection of penaeid shrimp using cationic liposomes. *Aquaculture*, Research 2014, 45, 106–112.

Amparyup, P., W. Charoensapsri and A. Tassanakajon, 2009. Two prophenoloxidases are important for the survival of *Vibrio harveyi* challenged shrimp *Penaeus monodon. Dev Comp Immunol* 33(2): 247-56.

Assavalapsakul, W., D. R. Smith and S. Panyim, 2006. Identification and characterization of a *Penaeus monodon* lymphoid cell-expressed receptor for the yellow head virus. *J Virol* 80(1): 262-9.

Astrofsky, K. M., M. M. Roux, K. R. Klimpel, J. G. Fox and A. K. Dhar, 2002. Isolation of differentially expressed genes from white spot virus (WSV) infected Pacific blue shrimp (*Penaeus stylirostris*). *Arch Virol* 147(9): 1799-812.

Attasart P, Kaewkhaw R, Chimwai C, Kongphom U, Namramoon O, Panyim S, 2010. Inhibition of Penaeus monodon densovirus replication in shrimp by double-stranded RNA. Archives of virology 155: 825-832.

Beckmann, J.S. and Soller, M., 1990. Toward a unified approach to genetic mapping of eukaryotes based on sequence tagged microsatellite sites. *Bio/ technology (Nature Publishing Company)*, *8*(10), pp.930-932.

Benzie JAH, 1998. Penaeid genetics and biotechnology. *Aquaculture,* 184:23–47.

Benzie, John AH, 2005. "Marine shrimp genomics." In Symposium of the 21st COE Marine Bio-manipulation Frontier for Food Production "Potential and Perspective of Marine Bio-Manipulation" in Hokkaido, Japan.

Bierne N, Beuzart I, Vonau V, Bonhomme F, Bédier E, AQUACOP, 2000. Microsatellite-associated heterosis in hatchery propagated stocks of the shrimp *Penaeus stylirostris. Aquaculture,* 184:203–219.

Brady, P., Elizur, A., Cummins, S. F., Ngyuen, N. H., Williams, R., and Knibb, W. 2013. Differential expression microarrays reveal candidate genes potentially associated with reproductive dysfunction of captive-reared prawn *Penaeus monodon. Aquaculture,* .400–401.

Brooker, A.L., Benzie, J.A.H., Blair, D., Versini, J.J., 2000. Population structure of the giant tiger prawn *Penaeus monodon* in Australia waters, determined using microsatellite markers. *Mar. Biol.* 136, 149–157

Browdy, C.L., 1992. A review of the reproductive biology of Penaeus species: perspective on controlled shrimp maturation system for high quality nauplii production. In: Wyban, J. (Ed.), Proceedings of the Special Session on Shrimp Farming, World *Aquaculture.* Society, Baton Rouge, Louisiana, USA, pp. 22–51

Cesar, J. R., B. Zhao and J. Yang. 2008. Analysis of expressed sequence tags from abdominal muscle cDNA library of the pacific white shrimp *Litopenaeus vannamei*. Animal 2(9): 1377-1383.

Chen, Y.Y., Chen, J.C., Lin, Y.C., Yeh, S.T., Chao, K.P. and Lee, C.S., 2014. White shrimp Litopenaeus vannamei that have received Petalonia binghamiae extract activate immunity, increase immune response and resistance against Vibrio alginolyticus. *Journal of Aquaculture Research & Development, 2014.*

Coman GJ, Arnold SJ, Jones MJ, Preston NP, 2007. Effect of rearing density on growth, survival and reproductive performance of domesticated *Penaeus monodon. Aquaculture,* 264:175 –183

Covi, J.A, Kim, H.W, Mykles, D.L., 2008. Expression of alternatively spliced transcripts for a myostatin-like protein in the blackback land crab, *Gecarcinus lateralis. Comp. Biochem. Physiol. A Mol. Integr. Physiol.* 150, 423–430

De la Vega E, Degnan BM, Hall MR, Cowley JA, Wilson KJ. 2004. Quantitative real-time RT-PCR demonstrates that handling stress can lead to rapid increases of gill-associated virus (GAV) infection levels in *Penaeus monodon. Diseas. of Aquat. Organisms* 59: 195-203.

De la Vega, E., B. M. Degnan, M. R. Hall and K. J. Wilson. 2007a. Differential expression of immune-related genes and transposable elements in black tiger shrimp (*Penaeus monodon*) exposed to a range of environmental stressors. *Fish Shellfish Immunol* 23(5): 1072-88.

De la Vega, E., M. R. Hall, K. J. Wilson, A. Reverter, R. G. Woods and B. M. Degnan. 2007b. Stress-induced gene expression profiling in the black tiger shrimp *Penaeus monodon. Physiol Genomics* 31(1): 126-38.

De la Vega, E., N. A. O'Leary, J. E. Shockey, J. Robalino, C. Payne, C. L. Browdy, G. W. Warr and P. S. Gross. 2008. Anti-lipopolysaccharide factor in *Litopenaeus vannamei* (LvALF): a broad spectrum antimicrobial peptide essential for shrimp immunity against bacterial and fungal infection. *Mol Immunol* 45(7): 1916-25.

De Lorgeril, J., D. Saulnier, M. G. Janech, Y. Gueguen and E. Bachere. 2005. Identification of genes that are differentially expressed in hemocytes of the Pacific blue shrimp (*Litopenaeus stylirostris*) surviving an infection with *Vibrio penaeicida. Physiol Genomics* 21(2): 174-83.

Dechklar, M., A. Udomkit and S. Panyim. 2008. Characterization of Argonaute cDNA from Penaeus monodon and implication of its role in RNA interference. *Biochem Biophys Res Commun* 367(4): 768-74.

Dhar, A. K., A. Dettori, M. M. Roux, K. R. Klimpel and B. Read. 2003. Identification of differentially expressed genes in shrimp (*Penaeus stylirostris*) infected with White spot syndrome virus by cDNA microarrays. *Arch Virol* 148(12): 2381-96.

Dixon, T.J, Coman, G.J.Arnold, S.J, Sellars, M. J, Lyons, R.E, Dierens, L, Preston, N.P. and LI, Y. 2008. Shifts in genetic diversity during domestication of Black Tiger shrimp, *Penaeus monodon*, monitored using two multiplexed microsatellite systems. *Aquaculture*, 283, 1-4, p. 1-6.

Dunham R A , 2004. *Aquaculture,* and fisheries biotechnology: genetic approaches: CABI

F.A.O. (Food and Agriculture Organization of the United Nations) 2010. The state of world fisheries and *Aquaculture,* .www.fao.org/docrep/013/i1820e/i1820e01.pdf . Flegel TW, 2007. The right to refuse revision in the genus *Penaeus. Aquaculture,* 264:2 –8

Fagutao, F. F., M. Yasuike, C. M. Caipang, H. Kondo, I. Hirono, Y. Takahashi and T. Aoki. 2008. Gene expression profile of hemocytes of kuruma shrimp, *Marsupenaeus japonicus* following peptidoglycan stimulation. *Mar Biotechnol* (NY) 10(6): 731-40.

Freitas PD, Calgaro MR, Galetti PM, 2007. Genetic diversity within and between broodstocks of the white shrimp *Litopenaeus vannamei* (Boone, 1931) (Decapoda, Penaeidae) and its implication for the gene pool conservation. *Braz J Biol* 67:939–943.

Guegen Y, De Lorgeril J, Robert L, Romestand B, Bachere E. 2004. Anti-infectious immune geneexpression in penaeid shrimp. *Mar. Biotechnol.* 6: S112-S117.

Gitterle T, Rye M, Salte R, Cock J, Johansen H, Lozano C, 2005a. Genetic (co)variation in harvest body weight and survival in *Penaeus* (*Litopenaeus*) *vannamei* understandard commercial conditions. *Aquaculture,* 243:83 –92.

Gao H, Kong J, 2005. The microsatellites and minisatellites in the genome of *Fenneropenaeus chinensis. DNA Seq,* 16:426-436.

Gitterle T, Salte R, Gjerde B, Cock J, Johansen H, Salazar M, 2005b. Genetic (co)variation in resistance to white spot syndrome virus (WSSV) and harvest weight in *Penaeus* (*Litopenaeus*) *vannamei. Aquaculture,* 246:139 –149.

Goyard E, Penet L, Chim L, Cuzon G, Bédier E, 2001. Performance of *Penaeus stylirostris* after six generations of selection for growth. Global *Aquaculture,* Advocate 4:31–32.

Goyard, E., Patrois, J., Peignon, J., Vanaa, V., Dufour, R., Viallon, J.,Bedier, E., 2002. Selection for better growth of *Penaeus stylirostris* in Tahiti and New Caledonia. *Aquaculture,* 204, 461–468.

Goyard E, Goarant C, Ansquer D, Pierre BP, de Decker S, Dufour R, 2008. Cross breeding of different domesticated lines as a simple way for genetic improvement in small *Aquaculture,* industries: heterosis and inbreeding effects on growth and survival rates of the Pacific blue shrimp *Penaeus* (*Litopenaeus*) *stylirostris*. *Aquaculture,* 278:43–50.

Gross, P. S., T. C. Bartlett, C. L. Browdy, R. W. Chapman and G. W. Warr. 2001. Immune gene discovery by expressed sequence tag analysis of hemocytes and hepatopancreas in the Pacific White Shrimp, *Litopenaeus vannamei*, and the Atlantic White Shrimp, *L. setiferus*. *Dev Comp Immunol* 25 (7): 565-77.

Harrison, K. 1990. The role of nutrition in maturation, re-production and embryonic development of decapod crustaceans, A review. *Journal of Shellfish Research*. 9, 1-28

He, N., H. Liu and X. Xu. 2004. Identification of genes involved in the response of haemocytes of *Penaeus japonicus* by suppression subtractive hybridization (SSH) following microbial challenge. *Fish Shellfish Immunol* 17(2): 121-8.

He, N., Q. Qin and X. Xu. 2005. Differential profile of genes expressed in hemocytes of White Spot Syndrome Virus-resistant shrimp (*Penaeus japonicus*) by combining suppression subtractive hybridization and differential hybridization. *Antiviral Res* 66(1): 39-45.

Hetzel DJS, Crocos PJ, Davis GP, Moore SS, Preston NC, 2000. Response to selection and heritability for growth in the kuruma prawn, *Penaeus japonicus*. *Aquaculture,* 181:215–223.

Huang Shiao-Wei, You-Yu Lin, En-Min You, Tze-Tze Liu, Hung-Yu Shu, Keh-Ming Wu, Shih-Feng Tsai, Chu-Fang Lo, Guang-Hsiung Kou, Gwo-Chin Ma, Ming Chen, Dongying Wu, Takashi Aoki, Ikuo Hirono and Hon-Tsen Yu, 2011. Fosmid library end sequencing reveals a rarely known genome structure of marine shrimp *Penaeus monodon*. *BMC Genomics,* **12**:242.

Hui, J. H., S. S. Tobe and S. M. Chan. 2008. Characterization of the putative farnesoic acid O-methyltransferase (LvFAMeT) cDNA from white shrimp, *Litopenaeus vannamei*: Evidence for its role in molting. *Peptides* 29(2): 252-60.

Ibarra AM, Pérez-Rostro CI, Ramirez JL, Ortega-Estrada E, 2007. Genetics of the resistance to hypoxia in postlarvae and juveniles of the Pacific white shrimp *Penaeus* (*Litopenaeus*) *vannamei* (Boone 1931). *Aquac Res* 38:838–846

Karlin S, 1998. Global dinucleotide signatures and analysis of genomic heterogeneity. *CurrOpinMicrobiol*, 1:598-610.

Kashi Y, King D, Soller M, 1997. Simple sequence repeats as a source of quantitative genetic variation. *Trends Genet*, 13:74-78.

Khayat M, Babin PJ, Funkenstein B, Sammar M, Nagasawa H, Tietz A, Lubzens E. 2001. Molecular characterization and high expression during oocyte development of a shrimp ovarian cortical rod protein homologous to insect intestinal peritrphin. *Biol. of Reprod.* 64: 1090-1099.

Kim, C. S., Z. Kosuke, Y. K. Nam, S. K. Kim and K. H. Kim. 2007. Protection of shrimp (*Penaeus chinensis*) against white spot syndrome virus (WSSV) challenge by double-stranded RNA. *Fish Shellfish Immunol* 23(1): 242-6.

Kim, B.K., Kim, K.S., Oh, C.W., Mykles, D.L., Lee, S.G., Kim, H.J., Kim, H.W., 2009a. Twelve actin-encoding cDNAs from the American lobster, Homarus americanus: cloning and tissue expression of eight skeletal muscle, one heart, and three cytoplasmic isoforms. Comp. Biochem. *Physiol. B Biochem. Mol. Biol.* 153, 178–184.

Kim, K.-S., Jeon, J.-M., Kim, H.-W., 2009b. A myostain-like gene expressed highly in the muscle tissue of Chinesemitten crab, Eriocheir sinensis. *J. Fish. Aquat. Sci.* 12, 185–193.

King DG, Soller M, Kashi Y: Evolutionary tuning knobs. *Endeavour* 1997, 21:36-40.

Klinbunga S, Kanchana Sittikankaew, Napaporn Jantee, Sayan Prakopphet, Sirithorn Janpoom, Rachanimuk Hiransuchalert, Piamsak Menasveta, Bavornlak Khamnamtong, 2015. Expression levels of vitellogenin receptor (Vtgr) during ovarian development and association between its single nucleotide polymorphisms (SNPs) and reproduction-related parameters of the giant tiger shrimp *Penaeus monodon.* Aquaculture 435:18–27.

Künzler P, Matsuo K, Schaffner W: Pathological, physiological, and evolutionary aspects of short unstable DNA repeats in the human genome. *Biol Chem Hoppe Seyler* 1995, 376:201-211.

Lee Ji-Hyun , Jalal Momani, YoungMog Kim, Chang-Keun Kang , Jung-Hwa Choi , Hae-Ja Baek , Hyun-Woo Kim, 2015. Effective RNA-silencing strategy of Lv-MSTN/GDF11 gene and its effects on the growth in shrimp, *Litopenaeus vannamei. Comparative Biochemistry and Physiology*, Part B 179, 9–16.

Leelatanawit R, Klinbunga S, Puanglarp N, Tassanakajon A, Jarayabhand P, Hirono I, Aoki T, Menasveta P. 2004. Isolation and characterisation of differentially expressed genes in ovaries and testes of the giant tiger shrimp *Penaeus monodon. Mar. Biotech.* 6: S506-S510.

Leelatanawit, R., K. Sittikankeaw, P. Yocawibun, S. Klinbunga, S. Roytrakul, T. Aoki, I. Hirono and P. Menasveta. 2009. Identification, characterization and expression of sex-related genes in testes of the giant tiger shrimp *Penaeus monodon. Comp Biochem Physiol A Mol Integr Physiol* 152(1): 66-76.

Lehnert, S. A., K. J. Wilson, K. Byrne and S. S. Moore. 1999. Tissue-Specific Expressed Sequence Tags from the Black Tiger Shrimp *Penaeus monodon. Mar Biotechnol* (NY) 1(5): 465-0476.

Leu, J. H., C. C. Chang, J. L. Wu, C. W. Hsu, I. Hirono, T. Aoki, H. F. Juan, C. F. Lo, G. H. Kou and H. C. Huang. 2007. Comparative analysis of differentially expressed genes in normal and white spot syndrome virus infected *Penaeus monodon. BMC Genomics* 8: 120.

Li Y, Byrne K, Miggiano E, Whan V, Moore S, Keys S, Crocos P, Preston N, Lehnert S. 2003. Genetic mapping of the kuruma prawn *Penaeus japonicus* using AFLP markers. *Aquaculture,* 219: 143–156.

Li J, Liu P, He Y, Song Q, Mu N, Wang Y, 2005. Artificial selection in the new breed of *Fenneropenaeus chinensis* named "Yellow Sea1"based on fast growth trait. *J Fish China* 29:1–5.

Li Z, Li J, Wang Q, He Y, Liu P, 2006b. The effects of selective breeding on the genetic structure of shrimp *Fenneropenaeus chinensis* populations. *Aquaculture,* 258:278–282.

Li, D. F., M. C. Zhang, H. J. Yang, Y. B. Zhu and X. Xu. 2007. Beta-integrin mediates WSSV infection. *Virology* 368(1): 122-32.

Luan S, Kong J, Wang QY, 2006. Genetic variation in wild and cultured populations of the kuruma prawn *Marsupenaeus japonicus* (Bate 1888) using microsatellites. *Aquac Res* 37:785–792.

Lugo, J. M., Y. Morera, T. Rodriguez, A. Huberman, L. Ramos and M. P. Estrada. 2006. Molecular cloning and characterization of the crustacean hyperglycemic hormone cDNA from *Litopenaeus schmitti*. Functional analysis by double-stranded RNA interference technique. *FEBS J* 273(24): 5669-77.

Luo, T., X. Zhang, Z. Shao and X. Xu. 2003. PmAV, a novel gene involved in virus resistance of shrimp *Penaeus monodon. FEBS* Lett 551(1-3): 53-7.

Maningas, M.B.B., Kondo, H., Hirono, I., Saito-Taki, T. and Aoki, T., 2008. Essential function of transglutaminase and clotting protein in shrimp immunity. *Molecular immunology*, *45*(5), pp.1269-1275.

Macbeth M, Kenway M, Salmon M, Benzie J, Knibb W, Wilson K, 2007. Heritability of reproductive traits and genetic correlations with growth in the black tiger prawn *Penaeus monodon* reared in tanks. *Aquaculture,* 270:51–56.

Maneeruttanarungroj C, Pongsomboon S, Wuthisuthimethavee S, Klinbunga S, Wilson KJ, Swan J, Li Y, Whan V, Chu KH, Li CP, Tong J, Glenn K, Rothschild M, Jerry D, Tassanakajon A, 2006. Development of polymorphic expressed sequence tag-derived microsatellites for the extension of the genetic linkage map of the black tiger shrimp (*Penaeus monodon*).*Anim Genet*, 37:363-368.

Meehan D, Xu Z, Zunizg G, Alcivar-Warren A, 2003. High frequency and large number of polymorphic microsatellites in cultured shrimp, *Penaeus (Litopenaeus) vannamei* [Crustacea:Decapoda]. *Mar Biotechnol*, 5:311-330.

Moore SS, Whan V, Davis GP, Byrne K, Hetzel DJS, Preston N, 1999. The development and application of genetic markers for the Kuruma prawn *Penaeus japonicus. Aquaculture,* 173: 19-32.

Mykles, D.L., 1997. Crustacean muscle plasticity: molecular mechanisms determining mass and contractile properties. *Comp. Biochem. Physiol. Part B Biochem. Mol. Biol.* 117,367–378.

Leary, N. A., H. F. Trent, III, J. Robalino, M. E. T. Peck, D. J. McKillen and P. S. Gross. 2006. Analysis of multiple tissue-specific cDNA libraries from the Pacific white leg shrimp, *Litopenaeus vannamei.* Integrative and Comparative Biology 46(6): 931-939.

Ongvarrasopone, C., M. Chanasakulniyom, K. Sritunyalucksana and S. Panyim. 2008. Suppression of PmRab7 by dsRNA inhibits WSSV or YHV infection in shrimp. *Mar Biotechnol* (NY) 10(4): 374-81.

Pan, D., N. He, Z. Yang, H. Liu and X. Xu. 2005. Differential gene expression profile in hepatopancreas of WSSV-resistant shrimp (*Penaeus japonicus*) by suppression subtractive hybridization. Dev Comp Immunol 29(2): 103-12.

Perez-Rostro, C.I., Ibarra, A.M., 2003a. Quantitative genetic para-meter estimates for size and growth rate traits for Pacific white shrimp *Penaeus vannamei. Boone. Aquac. Res.* 30, 1–13.

Perez-Rostro, C.I., Ibarra, A.M., 2003b. Heritabilities and geneticcorrelations of size traits at harvest size in sexually dimorphic Pacific white shrimp (*Litopenaeus vannamei*) grown in two environments. *Aquac. Res.* 34,1079–1085.

Perez F, Erazo C, Zhinaula M, Volckaert F, Calderon J. 2004. A sex specific linkage map of the white shrimp *Penaeus (Litopenaeus) vannamei* based on AFLP markers. *Aquaculture,* 242: 105-118.

Pongsomboon, S., Whan, V., Moore, S.S., Tassanakajorn, A., 2000. Characterization of tri- and tetranucleotide microsatellites in the black tiger prawn, *Penaeus monodon. Science Asia* 26, 1–8

Pongsomboon S, Wongpanya R, Tang S, Chalorsrikul A, Tassanakajon A, 2008a. Abundantly expressed transcripts in the lymphoid organof the black tiger shrimp, *Penaeus monodon,* and their implication in immune function. Fish Shellfish Immunol 25:485–493.

Pongsomboon S, Tang S, Boonda S, Aoki T, Hirono I, Yasuike M, Tassanakajon A, 2008b. Differentially expressed genes in *Penaeus monodon* hemocytes following infection with yellow head virus. BMB Rep 41:670–677

Preechaphol, R., R. Leelatanawit, K. Sittikankeaw, S. Klinbunga, B. Khamnamtong, N. Puanglarp and P. Menasveta. 2007. Expressed sequence tag analysis for identification and characterization of sex-related genes in the giant tiger shrimp *Penaeus monodon. J Biochem Mol Biol* 40(4): 501-10.

Racotta, I.S., Palacios, E., Ibarra, A.M., 2003. Shrimp larval quality in relation to broodstock condition. *Aquaculture,* 227, 107–130

Rijiravanich, A., C. L. Browdy and B. Withyachumnarnkul. 2008. Knocking down caspase-3 by RNAi reduces mortality in Pacific white shrimp *Penaeus* (*Litopenaeus*) *vannamei* challenged with a low dose of white-spot syndrome virus. *Fish Shellfish Immunol* 24(3): 308-13.

Robalino, J., J. S. Almeida, D. McKillen, J. Colglazier, H. F. Trent, 3rd, Y. A. Chen, M. E. Peck, C. L. Browdy, R. W. Chapman, G. W. Warr and P. S. Gross. 2007. Insights into the immune transcriptome of the shrimp *Litopenaeus vannamei*: tissue-specific expression profiles and transcriptomic responses to immune challenge. *Physiol Genomics* 29(1): 44-56.

Robalino, J., T. Bartlett, E. Shepard, S. Prior, G. Jaramillo, E. Scura, R. W. Chapman, P. S. Gross, C. L. Browdy and G. W. Warr. 2005. Double-stranded RNA induces sequence-specific antiviral silencing in addition to nonspecific immunity in a marine shrimp: convergence of RNA interference and innate immunity in the invertebrate antiviral response? *J Virol* 79(21): 13561-71.

Robalino Chapman Robert W., De La Vega Enrique, O'Leary Nuala A., Gorbach Danielle M., Du Zhi-Qiang, Rothschild Max F., Browdy Craig L., Warr Gregory, 2012. Advances in genomics and genetics of Penaeid Shrimp. In *Aquaculture,* biotechnology (Wiley- Blackwell, Oxford, UK.

Rojtinnakorn, J., I. Hirono, T. Itami, Y. Takahashi and T. Aoki. 2002. Gene expression in haemocytes of kuruma prawn, *Penaeus japonicus*, in response to infection with WSSV by EST approach. *Fish Shellfish Immunol* 13(1): 69-83.

Sarathi, M., M. C. Simon, V. P. Ahmed, S. R. Kumar and A. S. Hameed. 2008a. Silencing VP28 gene of white spot syndrome virus of shrimp by bacterially expressed dsRNA. *Mar Biotechnol* (NY) 10(2): 198-206.

Sarathi, M., M. C. Simon, C. Venkatesan and A. S. Hameed. 2008b. Oral administration of bacterially expressed VP28dsRNA to protect *Penaeus monodon* from white spot syndrome virus. *Mar Biotechnol* (NY) 10(3): 242-9.

Saski CA, Hlederman R, Chapman RW, Benzie JAH, 2009. Advancing shrimp genomics. Plant and Animal Genome XVII Conference, San Diego, USA

Sellars, M. J., Degnan, B. M. & Preston, N. P. 2006. Production of triploid Kuruma shrimp, *Marsupenaeus (Penaeus) japonicus* (Bate) nauplii through inhibition of polar body I, or polar body I and II extrusion using 6-dimethylaminopurine. *Aquaculture,* 256, 337–345.

Shen, Y. Q., J. H. Xiang, B. Wang, F. H. Li and W. Tong. 2004. Discovery of immune related factors in *Fenneropenaeus chinensis* by annotation of ESTs. Progress in Natural Science 14(1): 47-54.

Shockey, J. E., N. A. O'Leary, E. de la Vega, C. L. Browdy, J. E. Baatz and P. S. Gross. 2008. The role of crustins in *Litopenaeus vannamei* in response to infection with shrimp pathogens: An in vivo approach. *Dev Comp Immunol.* 33(5):668-73.

Somboonwiwat, K., P. Supungul, V. Rimphanitchayakit, T. Aoki, I. Hirono and A. Tassanakajon. 2006. Differentially expressed genes in hemocytes of *Vibrio harveyi*-challenged shrimp *Penaeus monodon. J Biochem Mol Biol* 39(1): 26-36.

Staelens J, Rombaut D, Vercauteren I, Argue B, Benzie J, Vuylsteke M: High-density linkage maps and sex-linked markers for the black tiger shrimp (*Penaeus monodon*). *Genetics* 2008, 179:917-925

Su, J., D. T. Oanh, R. E. Lyons, L. Leeton, M. C. van Hulten, S. H. Tan, L. Song, K. V. Rajendran and P. J. Walker. 2008. A key gene of the RNA interference pathway in the black tiger shrimp, *Penaeus monodon*: identification and functional characterisation of Dicer-1. *Fish Shellfish Immunol* 24(2): 223-33.

Supungul, P., S. Klinbunga, R. Pichyangkura, I. Hirono, T. Aoki and A. Tassanakajon. 2004. Antimicrobial peptides discovered in the black tiger shrimp *Penaeus monodon* using the EST approach. *Dis Aquat Organ* 61(1-2): 123-35.

Supungul, P., S. Klinbunga, R. Pichyangkura, S. Jitrapakdee, I. Hirono, T. Aoki and A. Tassanakajon. 2002. Identification of immune-related genes in hemocytes of black tiger shrimp (*Penaeus monodon*). *Mar Biotechnol* (NY) 4(5): 487-94.

Tassanakajon, A, Tiptawonnukul A, Supungul P, Rimphanitchayakit V, Cook D, Jarayabhand P, 1998. Isolation and characterization of microsatellite markersin the black tiger prawn *Penaeus monodon*. *Mol. Mar. Biol. Biotechnol.* 7,55–61.

Tassanakajon, A., Klinbunga, S., Paunglarp, N., Rimphanitchayakit, V., Udomkit, A., Jitrapakdee, S., Sritunyalucksana, K., Phongdara, A., Pongsomboon, S., Supungul, P., Tang, S., Kuphanumart, K., Pichyangkura, R. and Lursinsap, C. 2006. *Penaeus monodon* gene discovery project: the generation of an EST collection and establishment of a database. *Gene* 384: 104-112.

Tirasophon, W., Y. Roshorm and S. Panyim. 2005. Silencing of yellow head virus replication in penaeid shrimp cells by dsRNA. *Biochem Biophys Res Commun* 334(1): 102-7.

Tirasophon, W., S. Yodmuang, W. Chinnirunvong, N. Plongthongkum and S. Panyim. 2007. Therapeutic inhibition of yellow head virus multiplication in infected shrimps by YHV-protease dsRNA. *Antiviral Res* 74(2): 150-5.

Tiu, S. H., J. Benzie and S. M. Chan. 2008. From hepatopancreas to ovary: molecular characterization of a shrimp vitellogenin receptor involved in the processing of vitellogenin. *Biol Reprod* 79(1): 66-74.

Tiu, S. H. and S. M. Chan. 2007. The use of recombinant protein and RNA interference approaches to study the reproductive functions of a gonad-stimulating hormone from the shrimp *Metapenaeus ensis*. *FEBS* J 274(17): 4385-95.

Tiu, S. H., J. G. He and S. M. Chan. 2007. The LvCHH-ITP gene of the shrimp (*Litopenaeus vannamei*) produces a widely expressed putative ion transport peptide (LvITP) for osmo-regulation. *Gene* 396 (2): 226-35.

Treerattrakool, S., S. Panyim, S. M. Chan, B. Withyachumnarnkul and A. Udomkit. 2008. Molecular characterization of gonad-inhibiting hormone of *Penaeus monodon* and elucidation of its inhibitory role in vitellogenin expression by RNA interference. *FEBS J* 275(5): 970-80.

Uawisetwathana, U., Leelatanawit, R., Klanchui, A., Prommoon, J., Klinbunga, S. and Karoonuthaisiri, N., 2011. Insights into eyestalk ablation mechanism to induce ovarian maturation in the black tiger shrimp. *PloS one*, 6(9), p.e24427.

Wang H-C, Chang Y-S, Kou G-H, Lo C-F. 2004. White spot syndrome virus: Molecular characterization of a major structural protein in a baculovirus expression system and shrimp hemocytes. *Mar. Biotechnol.* 6: S95-S99.

Wang, B., F. Li, B. Dong, X. Zhang, C. Zhang and J. Xiang. 2006. Discovery of the genes in response to white spot syndrome virus (WSSV) infection in *Fenneropenaeus chinensis* through cDNA microarray. *Mar Biotechnol* (NY) 8(5): 491-500.

Wang, L., B. Zhi, W. Wu and X. Zhang. 2008b. Requirement for shrimp caspase in apoptosis against virus infection. *Dev Comp Immunol* 32(6): 706-15.

Westenberg, M., B. Heinhuis, D. Zuidema and J. M. Vlak. 2005. siRNA injection induces sequence-independent protection in *Penaeus monodon* against white spot syndrome virus. *Virus Res* 114(1-2): 133-9.

Witchulada Talakhun, Bavornlak Khamnamtong, Pachumporn Nounurai, Sirawut Klinbunga,Piamsak Menasveta, 2014. Characterization, expression and localization of valosin-containing protein in ovaries of the giant tiger shrimp *Penaeus monodon. Gene* 533:188–198

Wilson K, Li Y, Whan V, Lehnert S, Byrne K, Moore S, Pongsomboon S, Tassanakajon A, Rosenberg G, Ballment E, Fayazi Z, Swan J, Kenway. M, Benzie JAH. 2002. Genetic mapping of the black tiger shrimp *Penaeus monodon* with amplified fragment length polymorphisms. *Aquaculture,* 204: 297-309.

Wolfus G M, DK Garcia, A Alcivar-Warre, 1997. Application of the microsatellite technique for analyzing genetic diversity in shrimp breeding programs. *Aquaculture,* v.152, p. 35-47, 1997

Wong, Q. W., W. Y. Mak and K. H. Chu. 2008. Differential gene expression in hepatopancreas of the shrimp *Metapenaeus ensis* during ovarian maturation. *Mar Biotechnol* (NY) 10(1): 91-8.

Wouters, R., Lavens, P., Nieto, J., Sorgeloos, P., 2001. Penaeid shrimp broodstock nutrition: an updated review on research and development. *Aquaculture*, 202, 1–21.

Wu, W., R. Zong, J. Xu and X. Zhang. 2008. Antiviral phagocytosis is regulated by a novel Rab-dependent complex in shrimp *Penaeus japonicus*. *J Proteome Res* 7(1): 424-31.

Wu, Y., L. Lu, L. S. Yang, S. P. Weng, S. M. Chan and J. G. He. 2007. Inhibition of white spot syndrome virus in *Litopenaeus vannamei* shrimp by sequence-specific siRNA. *Aquaculture*, 271: 21-30.

Wuthisuthimethavee S, Lumubol P, Vanavichit A, Tragoonrung S. 2003. Development of microsatellite markers in black tiger shrimp (*Penaeus monodon* Fabricius). *Aquaculture*, 224: 39-50

Xu Z, Dhar AK Wyrzykowski, Alcivar-Warren A. 1999. Identification of abundant and informative microsatellites from shrimp (*Penaeus monodon*) genome. *Animal Genetics* 30: 150-156.

Xu, J., F. Han and X. Zhang. 2007. Silencing shrimp white spot syndrome virus (WSSV) genes by siRNA. *Antiviral Res* 73(2): 126-31.

Yodmuang, S., W. Tirasophon, Y. Roshorm, W. Chinnirunvong and S. Panyim. 2006. YHV-protease dsRNA inhibits YHV replication in *Penaeus monodon* and prevents mortality. *Biochem Biophys Res Commun* 341(2): 351-6.

Yamano K, Qiu G-F, Unuma T. 2004. Molecular cloning and ovarian expression profiles of thrombospondin, a major component of cortical rods in mature oocytes of penaeid shrimp, *Marsupenaeus japonicus*. *Biol. Reproduc.* 70: 1670-1678.

Yang Shi Ping , Jian-Guo He , Cheng Bo Sun , Siuming Francis Chan, 2014. Characterization of the shrimp neuroparsin (MeNPLP): RNAi silencing resulted in inhibition of vitellogenesis. *FEBS Open Bio* 4, 976–986.

Zhao, Z. Y., Z. X. Yin, S. P. Weng, H. J. Guan, S. D. Li, K. Xing, S. M. Chan and J. G. He. 2007. Profiling of differentially expressed genes in hepatopancreas of white spot syndrome virus-resistant shrimp (*Litopenaeus vannamei*) by suppression subtractive hybridisation. *Fish Shellfish Immunol* 22(5): 520-34.

CHAPTER 2

BASICS OF PROTEOMICS IN SHRIMP

2.1 INTRODUCTION

Proteomics is the large-scale study of proteins, particularly their structures and functions. Proteins are vital parts of living organisms, as they are the main components of the physiological metabolic pathways of cells. The word "proteome" is a blend of "protein" and "genome". The proteome is the entire complement of proteins, including the modifications made to a particular set of proteins, produced by an organism or system. This will vary with time and distinct requirements, or stresses, that a cell or organism undergoes. The word "proteome" is derived from Proteins expressed by a genome, and it refers to all the proteins produced by an organism, much like the genome is the entire set of genes. Proteomics is the large-scale study of protein, particularly their structures and functions. Most importantly, unlike genome, the proteome differs from cell to cell and is constantly changing through its biochemical interactions with the genome and surrounding environment. One organism has radically different protein expression in different parts of its body, in different stages of its life cycle and in different environmental conditions. Since proteins play a central role in the life of an organism; proteomics is instrumental in discovery of biomarkers. With completion of human genome project, many researchers are now looking at how genes and proteins interact to form other proteins. It is anticipated that the creation of adequate methodologies for the rapid and parallel analysis of proteins will accelerate the 'functionalization' of these biomolecules and develop new biomarkers and therapeutic targets for the diagnosis and treatment of diseases in organisms and also increase our understanding of biological processes.

Over the last forty years global aquaculture presented a growth rate of 7.0% per annum with an amazing production of 90.4 million tonnes in 2012, and a contribution of 51% of aquatic animal food for human consumption. In order to meet the world's health requirements of fish protein, a continuous growth in production is still expected for decades to come. Though aquaculture is a very competitive market, global awareness regarding the use of scientific knowledge and emerging technologies to obtain a better farmed organism through a sustainable production has enhanced the importance of proteomics in seafood biology research. Proteomics, as a powerful comparative tool, has therefore been increasingly used over the last decade to address different issues in aquaculture, particularly better methods of breeding and spawning, production of quality eggs and larvae, disease resistant shrimps, faster growth, capacity to with stand to biotic and abiotic stresses, nutrition, health, quality and safety.

2.2 KEY TECHNOLOGIES USED IN PROTEOMICS

Proteomics is a rapidly emerging set of key technologies that are being used to identify proteins and map their interactions in a cellular context. With the sequencing of the human genome, the scope of proteomics has shifted from protein identification and characterization to include protein structure, function and protein-protein interactions. Technologies used in proteomic research include X-ray crystallography, two dimensional gel electrophoresis, mass spectrometry, yeast two-hybrid screens, and computational prediction programs. While some of these technologies have been in use for a long time, they are currently being applied to study physiology and cellular processes in high-throughput formats. It is the high-throughput approach that defines and characterizes modern proteomics. Following are the key technologies used in proteomics. One- and two-dimensional gel electrophoresis is used to identify the relative mass of a protein and its isoelectric point (Fig 1).

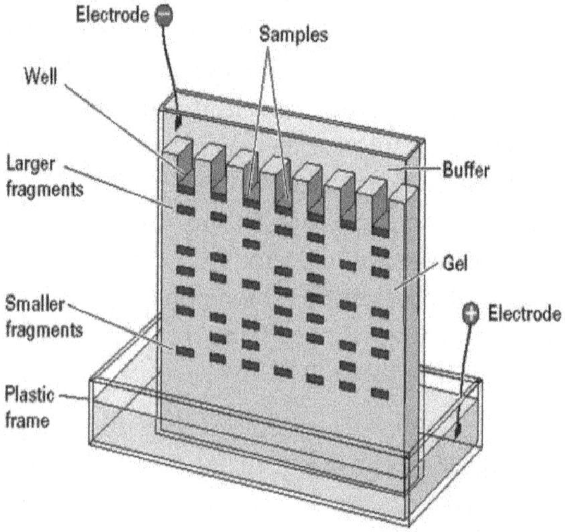

Fig. 1. 2D Gel electrophoresis

X-ray crystallography (Fig.2) and Nuclear Magnetic Resonance (NMR) are used to characterize the three-dimensional structure of peptides and proteins. However, low resolution techniques such as circular dichroism, Fourier transform infrared spectroscopy and small angle x-ray scattering can be used to study the secondary structure of proteins.

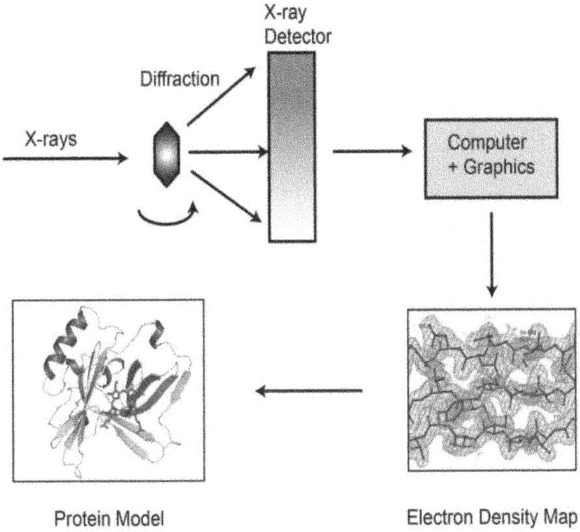

Fig. 2: X-ray crystallography

Tandem mass spectrometry combined with reverse phase chromatography or 2-D electrophoresis is used to identify (by de novo peptide sequencing) and quantify all the levels of proteins found in cells.

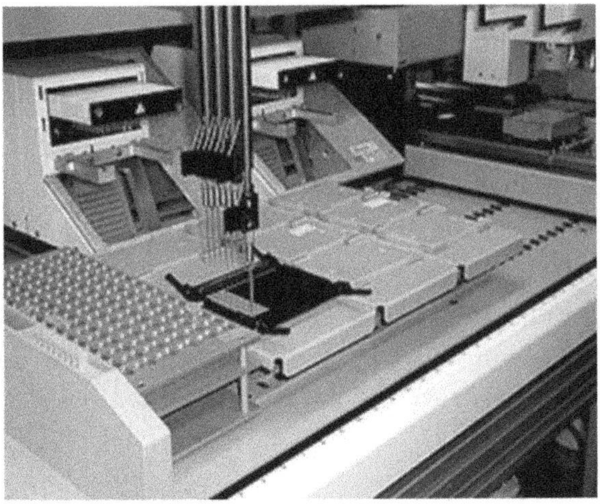

Fig. 3. Mass spectrometry

Mass spectrometry (no-tandem), often MALDI-TOF (Fig 3), is used to identify proteins by peptides mass finger printing. Less commonly this approach is used with chromatography and/or high resolution mass spectrometry.

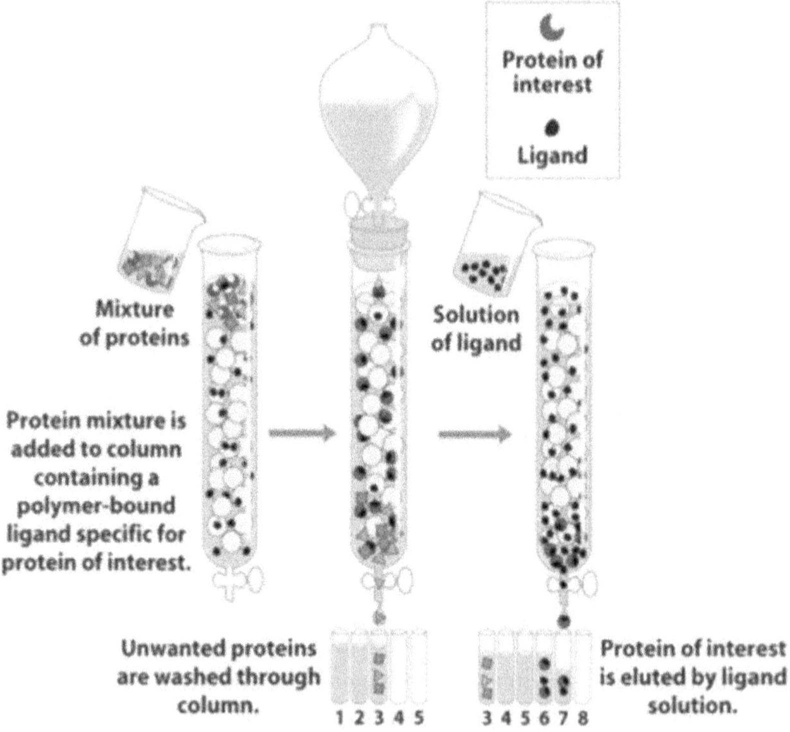

Fig. 4. Affinity chromatography

Affinity chromatography (Fig. 4), yeast two hybrid techniques, fluorescence resonance energy transfer (FRET), and Surface Plasmon Resonance (SPR) are used to identify protein-protein and protein-DNA binding reactions.

X-ray Tomography used to determine the location of labelled proteins or protein complexes in an intact cell, frequently correlated with images of cells from light based microscopes.

Software based image analysis is utilized to automate the quantification and detection of spots within and among gels samples. While this technology is widely utilized, the intelligence has not been perfected yet.

2.3 PROTEOMICS AND OVARIAN DEVELOPMENT AND MATURATION PROCESS

To understand molecular mechanisms of development and maturation of ovaries and oocytes in shrimps it is important to identify and characterise reproductive related genes/proteins expressed in ovaries. Differential expression of proteins during ovarian development of the shrimps could be inferred from the protein spectra. In recent years work on understanding the molecular mechanisms for development and maturation of oocytes/ovaries in penaeid shrimp have received great attention in the hope that it may lead to innovative ways of controlling reproductive maturation. Witchulada *et al.*, (2014) in their review paper it is mentioned that they found the induction effects of 5HT on transcription of reproductive related genes in several pathways in the shrimps, however it is also mentioned that there is no work on receptors for neurotransmitters based on EST analysis or on proteomic 2-DE analysis of genes and proteins expressed in the ovaries of *P. monodon* (Leelatanawit *et al.*, 2004, 2009, Witchulada *et al.*, 2012). The full length cDNA of 5-hydroxytryptamine (5-HT) receptor has been identified, isolated and characterised from *Metapenaeus ensis* (Tiu *et al.*, 2005) and from *P. monodon* (Ongvarrasopone *et al.*, 2006). Witchulada *et al.*, (2012) mentioned that the most abundantly expressed proteins in the ovaries of *P. monodon* belong to the protein disulphide isomerase family and they are functionally involved in chaperone activity and cell redox homeostasis. While carrying out proteomics studies in relation to ovarian development and maturation process in shrimp, number of workers have reported that compared to the wild shrimp *P. monodon*, reproductive maturation of captive tiger shrimp is reduced and probably this limits the species potential for growth of aquaculture as well as for domestication and selective breeding programs (Preechaphol *et al.*, 2007; Hiransuchalert *et al.*, 2013, Witchulada *et al.*, 2014). To overcome this problem in shrimp aquaculture, unilateral eyestalk ablation technique is being used to accelerate reproductive maturation. This is effective because eyestalks contain the X organ-sinus gland that is an important neuroendocrine complex for secretion and storage of various hormones, including gonad-inhibiting hormone (GIH), molt-inhibiting hormone (MIH), crustacean hyperglycemic hormone (CHH) and various pigment-concentrating and dispersing hormones (Fingerman and Nagabhushanam, 1992; Huberman, 2000; Okumura, 2004; Diwan, 2005, Subramoniam, 2011). Although unilateral eyestalk ablation induces reproductive maturation in penaeid shrimp (Okumura, 2004; Diwan, 2005, Marsden *et al.*, 2007), it also affects the balance of other physiological processes which directly or indirectly affects reduced fecundity and also egg quality and later the mother dies (Benzie, 1998). Therefore, Browdy, (1998) and Quackenbush, (2001) suggested for predictable and sustainable aquaculture, with induction of ovarian maturation and spawning of captive penaeid shrimp, without the use of unilateral eyestalk ablation. Preechaphol *et al.*, (2007, 2010), identified and characterised genes differentially expressed at different ovarian stages of *P. monodon* by EST analysis and by suppression subtractive hybridization (SSH) technique. More recently, Witchulada *et al.*, (2012) carried out cellular proteomics studies for the identification of reproduction-related proteins of *P. monodon* using two-dimensional gel electrophoresis (2-DE) and nano-electrospray ionization tandem mass spectrometry. By using this process, a total of 183 and 167 protein spots from vitellogenic (stage II) and 152 and 103 protein spots from mature (stage IV) ovaries of intact and eyestalk-ablated brood stock were examined. Differentially expressed proteins identified included, ubiquitin carboxyl-terminalhydrolase L5-like protein, 26S proteasome non-ATPase regulatory isoform A subunit 14, cysteine protease, valosin-containing protein and vitellogenin that are functionally implicated in the development and maturation of ovaries. Witchulada *et al.*, (2014) analysed the proteins in ovaries of domesticated and intact wild brood stock of *P. monodon* by using the technique of one-dimensional gel electrophoresis (SDS–PAGE) and further by nanoESI-LC-MS/MS. By this process, several different protein families were characterized. In addition, full-length cDNAs encoding β-thymosin (PmTmsb) and Rac-GTPase-activating protein 1 (PmRacgap1) were also further characterized and the expression profiles of these proteins were also examined by them during the course of ovarian development. It was reported that out of 1638 proteins identifies, 1253 of these significantly matched with the known proteins that were previously deposited in the data bases. Of these several reproduction related proteins, it is reported that the protein PmTmsb is expressed in all the tissues whereas PmRacgap 1 is more abundantly expressed in the gonadal tissue. Further it is reported that the expression levels of PmTmsb and PmRacgap 1 in ovaries of wild adult shrimp were significantly greater than those in ovaries of juveniles (Witchulada *et al.*, 2014). In the eyestalk ablated shrimp it was observed that there was significant reduction in PmTmsb expression at stages I and III of ovarian development, however on the contrary there was no effect in the expression of PmRacgap 1 during the same stages of ovarian development. From these results it was concluded that PmTmsb and PmRacgap 1 may act as negative effectors during ovarian development in *P. monodon* (Fig. 5, 6).

In penaeid shrimp, during maturation process of oocytes, development of oocytes is arrested at the meiotic prophase and reaches metaphase 1 after ovulation (Yano, 1998). It is still not known whether crustaceans possess a gonadotropin hormone that can trigger the meiotic resumption and final oocyte maturation as in most vertebrates (Miura *et al.*, 2006 and Thomas, 2008). Accordingly, understanding the molecular function of reproduction–related genes that are

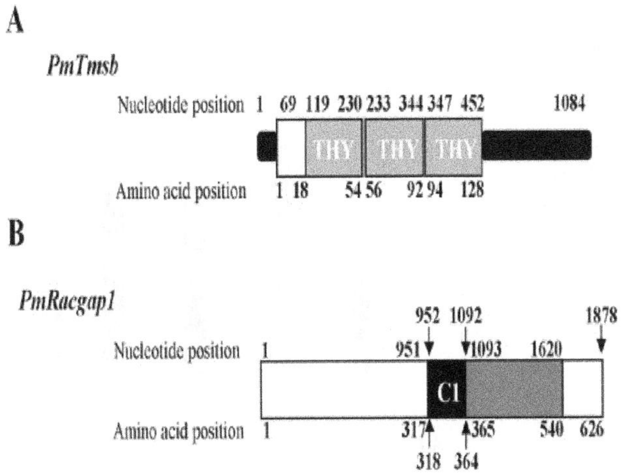

Fig. 5. Diagrams representing of the full-length cDNA of PmTmsb (A) and the complete ORF of PmRacgap1 (B). (**Source:** Witchulada *et al.*, 2014)

differentially expressed during ovarian maturation will be useful to increase the culture efficiency of many shrimps which are commercially important. Hiransuchalert *et al.*, (2013) isolated and characterised full length cDNA and genomic organization of small androgen receptor-interacting protein 1 which is called SARIP-1 in tiger shrimp *P. monodon*. Their studies indicated that SARIP 1 play an important role during ovarian maturation of *P. monodon*. The characterization of genes differentially expressed in ovaries is necessary for understanding the mechanisms involving ovarian developmental processes of the shrimp. Phinyo *et al.*, (2014) did the work on characterization and expression analysis of Cyclin-dependent kinase 7 gene (PmCdk7) and protein in ovaries of the tiger shrimp P monodon. The expression Cdk 7 in the ovaries has been reported to be greater than testes. The expression levels of Cdk 7 during ovarian development (stages I-IV) in the eyestalk ablated brood stock have been observed greater than that of the same ovarian stages in the intact brood stock. These results indicated that PmCdk 7 seems to play functional roles in the development and maturation of oocytes/ ovaries in *P. monodon*. Preechaphol *et al.*, (2010) isolated and characterized genes functionally involved in ovarian development of the shrimp *P. monodon* by suppression subtractive hybridization (SSH). They have identified several reproductive related transcripts and among them selenoprotein M precursor (PmSePM) transcript has been reported to be diminishing throughout the ovarian development. Their studies confirmed that the expression profiles of PmSePM and other transcripts like keratinocyte-associated protein 2 can be used as biomarkers for evaluating the degree of reproductive maturation in domesticated *P. monodon* (Preechaphol *et al.*, 2010).

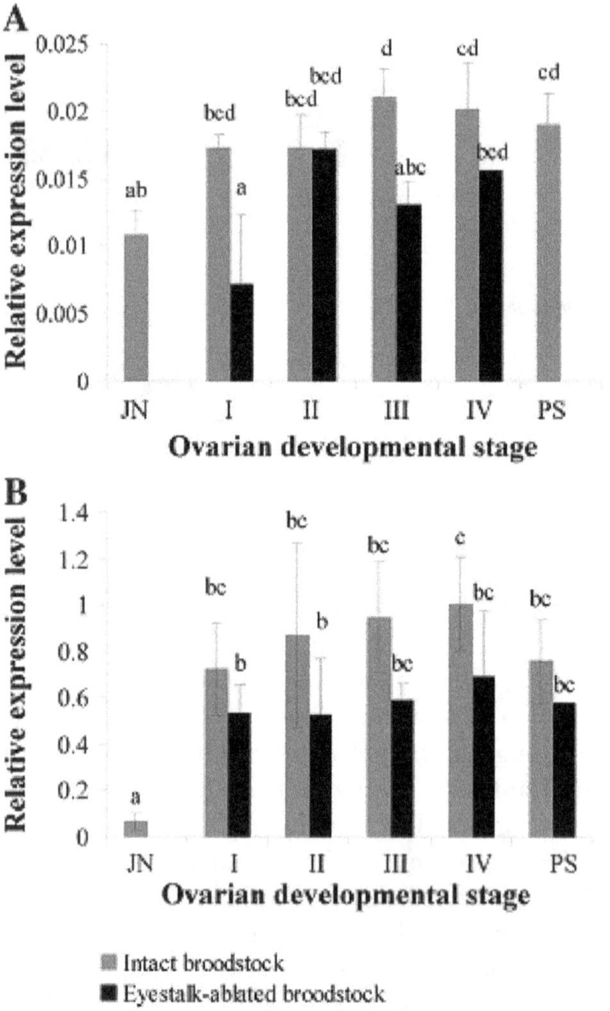

Fig. 6. Mean relative expression levels of PmTmsb (A) and PmRacgap1 (B) during ovarian development of wild intact and unilateral eyestalk-ablated broodstock of *P. monodon*. The same letters above bars indicate non-signiûcant differences between relative expression levels of different groups of samples (P N 0.05). JN = juvenile ovaries; I–IV = previtellogenic, vitellogenic, late vitellogenic, and mature ovaries, respectively; PS = ovaries of intact adults immediately collected after spawning (stage V). (**Source:** Witchulada *et al.*, 2014)

Cui *et al.*, (2013) carried out two-dimensional electrophoresis and mass spectrometry studies to identify proteins that are differentially expressed during ovarian maturation in *Metapenaeus ensis*. 87 spots with consistently significant quantitative differences among stage I, III and V ovaries were chosen for MS/MS analysis. 45 spots have been reported significantly to match to the known proteins in the databases. Functionally, these identified proteins further they have classified into five major groups, including cytoskeleton (11 %), metabolism (18 %), signal transduction (32 %), gene expression (14 %) and immune response (7 %). Among the differentially expressed reproduction-related proteins, the mRNA expression level of cellular retinol binding protein in *M. ensis* during ovarian maturation was further characterized by quantitative real-time PCR and it was reported that during ovarian maturation retinol binding protein transcript was found in ooplasm of previtellogenic oocytes but not in vitellogenic oocytes. These results demonstrate the application of proteomic analysis for identification of proteins involved in shrimp ovarian maturation and they provide new insights into ovarian development. Cui *et al.,* (2013) while carrying out studies on expression profiles and localization of vitellogenin mRNA and protein during ovarian development of the giant tiger shrimp *P monodon* found that this protein was expressed only in the ovaries and hepatopancreas but not in other tissues of female and testes of male brood stock. In the wild intact brood stock this protein was expressed at a low levels in the stage I ovaries up regulated in the stages II and III being significantly reduced in the stage IV ovaries. Further they have reported that in eyestalk ablated brood stock , the expression levels of this protein increased to nearly peak levels in vitellogenic ovaries (stage II) and peaked in cortical rods ovaries (stage III). The level of this protein in the hepatopancreas of intact shrimp with stages II and III ovaries was significantly greater than those with stages IV and V ovaries. Interestingly, it is mentioned that this protein in hepatopancreas has increased 25-40 times higher than that in the ovaries of intact shrimp with stages I-III of ovarian development.

Simulation of ovarian development in white shrimp *F. merguiensis* with a recombinant ribosomal protein (L10a) has been reported by Palasin *et al.,* (2014). In their findings they mentioned that a dose of 180µg of recombinant ribosomal protein L10A /shrimp was effective for inducing the ovaries to reach stages of I and II of vitellogenesis within 7 days after the first injection (Fig. 7). Even the levels of methyl farnesoate in the haemolymph of the treated shrimp are found to be increased in the stage II of vitellogenesis. Many earlier reports also suggested that differential expression of ovarian genes occurs at various stages of vitellogenesis in shrimp *P. monodon* (Karoonuthaisiri *et al.,* 2009, Preechaphol *et al.,* 2007, Uawisetwathana *et al.,* 2011), *F. chinensis* (Xie *et al.,* 2010), *M. japonicus* (Callaghan *et al.,* 2010), *M. ensis* (Lo *et al* ., 2007) and *F. merguiensis* (Wonglapsuwan *et al.,* 2009). The expression of ribosomal protein in relation to ovarian development in *F. merguiensis* has also been reported by Wonglapsuwan *et al.,* (2009). Perhaps the most important findings that have been put forth in this context is that of thrombospondin (TSP) and peritrophin are the ovarian development related genes, both of which exhibit their most abundant expression during vitellogenesis. Karoonuthaisiri *et al.,* (2009) reported a nuclear auto antigenic sperm protein which is involved in late stages of ovarian development in *P. monodon*. In recent years some early expressed genes have been suggested as molecular markers that may be used for detecting ovarian development in the shrimps (Makkapan *et al.*, 2011, Navakanitworakul *et al.*, 2012).

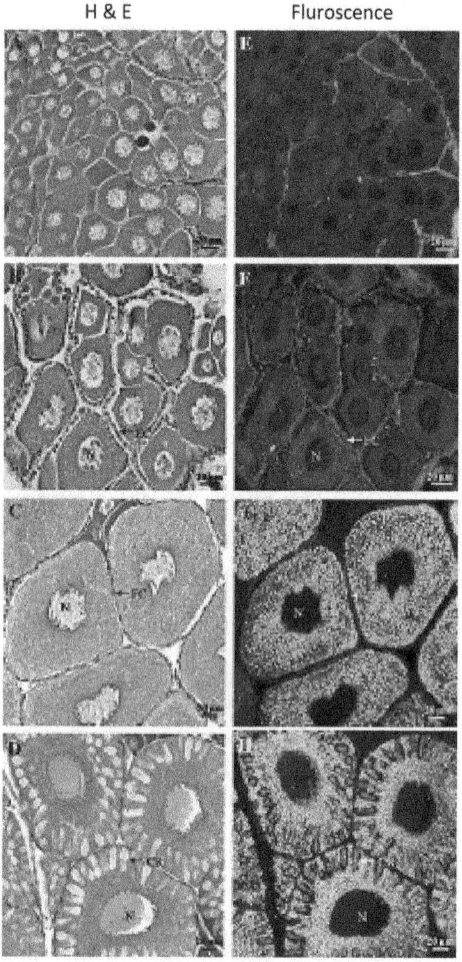

Fig. 7. Ovarian developmental stages (Vitellogenic I & II) in *F. merguiensis* after administration of recombinant ribosomal protein L10A. (Hematoxylin and Eosin stained sections and Fluroscent dye sections) (**Source:** Palasin *et al.*, 2014)

2.4 PROTEOMICS AND IMMUNOLOGICAL RESPONSES

In shrimp immune system the haemocytes are the main defence components which protect the shrimps against pathogenic microbes. Besides, antimicrobial peptides which are produced in the body are also considered as the first line of defence against pathogen infections. The other important shrimp defence systems include production of enzymes and proteins in the prophenoloxidase activation form and blood coagulation systems. Iwanaga *et al.,* (1998) in their review paper mentioned that in crustaceans the haemocytes upon endotoxin activation produce two types of protein molecules, one type helps in clotting for haemolymph coagulation and other type helps in antimicrobial activity. A septic injury induces the rapid and transient transcription of several genes encoding potent antibacterial and antifungal peptides that are released in to the blood where they act to destroy the invading microorganisms. In penaeid shrimp *L. vannamei,* three antimicrobial peptides have been purified form haemocytes and the plasma. These peptides have been characterized thoroughly and their cDNA cloned and research is being carried out to investigate the role of these peptides in the immune response against pathogens as well as with the characterization of other peptides in shrimp (Destoumieux *et al.,* 1997).

Recent approaches to fight against pathogens are Genomic and proteomic methods, such as simple gene cloning, high throughput expressed sequence tag analysis, suppression subtractive hybridization, and others, are some of the important tools for the identification of candidate genes involved in shrimp immunity (Dong and Xiang, 2007, Gross *et al.,* 2001, Leu *et al.,* 2007, Robalino *et al.,* 2009, Rojtinnakorn *et al.,* 2002, Supungul *et al.,* 2004, Tassanakajon *et al.,* 2006). Nevertheless, the full-genome sequence data of shrimps is seemingly necessary for a more complete search of immune-related proteins. Recently, a few reports have indicated that proteomic based techniques are a useful alternative method of choice in the identification of shrimp immune-related proteins (Robalino *et al.,* 2009, Jiang *et al.,* 2009, Wang *et al.,* 2007, Yao *et al.,* 2005). In all these studies, it is reported that differentially expressed proteins from shrimp tissues under hypoxia stress and various pathogenic conditions have been examined and compared to those found in shrimps under normal states. For example, the expression profile of proteins from the stomach of WSSV-infected *L. vannamei* was examined using 2D-gel electrophoresis (2-DE) and mass spectrometry (Wang *et al.,* 2007). Wu *et al.* (2007) used two proteomic approaches, a shotgun 2D-LC-MS/MS and a cleavable isotope-coded affinity tag, to study the differentially expressed cellular proteins from WSSV-infected shrimp epithelium. A 2D-LC-MS/MS approach has also been applied to explore the response of shrimps to WSSV infection in gill tissues (Robalino *et al.,* 2009). Somboonwiwat *et al.,* (2010) carried out, a proteomic analysis of *P. monodon* haemocytes to elucidate the shrimp immune responses at the translational level upon bacterial challenge. The haemocyte proteins whose expression changed upon *V. harveyi* infection were identified in order to explore the shrimp immune responses against bacterial infection. Then the expression patterns during the course of *V. harveyi* infection of some differentially expressed proteins were further assessed by western analysis and real-time RT-PCR. From the results obtained they have concluded the usefulness of a proteomic approach to the study of shrimp immunity and revealed hemocyte proteins whose expression were up regulated upon *V. harveyi* infection such as hemocyanin, arginine kinase and down regulated such as alpha-2-macroglobulin, calmodulin and 14-3-3 protein epsilon. The information is useful for understanding the immune system of shrimp against pathogenic bacteria.

Jung *et al.*, (2003) by utilising the knowledge of proteomics they could identify a novel allergen which will be useful in allergy diagnosis and in the treatment of Crustacean-derived allergic disorders. A novel allergen from *P. monodon*, designated Pen m 2, was identified by two-dimensional immunoblotting using sera from subjects with shrimp allergy, followed by matrix-assisted laser desorption ionization time-of-flight mass spectrometry analysis of the peptide digest. . This novel allergen was then cloned and the amino acid sequence deduced from the cDNA sequence. The cloned cDNA encoded a 356-aa protein with an acetylated N terminus at Ala2, identified by post source decay analysis. Comparison of the Pen m 2 sequence with known protein sequences revealed extensive similarity with arginine kinase from crustaceans. Pen m 2 was purified by anion exchange chromatography and shown to have arginine kinase activity and to react with serum IgE from shrimp allergic patients and induce immediate type skin reactions in sensitized patients. Using Pen m 2-specific antisera and polyclonal sera from shrimp-sensitive subjects in a competitive ELISA inhibition assay, Pen m 2 was identified as a novel cross-reactive Crustacean allergen. Huang *et al.*, (2002) carried out proteomic analysis of shrimp white spot syndrome viral proteins and characterization of a novel envelope protein VP4. It is well known that White spot syndrome virus (WSSV) is at present one of the major pathogens in shrimp culture worldwide and the complete genome of this virus has been sequenced recently. It has been reported that eighteen proteins matching frames of WSSV genome were identified and except for three known structural proteins and collagen, the functions of the remaining 14 proteins were not known. Results in this investigation thus proved the effectiveness of proteomic approaches for discovering new proteins of WSSV.

Zhang *et al.*, (2004) while doing the studies on identification of structural proteins from shrimp white spot syndrome virus (WSSV) by 2DE-MS found that this virus has 305-kb double-stranded DNA genome and has the capacity to encode 181 presumptive proteins. In an attempt to identify the viral proteins from the 181 theoretical proteins, proteins of the purified WSSV were separated by two-dimensional electrophoresis (2-DE) and mass spectrometry. By this kind of analysis with the proteomic approach, these viral protein and their corresponding genes were further confirmed by reverse transcription-polymerase chain reaction (RT-PCR). Two of them were characterized to be WSSV envelope proteins using immuno-electron microscopy. These studies showed that the proteomic approach is a powerful method for discovering the viral structural proteins and their corresponding genes. Zhengjun *et al.*, (2007) while carrying the studies on structural proteome of shrimp white spot syndrome virus found that this viral genome of WSSV contains a 305-kb double-stranded circular DNA, which encodes 181 predicted open reading frames. In their studies they used shotgun proteomics using off-line coupling of an LC system with MALDI-TOF/TOF MS/MS as a complementary and comprehensive approach to investigate the WSSV proteome. This approach has led to the identification of 45 viral proteins; 13 of them are reported for the first time. Seven viral proteins they reported to have acetylated N termini. RT-PCR confirmed the mRNA expression of these 13 newly identified viral proteins. Further by using the technique of quantitative proteomics, they could distinguish envelope proteins and nucleocapsid proteins of WSSV. Such studies on identification of WSSV structural proteins and their localization should facilitate further research on the discovery of mechanism of viral infection. Though crustaceans lack true immune system, potential to vaccinate *P. monodon* shrimp against white spot syndrome virus (WSSV) using the WSSV envelope proteins VP19 and VP28 was evaluated by Witteveldt *et al.*, (2004). Shrimp were vaccinated by intramuscular injection of the purified WSSV proteins and challenged 2 and 25 days after vaccination to assess the onset and duration of protection. It is reported that VP19-vaccinated shrimp showed a significantly better survival as compared to the control shrimp with a relative percent survival of 33% and 57% at 2 and 25 days after vaccination, respectively. Also, the groups vaccinated with VP28 and a mixture of VP19 and VP28 showed a significantly better survival when challenged two days after vaccination but not after 25 days. From the results it was concluded that shrimps can be protected against WSSV using its structural proteins as a subunit vaccine. This study further shows that vaccination of shrimp may be possible despite the absence of a true adaptive immune system, opening the way to new strategies to control viral diseases in shrimp and other crustaceans. Westenberg *et al.*, (2005) studied the effect of siRNA injection against White spot syndrome virus (WSSV) in P. monodon and it was reported that after siRNA injection mortality rate was reduced in shrimps. With this they concluded that siRNA induces a sequence-independent anti-viral immunity when injected in shrimp. Tsai *et al.*, (2006) also carried out studies on identification of Nucleocapsid and Envelope proteins of the shrimp white spot syndrome virus using proteomic approach and reported that there are at least 39 structural proteins are currently known for this virus. However, several details of the virus structure and assembly remain controversial. In order to understand the pathogenesis of white spot syndrome virus (WSSV) and to determine which cell pathways might be affected after WSSV infection, two-dimensional gel electrophoresis (2-DE) was done by Wang *et al.*, (2007) to study protein expression profiles from samples taken at 48 h post-infection from the stomachs of *L. vannamei* that were either specific pathogen free or else infected with WSSV. They identified 53 proteins, with functions that included energy production, calcium homeostasis, nucleic acid synthesis, signalling/communication, oxygen carrier/transportation etc. It is mentioned that 2-DE results observed were more consistent with relative EST database data from a previously developed EST database of two *P. monodon* cDNA libraries. It is further discussed that such analysis of differentially expressed proteins in WSSV-infected shrimp, would increase our understanding of the molecular pathogenesis of this virus-associated shrimp disease.

Robalino *et al.*, (2009) while working on functional genomics and proteomics in response to immune system in the shrimp *L. vannamei* emphasized the need for better control of infectious diseases in shrimp aquaculture which has prompted to develop interest in the study of crustacean immune systems. Further they emphasized that in order to understand immune system properly, functional genomic and proteomic approaches are better ways for obtaining molecular information regarding immune responses in these organisms. In their review article a series of results derived from transcriptomic and proteomic studies in shrimp (*Litopenaeus vannamei*) are discussed. They have used different techniques viz. Expressed Sequence Tag analysis, differential expression cloning through Suppression Subtractive Hybridization, expression profiling using microarrays, and proteomic studies using mass spectrometry, to analyse immune responses to infectious diseases which have provided a wealth of useful data and opportunities for new avenues of research. Examples of new research directions arising from these studies in shrimp include the molecular diversity of antimicrobial effectors, the role of double stranded RNA as an inducer of antiviral immunity, and the possible overlap between antibacterial and antiviral responses in the shrimp.An antimicrobial protein, crustin, is involved in the innate immunity of crustacean by defending the host directly against the microbial pathogens. Donpudsa *et al.*, (2014) while working on crustins from *P. monodon* identified different types of crustins by data mining from EST database. The abundant crustins reported were crustin pm1, 4 and 7, each with variation in the length of Gly-rich repeat among its members. The crustin pm4 which was characterized and later expressed from the hemocytes, its expression was up-regulated readily by WSSV infection and gradually decreased to normal level afterwards. It is reported that the recombinant crustin pm4 was able to inhibit the growth of a Gram-positive bacterium so also Gram –negative bacteria *E. coli* and vibrio at lower potency. Crustin which is identified as antimicrobial protein, has also been correlated with important function of defending immune system in number of crustaceans (Bartlett *et al.*, 2002, Jiravanichpaisal *et al.*, 2007, Smith *et al.*, 2008, Tassanakajon *et al.*, 2011).

2.5 PROTEOMICS AND BIOTIC AND ABIOTIC STRESSES

In shrimp culture operations the animals are continuously exposed to different kinds of biotic and abiotic stresses. These stresses when they persist continuously can cause imbalance in the physiological system, weakening of immune system, and spread of viral and bacterial diseases and mortality of the shrimps. In shrimps, the relationship between various environmental factors and medication physiological functions particularly that of immune system is largely unknown (Direkbusarakom and Danayadol, 1998; Hall and van Ham, 1998; Le Moullac, and Haffner, 2000; Le Moullac *et al.*, 1998; Perazzolo *et al.*, 2002). Survey of the literature shows that so far research has been focused mainly on measuring susceptibility to bacterial pathogens and quantitative analysis of a limited number of immune parameters (Le Moullac, and Haffner, 2000; Hall de la Vega, 2003). Similarly work on the relationship between environmental stressors and susceptibility to viral infections in shrimp is limited except the works of (Bray WA *et al.*, 1994) on *L. vannamei* on salinity and infectious hypodermal hematopoietic necrosis virus, another report on effect of acute salinityon WSSV out breaks in *F. chinensis* (Liu *et al.*, 2006) few reports on the effect of water temperature on the pathogenicity and replication of WSSV in *L. vannamei* (Granja, 2003; Guan, 2003; Vargas, 1993). Though there are few studies on number of genes involved in the shrimp immune system, their identification and characterization (Huang, 2004; Roux, 2002; Sotelo, 2003; Sritunyalucksana , 1999) in relation to environmental factors, the influence of different stressors on gene expression in the shrimp has not been explored and information is limited.

Studies have shown that stress could reduce shrimp resistance to pathogenic diseases, resulting in deceased growth rate and mass mortality (Moullac *et al.*, 1998, Moullac and Haffner, 2000, Burgents *et al.*, 2005, Zhang *et al.*, 2014). Costas *et al.*, (2008) mentioned that crowding stress could affect amino acid metabolism in fishes. Zhang *et al.*, (2014) studied effect of crowding on muscle protein, as a result of high stocking densities in shrimp culture operations and found that structural profile of muscle protein in the shrimp *F. chinensis* gets altered leading to cause muscle flesh quality degradation. These studies were carried out using two dimensional gel electrophoresis coupled with matrix-assisted laser desorption/ ionization time of flight/ time of flight (MALDI-TOF/ TOF) mass spectrometry (MS) to analyse differentially expressed muscle protein between shrimp under crowded and non-crowded conditions. Jiang *et al.*, (2009) carried out proteomic studies of the hepatopancreatic tissues in *F. chinensis* in response to hypoxic stress. Hypoxia, as one suboptimal environmental condition, can affect the physiological state of shrimp during pond aqaculture. To better understand the mechanism of response to hypoxic stress proteome research approach was utilized. Differentially expressed proteins of hepatopancreas in adult Chinese shrimp between the control and hypoxia-stressed groups were screened. It is reported that by 2-DE analysis, 67 spots showed obvious changes after hypoxia *and u*sing LC-ESI-MS/MS, 51 spots representing 33 proteins were identified. This study is the first analysis of differentially expressed proteins in the hepatopancreas of shrimp after hypoxia and provides a new insight for further study in hypoxic stress response of shrimp at the protein level.

Silvestre *et al.,* (2010) carried out the studies on assessment of the effects of chemotherapeutics and production management of the shrimp *P. monodon* by using proteomic method. For the purpose of assessing the stress level a proteomic analysis (2D-DE) was performed on haemolymph after exposure of the shrimp for 7 days to the antibiotics enrofloxacin or furazolidone via feed (4 g $kg^{"1}$) under laboratory conditions and also under field condition. It has been reported that no significant different protein abundance pattern was induced by antibiotics under laboratory conditions, while only one protein spot displayed after exposure to enrofloxacin in improved field conditions. These results showed that the very subtle effects of tested antibiotics on patterns of haemolymph protein expression in different production management systems.

DNA microarrays have become an important tool to measure global gene expression changes and genetic pathways involved in response to environmental stressors and toxicants. Li and Brouwer (2009) studied the gene expression profile of the grass shrimp *Palaemonetespugia* exposed to chronic hypoxia. In this study they have used a cDNA microarray which was designed and constructed from six libraries of expressed sequence tags generated in their previous study (Li and Brouwer, 2009). From the results obtained they concluded that cDNA microarray is a valid and useful tool to investigate the changes in gene expression of grass shrimp during chronic hypoxia exposure. Some genes, such as those coding for hemocyanin, ATP synthase, phosphoenolpyruvate carboxykinase, vitellogenin, trachealess, cytochrome *c* oxidase subunit I, lysosomal thiol reductase, and C-type lectin, could be used as molecular indicators of chronic hypoxia at specific time points. Vaga *et al.,* (2007) worked on the effect of environmental stress on gene expression profiling in the shrimp *P. monodon.* In their studies they investigated changes in temporal gene expression from shrimp exposed to hypoxic, hyper thermic and hypo osmotic conditions. From their findings, it is reported that hypoxic and hyper thermic stressors induced the most severe short-term response in terms of gene regulation, and the osmotic stress had the least response in gene expression profile relative to the control shrimps. It is further mentioned that these expression data agree with observed differences in shrimp physical appearance and behavior following exposure to stress conditions.

2.6 CHALLENGES IN PROTEOMICS WITH RESPECT TO SHRIMP CULTURE

The aim of proteomics is to simultaneously resolve all potential proteins expressed by a cell, tissue or organism in a specific physiological condition. It will allow scientists to build and test better hypotheses, with the ultimate goal to find better solutions to challenges in agricultural sciences, medicine and environmental management. Recent and very promising applications of proteomics have also been provided in the field of aquaculture, such as the search for antigenic proteins, the detection of differentially regulated proteins and the characterization of biologically active proteins primarily to investigate the physiology, development biology and impact of contaminants in aquatic organisms.

References

Bartlett, T.C., Cuthbertson, B.J., Shepard, E.F., Chapman, R.W., Gross, P.S., Warr, G.W., 2002. Crustins, homologues of an 11.5-kDa antibacterial peptide, from two species of penaeid shrimp, *Litopenaeus vannamei* and *Litopenaeus setiferus*. *Mar. Biotechnol.* 4, 278–293.

Benzie JAH, 1998. Penaeid genetics and biotechnology. *Aqaculture* 184:23–47.

Bray WA, Lawrence AL, Leung-Trujillo JR, 1994. The effect of salinity on growth and survival of *Penaeus vannamei*, with observations on the interaction of IHHN virus and salinity. *Aqaculture* 122: 133–146.

Browdy, C.L., 1998. Recent developments in penaeid broodstock and seed production technologies: improving the outlook for superior captive stocks. *Aqaculture* 164, 3–21.

Burgents Joseph, karen Burnett, and louis Burnett, 2005. Effects of Hypoxia and Hypercapnic Hypoxia on the Localization and the Elimination of Vibrio campbelliiin *Litopenaeus vannamei*, the Pacific White Shrimp. *Biol. Bull.* 208:159–168.

Callaghan, T.R., Degnan, B.M., Sellars, M.J., 2010. Expression of Sex and Reproduction-Related Genes in *Marsupenaeus japonicus*. *Mar. Biotechnol.* 12, 664–677.

Costas Benjamín, Cláudia Aragão, Juan Miguel Mancera, Maria Teresa Dinis and Luís E C Conceição, 2008. High stocking density induces crowding stress and affects amino acid metabolism in Senegalese sole *Solea senegalensis* (Kaup 1858) juveniles. *Aqaculture* Research 39, 1: 1–9.

Cui J, Wu, L., Chan, S.M. and Chu, K.H., 2013. cDNA cloning and mRNA expression of retinoid-X-receptor in the ovary of the shrimp *Metapenaeus ensis*. *Molecular biology reports*, 40(11), pp.6233-6244.

Destoumieux D, Bulet, P., Loew, D., Van Dorsselaer A., Rodriguez, J., Bachere, E, 1997. Penaedins: A new family of anti microbial peptides in the shrimp *Penaeus vannamei* (Decapoda) *J. Biol. Chem.* 272, 28398-28406.

De la Vega E, Michael R. Hall, Kate J. Wilson, Antonio Reverter, Rick G. Woods and Bernard M. Degnan, 2007. Stress-induced gene expression profiling in the black tiger shrimp *Penaeus monodon*. *Physiol. Genomics* 31:126-138.

Diwan A D, 2005. Current progress in shrimp endocrinology- A review. *Indian Journal of Experimental Biology*. 43, 209-223.

Direkbusarakom S, Danayadol Y, 1998. Effect of oxygen depletion on some parameters of the immune system in black tiger shrimp (*Penaeus monodon*). In: *Advances in Shrimp Biotechnology*, edited by Flegel TW. Bangkok, Thailand: National Center for Genetic Engineering and Biotechnology.

Dong B and Xiang JH, 2007. Discovery of genes involved in defense/immunity functions in a haemocytes cDNA library from *Fenneropenaeus chinensis* by ESTs annotation. *Aqaculture*, 272:208-215.

Donpudsa Suchao, Suwattana Visetnan, Premruethai Supungul, Sureerat Tang, Anchalee Tassanakajon, Vichien Rimphanitchayakit, 2014. Type I and type II crustins from *Penaeus monodon*, genetic variation and antimicrobial activity of the most abundant crustinPm4. *Developmental and Comparative Immunology* 47: 95–103.

Fingerman, M., Nagabhushanam, R., 1992. Control of the release of crustacean hormones by neuroregulators. *Comp. Biochem. Physiol.* C 102, 343–352.

Granja CB, Aranguren F, Vidal OM, Aragon L, Salazar M, 2003. Does hyperthermia increase apoptosis in white spot syndrome virus (WSSV) infected *Litopenaeus vannamei*? *Dis Aquat Organ* 54: 73–78, 2003.

Gross PS, Bartlett TC, Browdy CL, Chapman RW, Warr GW, 2001. Immune gene discovery by expressed sequence tag analysis of hemocytes and hepatopancreas in the Pacific White Shrimp, *Litopenaeus vannamei*, and the Atlantic White Shrimp, *L. setiferus*. *Dev Comp Immunol*, 25:565-577.

Guan Y, Yu Z, Li C, 2003. The effect of temperature on white spot syndrome infections in *Marsupenaeus japonicus*. *J Invertebr Pathol* 83: 257–260.

Hall MR, van Ham EH, 1998. The effects of different types of stress on blood glucose in the giant tiger prawn, *Penaeus monodon*. *J World Aquac Soc* 29: 290–299.

Hall MR, de la Vega E, 2003. Physiological response to stress and health implications in Crustacea. In: Styli: Thirty Years of Shrimp Farming in New Caledonia, edited by Goarant C, Harache Y, Herbland A, and Mugnier C. Plouzane, France: INRA Editions, 2004, p. 38 –56.

Hiransuchalert, R., Yocawibun, P., Klinbunga, S., Khamnamtong, B., Menasveta, P., 2013. Isolation of cDNA, genomic organization and expression of small androgen receptor interacting protein 1 (PmSARIP1) in the giant tiger shrimp *Penaeus monodon*. *Aqaculture*. 412-413, 151–159.

Huang Canhua, 2002. Proteomic Analysis of Shrimp White Spot Syndrome Viral Proteins and Characterization of a Novel Envelope Protein VP466 **Molecular & Cellular Proteomics*, 1, 223-231.

Huang CC, Sritunyalucksana K, Soderhall K, Song YL, 2004. Molecular cloning and characterization of tiger shrimp (*Penaeus monodon*) transglutaminase. *Dev Comp Immunol* 28: 279–294.

Huberman, A., 2000. Shrimp endocrinology. A review. *Aqaculture* 191, 191–208.

Iwanaga, S, Kawabata S. I., Muta, T., 1998. New types of clotting factors and defence molecules found in horseshoe crab haemolymph: their structures and functions. *J. Biochem*. 123, 1-15.

Jiravanichpaisal, P., Lee, S.Y., Kim, Y.A., Andrén, T., Söderhäll, I., 2007. Antibacterial peptides in hemocytes and hematopoietic tissue from freshwater crayfish *Pacifastacus leniusculus*: characterization and expression pattern. *Dev. Comp. Immunol.* 31, 441–455.

Jiang Hao, Fuhua Li, Yusu Xie, Bingxin Huang, Jinkang Zhang, Jiquan Zhang, Chengsong Zhang, Shihao Li and Jianhai Xiang, 2009. Comparative proteomic profiles of the hepatopancreas in *Fenneropenaeus chinensis* response to hypoxic stress. 12:3353–3367.

Jung Yu Chia; Lin YuFen; Chiang BorLuen; Chow LuPing, 2003: Proteomics and immunological analysis of a novel shrimp allergen, Pen m 2. *Journal of Immunology* 170(1): 445-453.

Karoonuthaisiri, N., Sittikankeaw, K., Preechaphol, R., Kalachikov, S., Wongsurawat, T., Uawisetwathana, U., Russo, J.J., Ju, J., Klinbunga, S., Kirtikara, K., 2009. ReproArrayGTS: A cDNA microarray for identification of reproduction-related genes in the giant tiger shrimp *Penaeus monodon* and characterization of a novel nuclear autoantigenic sperm protein (NASP) gene. *Comp. Biochem. Physiol.* D 4, 90–99.

Leelatanawit, R., Klinbunga, S., Puanglarp, N., Tassanakajon, A., Jarayabhand, P., Hirono, I.,

Aoki, T., Menasveta, P., 2004. Isolation and characterization of differentially expressed genes in ovaries and testes of the giant tiger shrimp (*Penaeus monodon*). *Mar. Biotechnol.* 6, S506–S510.

Leelatanawit, R., Sittikankeaw, K., Yocawibun, P., Klinbunga, S., Roytrakul, S., Aoki, T., Hirono, I., Menasveta, P., 2009. Identification, characterization and expression of sex-related genes in testes of the giant tiger shrimp *Penaeus monodon. Comp. Biochem. Physiol.* A 152, 66–76.

Leu JH, Chang CC, Wu JL, Hsu CW, Hirono I, Aoki T, Juan HF, Lo CF, Kou GH, Huang HC, 2007. Comparative analysis of differentially expressed genes in normal and white spot syndrome virus infected *Penaeus monodon. BMC Genomics*, 8:120

Li, T., and Brouwer, M., 2009. Bioinformatic analysis of expressed sequence tags from grass shrimp Palaemonetespugio exposed to environmental stressors. *Comp. Biochem. Physiol.* Part D Genomics Proteomics.

Liu B, Yu Z, Song X, Guan Y, Jian X, He J, 2006. The effect of acute salinity change on white spot syndrome (WSS) outbreaks in *Fenneropenaeus chinensis. Aqaculture* 253: 163–170,.

Lo, T.S., Cui, Z., Mong, J.L., Wong, Q.W., Chan, S.M., Kwan, H.S., Chu, K.H., 2007. Molecular Coordinated Regulation of Gene Expression During Ovarian Development in the Penaeid Shrimp. *Mar. Biotechnol.* 9, 459–468.

Makkapan, W., Maikaeo, L., Miyazaki, T., Chotigeat, W., 2011. Molecular mechanism of serotonin via methyl farnesoate in ovarian development of white shrimp: *Fenneropenaeus merguiensis* de Man. *Aqaculture* 321, 101–107.

Marsden, G., Mather, P., Richardson, N., 2007. Captivity, ablation and starvation of the prawn *Penaeus monodon* affects protein and lipid content in ovary and hepatopancreas tissues. *Aqaculture* 271, 507–515.

Miura, T., Higuchi, M., Ozaki, Y., Ohta, T., Miura, C., 2006. Progestin is an essential factor for the initiation of the meiosis in spermatogenetic cells of the eel. *Proc. Natl. Acad. Sci. U. S. A.* 103, 7333–7338.

Moullac Le G, Soyez C, Saulnier D, Ansquer D, Avarre JC, Levy P, 1998. Effect of hypoxic stress on the immune response and the resistance to vibriosis of the shrimp *Penaeus stylirostris*. *Fish Shellûsh Immunol* 8: 621–629.

Moullac Le G, Haffner P, 2000. Environmental factors affecting immune responses in Crustacea. *Aqaculture* 91: 121–131.

Navakanitworakul, R., Deachamag, P, Wonglapsuwan, M., Chotigeat, W., 2012. The roles of ribosomal protein S3a in ovarian development of *Fenneropenaeus merguiensis* (De Man). *Aqaculture* 338–341, 208–215.

Ongvarrasopone, C., Roshorm, Y., Somyong, S., Pothiratana, C., Petchdee, S., Tangkhabuanbutra, J., Sophasan, S., Panyim, S., 2006. Molecular cloning and functional expression of the *Penaeus monodon* 5-HT receptor. *Biochim. Biophys. Acta* 1759, 328–339.

Okumura, T., 2004. Perspectives on hormonal manipulation of shrimp reproduction. *Jpn. Agric. Res. Q.* 38, 49–54.

Palasin, K., Makkapan, W., Thongnoi, T. and Chotigeat, W., 2014. Stimulation of ovarian development in white shrimp, *Fenneropenaeus merguiensis* De Man, with a recombinant ribosomal protein L10a. *Aqaculture*, 432, pp.38-45.

Perazzolo LM, Gargioni R, Ogliari P, Barracco MA, 2002. Evaluation of some hemato-immunological parameters in the shrimp *Farfantepenaeus paulensis* submitted to environmental and physiological stress. *Aqaculture* 214: 19–33.

Phinyo M, V. Visudtiphole, S. Roytrakul, N. Phaonakrop, P. Jarayabhand, S. Klinbunga, 2014. Characterization and expression of *cell division cycle 2* (*Cdc2*) mRNA and protein during ovarian development of the giant tiger shrimp *Penaeus monodon Gen. Comp. Endocrinol.*, 193, pp. 103–111.

Preechaphol, R., Leelatanawit, R., Sittikankeaw, K., Klinbunga, S., Khamnamtong, B., Puanglarp, N., Menasveta, P., 2007. Expressed Sequence Tag Analysis for Identification and Characterization of Sex-Related Genes in the Giant Tiger Shrimp *Penaeus monodon*. J. *Biochem. Mol. Biol.* 40, 501–510.

Preechaphol, P., Klinbunga, S., Khamnamtong, B., Menasveta, P., 2010. Isolation and characterization of genes functional involved in ovarian development of the giant tiger shrimp *Penaeus monodon* by suppression subtractive hybridization (SSH). *Genet. Mol. Biol.* 29, 667–675.

Quackenbush, L.S., 2001. Yolk synthesis in the marine shrimp, *Penaeus vannamei*. *Am.Zool.* 41, 458–464.

Robalino, Javier, Ryan B. Carnegie, O'Leary Nuala, Severine A. Ouvry-Patat, Enrique de la Vega, Sarah Prior, Paul S. Gross, Craig L. Browdy, Robert W. Chapman, Kevin L. Schey, Gregory Warr, 2009."Contributions of functional genomics and proteomics to the study of immune responses in the Pacific white leg shrimp *Litopenaeus vannamei.*" *Veterinary immunology and immunopathology* 128, no. 1: 110-118.

Roux MM, Pain A, Klimpel KR, Dhar AK, 2002. The lipopolysaccharide and B-1,3-glucan binding protein gene is upregulated in white spot virus infected shrimp (*Penaeus stylirostris*). *J Virol* 76: 7140–7149.

Rojtinnakorn J, Hirono I, Itami T, Takahashi Y, Aoki T, 2002. Gene expression in haemocytes of kuruma prawn, *Penaeus japonicus*, in response to infection with WSSV by EST approach. *Fish Shellfish Immunol*, 13:69-83.

Silvestre Frédéric, Huynh ThiTu, Amandine Bernard, Jennifer Dorts, Marc Dieu ,Martine Raes, Nguyen ThanhPhuong, Patrick Kestemont, 2010. A differential proteomic approach to assess the effects of chemotherapeutics and production management strategy ongiant tiger shrimp *Penaeus monodon.* comparative Biochemistry and Physiology Part D: Genomics and Proteomics. 5, 3: 227–233.

Smith, V.J., Fernandes, J.M.O., Kemp, G.D., Hauton, C., 2008. Crustins: enigmatic WAP domain-containing antibacterial proteins from crustaceans. *Dev. Comp. Immunol.* 32, 758–772.

Somboonwiwat Kunlaya, Vorrapon Chaikeeratisak, Hao-Ching Wang, Chu Fang Lo and Anchalee Tassanakajon, 2010. Proteomic analysis of differentially expressed proteins in *Penaeus monodon* hemocytes after *Vibrio harveyi* infection. *Proteome Science*, 8:39.

Sotelo-Mundo RR, Islas-Osuna MA, de la Re Vega E, HernandezLopez J, Vargas-Albores F, Yepiz-Plascencia G, 2003. cDNA cloning of the lysozyme of the white shrimp *Penaeus vannamei. Fish Shellûsh Immunol* 15: 325–333.

Supungul P, Klinbunga S, Pichyangkura R, Hirono I, Aoki T, Tassanakajon A, 2004. Antimicrobial peptides discovered in the black tiger shrimp *Penaeus monodon* using the EST approach. *Dis Aquat Organ*, 61:123-135.

Sritunyalucksana K, Cerenius L, Soderhall K, 1999. Molecular cloning and characterization of prophenoloxidase in the black tiger shrimp, *Penaeus monodon. Dev Comp Immunol* 23: 179–186.

Subramoniam, T., 2011. Mechanisms and control of vitellogenesis in crustaceans. *Fish. Sci.* 77, 1–21.

Tassanakajon A, Klinbunga S, Paunglarp N, Rimphanitchayakit V, Udomkit A, Jitrapakdee S, Sritunyalucksana K, Phongdara A, Pongsomboon S, Supungul P, Tang S, Kuphanumart K, Pichyangkura R, Lursinsap C,2006. *Penaeus monodon* gene discovery project: the generation of an EST collection and establishment of a database. *Gene*, 384:104-112.

Tassanakajon, A., Amparyup, P., Somboonwiwat, K., Supungul, P., 2011. Cationic antimicrobial peptides in penaeid shrimp. *Mar. Biotechnol.* 13, 639–657 .

Tiu, S.H.K., He, J.G., Chan, S.M., 2005. Organization and expression study of the shrimp (*Metapenaeus ensis*) putative 5-HT receptor: up-regulation in the brain by 5-HT. *Gene* 353, 41–52.

Thomas, P., 2008. Characteristics of membrane progestin receptor alpha (mPRα) and progesterone membrane receptor component 1 (PGMRC1) and their roles in mediating rapid progestin actions. Frontiers in Neuroendocrinology 29, 292–312.

Tsai JM, Wang HC, Leu JH, Wang AH, Zhuang Y, Walker PJ, Kou GH, Lo CF.. 2006. Identification of the nucleocapsid, tegument, and envelope proteins of the shrimp white spot syndrome virus virion. *J Virol.* 80 (6): 3021-9

Uawisetwathana U, Leelatanawit R, Klanchui A, Prommoon J, Klinbunga S, Karoonuthaisiri N, 2011. Insights into Eyestalk Ablation Mechanism to Induce Ovarian Maturation in the Black Tiger Shrimp. *PLoS ONE* 6 (9): e24427.

Vargas-Albores F, Guzman-Murillo MA, Ochoa JL, 1993. An anticoagulant solution for haemolymph collection and prophenoloxidase studies of Penaeid shrimp (*Penaeus californiensis*). *Comp Biochem Physiol A* 106: 299 –303.

Vidal OM, Granja CB, Aranguren F, Brock JA, Salazar M, 2001. A profound effect of hyperthermia on survival of *Litopenaeus vannamei* juveniles infected with white spot syndrome virus. *J World Aquac Soc* 32: 364–372.

Wang HC, Leu JH, Kou GH, Wang AH, Lo CF, 2007. Protein expression profiling of the shrimp cellular response to white spot syndrome virus infection. *Dev Comp Immunol*, 31:672-686.

Westenberg Marce l, Bas Heinhuis, Douwe Zuidema, 2005. siRNA injection induces sequence-independent protection in *Penaeus monodon* against white spot syndrome virus. *Virus research* 114, 1–2: 133–139.

Witteveldt Jeroen, Just M. Vlak, Mariëlle C.W. van Hulten, 2004. Protection of *Penaeus monodon* against white spot syndrome virus using a WSSV subunit vaccine. *Fish & shellfish immunology*. 16, 5:571–579.

Witchulada Talakhun, Roytrakul, S., Phaonakrop,N., Kittisenachai,S., Khamnamtong, B., Klinbunga, S., Menasveta, P., 2012. Identification of reproduction-related proteins and characterization of the protein disulfide isomerase A6cDNA in ovaries of the giant tiger shrimp *Penaeus monodon. Comp. Biochem. Physiol.* D 7, 180–190.

Witchulada Talakhun, Narumon Phaonakrop, Sittiruk Roytrakul, Sirawut Klinbunga, Piamsak Menasveta, Bavornlak Khamnamtong, 2014. Proteomic analysis of ovarian proteins and characterization of thymosin-β and RAC-GTPase activating protein 1 of the giant tiger shrimp *Penaeus monodon. Comparative Biochemistry and Physiology*, Part D.

Witchulada Talakhun, Bavornlak Khamnamtong, Pachumporn Nounurai, Sirawut Klinbunga, Piamsak Menasveta, 2014. Characterization, expression and localization of valosin-containing protein in ovaries of the giant tiger shrimp *Penaeus monodon. Gene* 533:188–198.

Wonglapsuwan M., Phongdara, A., Chotigeat, W., 2009. Dynamic changes in gene expression during vitellogenic stages of the white shrimp: *Fenneropenaeus merguiensis* de *Man. Aquac. Res.* 40, 633–643.

Wu Jinlu, Qingsong Lin, Teck Kwang Lim, Tiefei Liu, and Choy-Leong Hew, 2007. White Spot Syndrome Virus Proteins and Differentially Expressed Host Proteins Identified in Shrimp Epithelium by Shotgun Proteomics and Cleavable Isotope-Coded Affinity Tag. *Virol*; 81(21): 11681–11689.

Yao CL, Wu CG, Xiang JH, Dong B, 2005. Molecular cloning and response to laminarin stimulation of arginine kinase in haemolymph in Chinese shrimp, *Fenneropenaeus chinensis. Fish Shellfish Immunol*, 19:317-329.

Yano, I, 1998. Hormonal control of vitellogenesis in penaeid shrimp. In: Flegel, T.W. (Ed.), Advances in Shrimp Biotechnology. Proceedings to the Special Session on Shrimp Biotechnology 5th Asian Fisheries Forum, November 11–14. National Center for Genetic Engineering and Biotechnology, Bangkok, Chiengmai, Thailand. 29–31.

Xie, Y., Li, F., Wang, B., Li, S., Wang, D., Jiang, H., Zhang, C., Yu, K., Xiang, J., 2010. Screening of genes related to ovary development in Chinese shrimp *Fenneropenaeus chinensis* by suppression subtractive hybridization. *Comp. Biochem. Physiol.* D 5, 98–104.

Zhang Xiaobo, Canhua Huang, Xuhua Tang, Ying Zhuang and Choy Leong Hew, 2004. Identification of structural proteins from shrimp white spot syndrome virus (WSSV) by 2DE-MS. *PROTEINS: Structure, Function, and Bioinformatics* 55:229 –235.

Zhang S, Linglin Fu, Yanbo Wang, Junda Lin, 2014. Alterations of protein expression in response to crowding in the Chinese shrimp (*Fenneropenaeus chinensis*). *Aqaculture* 428–429,135–140.

Zhengjun Li, Qingsong Lin, Jing Chen, Jin Lu Wu, Teck Kwang Lim, Siew See Loh, Xuhua Tang, and Choy-Leong Hew, 2007. Shotgun Identification of the Structural Proteome of Shrimp White Spot Syndrome Virus and iTRAQ Differentiation of Envelope and Nucleocapsid Subproteomes. *Molecular & Cellular Proteomics*, 6, 1609-1620.

SECTION-II

REPRODUCTIVE BIOTECHNOLOGY

CHAPTER 3

PHYSIOLOGY OF REPRODUCTION

3.1 INTRODUCTION

One of the major problems preventing optimization of the commercial culture of shrimp is the control of female reproduction, which is highly complex in penaeid shrimps (Chang 1992). It appears that a number of environmental signals can influence difference hormonal factors, which in turn regulate reproduction. The understanding of regulatory process of reproduction is an area of intense research (Adiyodi, 1985 and Charniaux- Cotton and Payen, 1988). Each animal species uses distinct environmental cues for timing its reproduction, has a full range of neuronal structures for perception of signals, and uses a complex neuroendocrine system for transduction of the messages to the endocrine organs, which themselves produce factors regulating the activity of the organs involved in reproduction (Van Herp 1992). Success depends on the optimal completion of each step in the reproductive cycle of the animals, but it is evident that the neuroendocrine factors play a predominant interactive role between the external factors such as photoperiodicity, temperature, nutrition, stress and internal organization of the animal. According to Van Herp (1992), the crustacean Aquaculture in the future needs the support of precise knowledge of the role and mode of action of external and internal factors controlling reproduction, in order to create a strong basis for the development of intensive Aquaculture. With the increasing knowledge of endocrine activity and its control over the gonadal development in crustaceans, the various techniques of endocrine manipulations for induced maturation of gonads are receiving great attention. In the past decade a vast amount of research has been directed towards the understanding of female reproduction in crustaceans. Studies on the breeding cycles of various penaeids have been extensively done by Cummings (1961), Subramoniam (1965), Rao (1968), Brown and Patlan (1974), Thomas (1974), Penn (1980), Kennedy and Barber (1981), Motoh (1981, 1985), Manasveta *et al.*,(1993), Qunitio *et al.*,(1993), and Susan *et al.*,(1993). Most of the investigators, especially on the female reproduction, have used morphometric characters like colour changes in the ovary (King, 1948; Cummings, 1961; Brown and Patlan, 1974; Primavera, 1980; Motoh, 1985; Tanfermin and Pudadera, 1989) or gonadosomatic index (GSI) (Pillai and Nair, 1971; Thomas, 1974; O'connor, 1979; Quinn and Barbara, 1987; Lawrence and Castle, 1991) for the assessment of ovarian maturation. Histological investigations on the reproductive organs during the ovarian maturation in various penaeid shrimps have been done by various workers like Hudinaga (1942) in *Penaeus japonicus*, King (1948) in *P. setiferus*, Sheikhmahmud and Tambe (1958) in *Parapenaeopsis stylifera*, Cummings (1961) in *Penaeus duorarum*, Duronslet *et al.*,(1975) in *P. aztecus* and *P. setiferus*, Anderson *et al.*,(1984) in *S. injentis*, Tom *et al.*,(1987) in *P. longirostris*, Yano (1988) in *P. japonicus*, Tanfermin and Pudadera (1989), and Qunitio *et al.*,(1993) in *P. monodon*, and Mohammed and Diwan (1994) in *P. indicus*. However, ultrastructural investigations during the process of ovarian maturation are restricted to the studies of Duronslet *et al.*, (1975) in *P. aztecus* and *P. setiferus* and Mohammed and Diwan (1994) in *P. indicus*. Vitellogenesis is the most intensely studied aspect in oogenesis in various crustaceans during the past two decades (Hinch and Cone, 1969; Lui *et al.*,1974; Komm and Hinch, 1985, 1987; Tom *et al.*,1987; Quackenbush, 1989; Young *et al.*,1993; Chang *et al.*,1994; Mohammed and Diwan, 1994; Chang and Shih, 1995 and Sagi *et al.*,1995). Still the site of vitellogenesis is a fact of great controversy among various crustaceans. According to Komm and Hinch (1987), an accurate evaluation of oogenesis and vitellogenesis requires a complete ultrastructural and biochemical investigation. However ultrastructural studies on the process of vitellogenesis are still fragmentary in crustaceans, and particularly in penaeids.

Until now studies on reproduction and its control in crustaceans have dealt with the female reproduction and not much attention has been paid to the males (Van Herp, 1992). Recent studies on the male reproduction in captivity revealed deterioration in the quality of semen (Leung-Trujillo and Lawrence, 1985, 1987; Alfaro and Lozano, 1993; Pratoomchat *et al.*,1993). In this respect full understanding of reproductive mechanism of male penaeids is essential for any manipulation of sperm. Reproductive anatomy of several penaeids have been described and figured by many investigators (Hudinaga, 1942; Eldred, 1958; Tirmizi and Khan, 1970; Huq,1980; Vasudevappa, 1992; Mohammed and Diwan, 1994).

However, studies on the spermatogenesis of penaeid shrimps are less (King, 1948; Lu *et al.*,1973; Pochon- Masson, 1983; Adiyodi, 1983; Vasudevappa, 1992; Mohammed and Diwan, 1994) compared to other crustaceans. Many of such studies focus only on light microscopic investigations and attempts on ultrastructural studies are, however, meager. Similarly though few investigators have described the mechanism of spermatophore formation in various penaeids like *P. kerathurus* by Malek and Bawab (1974 a, b); *P. indicus* by Champion (1987) and Mohammed and Diwan (1994) and in *M. dobsoni* by Vasudevappa (1992) through light microscopy, there has been little published work at ultrastructural level (Ro *et al.*, 1990) in *P. setiferus* and Chow *et al.*,(1991) in *P. setiferus* and *P. vannamei*. It has been demonstrated that in decapod crustaceans, reproduction is under the control of hormonal factors viz. the GIH and GSH produced from the various endocrine centres. The neuroendocrine regulation of crustacean reproduction has been reviewed by Legrand *et al.*,(1982), Payen (1986), Fingerman (1987), Charniaux-Cotton and Payen (1988), Meusy and Payen (1988), Van Herp and Payen (1991), Diwan (2005) and Shweta *et al.*,(2011). The morphology and histology of crustacean endocrine system have been studied by Carlisle and Knowles (1959), Nagabhushanam *et al.*,(1992) and Mohammed *et al.*,(1993) and X-organ sinusgland complex by Dall (1965 a), Nakamura (1974), Madhystha and Rangnekar (1976), Gynanath and Sarojini (1985), Nanda and Ghosh (1985) and Mohammed *et al.*,(1993). Different types of neurosecretory cells and different stages of secretion have been mentioned in the reports of Durand 1956, Peryman, (1969), Diwan and Nagabhushanam *et al.*,(1974) Ro *et al.*,(1981), Decaraman and Subramoniam (1983) and Joshi (1987) and Mohammed *et al.*,(1993) and X-organ sinus gland complex by Dall (1965 a), Nakamura (1974), Madhystha and Rangekar (1976), Gynanath and Sarojini (1985), Nanda and Ghosh (1985) and Mohammed *et al.*,(1993). However, a detailed description of the various phases of neurosecretion in relation to reproduction has been there only in very few reports (Gynanath and Sarojini 1985 and Mohammed *et al.*, 1993).

Endocrine control of reproduction has been investigated with a wide variety of crustaceans. However, the actual mechanism and hormones working, behind reproduction have not been properly revealed in any crustaceans (Fingerman, 1987), but eyestalk factors have been held responsible by many (Aoto and Nishida, 1956; Demeusy, 1967; Rengnekar and Deshmukh, 1968; Diwan and Nagabhushanam, 1974; Quackenbush and Herrkind, 1981; Fingerman, 1987; Mohammed and Diwan, 1991; Van Herp, 1992) in controlling this process. However, a second reproductive hormone found in the brain and thoracic ganglia has been attributed the role of gonad stimulation by Otsu 1960, Gomez (1965), Oyama (1968), Nagabhushanam *et al.*, (1982), Eastman-Recks and Fingerman (1984), Mohammed and Diwan (1991), Yano (1992) and Yano and Wyban (1992). In this regard cytological ultrastructural studies of gonads and neuroendocrine masses can give some clues to unravel the mechanism of reproduction. But, unfortunately such studies are meager in crustaceans and lacking in penaeids. The most successful technique for inducing maturation in captivity has been the removal of the eyestalk which is the site of GIH and this has been applied to various penaeids by many investigators with varying degrees of success (Arnstein and Beard,1975; Alikunhi *et al.*,1975; Aquacop, 1975; Muthu and Laxminarayana, 1979; Lumare, 1979; Emmerson, 1980; Kulkarni and Nagabhushanam, 1980; Primavera, 1982; Chamberlain and Gervais, 1984; Chow, 1987; Wyban, *et al.*,1987 and Mohammed and Diwan, 1991). Similarly advancement in the ovarian stages were demonstrated by the administration of thoracic ganglion extracts by Otsu (1960), Oyama (1968), Hinch and Bennet (1979), Nagabhushanam and Kulkarni (1982), Eastman-Recks and Fingerman (1984), Takayanagi *et al.*,(1986) and Yano (1992), by the administration of brain extract by Nagabhushanam and Kulkarni (1982) and Yano and Wyban (1992) and by extract of central nervous organs (brain and thoracic ganglion) by Mohammed and Diwan, (1991). Junera *et al.*,(1977) first proposed that a vitellogenic stimulating ovarian hormone was present in the ovary of amphipod, *Orchestia gammarella*. Since then, progesterone (Kulkarni *et al.*,1979; Yano, 1985) and 17-alpha hydroxy progesterone (Nagabhushanam *et al.*,1980; Yano, 1987; Tsukimura and Kamemoto, 1988 and Koskela *et al.*,1992) have been tried to accelerate gonad development with various degrees of success.

From the foregoing literature it was found that among these various techniques for the induction of gonads in captive shrimps, only eyestalk ablation has been tried successfully in *P. monodon* by various investigators. Further it is needed that the percentage of animals getting matured and spawned by this technique of eyestalk ablation in captive *P. monodon* are still low in our country and problems are more acute especially in areas where the natural population of this species is comparatively less. The successful cryopreservation of viable gametes would open new perspectives in any culture operations. Many attempts have been made to cryopreserve the spermatozoa of aquatic vertebrates, mainly fishes. However, less attention has been paid on invertebrates. The various attempts on cryopreservation in invertebrates include, the reports of Sawada and Chang (1964), Dunn and Mc Lachlin (1973), Zell *et al.*,(1979); Behlmer and Brown (1984), Chow *et al.*,(1982, 1985), Ishida *et al.*,(1986), Anchordoguy *et al.*,(1987, 1988), Jayalectumie and Subramoniam (1989), Renard (1992), Bury and Olive (1993) and Joshi and Diwan (1992). In spite of the fact that among invertebrates, decapod crustaceans are the most economically important group of animals, very little attention has been paid on the freezing and preservation of gametes. A few attempts made to cryopreserve the spermatozoa of decapod crustaceans including shrimps and crabs are by Behlmer and Brown (1984), in *Limulus polyphemus*, Chow *et al.*, (1982, 1985) in *Macrobrachium rosenbergii*, Anchordoguy *et al.*,(1987, 1988) in *Sicyonia injentis*, Jayalectumie and Subramoniam (1989) in *Scylla serrata*, Joshi and Diwan, (1992) in *Macrobrachium idella*. However, such studies are not done with the gametes in *P. monodon* till now.

Shrimp farming has become one of the most important food production industries of the world, most lucrative and widely traded Aquaculture products. This has expanded rapidly since early 1980 and in 2014 the global yield of farmed shrimp exceeded 4.33 mmt contributing more than 30% of global shrimp production. Shrimp farming is highly beneficial to local community as well as national economics of developing countries. In India, it has witnessed unprecedented growth in eighties and nineties and has become one of the most important food production industries of the country. Total farm shrimp production has increased from 30,000 mt in 1990 to 1, 30,000 mt in 2005 (Anonymous 2005) and further in 2014 this has gone upto 4.33 mmt. Though the growth of the industry and its importance in the coastal economy is impressive, the Aquaculture of shrimp has not certainly been without problems. Catastrophic infectious disease hit shrimp farming since mid nineties causing great losses. The enormous concentration of animals and their coprophagous behaviour imposed by an intensive culture have triggered the development of disease out break, which are often explosive and sometimes leads to the loss of complete stock. The way of disease outbreaks and mass mortality witness in the Indian shrimp farming sector was mainly dominated by white spot disease, caused by a lethal viral pathogen, White Spot Syndrome Virus (WSSV). Being epizootic caused by a viral pathogen, till date there is no therapeutic and treatment measures available. Because of their ability to trigger sudden epizootics and their potential for rapid propagation, epizootics caused by WSSV have become a major limiting factor in the growth of Aquaculture in India and other shrimp farming countries, across the globe.

Therefore, the existing and emerging disease in shrimp production systems would require the adoption of a comprehensive health management approach. Control of shrimp diseases will have to be addressed by importing or by developing SPF/SPR broodstock following Good Management Practices (GMP), cluster approach and bio-security measures. One of the issues in health management is the use of various drugs and chemicals for the control of diseases by the farmers without any idea about their efficiency. Hence, there is an urgent need for development of effective therapeutants, probiotics and vaccines. Awareness campaigns on an extensive scale are also needed to educate the large segment of small scale farmers. The sector is dependent on wild spawners and more than 60% of the wild spawners are infected with white spot virus. On a short term, import of specific pathogen free broodstock through appropriate quarantine mechanisms and adoption of biosecurity protocols in hatcheries is one solution. On a long term basis, development of captive broodstock and domestication is a permanent solution to ensure production of disease free seed. By combining this with a genetic selection programme with Biotechnology inputs, it would be possible to produce fast growing and pathogen resistant broodstock which can lead to higher production. Therefore, development of high-health and high growth shrimps through application of genetic and biotechnological tools is essential for increasing production in a sustainable manner. The world over, there has been very few programmes for genetic improvement of tiger shrimp. This is basically due to the fact that of all the penaeid shrimps, tiger shrimp is the most difficult to mature and breed in captivity. There are reports from Thailand which indicate that pond-reared tiger shrimps have a very low reproductive ability in that they are difficult to mature and spawn. In cases where such pond-reared specimens had matured and spawned, the number of eggs spawned was very low and the viability of eggs was also low. In certain cases the eggs were found to be unfertilized. The only way out of this dilemma is to initiate a domestication programme followed by genetic improvement through selection.

3.2 FEMALE REPRODUCTIVE PHYSIOLOGY

The extent to which any biological system can be successfully manipulated is a function of understanding the mechanisms through which that system operates. As reproductive biology is central to all biological sciences a proper understanding of it is vital for the management of the species concerned through successive generations (Adiyodi, 1982, Cuzon *et al.*,2004). Even though the breeding cycles based on the morphology of reproductive organs of crustaceans have been well studied, in-depth knowledge about the sexual physiology of crustaceans is still meagre in comparison to what is known of their terrestrial equivalents, the insects and many other animal groups (Huberman, 2000). Reproduction in many decapods in general and penaeidean shrimps in particular is significantly different from that of most aquatic invertebrates as well as vertebrates (Bauer and Martin, 1991, Peixoto *et al.*,2005).

The fundamental problem in shrimp mariculture industry is the lack of predictably abundant supplies of offspring's of known heritage. Development of a suitable biotechnology for controlled induction of reproduction in the candidate animal is the prime requisite to overcome this problem. In order to develop such a technology, basic information on morphology, anatomy, physiology, cytology and molecular biology of the reproductive systems of the species is most essential. The cytological studies correlated with the morphology during the process of ovarian maturation form a concrete framework for the assessment of maturity. Moreover, such studies are more important in the correct staging of reproductive phase of the animals in the hatcheries for their better management. The production of egg, well equipped with the necessary reserve food for the developing embryo takes place through a dynamic as well as dramatic process within the developing germ mother cells of the ovary. Cytological and electron microscopical studies in the oocytes during oogenesis revealed the clear picture of the cells during their development. Nevertheless, relatively little emphasis has been laid on the bio-molecular changes in the ovaries of the penaeids during different maturity stages. In order to probe into the complicated dynamic process of oocytes, it is evident that a detailed knowledge of the ultrastructural changes that take place during the various stages of maturation is needed. Over the past decade a great deal of useful information on the light microscopic features of reproductive organs particularly on oogenesis has been published. Many workers have studied reproductive cycles of decapods (Diwan and Nagabhushanam 1974, Hill, 1975, Haefner, 1977, Aiken and Waddy, 1980, Joshi and Khanna, 1982, Damrongphol *et al.*,1991, Janine *et al.*,1991, Cavalli *et al.*,1997, Browdy, 1998, Ayub & Ahmed, 2002, Alfaro, *et al.*,2004, Coman *et al.*,2005 and 2006.). Similarly studies on the breeding cycles of penaeid shrimps are reported by Cummings (1961), Subramoniam (1965), Rao (1968), Brown and Patlan (1974) Thomas (1974) Penn (1980), Kennedy and Barber (1981), Motoh (1981), & (1985), Menasveta *et al.*,(1993), Qunitio *et al.*,(1993) and Susan *et al.*,(1993), Cavalli *et al.*,(1997), Shoji, (1997); Browdy, (1998), Ayub & Ahmed, (2002), Peixoto, *et al.*,(2003), Alfaro, *et al.*,(2004), Peixoto, *et al.*,(2005), Coman *et al.*,(2005) and (2006). Most of the investigations especially on the female reproductive biology are restricted to the morphometric characters of the reproductive organs, spawning season and spawning areas. Morphometric characters for the assessment of maturation are based either on the colour changes during the reproductive cycles (King, 1948; Cummings, 1961 and Brown and Patlan, 1974) or on the gonado somatic-index (GSI) (Pillai and Nair, 1971; O'conor,1979; Quinn and Barbara, 1987; Cavalli *et al.*,1997; Shoji, 1997; Browdy, 1998; Ayub & Ahmed, 2002; Peixoto, *et al.*,2003; Alfaro, *et al.*,2004; Peixoto, *et al.*,2005; Coman *et al.*,2005 and 2006).

Many investigators studied the anatomical features of the female reproductive organs of crustaceans through light microscopy. Chief contributions on the histo-morphological features of the female reproductive system in various crustacean groups are by Ryan (1967) in *P. sanguinolentus*, Chandran (1968) in *C. veriegata*, Pillai and Nair (1970) in *Uca annulipes, P. pelagicus* and *M. affinis*, Diwan and Nagabhushanam (1974) in *B. cunicularis*, Joshi and Khanna (1982) in *P. koolooense* Komm and Hinch (1985 & 1987) *C. clypeatus*. Major histological investigations on the reproductive organs during ovarian cycle in penaeid shrimps are by Cummings (1961), Duronslet *et al.*,(1975), Anderson *et al.*,(1984), Tom *et al.*,(1987), Tanfermin and Pudadera (1989), Dall *et al.*,(1990), Menasveta *et al.*,(1993), Qunitio *et al.*,(1993) and Mohammed and Diwan (1994) Cavalli *et al.*,(1997), Shoji, 1997, Browdy, (1998), Crocos *et al.*,(1999), Ayub & Ahmed,

(2002), Peixoto, *et al.*,(2003), Alfaro, *et al.*,(2004), Peixoto, *et al.*,(2005), Coman *et al.*,(2005) and (2006). These investigators describe the anatomical as well as cytological changes during oogenesis.

In crustacean oogenesis, as in a number of other animal taxa, one of the more conspicuous events is the gradual accumulation of nutritional materials collectively referred to as yolk (Anderson, 1984). The process of yolk synthesis (vitellogenesis) is the most intensely investigated aspect in oogenesis (Hinch and Cone, 1969; Lui *et al.*,1974; Komm and Hinsch, 1985, 1987; Tom *et al.*,1987; Quackenbush, 1989; Young *et al.*,1993; Chang *et al.*,1994; Mohammed and Diwan, 1994; Chang and Shih, 1995; and Sagi *et al.*,1995, Cavalli *et al.*,1997, Shoji, 1997, Browdy, 1998, Ayub & Ahmed, 2002, Peixoto, *et al.*,2003, Alfaro, *et al.*,2004, Peixoto, *et al.*,2005, Coman *et al.*,2005 and 2006). Still the site of vitellogenin synthesis is an unrevealed fact in crustaceans. Studies on egg yolk protein synthesis have been focused on representatives from only a few of the many families of crustaceans (Charniaux-Cotton, 1985, Shoji, 1997, Khayat *et al.*,1998, Wouters *et al.*,2001b). Available information from the few penaeid species that have been studied, however, is controversial with respect to the mode of synthesis and packaging of yolk products in other decapods. Vitellogenin is the precursor to egg yolk protein and is one of the two lipoproteins known in crustaceans (Kerr, 1969; and Quackenbush, 1991, Wouters *et al.*,2001b) and its concentration in various tissues is correlated with yolk accumulation in the oocytes (Quackenbush, 1989; Okamura *et al.*,1992). The origin of yolk in crustaceans specially in decapods has been investigated through biochemical, electrophoretic, immune flourescence, electron microscopical as well as hormone title assays by many investigators. Biochemical and electrophoretic studies are made in crayfish *P. clarkii* and crab, *P. crassipes* by Lui *et al.*,(1974), Lui and O'connor (1976 & 1977), for isolation of vitellogenin to find out the site of yolk synthesis. With the same objectives Tom *et al.*,(1987) conducted histological and immune flourescence studies in *P. longirostris*. Morphological and cytological investigations on these aspects include the studies of Hinch and Cone (1969) in *L. emarginata*, Zerbib (1980) in *O. gammarellus*, Duronslet *et al.*,(1975) in *P. setiferus* and *P. azeticus*, Komm and Hinch (1985, 1987) in *C. clypeatus*, Mohammed and Diwan (1994) in *P. indicus* and Shoji, (1997) in *P. monodon*. Yano and Chinzei (1987) studied the vitellogenesis in *P. japonicus* through *in vitro* incubation and immune flourescence study. These studies reveal that controversy exists regarding the significance of extra ovarian source of vitellogenin synthesis in crustaceans. Nevertheless in decapod crustaceans, several sites for yolk protein production have proposed. These sites include the haemocytes of the haemolymph, the hepatopancreas and sub epidermal adipose tissue of maturing female crustaceans (Kerr, 1969; Charinaux-Cotton, 1978; Adiyodi and Subramoniam, 1983;

Eastman Recks and Fingerman, 1985; Paulus and Laufer, 1987; Tom *et al.*,1987; Quackenbush and Keelay, 1988; Rankin *et al.*,1989; Bradfield *et al.*,1989; Quinitio *et al.*,1989 & 1990; Chang *et al.*,1993; Chang and Shih, 1995; Sagi *et al.*,1995, Shoji, 1997). Some studies suggest that most of the vitellin in vitellogenic females is produced in the ovary and only a small amount is produced in extra-ovarian tissue (Fainzilber *et al.*,1992; Eastman-Recks and Fingerman, 1985, Lui and O'Connor, 1976). Further *in vitro* experiments showed a low amount of vitellogenin synthesis in the hepatopancreas of penaeid shrimp and crabs (Quackenbush and Keeley, 1988; Quackenbush, 1989 a, 1989 b). Other researchers have been unable to identify vitellogenin synthesis in any extra-ovarian tissue of penaeid shrimp (Yano and Chinzei, 1987; Rankin *et al.*,1989). On the other hand there is evidence that extra ovarian sites do produce yolk proteins in crabs (Paulus and Laufer, 1987; Lee and Puppione, 1988). Shafir *et al.*,(1992a) demonstrated a role for the haemolymph in transporting vitellogenin between its processing and target sites. Histological as well as electron-microscopical investigations made by Hinsch and Cone (1969) in *L. emarginata*, Duronslet *et al.*,(1975) in *P. aztecus* and *P. setiferus*, Mohammed and Diwan (1994) in *P. indicus* and Shoji, (1997) in *P. monodon* demonstrated the extra ovarian synthesis of yolk in crustaceans.

Even though vitellogenesis has been investigated by several investigators in various crustaceans, due to the diverse assemblage of animals, the penaeids have received little attention. Qunitio *et al.*, (1990) and Chang *et al.*,(1993) studied the isolation and characterization of vitellin from the ovary of *P. monodon* and Chang *et al.*, (1994) from the haemolymph. These studies indicate that a controversy exists regarding the extra ovarian source of vitellogenin synthesis in penaeids. Quinitio *et al.*, (1990) and Chang *et al.*, (1993) isolated the yolk precursors from two different sources in *P. monodon* and hence the exact source of vitellogenin synthesis is not clear. In this situation the biomolecular investigations give clearer pictures of yolk synthesis. Information regarding the bio-molecular changes occurring in the oocytes during gonadal maturation as well as vitellogenesis is however, scanty as far as *P. monodon* is concerned. Hence in the present chapter in depth analysis of the process of oogenesis in *P. monodon* were made to give more details about the ovarian development especially the mechanism by which ovaries accumulate their reserve food.

In penaeid shrimps the female reproductive system consists of the internal organs such as the paired ovaries, oviducts and the external organ, the single thelycum. The ovaries are partly fused, bilaterally symmetrical bodies located dorsal to the hepato pancreas and ventral to the heart in the cephalo thoracic region and dorsal to the gut in the abdominal region. It extends entire length of the animal from base of rostrum in anterior portion to telson in the posterior region in the mature animals. Morphologically three main lobes are observed on each half of the ovary, anterior, middle and posterior (Fig. 1). The anterior lobes are situated close to the oesophagus and cardiac region and they bear a pair of somewhat elongated lobes, in the cephalic region. The six-paired fingers like lobules spread as a saddle surrounding the hepato pancreas constitute the middle lobe. From this a pair of lobe, one from each half of the ovary extends over the entire length of the abdomen in the posterior lobe. Pair of short narrow tubes namely the oviducts originates from the 6th lateral lobule of the middle lobe of each ovary, and descents to the gonopores on the coxae of the third pereiopods. The thelycum is of closed type and is located on the ventral side between the 4th and 5th pereiopod and serves as the receptacle for spermatophores. Morphologically the thelycum consists of an anterior and two lateral lobes. A cavity is present in the thelycum behind the lateral plates for the storage of spermatophores.

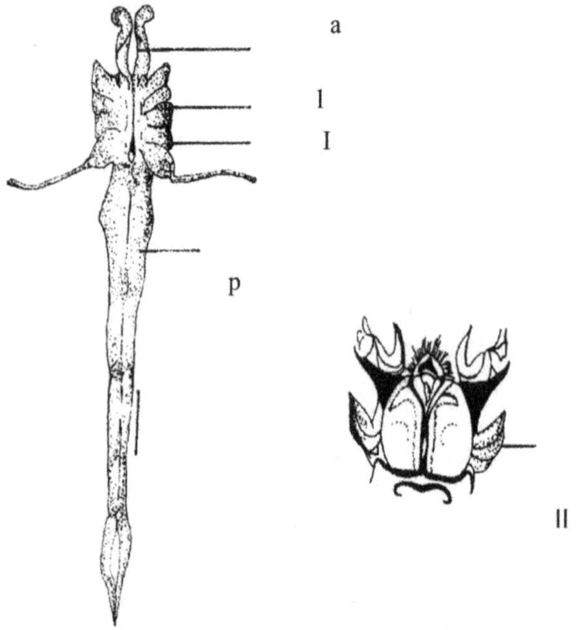

Fig.1. Diagramatic representation of the female reproductive system of *P. monodon.* I–Ovary, II–Thelycum a–anteriorlobe, l–laterallobe, p–posteriorlobe (**Source:** Diwan *et al.*, 2008)

3.3 ASSESMENT OF FEMALE REPRODUCTIVE MATURITY

Determination of reproductive state of a particular animal is very important for its fishery management and culture programmes. In many shrimps determination of ovarian maturation is possible through the evaluation of colour changes in the ovary during maturation and increase in the ovarian volume viz. (GSI) during maturation. The developmental changes in the external genital organ – the thelycum – in the maturing animals are important. Based on these changes the ovarian development is generally classified into 5 different stages (stage I to V) considering the immature stage as 0 stages. The different maturity stages include pre-vitellogenic, early-vitellogenic, late-vitellogenic, mature, and spent ovarian stages. Quantitative parameters of different reproductive stages in *Penaeus monodon* and *F. indicus* are presented in the table1 and table 2.

Table 1. Quantitative parameters of *Penaeus monodon* at different stages of ovarian maturation.

Stages	**CL (mm)**	**BL (mm)**	**BW (g)**	**GW (g)**	**GSI (%)**	**AOD (µm)**	**AND (µm)**	**Colour and appearance of ovary**
Immature	61.3	212	150	0.8	0.53	24.1	18.3	Transparent un pigmented and smooth
Pre-vitellogenic	63.1	208	147	1.9	1.3	61.4	34.0	Opaque, white & smooth
	62.4	215	140	2.38	1.7	98.9	47.8	Cream, smooth granular
Early-vitellogenic	64.3	228	151	8.8	5.8	189.5	69.2	Yellow firm & granular
	65.2	260	156	11.8	7.53	174.2	81.8	Greenish yellow, firm and granular
Late-vitellogenic	63.6	245	151	12.4	8.17	338.5	107.1	Green and highly granular
	68.5	265	158	14.7	9.3	406	122.5	Greenish black thick and firm
Spent	65.6	245	141	3.2	2.2	140.5	59.6	Pale yellow, granular, flaccid

CL – Carapace length

BL Body length

BW – Body weight

GW – Gonad weight

GSI – Gonado somatic index

AOD – Average oocyte diameter

AND – Average nuclear diameter

(**Source:** Diwan *et al.*,2008)

Table 2. Classification of maturity stages in female *F. indicus* based on colour of the ovary, GSI and oocyte diameter

Ovary Stage	Colour and appearance of ovary	Gonadosomatic index ± S.D	Oocyte diameter µ ± S.D	Nucleus diameter µ ± S.D
Stage I				
Immature ovary	Translucent Smooth	—	0.435 ± 0.272	—
Primary oogonial cells	—	—	7.2 ± 2.2	5.5 ± 0.8
Secondary oogonial cells	—	—	21.4± 4.3	13.2 ± 1.8
Primary oocytes			39.1 ± 10.0	20.9 ± 6.3
Stage II				
Early maturing ovary	Pale cream Smooth granular	2.207 ± 0.619	125.8 ± 22.6	44.2 ± 7.6
Stage III				
Late maturing ovary	Light green granular	4.527 ± 1.499	187.3 ± 19.4	60.0 ± 7.0
Stage IV				
Mature/Ripe ovary	Dark green granular	7.312 ± 1.059	241.3 ± 16.7	66.4 ± 3.6
Stage V				
Spent/spentre-covering ovary	Pale cream granular –flaccid	1.042 ± 0.413	25.1 ± 4.9	17.6 ± 2.6

(**Source:** Mohamed, 1989)

3.4 DEVELOPMENT OF OVARY

Immature or stage 0 ovaries are not visible externally through the dorsal exoskeleton. The ovaries in this stage appear as a thin, translucent, and un-pigmented band with small finger like linear lobules over the hepatopancreas. The posterior lobes are rudimentary. In stage-I (pre-vitellogenic) the ovary is not visible through the exoskeleton, as it is not developed. The paired ovaries at this stage lose its transparency and appear opaque. The colour of the ovarian lobules varied from white to cream and appears a bit granular especially in the anterior and middle lobes. The anterior and middle lobes appear larger in this stage but no increase in size in the posterior lobes. Early vitellogenic (stage II) ovaries are faintly visible through the exoskeleton as a thick linear solid band due to its expansion in the posterior thoracic and anterior abdominal regions. The diameter of the posterior lobes becomes longer than that of the intestine. The dissected ovaries appear firm and granular in texture with a light yellow to greenish yellow color due to the accumulation of yolk. The visibility of the ovary through the exoskeleton is due to the granular nature and the increased coloration in the ovarian lobules. A diamond shaped expansion of the ovary at the first abdominal segment characterizes the late vitellogenic or stage III ovaries. Ovaries are visible through the exoskeleton as a thick solid and dark linear band due to its further expansion of its lobules. The ovaries are light to dark green in color with a firm and granular texture. The mature (stage IV) ovaries are clearly visible through the exoskeleton as a thick band on the entire dorsal side of the animal. The fully diamond shaped or butterfly shaped expansion of the first abdominal segment is clearly visible. The ovarian lobes are considerably larger, and these fully developed ovaries fill up all the available space in the body cavity, both in the cephalo thoracic and abdominal region. The olive green to dark green ovaries is highly granular in texture. Spent (stage V) ovaries are not visible through the exoskeleton. Externally it is indistinguishable from those in the pre-vitellogenic stage. But the ovaries are flaccid and reduced in size but remain opaque, with less colouration, distinctive than the pre-vitellogenic stage. In incomplete spawns, portions of ovaries, particularly posterior lobes, retain the coloration of the matured ovary.

In maturing animals development of thelycum is closely related with the ovarian development. At stage 0 and stage I maturity stages the lateral plates of thelycum are widely separated leaving a "V" shaped notch in the middle. The cavity between the plates appears empty and not impregnated with spermatophores indicating that mating has not taken place. In stage II animals the lateral plates are somewhat broadened and start their overlapping with the lateral flanges of the median plate. Further, broadened lateral plates complete their overlapping with the lateral flanges of median plates in the stage III animals. There is only a "V" shaped small slit between these lateral plates and most of animals at this stage are impregnated with spermatophores. Thickened and intersected lateral plates are observed in mature animals. Later these plates become crescent shaped and the entire thelycum occupies almost the entire space between the sternite of the 5thpair of pereiopods. A concealed hollow space is present beneath the lateral plates where in the spermatophores are deposited. Similar structures are present in the thelycum of spent (stage-V animals).

3.5 PROCESS OF OOGENESIS

Reproductive cycle includes a series of events starting from activation of primordial germ cells to the differentiation of highly yolk-equipped ova. Light microscopical examinations of the ovaries at different stages of maturity indicated the chain of nuclear and cytoplasmic changes that occur inside the developing ovary. Oogonial cells developed from the primordial germ cells get transforms into the mature ova with sufficient yolk for the development of the embryo. A series of dramatic as well as complicated changes take place in the developing oocyte during its developmental phase. Based on the changes that occur inside the cytoplasm and nucleus of the growing oocytes, process of oogenesis is generally classified into six different phases, such as immature, pre-vitellogenic, early-vitellogenic, late vitellogenic, mature or gravid and spent oocytes. These oocytes phases correspond to the stages 0 to V as earlier based on morphological characters.

The light microscopical studies in the ovarian tissue of penaeid shrimp showed, a thin ovarian wall, encompass the ovary. It consists of 3 layers; a thin outer most pavement epithelium, an inner layer of germinal epithelium and a relatively thick layer of connective tissue in between. Blood capillaries are also present in the ovarian wall. A germinal zone (GZ) is found on the lateral periphery in the form of a thin band and this is the "zone of proliferation" from which the displacement of oogonial cells takes place. Invasion of this zone into the ovarian lobes is observed from the ventral portion of the ovary. The young oocytes moved farther from the germinal zone upon maturation so the developing oocytes and ova are found towards the center of each ovarian lobe.

1. Immature Stage

An active zone of proliferation with clusters of developing oogonial cells is the characteristic feature of immature ovary (Fig.2). The primary and secondary oogonial cells are arranged in a graded manner in the ovary so that the growing secondary oogonial cells are shifted to the interior. The nuclei of the primary oogonial cells are not conspicuous. These primary oogonial cells have a diameter of 11±1.35μ. These cells undergo mitotic division and gives rise to the secondary oogonia. The secondary oogonial cells possess a conspicuous nucleus. There is an increase in the number of diffused granule like nucleolus in the nucleoplasm of these oogonia, ranging from 8-20 in numbers. The secondary oogonia are bigger with a cell diameter of 24.1±3.5 μ and a nuclear diameter of 18.3±3.9 μ.

The oogonial cells appear round ultrastructurally and their large nucleus occupies approximately 80% of the cell volume. The oolemma is smooth and without any particular morphological specializations at this stage. The diffused electron-dense chromatin materials as well as small granule like round nucleoli are not present in the nucleolemma of these oogonial cells. Electron-loose cytoplasm in these oogonial cells contains only some small granules and filamentous materials (Fig.3). Other cell organelles are not at all visible at this stage of development.

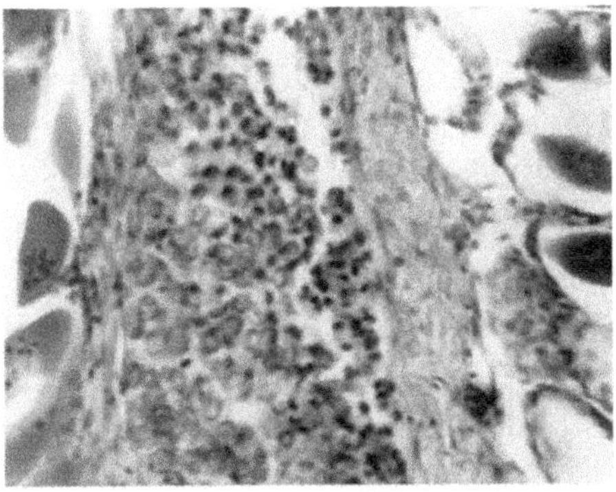

Fig. 2. Light micrograph of an immature ovary of *P monodon* showing germinal zone with developing oogonial cells (GZ – Germinal zone OG–Oogonial cells) X100 (**Source:** Diwan *et al.*, 2008)

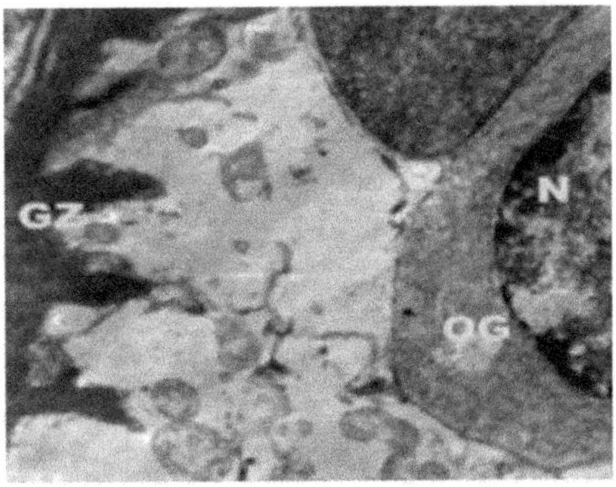

Fig. 3. Electronmicrograph of germinal zone and developing oogonial cells. (GZ – Germinal zone, OG – Oogonia, N – Nucleus, Nucleolus (Arrow). X8400 (**Source:** Diwan *et al.*, 2008)

2. Pre-vitellogenic ovary

The most striking feature of pre vitellogenic ovary is the presence of highly basophilic primary oocytes with much more increased cytoplasmic volume than that of the oogonial cells. These primary oocytes develop from the secondary oogonial cells through meiotic division. Two developmental phases are seen in the pre-vitellogenic oocytes i.e oocytes in the chromatin nucleolus stage and the oocytes in the perinucleolus stage. Former are smaller than the latter and appear as round to oval cells with a prominent nucleus. The nuclei contain 10 to 18 centrally located, deeply stained granules like nucleoli and prominent chromatin materials in their nucleoplasm. These oocytes vary in cell diameter and also in nuclear diameter and are devoid of individual follicle cell layers. The perinucleolar oocytes are characterized by the displacement of nucleoli towards the periphery of the nucleoplasm. The follicle cells in this stage are rectangular or cubical and a highly vacuolated conspicuous nucleus is also present (Fig.4).

Ultrastructurally the pre vitellogenic oocytes appear round, like oogonial cells but the cytoplasmic volume is more compared to the oogonial cells. The nuclei are round with a clear nucleolemma, which is interrupted by numerous nuclear pores through which the nuclear materials presumably passes into the perinucleolar ooplasm. The nucleoli of the pre vitellogenic oocytes are electron dense bodies along the inner periphery of the nuclear wall. Some membrane-free, electron-dense materials are present at the outer side of the nuclear wall associated with the nuclear pores (Fig.5 & 6).

Ultrastructurally both phases of previtellogenic oocytes i.e. the oocytes at the chromatin nucleolus stage and the oocytes at the perinucleolar stage exhibit the same sub-cellular structures. The cytoplasm contains ribosomes, mitochondria, golgi elements and cisternae of the granular endoplasmic reticulum. Variable sized vesicles filled with dense, osmio philic granules are present near oolema. Numerous long lamellae of parallel membranes are seen throughout the cytoplasm. These lamellar congregations possess some regular pore like structures and are identified as annulate lamellae. Membrane-free, electron-dense aggregates appear between dense lamellar membranes and distributed throughout the cytoplasm.

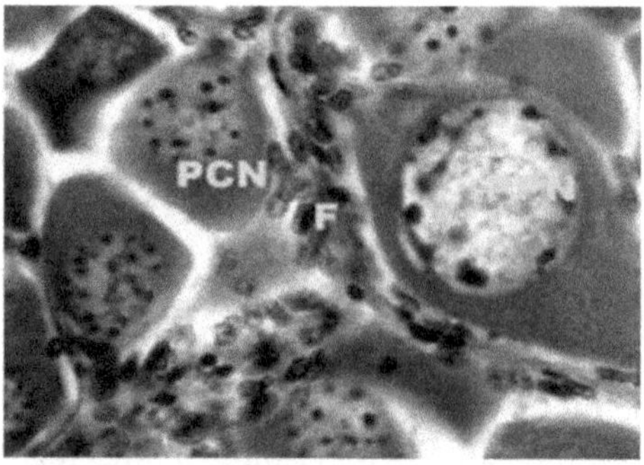

Fig. 4. Light micrograph of a pre-vitellogenic ovary containing oocytes at chromatin nucleolus and peri-nucleolus stages (PCN – Pre-vitellogenic oocyte at chromatin nucleolus stage, PPN-Pre- vitellogenic oocyte at peri-nucleolus stage,F–Follicle cells) X100 (**Source:** Diwan *et al.,* 2008)

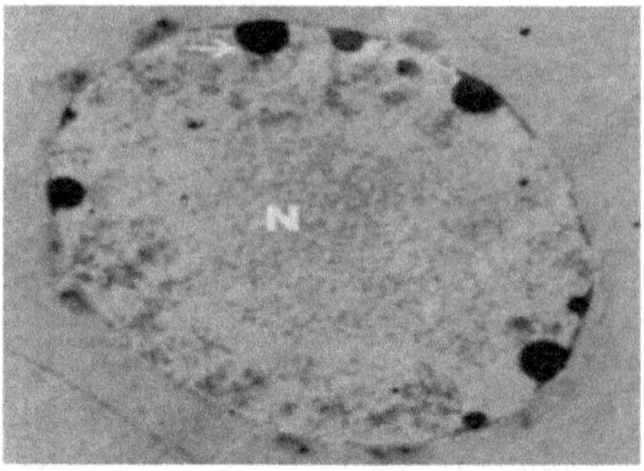

Fig. 5. Electron micrgraph of a pre-vitellogenic oocyte at peri- nucleolar stage. Centrally located Nucleus (N) contained several nucleoli (Arrows) X 660 (**Source:** Diwan *et al.,* 2008)

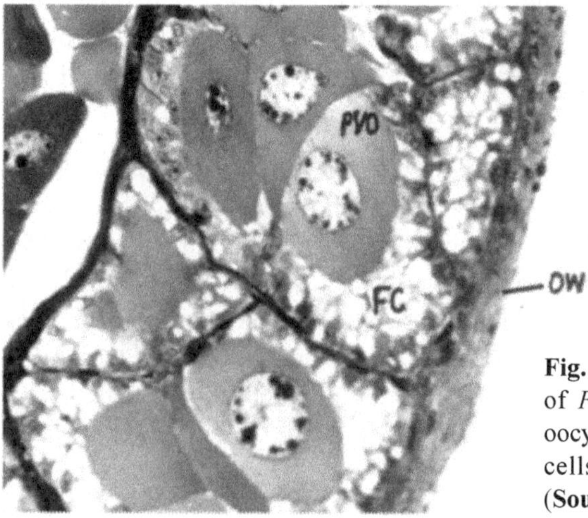

Fig. 6. Semithin section of a stage I ovary of *F. indicus* showing previtellogenic oocytes (PVO) surrounded by tall follicle cells (FC). OW ovarian wall. X 100. (**Source:** Mohamed and Diwan, 1991)

3. Early vitellognic ovary

Early vitellogenic ovary possesses oocytes in two sub-stages i.e. the oocytes at cisternal phase and oocytes at platelet phase (Fig.7). The cisternal phase early vitellogenic oocytes are round to oval with an oocytes diameter of 189.5±15.3μ and a nuclear diameter of 69.2 ±9.6μ. Here there is a sudden two fold increase in the size of the oocytes. The nature of the cytoplasm changes suddenly from homogenous to vesicular and little bit granular. From the granular nature of the cytoplasm and its sudden increase in the cytoplasmic volume it is seen that during this stage onwards the oocytes started its active accumulation of yolk. The granular nature is due to the accumulation of oil globules in the cytoplasm, which is the characteristic feature of the primary vitellogenic oocytes. Consequently the nucleolar materials made a halo around the nucleus due to their circular arrangements in the peripheral karyoplasms (Fig.8).

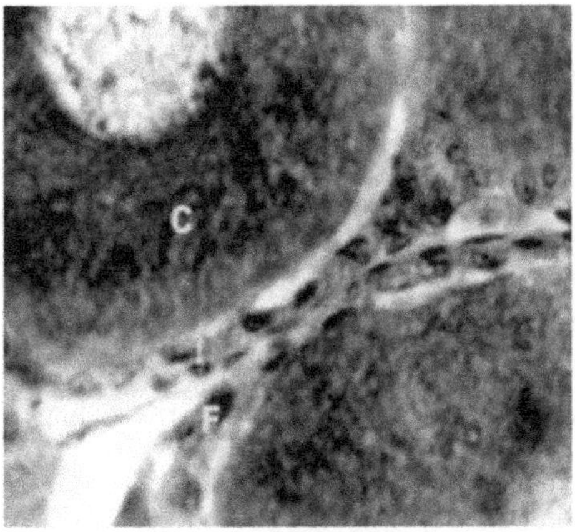

Fig. 7. Light micrograph of early vitellogenic ovary of *P Monodon* with plateletphase oocytes with centrally located palely stained nucleus (N), granular cytoplasm (C) and highly stretched folliclecells (F) X100 (**Source:** Diwan *et al.*, 2008)

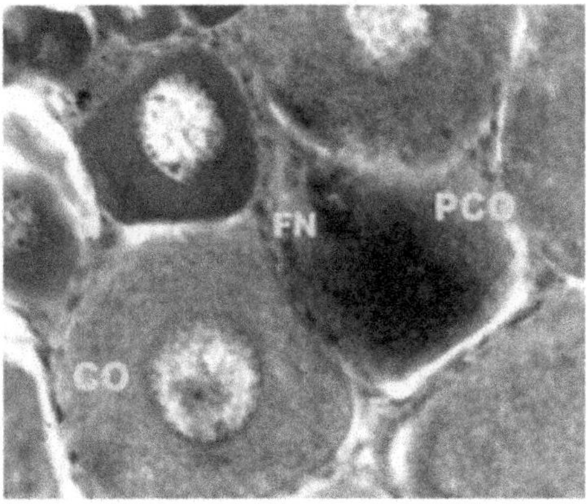

Fig. 8. Light micrograph of an earlyvitellogenic ovary of *P Monodon* with cisternal phase oocytes (CO) and a pseudooocyte (PSO) within a single follicle (F) covering (FN – Follicle cellnucleus) X100 (**Source:** Diwan *et al.*, 2008)

During this stage, the formation of follicle cells around each individual oocyte occurs. Because of the sudden increase in the cytoplasmic volume the follicle cells stretched considerably and consequently their thickness decreased. The hypertrophied nucleus as well as nucleolus of follicle cells becomes conspicuous during this stage. Some oocyte like cells but without a nucleus (Pseudo-oocytes) is also present among the growing oocytes. These cells and the oocytes in close proximity are found encompass by a common follicle cell. A reduction in the size of these pseudo-oocytes is seen as the neighboring oocytes grow (Fig. 9).

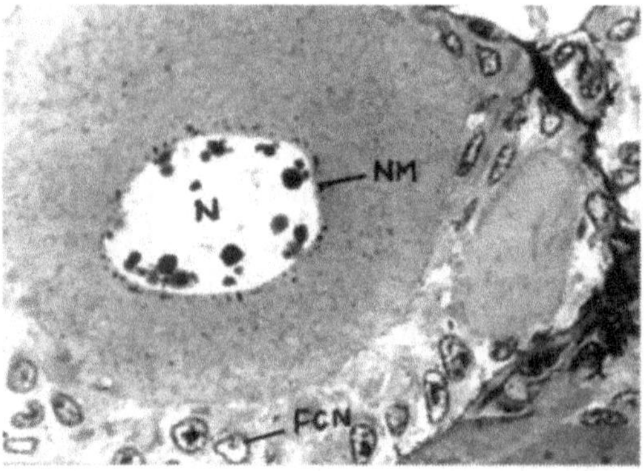

Fig. 9. Semithin section of early vitellogenic oocytes with *F. indicus* perinuclear halo of nucleolar material (NM) and granular cytoplasm. The follicle cells are flattened and have hypertrophied nucleus (FCN). X 200. (**Source:** Mohamed and Diwan, 1991a)

Platelete phase oocytes possess a highly vacuolated, rough and granular type cytoplasm with 274.2 ± 8.3µ oocyte diameter and 81.8 ±4.3µ nuclear diameter. The "pseudo-oocytes" reduced further in their size as the oocytes reach platelet phase. In the platelet phase oocytes, such cells appear as a small bit of granular cytoplasm just nearer to it inside the same follicular encompassment. Due to the increase in the volume of the oocytes the follicle cells encompassing them stretch further and appear as a narrow band of flattened cells around each oocytes.

The two different types of oocytes, the cisternal phase oocytes and platelet phase oocytes in the early vitellogenic ovary are clearly distinguished through ultrastructural studies. A lot of prominent ultrastructural changes occur in the perinuclear activity during its transition from the previtellogenic to early vitellogenic phase. Membrane free granular aggregations differing in their shapes and contents appear around the nuclear membrane (Fig.10 & 11).

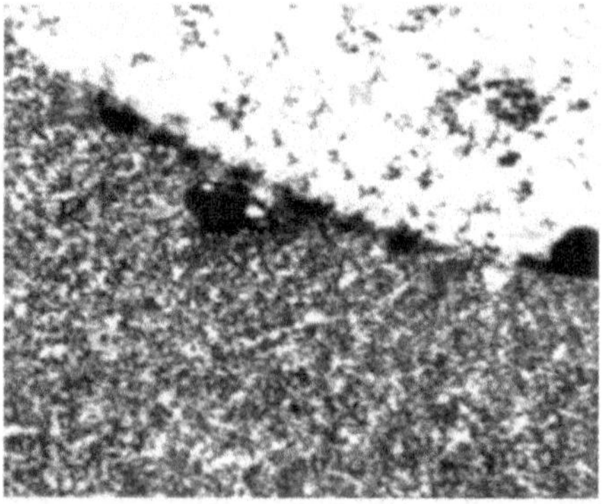

Fig.10. Electronmicrograph of the cisternal phase oocytes showing nucleus (N), membrane free aggregations & (Arrows). X 12,650 (**Source:** Diwan *et al.*, 2008)

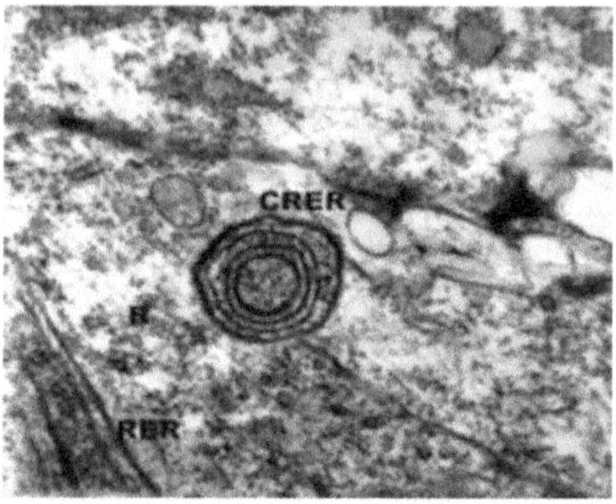

Fig.11. Electron-micrograph of cisternal phase-oocytes of *P monodon* showing active cell organelles like rough endoplasmic reticulum (RER), concentric rough endoplasmic reticulum and ribosomes (R) X 25,200 (**Source:** Diwan *et al.*, 2008

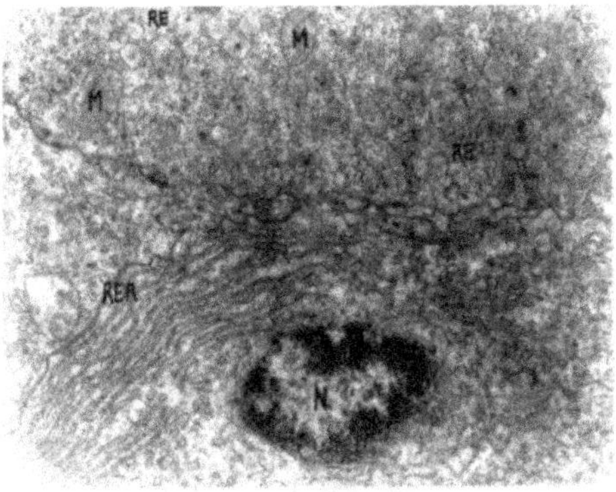

Fig. 12. Electron micrograph of the perinuclear region of an early vitellogenic oocyte of *F. indicus*d abundant mitochondria (M), rough endoplasmic reticulum (RER) swollen reticular elements (RE) and numerous free ribosomes. N nucleus. X 8000. (**Source:** Mohamed, 1989)

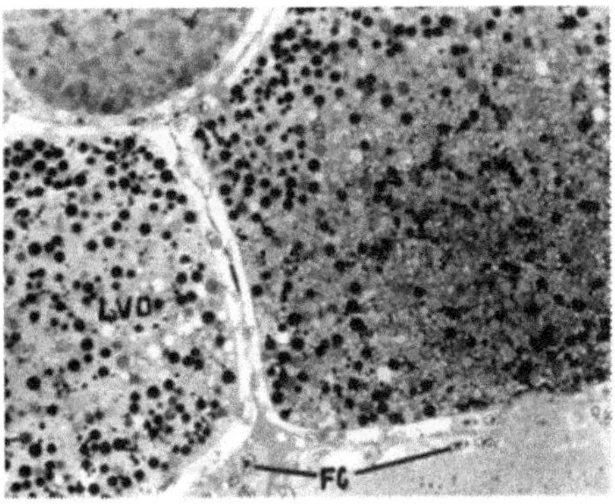

Fig. 13. Semithin section of late vitellogenic oocyte of *F. indicus* showing darkly stained yolk platelets, lightly stained yolk vesicles, unstained lipid globules and flattened follicle cells (FC). X 200. (**Source:** Mohamed and Diwan, 1991)

The perikaryon of the cisternal phase oocytes contains numerous active cell organelles and developing vesicles (Fig.12). All these granules contain electron-opaque flocculent materials. Electron-loose membrane-bound vesicles are apparent in the perinucleolar cytoplasm. These extra nuclear vesicles fuse to form large agranular vesicles. The cytoplasm of these cells contains numerous free ribosomes with minimal membrane association (Fig.13).

The perinucleolar cytoplasm of these oocytes is highly rough and granular in appearance due to the accumulation of both electrons dense as well as electron loose vesicles. Accumulation of membrane free and membrane bound granular bodies continues in this platelet phase oocytes. Masses of granular vesicles are apparent in the perinucleolar cytoplasm. These smaller granules fuse together giving rise to larger granules. Whorls of rough endoplasmic reticulum are present in association with these larger granules. In addition to these granules a variety of granules differing intheirsize, shape and electron density are seen in the perinucleolar cytoplasm. These include membrane free extremely electron dense large yolk bodies, rough endoplasmic reticulum bound developing yolk bodies and developing lipid vacuoles. Circular profiles of swollen rough endoplasmic reticular elements with an accumulation of condensing materials are also abundant in the perinucleolar cytoplasm. An increased numbers of round to elongate mitochondria with lamellar cristae of variable shape filled withelectron-densematrix, numerous free and attached ribosomes and highly active Golgi bodies are present in the cytoplasm of these oocytes (Fig 14, 15, 16 & 17).

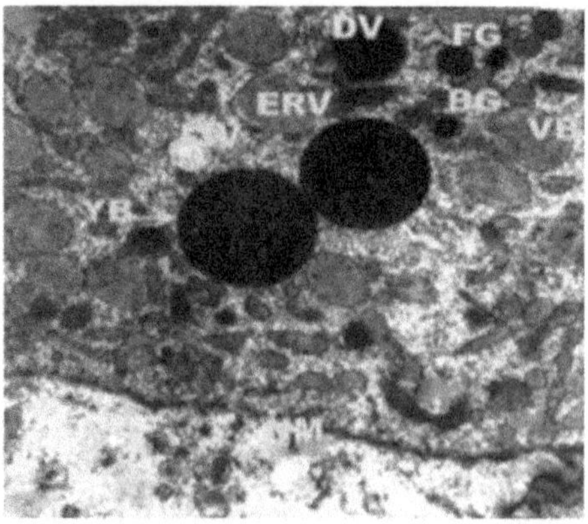

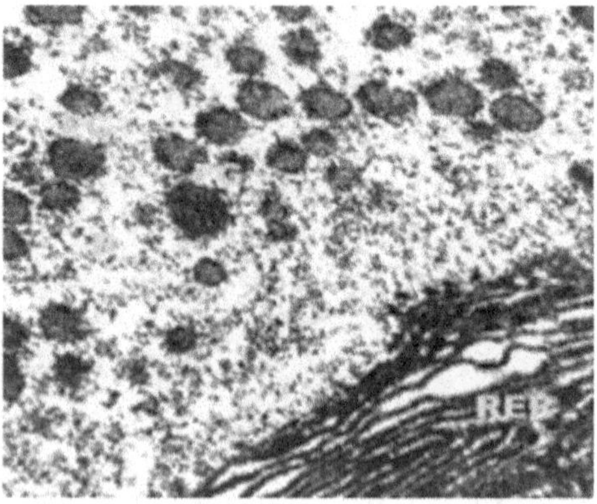

Fig. 14 and 15. Electron-micrograph of plateletl phase oocytes of *P. monodon* showing thickly packed yolk platelets (YP) and vesicular bodies (VS), highly porous and undulating nuclear membrane (NM), electron-dense (DV) and electron loose (LV) vesicles, membrane free (FG) and membrane bound (BG) granules, endoplasmic reticulum bound vesicles (ERV), large yolk bodies (YB), lipid vacuoles. (**Source:** Diwan *et al.*, 2008)

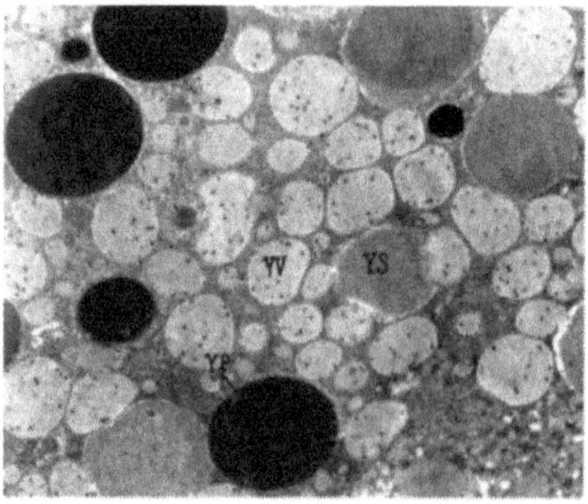

Fig. 16. Electron micrograph of the cytoplasm of late vitellogenic oocyte of *F. indicus* showing formation of yolk. The yolk vesicles (YV), Yolk spheres (YS) and fully formed yolk platelets (YP) are evident. X 4000. (**Source:** Mohamed and Diwan, 1991)

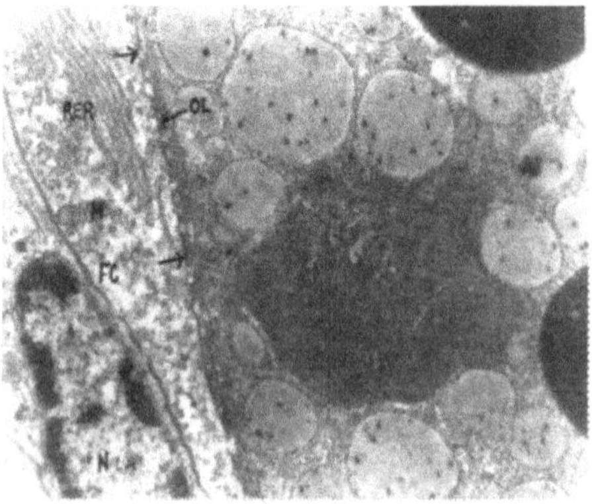

Fig. 17. Electronmicrograph of the late vitellogenic oocytes in *F. indicus* with oolemma (OL) showing micropinocytotic vesicles (arrows). The adjacent follicle cell (FC) has hypertrophied nucleus (N), mitochondria (M) and rough endoplasmic reticulum (RER). X 8000. (**Source:** Mohamed and Diwan, 1991)

Cytoplasm in the perinucleolar area to the oolemma contains numerous lipid vacuoles in association with Golgi complexes. Towards the periphery of the oocytes, RER vesicles as well as intra-cisternal granules are present along the forming face of the Golgi complex. Golgi complexes are prominent, in the peripheral cytoplasm with variable number of stacked cisternae. Small moderately electron-dense vesicles are pinched off from the cisternae of the active parallel stacks of rough endoplasmic reticulum with intra- cisternal granules the cytoplasm nearer to the oolemma. Developing yolk bodies are abundant in this peripheral cytoplasm in association with these reticular stacks. Numerous free ribosomes also are noticed around these bodies (Fig. 18 & 19). RER elements are also present in association with these vesicles. These granules in association with the RER bodies together suggested a source of yolk constituents. All these active organelles indicate the active role of the cytoplasm in the intra oocyte production of yolk. Biomolecular studies of the follicle cells around the platelet phase oocytes exhibit an increased sub cellular activity. It is seen that the flat follicle cells possess an oblong nucleus and a highly electron dense large nucleolus inside the nucleoplasm. Other cell organelles in the cytoplasm are round to oval mitochondriae, Golgi complex, smooth and rough reticular elements, ribosomes and vesicular bodies.

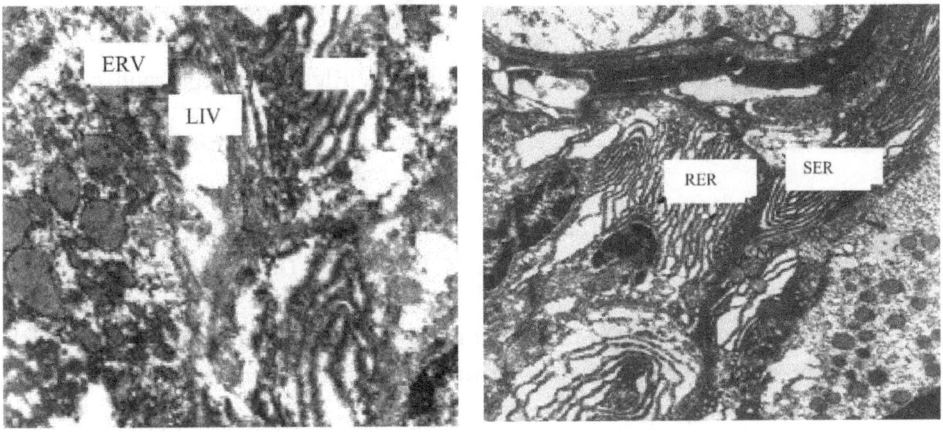

Fig. 18 and 19. Electron-micrograph of platelet phase-oocytes of *P. monodon* showing thickly packed yolk platelets (YP) and vesicular bodies (VS), highly porous and undulating nuclear membrane (NM), electron-dense (DV) and electron loose (LV) vesicles, membrane free (FG) and membrane bound (BG) granules, endoplasmic reticulum bound vesicles (ERV), large yolk bodies (YB), lipid vacuoles (LIV) and highly active cell organelles like mitochondria (M), Golgi body (G), Parallel stacks of rough endoplasmic reticulum (RER) and ribosomes. X 25, 200 (**Source:** Diwan *et al.*, 2008).

4. Late vitellogenic ovary

Late vitellogenic oocytes are characterized by the appearance of radially arranged well developed, club shaped structures the cortical rods in the peripheral ooplasm. Oocytes diameter was 338.5±9.7µ with a nuclear diameter of 107.7 ±7.3µ. Abundant numbers of yolk platelets are present in between the cortical rods of these oocytes. The cortical bodies measure 45.1±4.4µ in length and 20.7±3.4 in breadth. Ultrastructurally the cortical rods are moderately electron-dense with homogenous cytoplasm. Higher magnification revealed a distinct substructure in these cortical rods. The matrix of these bodies is packed with some "feathery" structures (Fig. 20 & 21).

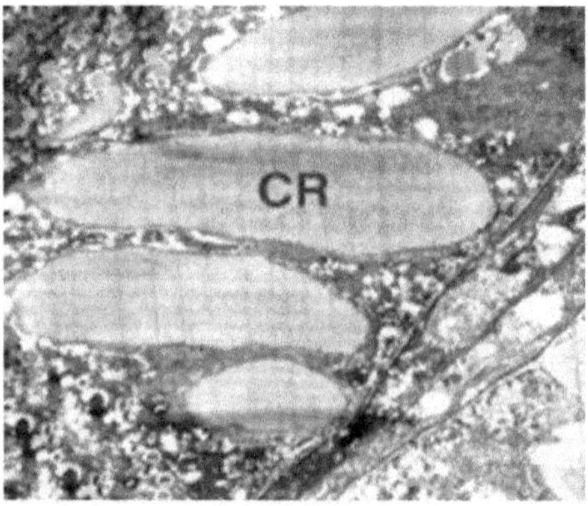

Fig. 20. Light micrograph of follicle cells around the platelet phase oocytes with active nuclei (FN) X100 (**Source:** Diwan *et al.*, 2008)

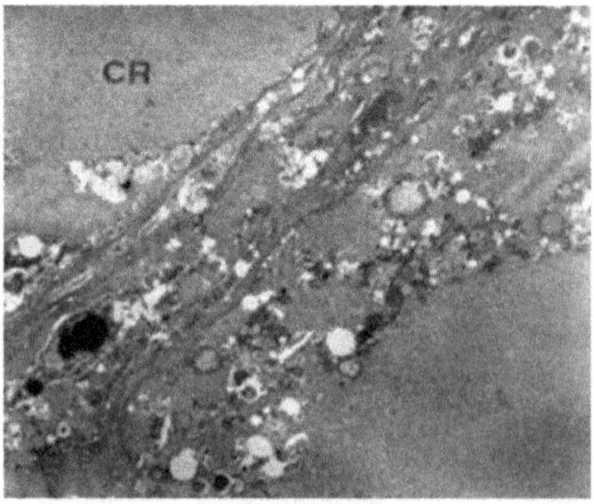

Fig. 21. Electron-micrograph of a late vitellogenic oocyte of *P. monodon* with cortical rods (CR), Cytoplasm in between these rods contained electron dense yolk bodies (YB) and moderately electron dense yolk platelets (YP). Fig. 28X430 and Fig. 29 X 4,600 (**Source:** Diwan *et al.*, 2008)

Each of these structures appears as an electron-dense central axis upon which numerous moderately electron-dense thin febrile like elements arranged as in a feather. The whole cytoplasm between these cortical rods is filled with similar type of yolk spheres. The moderately electron-dense developing yolk bodies in the cytoplasm of platelet phase oocytes are not apparent in this phase. Similar type of electron density of the newly forms cortical bodies and the sudden disappearance of these yolk bodies make to assume that those yolk bodies undergo some progressive changes and finally aggregate together to form the cortical bodies. The cell organelles like ER, Golgi bodies, and mitochondria are not prominent. The Oolemma at this stage appears broken with the presence of numerous pin cytotic vesicles (Fig 22).

Some concentric layers of reticular elements encompassing developing yolk granules are present (Fig. 16) in the peripheral ooplasm, and also in the cytoplasm of the adjacent follicle cells. As the granules enlarge, the circular layers change to semi-lunar shape and finally discharge the granules to the cytoplasm (Fig. 23). Concentric layers of reticular elements with developing yolk precursors are also present with abundant supply of free ribosomes. These types of vesicles appear in the oocytes wall in close apposition to the follicle cells. The oblongnucleus of the follicle cells at this stage is smaller than previous stage. Developing yolk bodies in these follicle cells are found surrounded by circular or semilunar endoplasmic reticular elements.

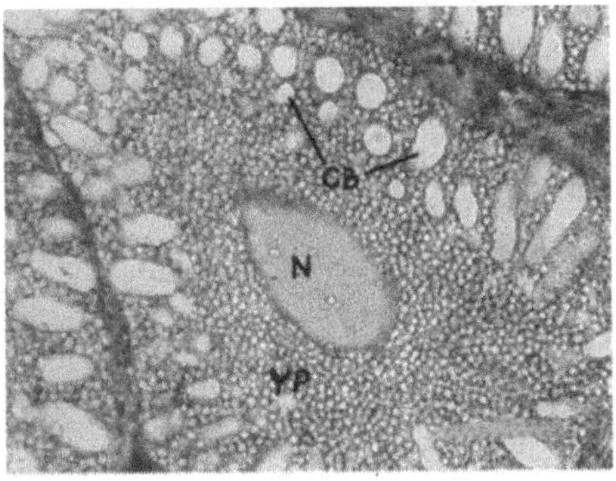

Fig. 22. Cross section of a vitellogenic/ mature oocyte of *F. indicus* with peripheral cortical bodies (CB) close to the oolemma. (N) nucleus. X 100. (**Source:** Mohamed and Diwan, 1991a)

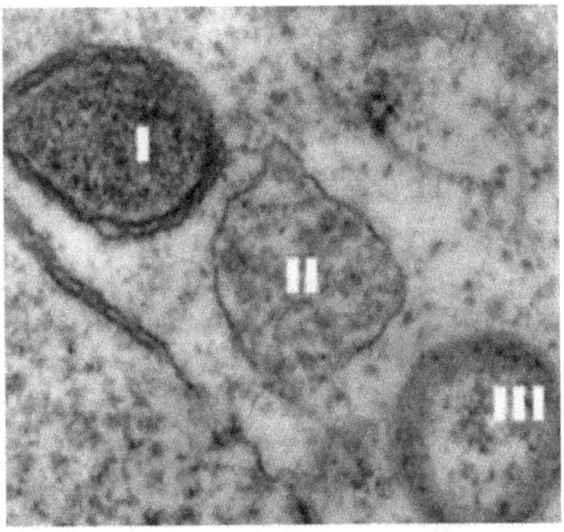

Fig. 23. Electron-micrgraph of the oolemma of the late vitellogenic oocytes of *P Monodon* showing pinocytotic up take of granular yolk materials (arrows) I –granule filled vesicle connected with the oolemma, II–newly separated vesicle, III-already separated vesicle in the ooplasm. X 15,220 (**Source:** Diwan *et al.*, 2008)

5. Mature ovary

The mature oocytes appear similar to the cortical rod phase oocytes, except the fact that the cortical rods moved to the extreme periphery of the ooplasm and lose its appearance. The fully matured oocytes appear as more elongated than circular with a very thin rim of follicle cells around it. The size of the nucleus much reduced, and starts its migration to the periphery and upon full maturation many of the cells are found without a nucleus as it is shifted to the cytoplasmic membrane (Fig. 24).

The follicle cells around each oocyte slowly get detached from the oolemma so that the ovulation becomes easy. Because of the more elongated appearance of the oocytes they measure upto 406.3±13.4μ in diameter and nuclei only 91.7±4.7μ in diameter. The greatly reduced nuclei lose their round shape due to the disintegration of the nuclear wall, but the nucleolus is prominent (Fig. 25). A prominent nucleus is present in the follicle cells around each oocyte.

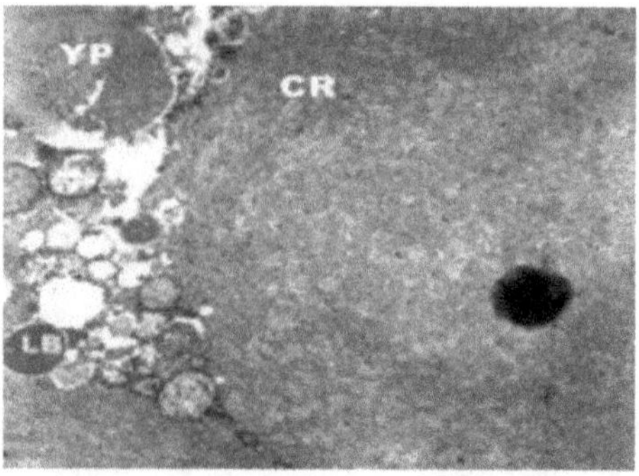

Fig. 24. Electron micrograph of the follicle cells and the adjacent oocytes of *P Monodon* with follicle cell nucleus (FN), part of the cortical rods (CR), yolk vesicles (YV) taken through pinocytosis and the membrane free yolk granules (YG) X8,400 (**Source:** Diwan *et al.*, 2008)

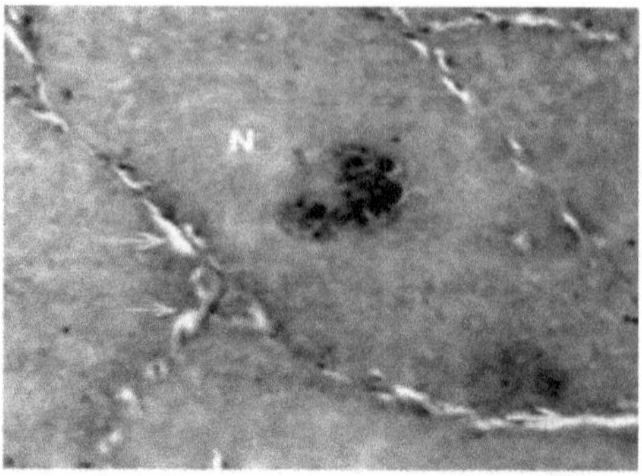

Fig. 25. Light micrograph of the fully matured (just before ovulation) ovary showing riped oocytes started its detachment from the oolemma (arrows). The disintegrated nucleus (N) can be seen in these oocytes X100 (**Source:** Diwan *et al.*, 2008)

Ultrastructural investigations reveal that mature oocytes are similar to that of cortical rod phase oocytes, except a few differences in their cytoplasm. Distinct oval cortical bodies are not apparent as they moved to the peripheral region of the oocyte at this stage. Almost the entire cytoplasm is packed with these cortical bodies and the same feather like appearance observed in the cortical rods; but the electron-dense axis of the feather like structures is not visible. Yolk platelets and lipid granules are found only in very restricted areas (Fig. 26). Most of the yolk bodies with similar contents empty their granules to the matrix of the cortical bodies. Cell organelles were not visible in the oocyte cytoplasm. The follicle cells at this phase get detached from the oolemma in most of the places. A clear space is visible in all oocyte between the oolemma and plasma membrane of the detached follicle cell. Follicle cell at this stage appear as a three layered structure without any distinct cell organelles.

6. Spent ovary

Oocytes of the spent ovary are mostly primary oocytes similar to those of pre-vitellogenic oocytes in the immature ovary but these can be distinguished by the presence of yolk substances at. Resorbing oocytes are also observed in between the empty thicker follicle cells caused by the retraction of the follicle cells from the oocytes. The zone of proliferation is active with irregular primary oogonial cells and developing oocytes at certain portions of the spent ovary.

In *P. monodon* the female genital organ consists of paired ovaries, which are partly fused, bilaterally symmetrical bodies extending from the cardiac region of the anterior portion to the telson, paired oviducts and the thelycum in the ventral side. Motoh (1981) and Solis (1988) have described similar type of ovarian morphology in the same species from the seas of Philippines. Other penaeids like *Penaeus setiferus* (King, 1948), *P. duorarum* (Cummings, 1961), *P. indicus* (Subramoniam 1965, and Mohammed and Diwan, 1994), *P. merguiensis* (Tuma, 1967), *P. setiferus* and *P. aztecus* (Duronslet *et al.*,1975) also have similar morphological structures.

Visual assessment of gonadal development is one of the effective methods of studying reproductive state of the marine organisms to determine the degree of gonadal growth. For females the common method has been the visual assessment of the ovaries on the basis of egg size and colour (King, 1948; Cummings, 1961; Subramoniam, 1965; Tuma, 1967; Rao, 1968; Villaluz *et al.*,1969; Primavera, 1980; Motoh, 1981; Yano, 1985; Tanfermin and Pudadera, 1989; Shoji, 1997; Peixoto, *et al.*,2003; Alfaro, *et al.*,2004; Peixoto, *et al.*,2005; Coman *et al.*,2005 and 2006). The gonadal maturation in female penaeids categorized by the external staging system developed by King (1948) for *P. setiferus* and Tuma (1967) for *P. merguiensis* based on the colour, size and nature of developing ovary issued for the classification of maturity stages in *P. monodon* in the present chapter. Another simple indicator of the reproductive state of the marine organism is the gonado somatic index (GSI). In this method of assessing gonadal growth, changes in the ratio between gonadial weight and body weight are taken into account. This has been variously used in penaeids (Lawrence, *et al.*,1979; Joshi, 1980; Lawrence and Castle, 1991; Shoji, 1997; Peixoto, *et al.*,2003; Alfaro, *et al.*,2004; Peixoto, *et al.*,2005; Coman *et al.*,2005 and 2006).

Histological characteristics are more reliable concrete evidence for the assessment of gonadal maturation. Macroscopic and histological characterization of oocyte development in several penaeid species have been carried out by different investigators like Subramonyam (1965) and Rao (1968) in *P. indicus*, Brown and Patlan (1974) in *P. aztecus*, *P. duorarum*, and *P. setiferus*, Duronslet *et al.* (1975) in *P. aztecus* and *P. setiferus*, Tom *et al.*,(1987) in *Parapenaeus longirostris*, Yano (1988) in *P. japonicus*, Mohammed and Diwan (1994) in *P. indicus* and Shoji (1997) in *P. monodon*. Although the evaluation criteria for the classification of gonadal development and number of ovarian developmental stages are arbitrary and vary in literature, the most commonly used system divides the ovarian maturation into five stages. On the basis of ovum size, gonad expansion and colouration, the process of maturation of ovary in *P. monodon* has been categorized into 5 stages by Villaluz *et al.*,(1969), Santiago (1977), Primavera (1980), Motoh (1981) and Solis (1988). But later TanFermin and Pudadera (1989) revised this classification into four stages based on the histological and histo-chemical studies for an easy identification of the maturity stages of spawners. Qunitio *et al.*, (1993) added 2 more stages and described the ovarian maturation of *P.monodon* in six stages for the practical use of brooders in the hatcheries. From these reports the classification is found confusing and therefore in the present chapter based on the morphological characters and histological and ultrastructural changes in oocytes, the maturity stages of *P. monodon* are classified into five main ovarian stages, immature (stage 0), pre-vitellogenic (stage I), early- vitellogenic (stage II), late-vitellogenic (stage III) mature (stage IV) and spent (stage V). The thin ovarian wall is composed of 3 layers. King (1948) in *P. setiferus* and Subramoniam (1965) in *P. indicus* described three layers in the wall of the ovary, a thin outermost layer of pavement epithelium, an inner layer of germinal epithelium and a layer of connective tissue in between. Mohammed and Diwan (1994) however, described only 2 layers in the ovarian wall of *P. indicus*. A zone of proliferation or germinal zone is present in the periphery of the ventro-lateral wall in *P. monodon* in almost all maturity stages. Invasion of this zone into the ovarian lobes is seen during early developmental stages. Similar types of observations have been in the ovary of *P. setiferus*, by King (1948), *P. indicus* by Subramoniam (1965), and Mohammed and Diwan (1994), *P. japonicus* by Yano (1988) and Shoji, (1997) in *P. monodon*.

Oogenesis in crustaceans involves two distinct processes, proliferative and differentiative. During the proliferative process in the germinal zone the number of oogonial cells increases by mitotic multiplication. The primary oocytes derived from the secondary oogonial cells transforms into typical egg cells by differentiative process (Adiyodi and Subramoniam, 1983). In *P. monodon* also the process of oogenesis get completed through these two phases. The proliferative phase of multiplication is seen in the immature as well as maturing ovaries. The primary oogonial cells derived from the germinal zone undergo rapid multiplication and form the secondary oogonial cells. During differentiative phase these oocytes accumulate its reserve yolk and finally develop into fully matured oocytes. Yano (1988) in *P. japonicus* observed a similar pattern of oocyte formation. The stages of mitotic division of primary oogonial cells are difficult as they occur in rapid succession (Adiyodi and Subramoniam, 1983). Description of the primary oogonial cells in crustaceans, especially in penaeids are scanty. Instead the secondary oogonial cells in the immature ovary as well as in their resting phase in maturing ovary have been often described in other crustaceans but still are rare in penaeids. In *P. monodon*, the primary oogonial cells appear as small round deeply stained cells around the germinal zone. These condary oogonial cells also appear as round cells with a prominent nucleus and a thin layer of cytoplasm around the nucleus. Diffused chromatin materials as well as small granular nucleoli become prominent in the karyoplasm. Weitzman, (1966) described the secondary oogonial cells in *Gecarcinus lateralis*as "germ nests" having large round vesicular nuclei with deeply stained chromatin materials and a thin rim of cytoplasm. In *P. indicus* Mohammed and Diwan (1994) reported that the oogonial cells possess a large, conspicuous nucleus and weakly eosinophilic cytoplasm. Qunitio *et al.*,(1993) reported that the secondary oogonial cells in *P. monodon* had a large nucleus with granular nucleoli, which fill up the entire cytoplasm.

The Trentini and Scanabissi, (1978) described similar types of observations in the oogonial cells of crab *Tripos cancriformis* but further they described the oogonial cells as four-celled groups (one oogonia and 3 nurse cells) resulted from incomplete cytokinesis. In *P. monodon* it is seen that after the proliferative divisions, the oocyte entered the differentiative phase during which a chain of nuclear and cytoplasmic changes and an appreciable increase in the size of the oocyte occur. These changes associated with the vitellogenesis are included here in the classification of different maturity stages of the ovary. The first differentiated oocytes are considered as the pre-vitellogenic oocytesin *P. monodon.* Tanfermin and Pudadera (1989) described two substages, viz. the chromatin nucleolus stage and perinucleolus stage based on the arrangement of nucleolus in the karyoplasms in the pre-vitellogenic oocytes of *P. monodon.* Again Qunitio *et al.*,(1993) described these two sub-stages in the stage III pre-vitellogenic ovary of *P. monodon.*

Ultrastructural studies of pre-vitellogenic oocytes of crustaceans have been over looked because most of the investigators have been primarily concerned with vitellogenic rather than pre-vitellogenic oocytes. Fine structural details on pre-vitellogenic oocytes of *P. monodon* shown that these oocytes are characterized by a large nuclear to cytoplasm ratio and are immediately recognizable by the dense nucleolus located on the peripheral nucleoplasm. The oocytes at this stage of pre-vitellogenesis are found accumulated with precursors, which are required for the complicated process of vitellogenesis in the developing ovary. The cytoplasmic organelles in the oocyte become active and recognizable in contrast to the oogonial cells. The nuclear walls at this stage appear interrupted by numerous gaps or nuclea rpores through which the electron dense nucleolar materials diffused in to the cytoplasm. It is also seen that the cytoplasm especially around the nucleusis supplied with abundant number of ribosomes, mitochondria, Golgi elements and RER in the pre-vitellogenic oocytes of *P. monodon.* Similar types of observations have been made in *L. emarginata*, (Hinch, 1969), *P. setiferus*, (Duronslet *et al.*,1975), *P. indicus*, (Mohammed and Diwan, 1994) and in *P. monodon*, (Shoji, 1997). In *P. monodon* it is seen that along with these organelles, the cytoplasm of the pre-vitellogenic oocytes possess an agranular membrane system (annulate lamellae) and nuages. This type of annulated lamellae and nuage has been reported in the gametes of many animals (review by Kessel, 1983). In decapod crustacean's occurrence of these annulate lamellae are first reported by Komm and Hinch (1985) in the pre-vitellogenic oocytes of *Coenobitaclypeatus*. Another characteristic feature observed in the pre-vitellogenic oocytes of *P. monodon* is the granule filled vesicles accumulated in the cytoplasm around the nucleus. The similar electron-density of these vesicles and nucleolar materials and its appearance near the nuclear pores indicate that these vesicles originate in the nucleus. Hinch (1980) and Komm and Hinch (1985) have reported similar types of granules in *C. clypeatus*.

Though the vitellogenesis is a largely explored area in crustaceans over the past two decades, the site of synthesis of crustacean yolk is still not yet known. The detailed studies of the early-vitellogeni coocytes of *P. monodon* revealed two distinct types of oocytes, oocytes at cisternal phase and oocytes at platelet phase. Duronslet *et al.*, (1975) described these two phases of oocytes in *P. setiferus* and *P. aztecus*. Early-vitellogenic oocytes have been investigated through electron microscope by Hinch (1969) in *Libinia emarginata*; Duronslet *et al.*,. (1975) in *P. setiferus* and *P. aztecus*, Komm and Hinch (1987) in *C. clypeatus* and Mohammed and Diwan, (1994) in *P. indicus*. Ultrastructural studies in the cisternal phase oocytes of *P. monodon* indicate an intense perinuleolar activity. The vesicles transferred from nucleus act as the prelude for further vitellogenesis inside the oocyte cytoplasm. Such nuclear emissions are reported in many early-vitellogenic crustacean oocytes by Adiyodi and Subramoniam (1983). Dark staining cytoplasmic bodies around the nuclear envelope and in the perinucleolar cytoplasm are reported in the cisternal phase oocytes of *P. setiferus* and *P. azeticus* by Duronslet *et al.*,. (1975) and in the developing oocytes of *P. indicus* by Mohammed and Diwan (1994).

As the oocyte of *P. monodon* reaches the cisternal phase, an increase in the concentration of free ribosomes and RER occur. Zerbib, (1980) attributed the growth of *Orchestia* oocytes along with increased ribosome concentration. The abundant RER in growing oocytes has been demonstrated by many crustacean investigators namely Beams and Kessel (1962 & 1963) in crayfish oocytes, Hinch and Cone (1969) in *Libinia emarginata*, Zerbib (1980) in *Orchestia gammaralla* and Komm and Hinch (1987) in *Coenobitaclypeatus*. In penaeids the abundant supply of both RER and free ribosomes has been reported by Duronslet *et al.*,(1975) and Mohammed and Diwan (1994). In *P. monodon*, the ultrastructure of early-vitellogenic oocytes show well developed Golgi bodies in association with some RER. It is seen that the vesicles blebs off from the ER, fuse with the cisternae of these Golgi bodies, condense into yolk bodies and accumulate within the confines of the oocyte. Such metabolic co-ordination has been extensively investigated and demonstrated (Caro and Palade, 1964, Tartakof and Vassalli, 1978 and Komm and Hinch, 1987). Micro pinocytotic vesicles also made their appearance in the plasma membrane of these oocytes at the end of this phase.

Platelet phase oocytes of *P. monodon* are packed with different types of granule filled vesicles and yolk platelets. The vesicles at this phase include membrane free electron- dense yolk bodies; RER bound developing yolk bodies and developing lipid vacuoles. Another characteristic feature is swollen circular profiles of RER with condensing materials, which are blebbed off from the reticular elements. This type of vesicles and endoplasmic reticular profiles are typically demonstrated by Beams and Kessel (1962 & 1963) in crayfish, Hinch and Cone (1969) in *Libinia emarginata*, Euurenius (1973) in *Cancer pagurus*, Zebib (1980) in *Orchestia gammarellus* and Komm and Hinch (1987) in *Coenobitaclypeatus*. Ultrastructural studies of Duronslet *et al.*, (1975) in *P. aztecus* and *P. setiferus* and Mohammed and Diwan (1994) in *P. indicus* documented this type of ER elements and yolk bodies. Very active Golgi complexes at this stage are associated with developing vesicles as well as large lipid vacuoles. Hinch and Cone (1969) in *Libinia emarginata* and Komm and Hinch (1987) in *C. clypeatus* reported this type of developing yolk bodies and lipid vacuoles associated with active Golgi bodies.

The ultrastructural studies on platelet phase oocytes in *P. monodon* reveal an irregular oocyte wall at this phase due to the dense accumulation of micro pinocytotic vesicles. These vesicles are gradually blebbed off into the peripheral cytoplasm of the oocytes. These vesicles together with the associated RER elements suggested a new source of yolk materials in these oocytes. In the oocytes of *P. monodon* these smaller granule filled vesicles, which originate in the ovary get fuse together, undergo some progressive modifications and finally transform into the larger yolk spheres. An intra oocytic yolk synthesis with an extra oocytic yolk supplementation is seen in these platelet phase oocytes. Ultrastructural studies by Hinch and Cone (1969) in *L. emarginata*, Euurenius (1973) in *C. pagurus*, Zerbib (1980) in *O. gammarellus*, Beams and Kessel (1980) in *Oniscusasellus*, Komm and Hinch (1987) in *C. clypeatus* all proposed an extra-oocytic source of yolk constituents, complementing the intra-oocytic source. In penaeids like *P. setiferus* and *P. aztecus* studies by Duronslet *et al.*, (1975) and in *P. indicus* by Mohammed and Diwan (1994) demonstrated a similar type of micro pinocytic vesicles and extra- ovarian uptake of yolk materials in the vitellogenic oocytes.

Late-vitellogenic oocytes in *P. monodon* are characterized by the presence of a distinct sub-structure the cortical bodies. Tanfermin and Pudadera (1989) and Qunitio *et al.*,. (1993) described these cortical rods in the cortical rod phase ovary of *P. monodon*. This is a unique feature in penaeids studied so far. Different investigators described the same structure in different terminologies. In *P. japonicus*, Hudinaga (1942) described these structures as jelly substances, King (1948) in *P. setiferus* and Rao (1968) in penaeids as peripheral bodies, Subramoniam (1965) in *P. indicus* as marginal bodies, Duronslet *et al.*,. (1975)

in *P. setiferus* and *P. aztecus* as cortical bodies, Anderson *et al.*,(1984) in *S. injentis* as cortical specializations and Yano (1988) in *P. japonicus* as cortical crypts.

Sheikhmahmud and Tambe (1958) did not find the cortical rods or related structures in *P. stylifera*. Ultrastructural investigations of Duronslet *et al.*,. (1975) in *P. setiferus* and *P.aztecus* and Mohammed and Diwan (1994) in *P. indicus* reveal that the matrix of these oocytes is packed with small feathery structures. The cortical bodies present in *P. monodon* exhibit a similar type of ultrastructure. The matrix is packed with small "feather" like structures and each of these structures possess a thick electron-dense axis and thin fibrillar elements. The cytoplasms between these bodies appear similar to that of platelet phase, with full of lipid bodies and yolk platelets. Micro pinocytotic yolk uptake continues in these late-vitellogenic oocytes. Cell organelles like ER, Golgi bodies and mitochondria are not apparent. Consequently the intra oocytic yolk production does not occur in these oocytes. In *P. indicus*, Mohammed and Diwan (1994) reported a similar type shift of in yolk synthesis, from intra oocytic to extra oocytic as the oocytes reached in the late- vitellogenic phase. Hinch and Cone (1969) in *Libinia emarginata* and Komm and Hinch (1987) in *C. clypaetus* has been described the similar type of shift during the late- vitellogenesis.

In *P. monodon* fully matured ovary (late-vitellogenic) oocytes are apparent shortly before ovulation as enlarged cells with eosinophilic cytoplasm. The distinct cortical rods are not apparent in these oocytes. In the histological investigations Tanfermin and Pudadera, (1989) observed similar type of oocytes in 5% of the fully matured *P. monodon*. They concluded that these oocytes with vacuolated cytoplasm are in another step of vitellogenesis/artesia. Ultrastructurally it is seen in *P. monodon* that highly enlarged cortical bodies shift further to the periphery and because of their larger size and shift to the periphery they are not visible at light microscopical level. Similar type of cortical rod movement to the peripheral cytoplasm has been reported in the mature oocytes of *P. setiferus* and *P. aztecus* (Duronslet *et al.*,1975). In *P. monodon*, it is seen that the highly expanded cortical rods fill the entire space of the ooplasm of these fully matured oocyte. Ultrastructurally the whole ooplasm of the mature oocytes has an appearance, which is similar to that of the matrix of the cortical bodies. Cytoplasm in restricted are as in between these cortical rods contains electron-dense lipid globules, lipid vacuoles and few yolk platelets as found in the late vitellogenic oocytes. It is found that the remaining yolk platelets also emptied their contents to the matrix of the cortical rods. From these observations it is concluded that these cortical bodies are forms by the fusion of the abundant yolk platelets found in the preceding stages.

Folliculogenesis or the investment of follicular cells around the oocytes in *P. monodon* is found initiated only at the end of pre-vitellogenic oocytes. During this phase the rectangular or cuboidal follicle cells are thick and possess a highly vacuolated cytoplasm and conspicuous nucleus. Similar type of follicle cells has been reported by Tanfermin and Pudedera, (1989) in *P. monodon* and by Mohammed and Diwan (1994) in *P. indicus*. The flattening of the follicle cells is observed during the progressive growth of the oocyte due to its increase in cell volume. The complete encompassment of individual oocytes with the follicular cells is observed in the early vitellogenic phase of the ovary. The follicle cells at this stage possess a hypertrophied nucleus with prominent nucleolus.

According to Charinaux-Cotton (1985), folliculogenesis in the early vitellogenic oocytes is a pre-requisite for the uptake of yolk proteins form external sources. Yano and Chinzei (1987) suggested that the follicular cells, which surround the oocytes, are the sites of exogenous egg yolk protein production in *P. japonicus*. So far the cytological changes in the follicle cells encompassing the oocytes of penaeids has not been a subject of detailed investigation and consequently very little is known about such changes.

The ultrastructural observations in the follicle cells of early vitellogenic oocytes of *P.monodon* show an active cytoplasm, with numerous RER and agranular membrane systems. These active organelles in the follicle cells suggested a role in yolk protein synthesis in these cells. Follicle cells around the platelet phase oocytes show an active protein synthesis than the preceding stage. At this phase along with the active cell organelles like Golgi bodies, RER, and mitochondria, granule filled vesicles are also apparent. The vesicular bodies found in the follicle cell cytoplasm include membrane- free electron-dense yolk bodies, developing membrane-bound yolk platelets and lipid vacuoles. According to Ratenu and Zerbib (1978) protein synthetic activity in the vitellogenic oocytes of *O. gammarella* is evident in the endoplasmic and Golgi complex region of the follicle cells. Highly active organelles in the follicle cells of vitellogenic oocytes have been reported in by Griffond and Gomot, (1979) and Shoenmakers *et al.*,(1981). The follicle cells surrounding the late-vitellogenic oocytes in *P. monodon* are characterized by the presence of concentric layers of RER elements with developing yolk granules. Trentini and Scanabicci (1978), has also reported similar type of yolk synthesis inside the concentric layers of smooth endoplasmic reticulum in the 4 celled follicle cells of *Triopscancriformes*. Abundant pino cytotic vesicles found apparent in the oolemma of these oocytes in close apposition to the follicle cells give added evidence for the protein synthetic activity of these cells during this phase. Similar type of pinocytotic vesicles has been described in *P. indicus* (Mohammed and Diwan, 1994). A dramatic change is observed in the follicle cells around the fully matured oocytes. Nucleus as well as other cell organelles is not at all observed in the follicle cells. A clear space is apparent between the plasma membrane of the follicle cell as well as the oolemma. According to Bray and Lawrence (1990) the final flush of colour change during ovarian maturation seems to occur within hours of spawning coupled with the breakdown of follicle cells surrounding the ova in preparation of spawning.

Preechaphol *et al.*,(2007) carried out the work on identification and characterization of sex related genes expressed in vitellogenic ovaries of tiger shrimp, *P. monodon* by using Expressed Sequence Tag (EST) technology. It was observed that out of 1051 clones sequenced from the 5-terminus nucleotide sequences, 743 EST significantly matched with the known genes of the tiger shrimp previously deposited in the gene bank where as 308 ESTs were regarded as newly unidentified transcripts. Preechaphol *et al.*,(2010) while studying maturation process ovarian tissues in *P. monodon* have isolated and characterized progesterone receptor related protein P23 differentially expressed during ovarian development. They have reported that their findings strongly suggest functionally important role of P23 gene products during vitellogenesis of *P. monodon* oocytes.

Arcos *et al.*,(2011) evaluated correlations of vitellogenin (Vg) in haemolymph with the number and diameter of oocytes in the gonads, as a measure of reproductive status of pacific white shrimp, *Litopenaeus vannamei.* A significant correlation has been reported between vitellogenin and oocyte diameter but not for Vtg and number of oocytes. From these findings it was concluded that Vtg is the first reproductive trait to predict reproductive capacity of this white shrimp before eyestalk ablation, and shrimp producers can utilize it to define a priority which females to introduce in to spawning tanks for breeding and spawning.

Kruevaisayawan *et al.*,(2010) while working on morphological changes of developing oocytes during oogenesis in the black tiger shrimp, reported the abundant presence of ribosomes and dilated rough endoplasmic reticulum in the developing oocytes, so also presence of yolk granules and lipid droplets indicating active synthesis of protein and lipid components. The main characteristic of the mature oocyte as reported by them was the presence of cortical rods which were composed of the tightly packed structural units each resembles a bottle brush. With the help of immunostaining further they showed that these structural units first get synthesized in the RER-Golgi complex of maturing oocytes and transported in to the extracellular crypts of the mature oocytes where they finally assemble in to cortical rods.

Reproduction in the female shrimp is characterized by active production of yolk protein vitellin which is accumulated in the oocytes. Unlike the vertebrates which synthesize vitellogenin precursor in the liver, the sources of the crustacean vitellogenin have been debated for many years. Using a molecular approach later it has been confirmed that the hepatopancreas is the major site of synthesis of vitellogenin (Rankin *et al.*,1989, Tom *et al.*,1992). On the contrary, there are also reports for ovarian synthesis of vitellogenin in *P. japonicus* (Yano and Chinzei, 1987) and *P. semisulcatus* (Browdy *et al.*,1990). Several hormones have been suggested to play a role in regulation of vitellogenesis. Role of eyestalk factors in the regulation vitellogenesis has been described by Chan *et al.*,(2003) and that of steroids and terpenoids have been shown by Laufer *et al.*,(1992). In recent past Shirley *et al.*,(2006) while working on vitellogenisis of *P. monodon* mentioned that both the ovary and hepatopancreas equally contribute to the production of vitelloginin transcript during reproductive maturation. Their study was based on the effect of farnesoic acid and 20-hydroxyecdysone in vitro hepatopancreas and ovarian tissues which stimulated the expression of vitellogenin.

Tahara *et al.*,(2005) developed a vitellogenin ELISA method to find out the relationship between hemolymph vitellogenin levels and spawning conditions in female kuruma shrimp, *Marsupeaneus japonicus.* It was observed that hemolymph vitellogenin levels did not fluctuate significantly during pre-maturation and maturation stages but during late maturation stages large accumulation of vitellin was seen in the developed oocytes. These results indicated that high levels of vitellogenin are required to the shrimp throughout the late maturation phases as well as during spent stages. Tinikul *et al.*,(2014) investigated the effects of two isoforms of gonadotropin-releasing hormone, namely octopus GnRH and lamprey GnRH, and a neurotransmitter, dopamine on the ovarian maturation and spawning in the pacific white shrimp, *L. vannamei.* It has been reported that both GnRH stimulated the ovarian maturation, while dopamine showed negative effect (Fig. 26)

The penaeid breeding programmes have been frequently associated with the improvement of quality traits in terms of disease resistance and better growth performance (Goyard *et al.*,2002, Gitterle *et al.*,2005 a, b). Several methods are being used to improve the traits either through genetic selection, wild stock selection, improving immune system of the domesticated shrimps by various means (Arcos *et al.*,2005, Ibarra *et al.*,2007). Peixoto *et al.*,(2008) carried out studies on comparison of reproductive performance, offspring quality, ovarian histology and fatty acid composition between similarly sized wild caught and domesticated shrimp, *Farfantepenaeus paulensis.* From the studies it was suggested that the reproductive performance, offspring quality, and ovarian maturation of 16th month old domesticated *F. paulensis* brood stock were equivalent to that of similarly sized wild caught brood stock from deep sea waters. Andriantahina *et al.*,(2012) conducted studies on comparison of reproductive performance and offspring quality of domesticated white shrimp *L.vannamei* in pond reared and tank reared conditions. They found that there was no significant difference in

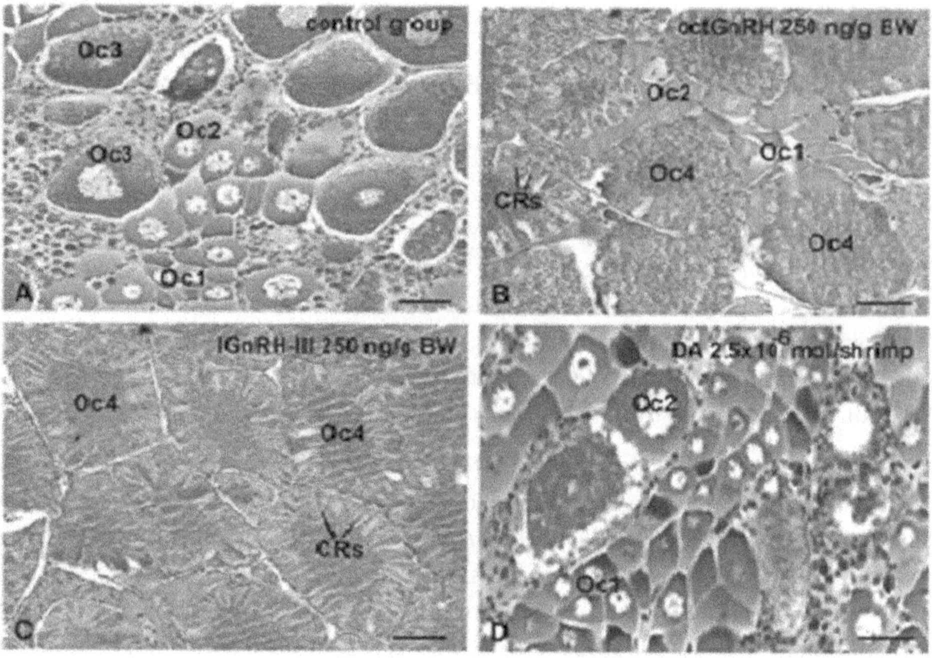

Fig. 26. Effects of oct GnRH, lGnRH-III, and DA, on ovarian histology. (A) Ovaries of the control group on day 21 after injection with normal saline, showing characteristics of early ovarian stages II–III. (B–C) Ovaries of groups injected with octGnRH-injected (250 ng/g BW), lGnRH-III-injected (250 ng/g BW), on day 21 demonstrating characteristics of stage IV, containing fully mature oocytes with cortical rods. (D) DA-injected group (2.5× 10 -6 mol/shrimp) showing characteristics of stage I contained mostly early previtellogenic oocytes. CRs, cortical rods; Oc1, early previtellogenic oocyte; Oc2, late previtellogenic oocyte; Oc3, early vitellogenic oocyte; Oc4, late vitellogenic oocyte/mature oocyte; Scale bars, 50 µm. (**Source:** Tinikul *et al.*, 2014)

the spawning frequency, fecundity, egg spawning success rate, number of eggs per spawning and egg hatchability between the pond reared and tank reared female shrimps. However, it was reported that the pond reared brood stock resulted in better offspring quality than the tank reared brood stock. Peixoto *et al.*,(2011) updated the review on research and development on brood stock maturation and reproduction of the pink shrimp, *F. paulensis*. In their review besides basic knowledge of the reproductive biology and maturation process of *F.pualensis*, mention has been made on basic procedures and environmental requirement for maturation and spawning of the shrimp in captive condition. Effect water temperature, water exchange rate, effect of ammonia, brood stock feeding etc. on maturation has been covered in this review. Impact of unilateral eye stalk ablation, artificial insemination studies on reproductive performance, brood stock source, size and age and their relationship with reproduction has been also discussed in detail. Mention has also been made on genetic selection studies and even the impact of reverse genomic technique to generate expression profiles for individual shrimp. Results from such studies represent an initial approach towards a better understanding of the genetic regulation of growth and breeding programmes for cultured shrimps.

The mechanism underlying vitellogenesis in shrimp is complex and many regulating factors are involved in the process. The expression of various genes in reproductive related tissues and their control by hormones are essential steps in the regulation of ovarian development. Many worker have suggested that differential expression of genes occurs at various stages of viltellogenesis in the shrimps, *P. monodon* (Karoonuthaisiri *et al.*,2009, Preechaphol *et al.*,2007, Uawisetwathana *et al.*,2011), *Fenneropenaeus chinensis* (Xie *et al.*,2010), *Marsupenaeus japonicus* (Callaghan *et al.*,2010). The most important hypothesis that has been conceptualized in this context is that thrombospondin (TSP) and peritrohin are ovarian development related genes. Kunwadee *et al.*, (2014) while working on simulation of ovarian development in white shrimp *F. merguiensis* with a recombinant ribosomal protein found that this protein is encoded by an essential gene related to the early stages of ovarian development. The recombinant protein was produced and the same was evaluated to determine its ability to stimulate ovarian development. Further they have mentioned that administration of recombinant protein created high levels of methyl farnesoate hormone in the haemolymph during stage II of vitellogenesis.

Studies on the existence of gonadotropin-releasing hormone (GnRH) in the ovary and effects of GnRH on the ovarian maturation was made in the black tiger shrimp *P. monodon* by Ngemsoungnem *et al.*,(2008). The findings suggested that GnRH may be highly conserved peptides that play an important role in inducing the ovarian maturation in the shrimp.

Arcos *et al.*, (2011) found that vitellogenin when measured in haemolymph of the shrimp, *L vannamei* reflect the degree of ovarian development before eyestalk ablation and to be useful predictor of ovarian development after ablation. This technique is quite useful for shrimp producers as it indicates which females to be introduced in to the spawning tanks for breeding and spawning in captive condition.

In penaeid shrimp egg maturation process is characterised by vitellogenic and cortical rod protein (CRP) synthesis. Kim *et al.*, (2005) while working on *Marshupenaeus japonicus* mentioned that CRP mRNA is highly expressed before the onset of vitellogenesis and that vitellogenin (Vg) mRNA exhibited high expression during intense vitellogenesis. These observations suggest that different genes are involved in ovarian synthesis of CRP and Vg proteins. In the same penaeid shrimp Okumura *et al.*,(2006) provided further evidence that Vg and

CRP synthesis get accelerated after eyestalk ablation. Earlier Khayat *et al.*,(1994) also demonstrated high levels of Vg mRNA in the vitellogenic ovary of *P. semisulcatus.* Unlike other decapods, where autosynthesis of yolk has been shown to occur within the oocytes (Lui and O'connor 1977), in penaeid shrimp ovarian synthesis of yolk probably takes place in the follicle cells. Tsutsui *et al.*,(2000) by using the technique of Northern blot analysis and *in situ* hybridisation revealed that mRNA encoding vitellogenin was expressed in the follicle cells of vitellogenic females of Kuruma shrimp *M. japonicus.* Tsang *et al.*,(2003) also showed the expression of vitellogenin gene in the ovary and hepatopancreas of *Metapenaeus ensis* suggesting equal contribution from both tissues.

In recent years, several gene expression studies using the technique of quantitative Real time PCR, have demonstrated that the ovary remains the principle organ which synthesizes yolk proteins in several Penaeid shrimp species. In species like *M. japonicus* and *P. semisulcatus* one and the same Vg is experessed in the ovary and hepatopancreas. However, in shrimps like *Leptopenaeus mergurensis*, *M. ensis* and *P. monodon* more than one Vg may be involved in the tissue specific expression of the gene in both the ovary and hepatopancreas. Phiriyangkul *et al.*,(2007) while working on *L. Merguiensis* mentioned that the pattern of Vg mRNA expression between the hepatopancreas and ovary differ in that the expression level in the hepatopancreas is much lower than that in the ovary at all the stages of ovarian development. Thus the relative contribution of the ovary and hepatopancreas to overall yolk production may differ among various shrimp species.

Considerable information is available on ovarian lipovitellin acting as precursor for transport of lipids into the ovary from the haemolymph during vitellogenesis process. Lipid transport through the haemolymph takes place by two high density lipids (HDL) and a very high density lipoprotein. Female specific vitelliginin is one of the HDLs with its production being correlated with ovarian development. It has been noticed that in *P. semisulcatus* the non sex specific haemolymph lipoprotein consist, one of the peptide units of vitellogenin whereas, the sex specific lipoprotein consists of three subunits of vitelligenin peptide. With regard to the location of site vitellogenin synthesis considerable work has been carried out. In this regard it has been observed that VG expression may occur at multiple sites according to the species. That one and the same gene for vitellin and vitellogenin can be simultaneously expressed both in the ovary and hepatopancreas was shown in *P. Semisulcatus.* Multiple genes may also show tissue specific expression of Vg in the ovary and hepatopancreas, are demonstrated in another penaeid shrimp *M. Ensis.* Evidently in the shrimp ovary is the primary site of yolk synthesis as indicated by gene expression studies whereas, in other decapods they largely rely on extra ovarian organs such as the hepatopancreas for synthesis of Vg.

Treerattrakool *et al.*,(2008) while studying molecular characterization of gonad inhibiting hormone (GIH) of *P. monodon* and its elucidation on inhibitory role in vitellogenesis expression by RNA interference found that after knocking down GIH expression both in the optic and abdominal ganglion resulted in a conspicuous increase in Vg transcript level in the ovarian tissue of the shrimp, although Vg expression in the hepatopancreas has been reported less significant. Tsutsui *et al.*,(2005) while studying the effects of crustacean hyperglycemic hormone on vitellogenin gene expression in Kuruma shrimp *M. japonicus* mentioned that the vitellogenin inhibiting hormone exerts its effect primarily through vitellogenin gene expression in the ovary whereas in the hepatopancreatic tissue this gene expression was not that significant. In a more recent study Tsutsui *et al.*,. (2007) isolated as many as six sinus gland peptides with vitellogenesis inhibiting hormonal (VIH) activities in the shrimp *L. vannamei.* These VIHs have been reperted to cause varying degrees of inhibition in Vg mRNA expression in ovarian tissues of *M. japonicus* incubated *in vitro.*

Considerable work has also been carried out on vitellogenin receptors and yolk protein uptake. In recent studies, the cloning and characterization of cDNA encoding a putative Vg receptor from the tiger shrimp *P. monodon* has been reported. Similar studies have been also carried out for the kuruma shrimp *M. japonicus.* In *Penaues monodon,* after binding Vg with vitellogenin receptor, it has been reported that this complex moves into the oocytes cytoplasm aided by internalization signals present in vitellogenin receptors. Subramoniam (2011) in his review article on mechanism and control of vitellogenesis in crustaceans has mentioned about yolk proteins, yolk processing and sites of synthesis of yolk in different crustaceans. Further he has also described endocrine regulation of vitellogenesis, vitellogenesis stimulating hormone and steroidal control of vitellogenesis in detail.

Iamsaard *et al.*,(2012) carried out studies on changes in oviduct structure in the black tiger shrimp, *Penaeus monodon,* during different stages of ovarian maturation. They examined the structure of the oviduct using light, transmission and scanning electron microscopy techniques. It is reported that the epithelium of the oviduct during initial phase of maturation is composed of tall simple columnar cells with their basal nuclei located on the basement membrane and its thick collagen fibers. In the second phase of maturation the oviduct seemed to produce some substances and their epithelial cells became transitional with centrally located nuclei and formed some vacuoles. In stage IV of the maturation the epithelial cells in the oviduct are found to be disorganized, disrupted, and shed accumulated spherical secretory substances including some cellular contents into the lumen. Prior to spawning, only the oviduct epithelium at ovary Stage IV produced and secreted a number of spherical secretion substances into the lumen. These substances may act as the oviductal lubricants to facilitate the spawning process. Talbot and Helluy (1995) and Yu and Lu (2006) reported that the oviduct of penaeid shrimp plays an important role not only in carrying the mature eggs from the ovary to the gonophore during spawning but also in the secretion of some molecules. The possible substances secreted from the oviduct may play various physiological roles in the female reproductive system, such as lubrication of the oviduct, participation in oocyte maturation, and induction of capacitation of sperm stored in the female seminal receptacle. King (1948) while working on *P. setiferus* observed that oviduct wall is folded in some portions of the oviduct. Similar observations have been made by Lu *et al.*,2006) while working on *M. nipponense*. Bell and Lightner, (1988) also studied the histological changes in the oviduct in *P. monodon* in relation to body weight and reported that the epithelium of the oviduct consists of a layer of tall simple columnar cells with a basal nucleus in smaller shrimps (2 g body weight) and in larger shrimp (80-100 g BW) the oviductal epithelial cells become disorganized. It has been suggested that the epithelial cells of the oviduct of lobsters undergo cyclical changes with ovarian development and spawning. It is believed that the tall columnar epithelial cells lining the oviductal wall secrete a lubricating fluid to facilitate the passage of the mature eggs along the oviduct, although lubricating substances have never been characterized. However, it has been proposed that substances secreted from the shrimp oviduct may be involved in fertilization during spawning (Lu *et al.*,2006).

3.6 REPRODUCTIVE PERFORMANCE OF DOMESTICATED BROOD STOCK

Improving the reproductive performance of domesticated brood stock remains a high priority for commercially important shrimps for growth of shrimp Aquaculture industry. One of the main issues with many domesticated shrimps is low rate of egg fertilization probably due to sperm quality (Pongtippatee *et al.*, 2007). Therefore a critical requirement for improving reproductive performance is the development of practical and reliable measures of male fertility, which can be used to determine culturing technologies that optimisefertility. There are different methods to determine the efficiency of male fertility in penaeid shrimps. Some of the methods such as spermatophore weight and morphology and sperm count are fairly coarse but are simple to perform (Gupta and Rao, 2000, Jiang *et al.*, 2009 and Pratoomchat *et al.*,1993). More complex method such as acrosome reaction assay is sometimes considered more reliable because they assess specific characteristics of sperm quality and its activation when it comes in contact with the eggs. Arnold *et al.*, (2012) developed a novel approach to evaluate the relationship between measures of male fertility and egg fertilization in *P. monodon.* In their study they employed a novel, pairwise comparison approach to discriminating the fertilization influence of males: female brood stocks were artificially inseminated with one spermatophore each from a pair of the males. It was reported 22 successful spawnings involving selected pairs from 33 males and estimated the proportion of embryos fertilized by each paired male by individual genotyping of embryos. The non -inseminated twin spermatophore has been used to estimate measure of male fertility, sperm number and the number of normal sperm of each inseminated spermatophore. Using this approach it was shown that the egg fertilization potential of males cannot be determined simply by total sperm number or the number of normal sperm. This novel approach provides a method that can be used to determine whether more complex measures of male fertility can reliably predict egg fertilization potential.

Chung *et al.*, (2011) while working on enhancing the reproductive performance of tiger shrimp, *P. monodon* by incorporating sodium alginate in the brood stock and larval diets found that WSSV free brood stock of tiger shrimp fed with sodium alginate-enriched polychaete worms at a concentration of 200 mg kg-1 led to significantly higher egg production per gram of spawner's body weight, total larval production per spawner and the larval hatching rate. Hence it was concluded that supplementation of food containing polychaete worms with sodium alginate enhanced brood stock reproductive performance, and also increased the larval survival and body size.

While working on reproductive performance of kuruma shrimp, *Marsupenaeus japonicus* Nguyen *et al.*,(2012) found that the three fractions extracted from polychaetes worms namely trichloroacetic-soluble, neutral lipids, and polar lipids, neutral lipids were most effective in advancing ovarian maturation process.

Polychaetes worms are being used extensively for shrimp brood stock maturation diet due to their qualities in enhancing shrimp reproductive performances (Lytle *et al.*,1990). This is because of presence of HUFAs component, particularly arachidonic acid content (Meunpol *et al.*,2005) as well as some reproductive hormones such as prostaglandin E2 and prostaglandin F2a (Poltana, 2005). Other hormones discovered in polychaetes are ecdysteroids (Laufer *et al.*,2002), osmoregulatory hormones (Andries, 2001), oxytocin/vasopressin hormones (Fujino *et al.*,1999), and sex pheromones (Zeeck *et al.*,1998). Meunpol *et al.*,(2007) identified progesterone and 17α-hydroxy progesterone in polychaetes and studied the effect of these hormones on oocyte development in vitro in penaeid shrimp, *P. monodon.* Twenty four hours in vitro incubation previtellogenic oocytes with progesterone or 17α-OHP4 or with synthetic hormones, significantly increased percentages of vitellogenic oocytes and oocytes with cortical rod compared to the control with no hormones. It was observed that P4 was more effective in enhancing the final maturation of oocytes while 1α-OHP4 had more effects on vitellogenic oocytes.

3.7 MALE REPRODUCTIVE PHYSIOLOGY

Achievement of full economic potential of the shrimp mariculture industry depends on the successful domestication of the species concerned, along with genetic selection for desired traits; such as rapid growth, larger size, and high tolerance capacity. The key to domestication lies in enhanced controlled reproduction by the broodstock animals. It can be achieved only through a better understanding of specific biological aspects, particularly related to the reproductive biology of the animal, to ensure adequate numbers of laboratory reared post larvae can be produced at higher efficiency levels.

Although significant achievements in controlled reproduction of the penaeids have been made in different species in the past 10 to 15 years, much of the interest has been focused on female maturation alone (Primavera, 1985) and knowledge about the male maturation, semen quality and maximum utilization of the high quality sperm etc. are still fragmentary (Leung-Trujillo and Lawrence, 1987, Velazquez, *et al.*,2001). Therefore, for the selection of better strains and production of offspring with the desired quality, not only require the knowledge about the reproductive biology of females but also of males. Recently, condition of the male gonad has also been observed to be an important variable in captivere production due to the higher number of unsuccessful attempts in captive maturation. Studies dealt are mainly on the quality of the semen stored in the spermatophores and its deterioration in captive animals (Leung- Trujillo and Lawrence, 1985, 1987; Bray *et al.*,1985; Talbot, *et al.*,1989; Alfaro, 1993; Alfaro and Lozano, 1993; Pratoomchat *et al.*,1993; Heitzman *et al.*,1993, Gomes and Primavera, 1994 and Velazquez, *et al.*,2001). These reports reveal that the reproductive quality of the males is being decrease during captivity in penaeids. The exact reason for this is not clear and therefore it is emphasized to gain full understanding of the male reproductive mechanism, for the development of sound captive breeding techniques in penaeids. Therefore, in this chapter, physiology of male reproduction at its fine structural level is explained through different aspects, like spermatogenesis and spermatophore formation in the penaeid shrimps.

Reproductive anatomy of male reproductive system in several penaeids has been described by a number of workers (Hudinaga, 1942; King, 1948; Eldred, 1958; Subramoniam, 1965; Tirmizi and Khan, 1970; Huq, 1980; Mohammed and Diwan, 1994 and Gomes and Primavera 1994; Cavalli, *et al.*,1997; Crocos and Coman, 1997; Crocos, *et al.*,1999; Velazquez, *et al.*,2001; Peixoto *et al.*,2003;

Coman, *et al.*, 2003, 2004, 2005, 2006, 2007; Cuzon, *et al.*, 2004). Similarly, there is voluminous information available on the spermatogenesis in various other crustaceans (King, 1948; Langreth, 1969; Koebler, 1979; Hinsch, 1980; Jesperson, 1983; Sagi *et al.*, 1988; Mohammed and Diwan, 1994; Gomes, and Primavera 1994; Cavalli, *et al.*,1997; Crocos, and Coman, 1997; Crocos, *et al.*, 1999; Velazquez, *et al.*, 2001; Peixoto *et al.*, 2003; Coman, *et al.*, 2003, 2004, 2005, 2006, 2007; Cuzon, *et al.*, 2004). However, studies on the spermatogenesis of penaeids are less compared to other crustaceans (King, 1948; Pochon-Masson, 1983; Adiyodi, 1985; Vasudevappa, 1992; Mohammed and Diwan, 1994; Gomes and Primavera 1994; Cavalli, *et al.*, 1997; Crocos, *et al.*, 1999; Velazquez, *et al.*, 2001; Coman, *et al.*, 2003; Coman, *et al.*, 2005; Coman *et al.*,2006; Coman, *et al.*, 2007). Many of such studies focus only on light microscopical investigations while ultrastructural studies on spermatogenesis of penaeids are meager. The male reproductive system of shrimps of the genus *Penaeus* consists of paired testes and paired vas deferentia (King, 1948; Subramoniam, 1965; Tuma, 1967; Malek and Bawab, 1974 a; Tirmizi and Javed, 1976; Motoh, 1978; Motoh and Buri, 1980; Vasudevappa, 1992 and Mohammed and Diwan, 1994, Velazquez, *et al.*, 2001).

Spermatogenesis begins with the peripheral germinative layer of the testicular lobes, when spermatogonia enter into the prophase of meiosis (King, 1948).These spermatogonia after repeated divisions and transformations develop into spermatozoa in the testicular acini (Hinch, 1980). Most of these studies, particularly in the earlier stages of spermatogenesis, in various crustaceans are only through light-microscopy. Spermatid differentiation has been studied electron-microscopically in *P. setiferus* (Lu *et al.*,1973) and in *Sicyonidae* (Shigekawa and Clark, 1986). The spermatozoa in Decapods is considerably modified and usually consists of a number of non-motile arms surrounding a central nuclear region (Pochon-Masson, 1965). Clark *et al.*,(1973) studied the ultrastructure of spermatozoa in *P. aztecus*; Oyama and Kakuda (1987) in *P. japonicus*; and Mohammed and Diwan (1994) in *P. indicus*, Velazquez, *et al.* (2001) in *L. Vannamei*.

Several studies have dealt with the formation and/or structure of the spermatophores in decapods (Hudinaga, 1942; King, 1948; Eldred, 1958; Tirmizi, 1958; Subramoniam, 1965; Tirmizi & Khan, 1970; Malek and Bawab, 1974 a & b; Farfante, 1975; Haley, 1984; Champion, 1987; Ro *et al.*,1990; Chow *et al.*,1991; Vasudevappa, 1992; Mohammed and Diwan 1994; Velazquez, *et al.*,2001). The spermatophore formation has been described histologically in detail, in *P. kerathurus* by Malek and Bawab (1974 a, b) in *P. indicus* by Champion (1987), and Mohammed and Diwan (1994) and in *M. dobsoni* by Vasudevappa (1992). Despite the importance of the spermatophore in decapods reproduction, little is known about its formation at the ultrastructural level. Hinch and Walker (1974) have described the basic ultrastructure of the secretory cells as well as different stages of spermatophore formation in the Vas deferens of *L. emarginata.* Kooda-Cisco and Talbot (1986) described the production of the primary and intermediate spermatophore layers in the proximal Vas deferens of lobster (*Homarus*). Talbot and Beach (1989) used light microscopy and transmission electron microscopy to examine the structure of the vas deferens and its role in the formation of the spermatophore wall in the cray-fish *Cheraxalbidius.* In *P. indicus* Mohammed and Diwan (1994) have given a detailed description of male reproductive system, spermatogenesis and spermatophore formation through light microscopy.

At ultrastructural level there has been very little published work on the mechanism of spermatophore development in penaeid shrimps. Ro *et al.*,(1990) described the role of segments 1 and 2 in the spermatophore production of penaeid shrimp *P. setiferus.* Similarly the structure of Vas deferens, terminal ampoule and compound spermatophore, the process of spermatophore matrix deposition and the role of Vas deferens in spermatophore formation etc. are studied through electron-microscopy in two American white shrimp *P. setiferus* and *P. vannamei* by Chow *et al.,* (1991). The spermatophore not only functions as a vehicle for sperm transportation but also the acellular layers protect the sperm during their transfer, and storage by female. Therefore, it is very much essential to study their development and structure for the fruitful manipulations of their gametes during biotechnological applications for augmenting shrimp seed production.

3.8 BIOLOGICAL CHARACTERIZATION OF THE MALE REPRODUCTIVE SYSTEM

The male reproductive system of *P. monodon* is composed of both internal and external organs. The internal organs consist of paired testis and paired vas deferentia terminating in ampoules, containing spermatophores or sperm packets, and opens out through the gonopores, at the fifth waglkin leg. The external organs are the petasma, forms by the modified endopods of the 1st pleopod and a pair of appendix masculine, the modified endopods of the 2nd pleopod. The six lobed translucent testes are located in the cardiac region. These testicular lobes are connected to each other at their inner ends and lead to the vas deferentia (Fig. 27).

Vas deferens or testicular duct is the collecting tubule of the fully matured spermatozoa from the testis to the exterior, sperm-packets passes through gonopores. The paired Vas differentia of *P. monodon* originates from the posterior margin of the main axis of the testicular lobes, and descends ventro-laterally to terminate in the gonopore at the coxopodite of the 5th pereiopod. Each Vas deferens consists of four portions, the short and narrow proximal vas deferens (PVD) located nearer to the testes, the thickened and long median vas deferens (MVD) located in the middle region which consists of a blind pouch at the junction of PVD and MVD, an ascending limb and a descending limb, the long narrow distal Vas deferens (DVD) located at the farther end of the testicular duct just prior to the terminal ampoule, and the greatly dilated muscular terminal *ampoule* (TA) which is embedded in the coxal muscles of 5th pereiopod. The fully developed sperm packets or spermatophores are found inside this muscular organ. It opens to the outside through the gonopores situated on the coxae of the 5th pereiopod.

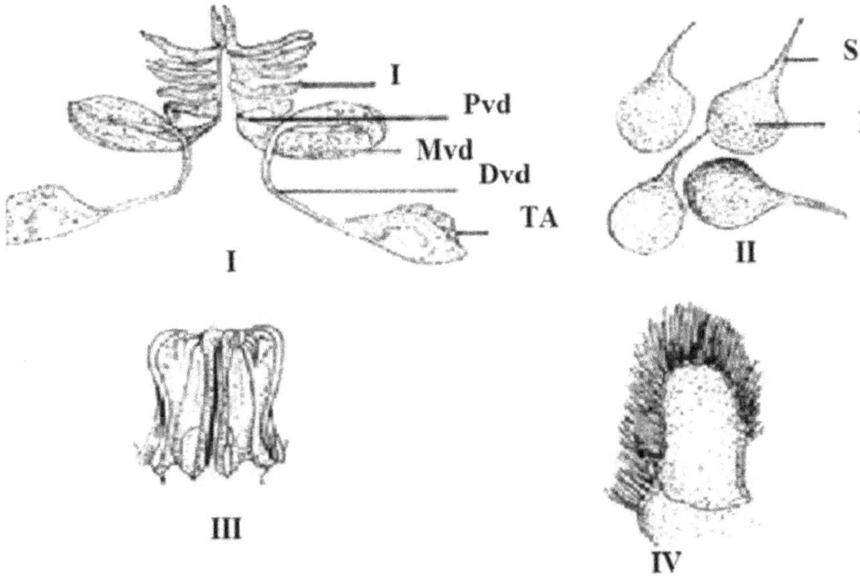

Fig. 27. Digramatic representation of the male reproductive system of *P.monodon.* I–Tests, II–Sperm, III–Petasma,IV–Appendix musculina B–body, l–lobules, S–spike, Pvd–Proximal Vas deferens, Mvd–Median vasdeferens, Dvd–Distal vasdeferens, TA–Terminal ampoule (**Source:** Diwan *et al.*, 2008)

3.9 DEVELOPMENT OF MALE REPRODUCTIVE ORGANS-MORPHOLOGICAL CHARACTERIZATION.

Based on the morphological as well as histological investigations, the maturity stages of male shrimp are classified into three stages, viz. immature, maturing and matured. In immature males, the testis is a thin translucent and extremely delicate organ in the cardiac region and the testicular. The extremely delicate Vas deferens appears as a thin translucent thread. Different parts are not distinguishable morphologically. The terminal ampoule appears as a thin delicate bag in immature males. The secondary sexual characters are not developed and the endopods of the 1st pleopods are too simple, elongated and flattened structures.

Even though the testis is transparent and delicate in maturing animals it is much more developed than the immature animals. The testicular lobes in maturing animals are somewhat larger and distinguishable, and the Vas deferens is much thicker and tubular in structure. The terminal ampoule appears as a small slightly thickened bag like structure, in the cavity at the base of the 5th pereiopods. The endopods of the 1st pleopods are modified and get partially united to form the petasma in matur in gmales. Appendix masculine appear as small buds at the base of the 2nd pleopods.

However, in fully matured males the entire primary as well as the secondary reproductive organs is well developed and fully functional. The fully developed testis now appear as opaque to creamin colour. The testicular lobes are thick short and stumpy in appearance and the well developed Vas deferens is easily distinguishable externally. The fully developed terminal ampoule appears thick, muscular and white due to the presence of spermatophores inside. The white colour at the coxae of the 5th pereiopods, due to the fully developed spermatophores present inside the TA, can be used as a character for identifying mature males. The endopods of the 1st pleopod become more elaborated and closely linked by a series of minute hook like structure in the mature males. Well-developed appendix masculinais also clearly visible.

3.10 DEVELOPMENT OF MALE REPRODUCTIVE ORGANS–HISTOLOGICAL CHARACTERIZATION

Testis

Histological studies of the testes at different maturity stages reveal that each testicular lobe is in the form of a moderately convoluted collecting tubule, held together by connective tissue, which are arrayed with multiple follicles (acini). These follicles occupy bulk of the testicular volume. Each follicle is found to be a collecting tubule lined by an epithelium.

The testicular follicles are empty in immature animals. The follicular wall is thicker and contains a small germinal area with non-differentiated cells at one side. Whereas, pronounced germinal area with differentiating primordial germ cells becomes prominent in maturing animals. Most of the follicular area is filled with germinal zone. In the testes of fully matured males, the testicular follicles are lined by germinal epithelium, however, the germinal zone is shifted to one of its side and it contains a few developing spermatogonial cells (Fig. 28). Fully developed spermatozoa are found in the central most region of each follicle.

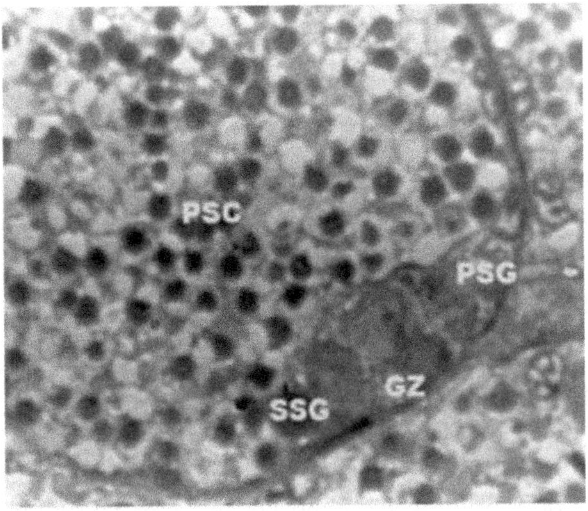

Fig. 28. Light micrograph of the maturing testes with primary spermatocytes (PSC), and shifted germinal zone with primary spermatogonia (PSG) and secondary spermatogonia (SSG) X 400 (**Source:** Diwan *et al.*, 2008)

Vas Deferens

Vas deferens of immature animals is a simple tube throughout its entire length whereas it is highly modified and specialized in fully matured animals. Different morphological divisions of the vas deferens are apparent in maturing as well as fully matured animals. Hence the corresponding stages in the maturing and matured animals are considered for the identification of different parts. The proximal portion of the testicular duct is identified as the PVD, middle part as the MVD and the distal part as the DVD respectively.

Proximal Vas Deferens (PVD)

The proximal vas deferens is a short slender tube leaving the testes and it is comprised of outer connective tissue sheath and an inner circular muscle layer. However, the duct of fully matured animals shows a highly modified internal anatomy. PVD is the site of completion of spermatogenesis, as well as the beginning of encapsulation of the sperm into spermatophores and the production of seminal fluid. PVD in general appears as a short tube in immature as well as matured animals. Cross-sections reveal that in immature animals, it is composed of an outer connective tissue layer and an inner layer of circular muscle fibers. In maturing animals besides outer connective tissue layer and inner circular muscle layer, an innermost layer of epithelial cells is also visible. These epithelial cells lining the inner lumen of the PVD secrete an amorphous matrix. Sperm cells are not present in the PVD from immature or maturing animals. The inner lining of low columnar epithelial cells is fully developed in the lumen of the PVD of fully matured animals. Even though the tube is empty most of the times, aggregations of sperms bathed in an amorphous matrix secreted by these epithelial cells are encountered occasionally.

Median Vas deferens (MVD)

The proximal vas deferens dilates posterior to form the median vas deferens (MVD). MVD is composed of two layers viz. the outer connective tissue layer and the inner lumen of MVD is found empty in immature animals. Whereas, the MVD of maturing animals is more developed and contains different morphologically distinct areas viz. blind pouch, ascending limb and descending limb. Blind pouch is the junction of PVD and the ascending limb of the MVD. A thin layer of columnar epithelial cells lines the inner lumen of the blind pouch in maturing animals. Even though sperm cells are absent, the lumen contains an amorphous material secreted by the epithelial cells lining the inner lumen. The blind pouch of MVD in fully matured animals exhibited structural features of PVD and MVD. Its lumen contains thickly packed sperm cells in a sticky amorphous material. The circular muscle layer of this blind pouch is thicker than that of PVD.

The blind pouch dilates into wider ascending limb of the MVD. In immature animals, MVD contains a single lumen and its ascending and descending limbs do not show any structural differences and its empty lumen does not possess an epithelial lining. However MVD of maturing animals shows that its lumen is partially compartmentalized and contains well developed columnar epithelial cells throughout its length. As it enters the ascending limb of the MVD, elongated as well as stunted typhlosoles are found inside its lumen, in the maturing animals. Typhlosoles are nothing but out growths of the epithelial cells into the lumen of the duct, in order to facilitate the increased secretory activity of these cells. The typhlosoles possess more elongated epithelial cells than that of the other area. These elongated typhlosoles make the partial division of the inner lumen and are composed of an inner layer of connective tissue lined by epithelial cells on its two sides. Sperm cells are not found in the ascending limb of the maturing animals and the lumen is found empty in most of the times. However, some acellular basophilic secretions are found occasionally in the lumen of this duct. The ascending limb continues as the descending limb, which also exhibit more or less a similar structure. Elongated typhlosoles are found in this descending limb of the MVD in maturing animals. Sperm cells are not present but the lumen contains an amorphous secretory material. The descending limb leads to the narrower and elongated distal Vas deferens. The distinction of MVD into blind pouch, ascending and descending limbs is seen clearly particularly in fully matured males. Unlike the previous cases, blind pouch of the matured animals contains a densely packed sperm mass in a sticky fibrillar-supporting matrix. Epithelial cells' lining this area probably releases the matrix, which give the sperm-supporting matrix in this region its cohesive property.

MVD has enlarged lumen, which is divided into two unequal independent ducts. The wider one is in continuation with the cavity of the blind pouch, and the sperm mass passes forward through this duct. It is the 'spermatophoric duct' (Fig 27) which deposits the different spermatophoric layers, around the sperm mass during its forward movement towards the terminal ampoule. The other narrower duct is called 'wingduct' (Fig.36) due to its role in the wing formation.Walls of these ducts are highly secretory in nature. All along the two ducts, innumerable numbers of glands are present, and more over they are in a state of supreme activity. The wall of the spermatophoric duct is thinner; even though it is composed of the three layers viz. outer connective tissue layer, middle circular muscle layer and the innermost columnar epithelial lining. Highly enlarged and secretory typhlosoles are particularly distinct in the duct. Amorphous materials of different nature surround the sperm mass inside the lumen. The wingduct is composed of the same three layers. Highly active epithelial cells are apparent also in this small duct. The secretory epithelial cells lining this duct secrete the amorphous materials of different natures. Only a small typhlosole is present in the wingduct.

The intervening septum separating the spermatophoric and wingduct is made of a thin strand of connective tissue with glandular epithelial cells on either of its side (Fig. 36). Invading this strand, are few muscle fibers scattered here and there and of more importance, blood islands for conveying the metabolites of these highly active epithelial cells. Despite the differences in size or thickness, both the spermatophoric duct and the wingduct consist of the same layers as indicated earlier. The massive secretions, discharged by the epithelial cells, occupy the cavities of both the spermatophoric as well as the wing ducts. These materials appear to form discrete layers around the sperm mass. The sperm mass along with its discrete a cellular secretions move towards the descending limb of the MVD.

The descending limb shows that, the same three layers already identified in the ascending limb are present here also. Almost similar structural configurations are seen in the descending limb of the MVD. The division of the lumen into the spermatophoric and wing duct persists to some distance and then gradually opens at one side and finally it become a common lumen before reaching the DVD. Compactly arranged sperm mass in the descending limb is packed in pouches with different layers inside the spermatophoric layer. Elongated and branched typhlosoles with long and highly secretory cells are found inside its lumen. The entire lumen is found packed with amorphous secretions of the epithelial cells. Descending limb of the MVD gets constricted and becomes narrower and continues as distal Vas deferens.

Distal Vas deferens (DVD)

In immature animals, distal vas deferens is a slender tube with a diameter of 0.15 to 0.2 mm. Two layers constitute it viz. the outer connective tissue layer and the inner circular muscle layer of the preceding parts and the inner most epithelial lining is absent. However, in maturing animals the distal Vas deferens contain all the three layers as in the MVD of this stage. The inner most epithelial lining is fully developed and contains well-developed typhlosoles. The diameter of the lumen is comparatively lesser than that of the MVD. The diameter of DVD varies between 0.5-0.8 mm in maturing animals.

In matured animals the distal as deferens is an elongated slender tube ending in the terminal ampoule. Highly elongated and branched typhlosoles are present in the lumen. The epithelial cells in the typhlosoles and other areas are particularly active in the secretion of the amorphous materials. The sperm mass in the DVD is encompassed in a covering made of these amorphous materials. Amorphous materials secreted by the epithelial cells in the DVD deposits over the spermatophore. The distal Vas deferens ends in a pear shaped muscular organ viz. the terminal ampoule.

Terminal Ampoule

The terminal portion of the vas deferens is a muscular organ in. It is located within the musculature of the fifth pereiopodal coxopodite. In immature animals it is not well developed and it appears as a small-elongated bag like organ at the terminal end of the vas deferens. No secretory material is present in this swollen bag like structure in the immature animal. Terminal ampoule of the maturing animals does not exhibit any conspicuous developments except in size, which has increased considerably than that of the immature animals. Its musculature is well developed, with distinct inner longitudinal and outer circular muscle layers. Epithelial lining is not well developed in the inner lumen.

TA of the matured animals is well developed and highly enlarged and has well developed musculature, with a distinct inner longitudinal and outer circular layer (Fig. 38). Highly developed and branched typhlosoles are present in the inner lumen of the terminal ampoule. However, the epithelial cells lining the inner lumen and on the typhlosoles are found to below columnar and matrix is secreted by these epithelial cells. The epithelium is thrown into longitudinal folds throughout its length.

Mechanism of Spermatogenesis

Spermatogenesis or production of mature spermatozoa takes place inside the lumen of the testicular acini. Spermatogenesis in each testicular follicle is cyclic, characterized by a distinct set of cytological transitions. In cross-section of the testis a strand like germinal zone is apparent in each follicular wall at one of its sides (Fig. 29 & 30). Germinal zone contains primary gonial cells and nurse cells, and the process of spermatogenesis initiates with the development of primary gonial cells in the germinal zone.The follicular epithelium always contains a few numbers of developing spermatogonial cells. Spermatogenesis is progresses towards the central lumen of the acini and therefore subsequent developmental stages are found towards the center in a graded manner. At the onset of spermatogenesis, the primary spermatogonial cells derived from the germinal zone undergo rapid reduction division and forms two secondary spermatocytes, which again divide mitotically into two and thus result in the formation of four spermatids, from each of the primary spermatocyte. These spermatids without further division get modified into spermatozoa (Fig. 31). The secondary spermatocytes, spermatids and spermatozoa are present in the center of each acinus in a sequence from the primary spermatocyte, which in turn are again developed from spermatogonial cells in the germinal area.

Since spermatogenesis involves the progressive reduction of cytoplasm and condensation of chromatin to produce the spermatozoa, the spermatogonial cells are seen to be the largest of all the cell types, followed by spermatocytes, which in turn are larger than spermatids and spermatozoa. The spermatogonial cells possess a large round vesicular nucleuswith diffused chromatin materials (Fig. 30 & 31). The nuclear diameter of these cells varies between 6-9μ. Small-elongated nurse cells are found in close association with the primary gonial cells. The nurse cells are seen to have prominent nuclei. By virtue of their close association with the spermatogonial cells, it is assumed that they have nutritive and supportiverole.

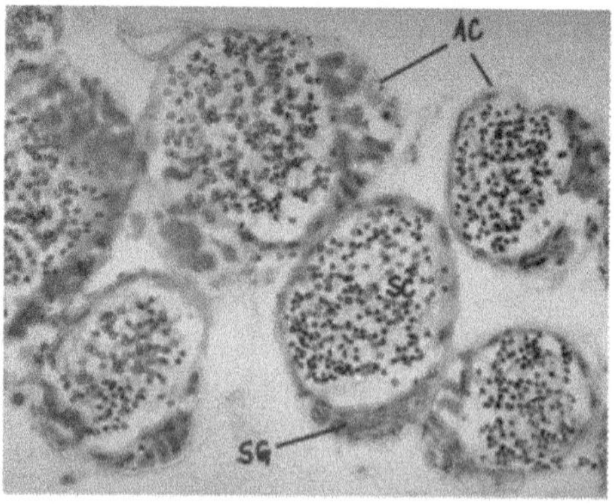

Fig. 29. Cross section of the testicular acini (AC) showing the strandlike germinal zone with spermatogonia (SG) and spermatocytes (SC) in the lumen of *F. indicus*. X 100. (**Source:** Mohamed and Diwan, 1994).

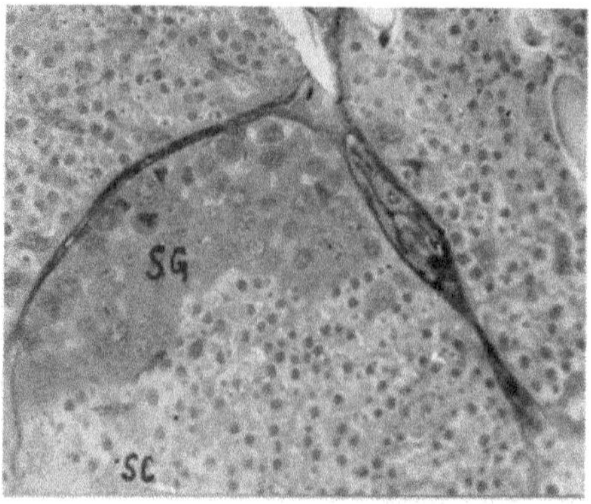

Fig. 30. Semithin section of the acini showing spermatogonia (SG), in the germinal zone and spermatocytes (SC) in the lumen X 100. (**Source:** Mohamed and Diwan, 1994).

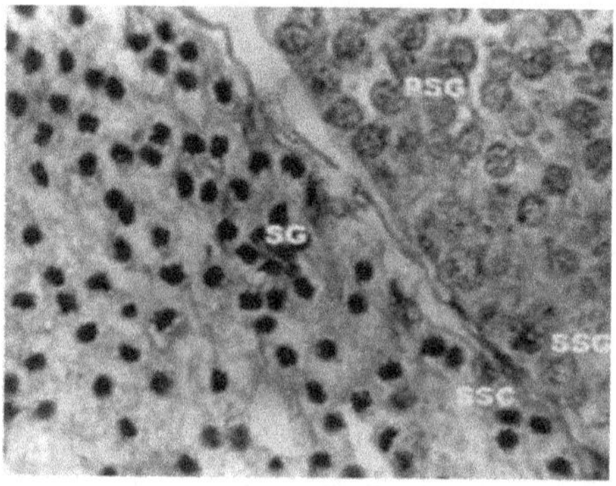

Fig. 31. Light micrographs of the mature testes with dividing (second meiotic division) secondary spermatocytes and few spermatids (ST) X 400 (PSG –Primary spermatogonium, SG – secondary spermatogonium, SSG – Secondary spermatocyte, ST-spermatids)
(**Source:** Diwan *et al.*, 2008)

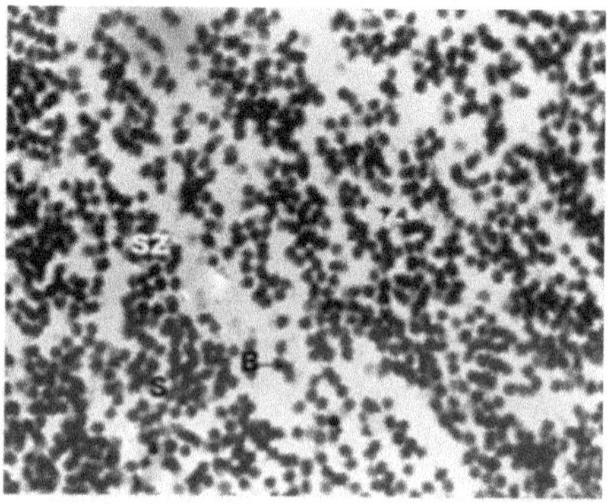

Fig. 32. Light micrograph of fully developed spermatozoa (SZ) in the proximal vas deferens with anterior spike (s) and round to oval body (B) x400
(**Source:** Diwan *et al.*, 2008)

Cell boundaries of spermatocytes, spermatids and spermatozoa are distinct. The primary spermatocytes have a cell diameter of 6-8μ and a nuclear diameter of 4-6μ. The nuclei of primary spermatocytes are dense in appearance, with heterochromatin material spread over homogeneously throughout. The secondary spermatocytes are smaller in size than the primary spermatocytes. The cytoplasm in them appears thin, but cell boundaries are clearly visible. These spermatocytes are usually the dividing cells in the testis, and diakinetic stages characteristic of dividing cells are frequently noticed among them. The spermatids are round structures with a very dense cluster of chromatin in their nucleus and are much smaller than the spermatocytes due to the reduction division and condensation of the chromatin materials. The diameter of the spermatids varies between 5-7 μ and its nuclei between 4-5 μ.

The spermatids undergo further structural modifications and later get differentiated into spermatozoa without any more divisions. Fully matured spermatozoa are found to be located at the central part of each follicle and appear to be almost spherical in outline, with condensed chromatin materials (Fig. 32). A short spike is apparent at the anterior region of the main body. The size of the body varies between 4-5μ. Ultrastructures of the testis indicate that the testicular wall is comprised of three layers, an outer layer of squamous epithelium, a middle layer of connective tissue and an inner layer of germinal epithelium. The spermatogonial cells are found to be roughly spherical to oval in outline. Small radial arms of the cytoplasm are apparent at the boundary of these gonial cells. The centrally located large nucleus occupies nearly one third of the cell volume. The vacuolar nucleus contains highly electron-dense condensed chromatin materials in their finely granular nucleoplasm. The nuclear boundary not clearly separates from the cytoplasmic materials. The highly electron-dense chromatin materials present at the outer boundary of the nucleus separated the nucleus from the cytoplasm. The cytoplasm of the spermatogonial cells contains numerous mitochondria and stacks of RER. Ribosomes are found in association with the RER as well as free in the cytoplasm. Large numbers of small-elongated non-germinative cells are apparent around each spermatogonial cells.

These non-germinative cells also possess cytoplasmic extensions, which extend out around and between the developing spermatogonial cells. The nucleus of these cells is not well developed, however some of these cells contains electron-dense chromatin materials. Nuclear membrane is not distinguishable and the cytoplasm is electron-denser than the adjoining germinal cells (Fig. 31).

Primary spermatocytes form the spermatogonial cells through mitotic division. The nucleus of such cells occupies almost 80% of the cell volume. The primary spermatocytes exhibit almost similar characteristics of late spermatogonial cells, except the fact that the nucleus is surprisingly larger in size and it contains highly electron-dense chromatin materials. Nucleolus is not apparent in the nucleus; however, patches of condensed chromatin materials distributed in the finely granular nucleoplasm become prominent in close association with the nuclear envelope. Nuclear pores interrupt the nuclear membrane of spermatocytes. The secondary spermatocytes form through the reduction division of the primary spermatocytes. During this phase, blebbing of the nuclear membrane become apparent in these dividing spermatocytes. Cytoplasm of the secondary spermatocytes contains sub-cellular organelles like RER, free and attached ribosomes, mitochondria, multi vesicular bodies and clear vesicles. RER is present in differentforms, like circular layers around the nucleus and near the plasma membrane of cells, segments of cisternae richly endowed with ribosomes containing little secretion products and swollen round to oval bodies with secretory products of medium electron density. Smooth ER is also seen in very less numbers. A few numbers of mitochondria are found in the outer cytoplasm, though their cristae are not clearly seen. Their outer profile varies from elongated to oval. A few clear vesicles have been observed in association with the RER. The early spermatids are found to originate from the secondary spermatocytes through simple mitotic division and get transferred into spermatozoa (Fig. 33). In the spermatids polarization of the cell contents get initiated. The nucleus migrates to one side of the cell while the cytoplasmic organelles move towards the other side. The shifted nucleus contains a prominent nucleolus. RER is abundant in circles around the nuclear membrane and also parallel to the plasma membrane of the cells. Vacuoles and granules filled vesicles become apparent throughout the cytoplasm.

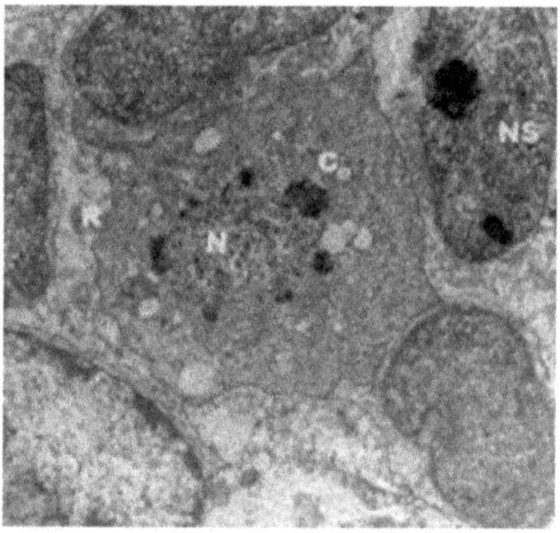

Fig. 33. Electron-micrograph of developing primary spermatogonia (PSG) with nursecells (Ns) around it (N-nucleus, C-chromatin materials, R-radialarms) X 4600
(**Source:** Diwan *et al.*, 2008)

Spermatids are found very much reduced in size due to the repeated divisions than the preceding stages of spermato genesis. The large elongated nuclei of the spermatocytes become half in size and roughly spherical in shape during the early stages of spermatid differentiation. The nuclear envelopes become increasingly convoluted and irregular (Fig. 34). The most striking characteristic noticed in the early spermatids, are the blebbing of the nuclear surface, which is more pronounced in this stage. As the blebbing progresses, small vesicles and dense granules come and get closely associated with these blebs. Clusters of small vesicles and dense granules get, shifted towards the nucleus and aligned themselves more or less perpendicularly with the nuclear surface and those which are closely associated with the nuclear blebs fused to the nuclear envelope. Generally outer membrane of the nucleus is involved in the blebbing, though both of them may also get involved occasionally.

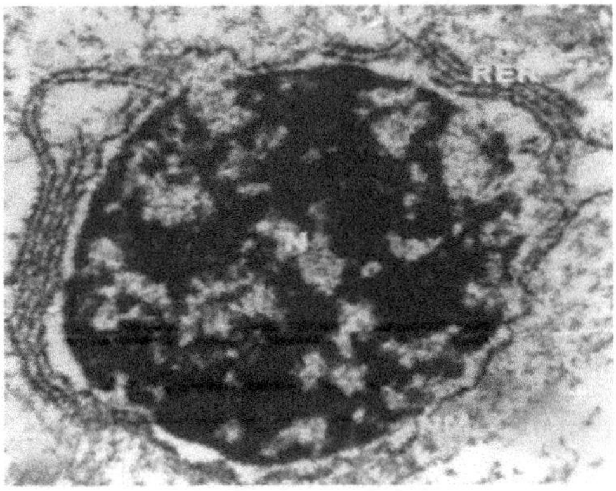

Fig. 34. Electron-micrograph of the mature spermatid showing the highly convoluted nuclear membrane (NM) and the circular layers of rough endoplasmic reticular elements (RER) around the nucleus (N)(Gr–Granules) X8600 (**Source:** Diwan *et al.*, 2008)

The cytoplasm of the early spermatids contains numerous small granules similar to those observed around the nucleus, freely and also on the outer surfaces of the granular endoplasmic reticulum (Fig. 34). In addition to this, it also contains numerous mitochondria, rough and smooth endoplasmic reticular elements, ribosomes and multi vesicular bodies. Ribosomes are present in association with the endoplasmic reticulum and in the cytoplasm. The mitochondria of these early spermatids are found to be varying somewhat in their outer profile. Transformation of the filamentous type of mitochondria to the oval spheroid is noticed in the spermatids, during the earlier stages. Vesicles similar to those seen in the secondary spermatocytes also occur in these spermatids. Some of these vesicles have smooth membranes, while some others are lined with ribosomes and a number of these vesicles contain faintly electron-opaque materials. It is also seen that these vesicles fused to form homogenous large vesicles surrounded by a single membrane in the sub-acrosomal region. Consequently, the acrosome also starts its development as the vesicular fusion takes place. It is seen that the acrosome has forms from the fused product of the small granule filled vesicles, which are found throughout the cytoplasm, especially around the nuclear membrane. Structures resembling lipid bodies sometimes appear between the acrosomal area and the nucleus of the spermatids. Spermatids during their intermediary stages of development, possess a comparatively electron dense round to oval nucleus. Concomitantly, a reduction in the total cell volume is occurs. During this stage, the outer nuclear membrane has an intended appearance, caused by the in foldings of the nuclear membrane. Concentric layers of highly active RER are found encompassing the nucleus (Fig. 34). Small mitochondria are apparent in the indentations of the nuclear membrane, as well as around these active RER. Areas of increased cytoplasmic activity are found between the cisternae of these RER. A few projections are noticed in the nuclear envelope and this out pocketing of the membrane collapses, leaving short flat sheets of closely apposed envelope membrane.

In the later stages of spermatid differentiation the active RER elements produced numerous electron-dense granular materials. Masses of the small vesicles and granules get excluded from the vicinity of the nucleus and are found packed near the cell periphery. Stacks of two to six annulated lamellae are also noticed in association with these vesicle- masses usually on the pole, away from the nucleus, near the clear area. These vacuoles and granules coalesce gradually and form the acrosomal vesicle. Condensation of the sub-acrosomal substance appears between the vesicle and the nucleus. The acrosomal vesicle gets filled with the condensed materials thereby forming the acrosomal complex. The vesicle appears cupshaped and the sub-acrosomal material underlined the acrosomal complex. The nucleus remains free of intrusion of vesicular and granular materials from the cytoplasm and is filled with clusters of fibrillar particles. During the final stages of spermatid differentiation the acrosomal vesicle along with the attached annular lamellae moved inwards and forms the acrosomal core, and it become a part of the acrosomal complex. In later spermatids, presumably by further invagination of the acrosomal vesicle along with the cell membrane of the spermatid and fusion with the small vesicles, formation of the limiting membrane get initiated.

During the transition of the late spermatids to the spermatozoa, the granular materials of the acrosomal vesicle completely set off from the rest of the cell and get incorporate into the acrosomal complex. The acrosomal granules get replaced by a denser imor cap, which lines the surface of the vesicle proximal to the nucleus. A dense material from the cytoplasm, accumulate at the distal part of the acrosomal vesicle and join with the acrosomal cap region, appearing as a highly condensed structure. At this stage the entire acrosomal complex protrude out from the spermatid and get differentiated into a highly condensed and integrated area at the anterior most regions of the spermatozoa. The formation of the anterior spike is also apparent in these developing spermatozoa. Small projections of the acrosomal membrane begin, to project across the periacrosomal space and become closely associated with plasma membrane. Extensions of the dense materials from the cap region get projected to some distance anteriorly in a slender and elongated form. This extension of the capregion develops into the anterior spike. The spike also contains the dense granules that accumulate from the cytoplasm of the spermatids, which forms the main component of the acrosomal vesicle. As maturation continues, an amorphous material originate from the cytoplasm at the peri acrosomal area, accumulate and filled the space between the acrosomal vesicle and plasma membranes, and along the sides of the acrosome to the middle of the spike. The acrosomal membrane that surrounds the spike at the proximal part of these developing spermatozoa does not appear to cover the outer edge, at the point where it approaches the plasma membrane. The acrosomal membrane, however, can be seen along the sides of this projection (Fig.35 and 36).

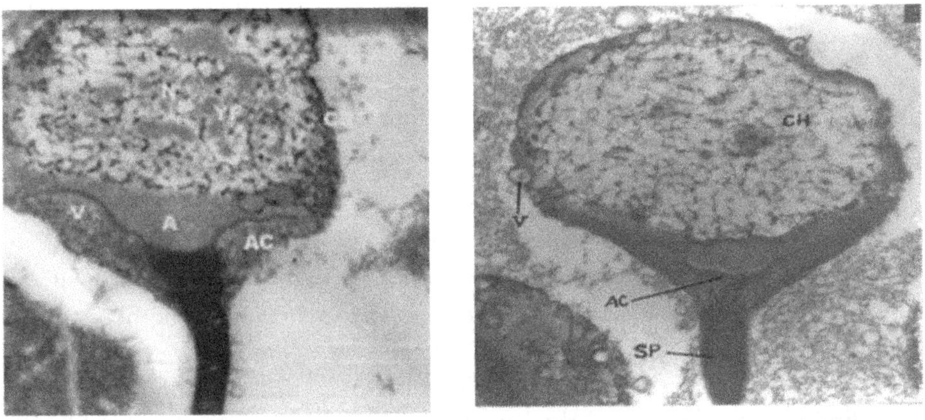

Fig. 35 and 36. Electron micrograph of the fully matured spermatozoa of *P. monodon* and *F. indicus* (A-acrosomal complex, AC-acrosomal, C-capregion,V -vesicle, YF-fibrilla rmaterials) X8,400 (**Source:** Diwan *et al.*, 2008).

The fully matured spermatozoa of penaeid shrimp is unistellate but non-motile. Ultrastructural observations show that the sperm is composed of a spherical main body, partially encompassed by a morphologically diverse cap region, containing acrosomal complex from which extended a single spike. The sperm's main body is found bean irregular sphere, one half of which encompassed by the cap region or the acrosomal cap and composed of structurally different components like the cytoplasmic band, small vesicles and the fibrillar mass. The cytoplasmic band, which contains anamorphous material, surrounds the posterior and lateral margins of the main body. A few small membrane bound vesicles are apparent around the periphery of this band. Loosely packed fibrillar materials constitute the majority of the cell body and it is composed of an electron loose matrix containing areticulum of fine putative electron-dense DNA fibrils. The fibrillar materials are not membrane bound at its outer region but the cytoplasmic band borders its lateralsides, and the acrosomal vesicle at its anterior side. The nuclear material is in direct contact with the cell membrane, as a discrete nuclear membrane is not present (Fig. 35).

The middle cap region is the most structurally complex region of the spermatozoa. It consists of distinct components viz. the central core of the acrosome, sub-acrosomal materials and pro-acrosomal granule. In cross-sections, the acrosomal cap region appears cup shaped with posterior surface partially encompassing the sperm nucleus. The components of the acrosomal complex, except the pro-acrosomal granule are radially symmetrical around the sperms central axis. The sub-acrosomal materials underline the acrosome and at the posterior end it separates the cap region from the nucleus. The fibrillar nuclear materials are get often inserted into the sub-acrosomal materials.The pro-acrosomal granule is present inside this sub-acrosomal material, usually in oneside, in close apposition to the plasma membrane. The thick saucers shaped acrosomal membrane separates the sub-acrosomal materials from the central core of the cap region. Extensions of this central core are found also towards the anterior spike region.

A common membrane is found covering the anterior regions of the sperm i.e. the acrosomal cap region and the spike. A group of somewhat thick and elongated filaments are found at the widest part of the dense acrosomal core between the two arms, which forms the cup. These elongated fine filaments are found extended anteriorly and constitutes the inner core. Structure of anterior spike is comparatively simple, and consists of two distinct structural components viz. a limiting membrane and the internal spike material. Fine elongated tubules extending from the central core of the acrosomal vesicle constitute majority of the spike materials.

3.11 MECHANISM OF SPERMATOPHORE FORMATION

Light microscopical as well as electron microscopical studies in the vas deferens of fully matured male *P. monodon* shows that, each part of this highly modified testicular duct has specific roles in the production of the highly evolved spermatophores. The formation of the spermatophore starts at the PVD and the process continues throughout the ductin a stepwise manner. Light microscopical observations reveals, the irregular distribution of multicellular glands in the inner lumen of the PVD are of highly secretory in nature. It is noticed that many of the secretory epithelial cells are packed with highly retractile granules, secreting copious materials spread over rather unevenly among the individually separated sperm cells and the same is found to convert the individual sperm cells into a coherent mass, inside the lumen of the PVD.

Ultrastructural studies of different parts of the Vas deferens show its significant role in the spermatophore production. It is found that PVD is the site of completion of spermatogenesis, as it receives the sperm cells, which are individually released from the testis. At the same time, it is the site for initiation of the spermatophore formation, as the acellular spermatophoric matrix is found secreted in this area by the glandular epithelial cells. Ultrastructural observations reveals that the low columnar epithelial cells are responsible for the secretion of the amorphous materials inside the duct and these cells are supported by a thick basal lamina of circular striated muscular fibers(Fig. 37). A holocrine mode of secretion is noted in these epithelial cells lining the proximal part of the Vas deferens. The columnar epithelial cells contain numerous mitochondria and cell organelles associated with protein synthesis and secretion. The basal region of the adjacent secretary epithelial cells is highly inter-digitated and contains numerous mitochondria.

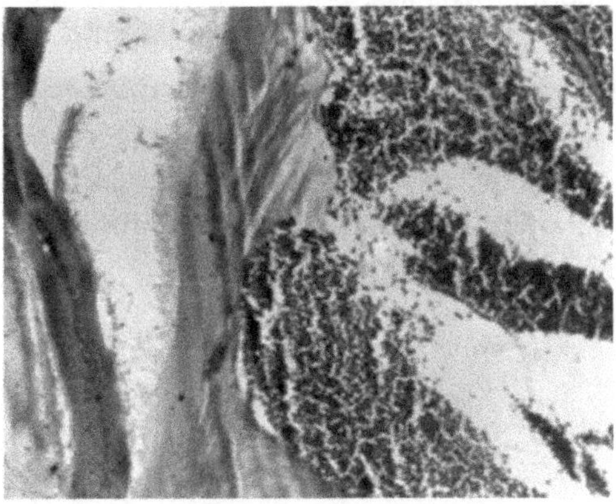

Fig. 37. Light micrograph of the blind pouch of *P Monodon* with sperm mass covered by the primary spermatophoric layer (S-sperm mass, PSL–primary spermatophoric layer) X100 (**Source:** Diwan *et al.*, 2008)

The nucleus is comparatively large and possesses rich supplies of euchromatin materials. Numerous large vesicles are found between the nucleus and basal plasma membrane and these vesicles contain numerous lamellar zones and low-density materials. RER occupies a large fraction of the cell volume. Ribosomes and poly ribosomes are observed free in the cytoplasm. Numerous small Golgi bodies are distributed throughout the cell, but most abundant in the perinuclear zone. Many small vesicles appear to originate from the Golgi bodies. A relatively thick layer of connective tissue covers the muscular layer.

In concomitance with the release of the sperm cells from the testis, the epithelial cells described above, secrete a flocculent moderately electron-dense matrix, which forms the sperm-supporting matrix in the PVD. The sperm-supporting matrix is comprised of low electron density components and linear strands of dense fibers, which together get converted with the individual sperm cells into a mass. As the sperm passes forward, another electron- dense material is also found to be secreted from the epithelial cells of the PVD, which forms the components of the basal layer or primary spermatophoric layer around the sperm mass and the supporting matrix. The proximal Vas deferens joins the MVD through blind pouch, which is a storage space for the sperm mass during its forward movement. Three small buds of typhlosoles found inside the blind pouch region of MVD exhibits active secretory activity. The sperm mass in the blind pouch is densely packed and enlarged than the same in the PVD. Irregularly distributed free sperm cells are not seenin the blind pouch region. The fibrillar nature of the sperm matrix is found increased tremendously during storage of sperm mass in the blind pouch for a short period. Later, these filaments often join together end to end and form an almost continuous but complicated net work, in which the spermatozoa are, bound closely forming a well oriented and thickly packed sperm mass. Electron-dense vesicles produced within the epithelial cells are found to release in to the lumen and distributed in abundance within the network of the sperm-supporting matrix. The spermcord during its transit through this blind pouch becomes highly convoluted and oriented into deep pouches of the spermatophores. Secretions of the primary spermatophoric layer are found to be continues in this area (Fig. 38).

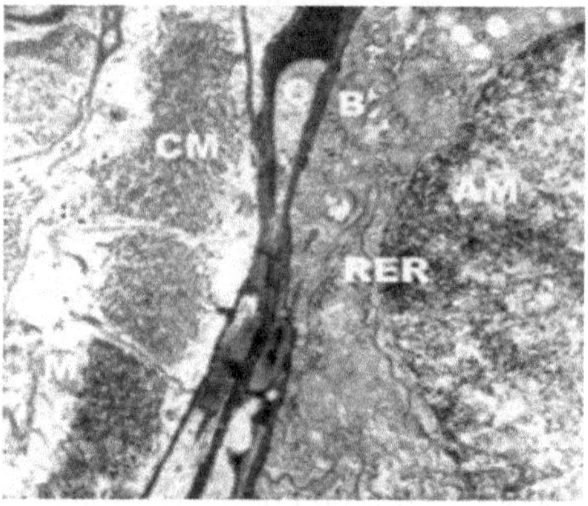

Fig. 38. Electron-micrograph of the basal lamina of the epithelial cells in the PVD of *P Monodon* showing the holocrine mode of secretion.(A–amorphous materials which are ready to separate, B-basal lamina, C-connective tissue, CM – Circular muscle layer, M-mitochondria, RER–rough endoplasmic reticulum) X15,500 (**Source:** Diwan *et al.*, 2008)

The primary spermatophoric layer appears heterogeneous and the region closest to the sperm matrix consists of a moderately electron dense fibrillar band, and outside this layer a much more electron dense zone is apparent in the distal end of the PVD. The sperm mass supported by the acellular matrix together with the primary spermatophoric layer forms a sperm cord, which initially fill the entire lumen before its transfer to the median Vas deferens.

The blind pouch contained two completely separated lumen are found separated from each other by a double layer of epithelial cells. Massive secretions discharged by these epithelial cells occupy the cavities of both the ducts. The wider spermduct contains densely packed and well-oriented sperm cord covered by the primary spermatophoric layer. The nature of the secretary material varies as the spermatophore passes forward through the ascending limb. The glandular epithelial cells in the ascending limb at its middle portion show an extraordinary secretory activity. And owing to the marked increase in the number of the secretory cells, they are projected into the lumen as branched typhlosoles. Copious out flow of the secretory material is apparent from the epithelial cells of the typhlosoles as well as the cells lining the inner lumen of this ascending limb. Here the glandular epithelial cells are packed with another type of granules, which are identical to the secretionin the lumen as small droplets here and there. These droplets fused to form large droplets and in certain areas the larger droplets are found diffused. These materials are found to constitute the secondary spermatophoric layer, which gets deposited over the primary spermatophoric layer.

At the same time, the wing duct also shows vigorous secretory activity. Like the spermatophoric duct, wingduct also contains a typhlosole with apparently active epithelial cells (Fig. 39). A frothy secretion is seen at the proximal most region of the wingduct. This secretion forms the third spermatophoric layer. It is found that a finely granular material is first secreted in wing duct, and later on, more and more materials of different characters are laid on these granular materials and finally get converted into a thick accumulation in the wing duct (Fig. 39). Meanwhile, the capacity of the wing duct increased at the expense of the delimiting septum towards the latter duct. Still it is found that the materials for the3rd layer continue to add up asit passed forward. Apart from the materials of the 3rd layer, another homogenous material constitutes the 4th layer of the spermatophore, which is secreted from the glandular cells of the wingduct at the father most end of the ascending limb. Immediately after this secretion, the materials are set around the free borders of the 3rd layer and appear as a sandwich. The 3rd layer sandwiched on either of its two flattened sides with the newly developed 4th layer. Secretion of the 4th layer is found to be continuing through the descending limb of the MVD to the terminal portions of the testicular duct.

The descending limb of the MVD shows separation of the lumen into spermatophoric duct and wing duct. Immediately after this area, it is observed that the septum intervening between the two duct, get detached at a point on the opposite end. This occurs when the newly formed material in the wing duct start flowing into the spermatophoric duct, in order to establish its connection on the spermatophore. The middle and terminal portions of the descending limb show that, the opening of the wing duct to the sperm duct become wider by the degeneration of the intervening septum. Epithelial cells in this region are found to be similar to that of the preceding region but the typhlosole is highly branched. The ascending limb at its farthest end and in the proximal parts of the descending limb shows the presence of a third layer in the enlarged wingduct. This material gets fixed around the 4th layer in the same manner as the 4th layer deposits around the 3rd layer. This newly formed layer constitutes the 5th and the outermost layer of the spermatophore. Epithelial cells in the sperm duct of the descending limb also secrete the materials responsible for this V layer. The fifth layer is thinner during its formation at the descending limb of the MVD and becomes thicker during its passage through the distal Vas deferens.

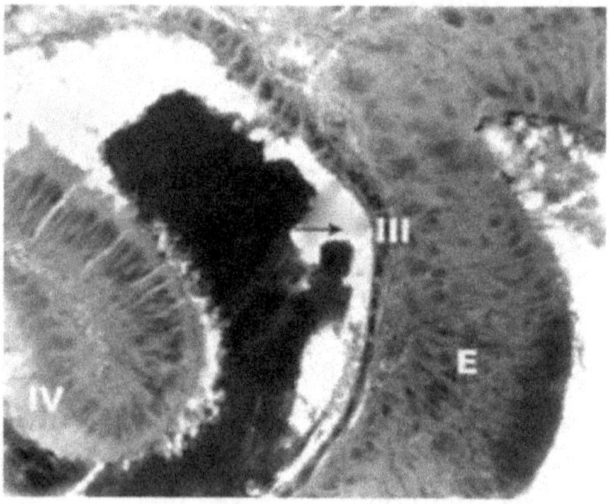

Fig. 39. Light micrograph of the wing duct of *P Monodon* in the ascending limb of MVD. (T-typhlosole, E-epithelialcells, A amorphous secretions of the IIIlayer) X100
(**Source:** Diwan *et al.*, 2008)

Electron microscopical studies reveal that the MVD differs from the PVD in many ways. At all columnar epithelial cell lines the inside of the MVD both inthespermatophore andwingduct. The surrounding muscle layer is thicker in this part than the PVD. The wider spermatophoric duct of MVD contains portions of the sperm cord and materials of the secondary spermatophoric layer. The secondary spermatophoric layer comprises of a matrix of low density, irregular zones of high electron-density and small and large electron dense granules. The basal plasma membrane of these epithelial cells is highly in folded and contains numerous mitochondria and small cisterns of RER, consistent with protein synthesis. At the apex of these cells, the plasma membrane is projected into blebs to the inner lumen (Fig. 40). These blebs of the epithelial cells contain materials, which appear similar to the secretion present inside the duct. It is noted that the microvilli extending off the basal portion of the bleb are smooth. Free ribosomes are noted throughout the cytoplasm of these cells. These free ribosomes intact presume to synthesize the secretory materials, remain in the cytoplasm and subsequently appear in the apical most region of the bleb.

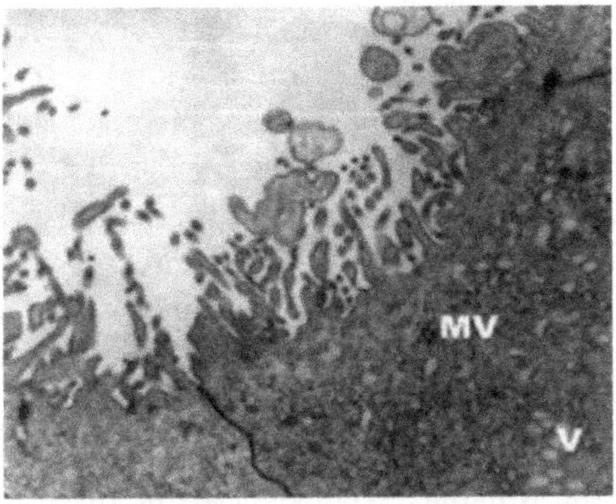

Fig. 40. Electron micrograph of the apical portion of the epithelial cells of the sperm duct showing the apocrine (large arrows) and exocytotic (small arrows) mode of secretion (MV–microvilli, V-vesicles) X8,400 (**Source:** Diwan *et al.*, 2008).

The portion of the bleb contains a distinct flocculent matrix, which is almost completely devoid of organelles (Fig. 40). The boundary between the organelle free and organelle-containing zone of the bleb is often surrounded by an electron dense material, which make the plasma membrane difficult to trace. However, at high magnifications it is clear that the apical most region of the bleb is devoid of the plasma membrane covering. The contents of the bleb and secondary spermatophoric layer appear to be continuous in these regions. Furthermore, it appears that the highly electron-dense contents of the bleb give rise by way of apocrine secretion to the low-density flocculent material of the secondary spermatophore layer.

The remainder of the secondary spermatophoric layer appears to be derived from the distal end of the ascending limb and descending limb of the MVD. The blebs of the epithelial cells contain electron-dense materials, which appears similar to the secretions present inside the duct and these cells release their contents from the blebs to the inner lumen by exocytosis. In addition to this, secretory materials processed through the Golgi apparatus and released from the microvillar portions of the epithelial cells are also added to this secondary spermatophore layer. In these region two types secretory vesicles are present near the Golgi bodies. Small vesicles with homogenous material of moderate electron-density form the first type and the second type includes few small electron dense granules. Both types of vesicles are frequently observed just beneath the plasma membrane, although they are not necessarily attached to each other. Contents of both of these granules are released to the space between microvilli by exocytosis. At the tips of the microvilli, the dense material appears to breakup and form the secondary spermatophore layer.

Relatively smaller lumen of the wing duct in the ascending limb of the MVD contains a thick core of moderately electron dense material, secreted by the epithelial cells lining the innermost area. This material appears as extremely granular and moderately electron- dense during the time of its formation. Later it become electron-dense and homogenous at the terminal portions of ascending limb after some structural modifications. Blebs are apparent at the apical most regions of the epithelial cells lining the inner lumen of this wingduct. The same apocrine mode of secretion is observed in these epithelial cells. But in certain cases, the blebs are found pinched off from the apical portion and join the secretory materials present in the lumen. Secretion of the 3rd layer is found in the entire lumen of the wingduct. Wingduct at the distal part of the ascending limb and the proximal part of the descending limb secretes and electron-dense and homogenous matrix, which is found deposited over the 4th layer, forming the 5th layer (Fig. 41). Formation of this layer leads to the detachment of the septum interveining the two ducts at one point. The materials from the wing start its displacement to the larger spermatophoric duct immediately after the detachment of this intervening septum. Excess materials of the 5th layer make its entry first in the sperm duct and get deposited over the spermatophore one it her of its sides to some extent. Secretion of the 5th layer is found till the terminal end of the descending limb.

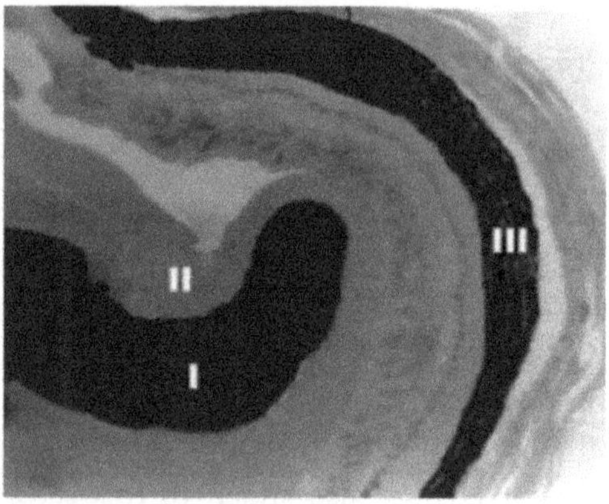

Fig. 41. Light micrograph of the terminal ampoule of *P Monodon* with spermatophore showing different spermatophoric layers (I-primary spermatophoric layer, II-secondary spermatophoric layer, III-third spermatophoric layer) X100 (**Source:** Diwan *et al.*, 2008).

Cross-sections of the long distal Vas deferens (DVD) show that, it had a thick layer of muscle and highly infolded epithelium, which create compartments, which convey the spermatophores to the terminal ampoule. These folds do not form complete partitions between compartments. Even though the typhlosoles are found lined with large accumulations of secretory epithelial cells, they do not contain secretory materials as in the case of the preceding parts (MVD) of the Vas deferens. The comparatively thick connective tissue layer found in the center of the wall of this duct indicated its role of conveyance of the spermatophores. The size of the spermatophore decreases as it passes through the slender DVD. The different layers get attached firmly against the sperm mass and distinctions of different layers are well identifiable at this stage. Ultrastructural studies reveals that the epithelial cells lining the inner lumen of the DVD become low columnar and the cell organelles like RER and Golgi bodies are less abundant, than the preceding parts of the vas deferens. Large moderately dense fibers of circular muscles are apparent in this part of the Vas deferens. Acellular materials morphologically similar to that of the 5th layer are observed in the lumen occasionally. The basal region of these epithelial cells is found highly in folded and numerous mitochondria are found among these folds. As the distal Vas deferens approaches the TA are duction in epithelial area is noticed and the cells do not contain any apical blebs.

The terminal ampoule reveals the presence of single lumen, which is found, divided into four to five interconnecting chambers or lumina. These interconnecting chambers are form by the branched typhlosoles. The lumen of the highly muscular organ contains a basophilic epithelium. This epithelium is thrown into longitudinal folds along with its inner lumina. These cells secrete some acellular eosinophilic matrix into the lumen of TA. No spermatophoric layers are found secreted by the epithelial cells. However, the acellular matrix and the longitudinal folds of the basal lamina along with its well-developed musculature are meant for the final molding of the spermatophore. Microscopical observations of the TA in different intervals after the manual extrusion of the spermatophore show that hardening of the spermatophoric materials takes place inside the terminal ampoule.

Ultrastructural observations reveal that the epithelial cells of terminal ampouleare not involved in the secretion of any spermatophoric layers. However, some electron- lucent materials secreted by these cells are apparent inside the lumen. This material is not found incorporated into any of the spermatophoric layers, instead is found inside the lumen as such and functions as a lubricant.

A perusal of the available literature on the reproductive biology of penaeid shrimps reveals that the process of gametogenesis has not been studied in male as thoroughly as that of the females. However, the morphology of the male reproductive system has been described in various penaeid shrimps like *P. Setiferus* (King, 1948); *P. stylifera* (Sheikhmahmud and Tambe, 1958; Rao, 1969); *P. indicus* (Subramoniam, 1965; Mohammed and Diwan, 1994); *P. merguiensis* (Tuma, 1967); *Metapenaeus stebbinal* (Tirmizi and Javed, 1976) *P. monodon*

(Motoh, 1978 & 1981 and Solis, 1988) and *M. dobsoni* (Vasudevappa, 1992). Apart from these Gomes and Primavera (1994) in *P. monodon*; Cavalli, *et al.*, (1997) in *Penaeus paulensis*; Crocos and Coman, (1997) in *Penaeus semisulcatus*; Crocos, *et al.*,(1999) in *P. monodon*; Peixoto *et al.*, (2002) in *Farfantepaulensis*; Peixoto *et al.*, (2003) in *Farfantepenaeus paulensis*; Coman, *et al.*, (2003) in *P. monodon*; Peixoto, *et al.*, (2004) in *Farfantapenaeus paulensis*; Coman, *et al.*, (2004) in *P japonicus*; Cuzon, *et al.*,(2004) in *Litopenaeus vannamei*; Coman, *et al.*, (2005) in *P. monodon*; Coman *et al.*,(2006) in *P. monodon*; Coman *et al.*, (2007) in *P. monodon*; Coman, *et al.*,(2007) in *P. monodon*; has also described the reproduction in penaeids.

General morphology of the male reproductive system of *P. monodon* is also more or less similar to that of the other penaeids described by these above mentioned workers. There are some variations in the number of testis lobes and the nature of fusion of the two halves of the testis. For an instance in *P. indicus* it is reported that only four lobes on either side of the testis are found (Subramoniam, 1965; Mohammed and Diwan, 1994). In *P. stylifera* testis has only three lobes on both the sides (Sheikhmahmud and Tambe, 1958). However, (King, 1948) in *P. setiferus* and in *M. dobsoni* (Vasudevappa, 1992) it has been reported that the testis of these shrimps possess an anterior lobe, six lateral lobes and one posterior lobe. Similarly an accessory gland is noticed in *P. stylifera* (Sheikhmahmud and Tambe, 1958 and Rao, 1969). However, no such gland is recognized in *P. monodon.*

The Vas deferens of penaeids is commonly distinguished into four regions viz. Proximal vas deferens, Median vas deferens, Distal Vas deferens and the Terminal ampoule by various investigators like King, (1948); Subramoniam, (1965); Malek and Bawab, (1974 a, b); Champion, (1987); Vasudevappa, (1992) and Mohammed and Diwan,(1994). Vas deferens of *P. monodon* also contains similar morphological regions with characteristic size and shape. However, some workers in certain species of the same genera have reported slight variations of this general pattern.

According to Subramoniam, (1965) the median vas deferens is the tubular portion of the testis in *P. indicus.* Similarly Tirmizi and Khan (1970) described a membranous bag filled with spermatic cones attached to the middle portion of the Vas deferens. These types of modifications are not seen in the Vas deferens of *P. monodon.*

Unlike female *P. monodon* males do not possess any clearly visible characteristics for the visual assessment of gonadal maturity. Therefore, less emphasis has been placed so far for the assessment of gonadal development in male penaeid shrimps. Moreover in terms of the productions of spermatozoa, males mature at a smaller size and earlier than females and information regarding the maturity stages in male shrimps is scanty. Based on the smaller changes in opacity and size of the testes in relation to the size of the animal, Subramoniam, (1965) had described five maturity stages in *P. indicus*. Whereas Castille and Lawrence, (1991) divided the maturity stages of the male penaeid shrimps *P. aztecus* and *P. setiferus* into three viz. immature, developing and mature, based on the size of the terminal ampoule. In case of *P. monodon* depending upon the entire morphological characters of the testis as well as the testicular duct or Vas deferens, the gonadal maturation, is classified into three stages viz. immature, maturing and mature. It is found that morphological changes are more pronounced in the vas deferens or testicular duct, rather than the testes during these transitional stages. Based on the morphological differences, histological characters of different regions of the reproductive system also differ during these maturity stages. Though structural variations are present in both the testes as well as testicular duct it is more pronounced in the highly modified vasdeference than the testis. In contrast to female reproductive system, there is a paucity of information in male reproductive system about the structural modifications that occur in different regions during its development. However, many have studied the complete structure of the fully matured vas deferens as well as its role during the process of spermatophore formation in the past two decades.

Four different regions proximal Vas deferens, median Vas deferens, distal Vas deferens and terminal ampoule are recognized in the Vas deferens of *P. monodon* as in other penaeid shrimps like *Penaeus kerathurus* (Malek & Bawab, 1974 a, b); *P. setiferus* (Chow *et al.*,1991a); *P. vannamei* (Chow *et al.*,1991 a) *M. dobsoni* (Vasudevappa, 1992); *P. indicus* (Mohammed and Diwan, 1994). Among the four regions, the median Vas deferens in *P. monodon* is again divided into three functional sub-regions as blind, pouch, ascending limb and descending limb, as in many of the above mentioned shrimps. These regions are distinct in maturing and fully matured males. Histologically, the structural modifications of the Vas deferens has been studied in other penaeids by different investigators in connection with the spermatophore formation in *P. kerathurus* (Malek and Bawab, 1974 a, b); *P. setiferus* (Ro *et al.*,1990); *P. setiferus* and *P. vannamei* (Chow *et al.*,1991 a); *M. dobsoni* (Vasudevappa, 1992); *P. indicus* (Mohammed and Diwan, 1994). However, none of them studied the structural variations in the duct during maturation.

Histologically the proximal vas deferens of *P. monodon* is a short unmodified tube composed of an outer connective tissue layer and inner circular muscle layer. Low columnar epithelial cells lining the inner lumen of PVD as in other penaeids like *P. setiferus* (King, 1948; Chow *et al.*,1991b; Ro *et al.*,1990); *P. vannamei* (Chow *et al.*,1991b) and *M. dobsoni* (Vasudevappa, 1992) are apparent in the maturing and mature animals but not in the immature animals. In the above-mentioned shrimps it is further reported that the columnar cells are responsible for the secretion of sperm supporting matrix present in the PVD. However, some investigators like Mohammed and Diwan, (1994) in *P. indicus* has not observed any glandular epithelial cells as well as sperm- supporting matrix in the PVD.

The MVD, the highly complicated region has been identified into three sub-regions in *P. monodon* based on their internal anatomical modifications, as in other penaeids. Histological investigations of the MVD in different maturity stages indicate that these structural modifications are apparent only in the mature animals and in immature animals the MVD is a simple tube. The blind pouch of the MVD, in the fully matured *P. monodon* encloses a single cavity lined by low columnar epithelial cells and always contains thickly packed sperm cells indicating its storage function. This blind pouch dilates in to the ascending limb of the MVD, the inner lumen of which divide into two unequal compartments, the wing duct and the sperm duct as in other penaeids. The lumen of the third portion of the MVD, the descending limb is also compartmentalized for a short distance and the separation is found gradually disappearing in the distal end. Similar characters are described for the MVD of penaeids like, *P. setiferus* (King, 1948; Ro *et al.*,1990, Chow *et al.*,1991); *P. kerathurus* (Malek and Bawab, 1974 a, b); *P. vannamei* (Chow *et al.*,1991 b); *M. dobsoni* (Vasudevappa, 1992) and *P . indicus* (Mohammed and Diwan, 1994).

The blind pouch of *P. monodon* contains a branched typhlosole in contrast to the previous reports. Presence of two to three typhlosoles in the sperm duct and one typhlosole in the wing duct are found in the MVD of *P. monodon* as in *M. dobsoni*. In all other penaeids studied, usually one typhlosole in each of the sperm duct and wing duct is reported (Malek and Bawab, 1974 a, b; Mohammed and Diwan, 1994).The three typhlosoles of the sperm duct developed in the ascending limb of the MVD disappear in the distal part of the descending limb of the MVD. The presence of rich blood supply totyphlosoles indicate high metabolic rate of the cells.

The distal vas deferens of fully matured male *P.monodon* is a highly complicated tube to which the MVD opens. It possesses only a single lumen throughout its length with the highly developed typhlosoles, which partially divide the lumen into different compartments. Almost similar structure has been reported in other penaeids (King, 1948; Malek and Bawb, 1974 a, b; Chow *et al.*,1991 a; Vasudevappa, 1992; Mohammed and Diwan, 1994). The musculature of the terminal ampoule is fully developed in the matured *P. monodon*. In immature shrimps it is a small sac like structure without musculature or secretory cells. Outgrowths of columnar epithelial cells as typhlosoles as described in many penaeids are found in *P. monodon.*

The process of spermatogenesis in *P. monodon* is almost similar to that reported in other decapods crustaceans (Pochon-Masson, 1983). There is a volume of information on spermatogenesis of various decapods crustaceans (King, 1948; Koebler, 1979; Hinch, 1980; Jesperson, 1983; Sagi *et al.*,1988; Vasudevappa, 1992 and Mohammed and Diwan, 1994). However, studies on this aspect in penaeids are less compared to other decapods.

Spermatogenesis begins in the peripheral germinative layer of the testicular tubules, when spermatogonia enter into the prophase of meiosis (King, 1948). A general feature of the process is that, these spermatogonial cells originate from the germinal area after passing through a period of quick growth will transform into primary spermatocytes and then undergo reduction division to become the secondary spermatocytes. These secondary spermatocytes divide mitotically and form the spermatid, which without further division will transform into the spermatozoa. This pattern of spermatogenesis has described in *P. setiferus* (King, 1948) through light microscopy and by Lu *et al.*, (1973) through electron microscopy. In *P. indicus* the whole process of spermatid differentiation has studied by Rao (1968) and Mohammed and Diwan (1994). Joshi *et al.*, (1982) and Vasudevappa, (1992) gametogenesis of *Parapenaeopsis stylifera* studied the spermatogenesis in *M. dobsoni* respectively through light microscopy. There is no much difference in the spermatogenesis in these penaeids according to them.

Apart from these, the process of spermatogenesis had been investigated through electron microscopy in many other decapod crustaceans like crabs *Eriocheir japonicus* (Yasuzumi, 1960); cray fishes, *P. clarkia* (Moses, 1961a & 1961b); and in penaeids, *S. injentis* (Kleve *et al.*,1980 and Shigikawa and Clark, 1986). However information on the ultrastructural details of the spermatogenesis in natantian shrimps is restricted only to *S. injentis* (Kleve *et al.*, 1980 and Shigikawa and Clark, 1986) and *P. setiferus* (Lu *et al.*,1973). However, these investigations reveals that during spermatogenesis the cytoplasm of the developing spermatogonial cells follows basically the same pattern of changes as seen in other crustaceans i.e. general reduction in organellar content. Early spermatogonial cells are located in the germinal zone and in *P. monodon* it is found towards the periphery of the follicular tubule. Similarly spermatogonia are grouped along the periphery of the tubules in *P. setiferus* (Lu *et al.*,1973). They are large cells with a centrally located spherical nucleus and during the early stages; the nuclei are not easily distinguishable. Its cytoplasm contains numerous cell organelles like aggregates of ribosome and lamellae of endoplasmic reticulum. Another interesting feature noticed during the early stages of spermatogenesis is the presence of large numbers of non-germinative cells among the germinal components. Even though Lu *et al.,*. (1973) did not mention about these types of cells, there are other reports of crustacean testis, which mention the presence of "nurse" or nutritive cells among the germinal components (Pillai, 1960, Pochon-Masson, 1968 a, b and Hinch, 1980). These cells function in providing nourishment, support and possibly hormones during Spermeogenesis (Hinch, 1980).

From earlier reports it is found that, during the ultrastructural investigations on spermatogenesis in decapod crustaceans, investigators have given more emphasize on the later stages of spermatogenesis i.e. the transition of spermatids to spermatozoa, which is more complicated and dramatically different than earlier stages. Therefore the information regarding the ultrastructural changes in the early stages of spermatogenesis is very limited in decapod crustaceans and it is lacking in penaeids.

Secondary spermatogonia transform into spermatocytes are distinguishable by certain characters. It is observed that the nuclear wall of the spermatocytes become more prominent and is easily demarcated from the surrounding cytoplasm. Highly electron dense chromatin materials are also apparent in the nucleoplasm and the cytoplasm contains organized ergastoplasm. Similar types of characteristics have reported in other crustaceans by Meusy (1972) and Lu *et al.,*(1973).

The cytoplasm of secondary spermatocytes is characterized by the abundant supply of membrane system around the nuclear wall. Lu *et al.,*. (1973) described similar type of secondary spermatocytes in *P. setiferus*. The cell organelles noticed in the cytoplasm are the abundant randomly distributed cisternae, rough endoplasmic reticulum (RER) and the granular bodies originating from the cisternae. Three types of membrane complex are described in the early spermatids of crab *Cancer sp* by Langreth, (1969) as well as in the cray fish *P. clarkia* by Moses (1958). Apart from the membrane systems oval to elongated mitochondria with feeble cristae are apparent in the secondary spermatocytes of *P. monodon*. Other prominent cell organelles found in *P. monodon* are, numerous ribosomes and vesicular systems. Transformation from the common filamentous type of mitochondria to be oblated spheroid is noticed as the meiosis progresses in the spermatocytes of the cray fish *P. clarkia* (Moses 1961b). Similarly, the role of mitochondria in the membrane complex of dividing spermatids has been reported in *Cancer* crabs by Langrenth (1969), in *A. tasmaniae* by Jesperson (1983) and in *S. chacei* by McKnight and Hinch (1986).

In *P. monodon*, another factor noticed in the secondary spermatocytes is, the out pocketing of the nuclear membrane as vesicles of varying size and shape. In cray fish *P. clarkii*, Moses, (1961 b) has described similar pattern of changes in nucleus as well as nuclear membrane during the transition of the spermatocytes to the spermatids. As the secondary spermatocytes transform into the spermatids by the 2nd division of the meiosis the size of the cells become considerably less. Drastic ultra structural changes are noticeable in the developing spermatids. The cell organelles become highly active and secretory. The most striking feature noticed in the early spermatids of *P. monodon* is the polarization of the cellular materials and the nucleus. Displacement of the ER elements to one side of the cells and the nucleus to the other pole has described in *P. Setiferus* by Lu *et al.*,. (1973). In *Cancer* crabs Langrenth (1969) reported that the cytoplasmic materials encircle the centrally located nucleus in early spermatids. A partial polarization of the cell contents along with the multivesicular bodies to one end of the cells has reported in cray fish *P. clarkia* by Moses, (1961b). McKnight and Hinch, (1986) reported that in *Scyllarus chacei* polarization of the cell components started in stage1 spermatids, as the nucleus moves to one side of the cell while the cytoplasmic organelles move to the other side. Similarly, in the early spermatids of *P. monodon* the RER is present in concentric layers around the nucleus at one side. While other cell organelles are shifted towards the opposite end of the cell. On the contrary, the polarization of the cells is found only in later spermatids in *Anaspidestasmaniae* (Jesperson, 1983).

The de-condensation of the nucleus as well as the blebbing of the highly convoluted nuclear wall is more pronounced in developing spermatids of *P. monodon*. Clusters of small vesicles and dense granules appear around the nucleus in close proximity to the nuclear blebs, indicating the highly active condition of the spermatid nuclei during its development as a pre-requisite for acrosome formation. De-condensation of the spermatid nucleus has been reported in many decapod crustaceans including penaeids. In *P. setiferus* Lu *et al.*, (1973) reported that the size of the nucleus diminishes in the spermatids due to the condensation of the nuclear materials. During the transformation of the spermatids to the spermatozoa in *P. clarkii*, the nucleus underwent considerable shrinkage and it represents only one eighth of the total volume of the cell (Moses, 1961b). McKnight and Hinch, (1986) reported that the nuclear e-condensation in *S. chacei* begins at the stage1 spermatid and gets completed in the stageIII spermatids. A reduction in the size of nucleus is reported in *Cancer* crabs (Langrenth, 1969) and in *C. clypeatus* (Hinch, 1980).

Another noticeable feature of the early spermatids in *P. monodon* is the blebbing of nuclear membrane and the highly active perinuclear cytoplasm (Diwan *et al.*, 2008). In *P. setiferus* Lu *et al.*,. (1973) reported that, the distinguishing structures of early spermatid are the dilated vesicles, which may have originate from the nuclear membrane and subsequently fuse to form a structure which caps the nucleus at one pole. Blebbing of the nuclear wall and a high perinuclear activity is also reported in the early stages of spermatid of *P. clarkii*, and blebs have been found recede during the later stages, but the surface of the nucleus is still irregular (Moses, 1961b). The vesicular formation as well as its fusion is observed throughout the spermatids in *S. chacei* (McKnight and Hinch, 1986). Similarly, smooth and round vesicles containing faintly electron opaque filamentous materials are described in the early spermatids of *Cancer*crabs (Langrenth, 1969), which are presumed to be the acrosome precursor because of their nature and subsequent fusion.

The most important event in the spermatid differentiation is the acrosome formation. In *P. monodon* it is observed that the acrosome formation starts in the early spermatids and completed only in the final stages of differentiation of the spermatozoa. The classical literature on decapod spermatogenesis is not clear in its descriptions of the origin of acrosome. Some earlier workers (Nath, 1937 and 1965) attributed its formation to Golgi elements and some to mitochondria (Mc Croan, 1940) and Nath (1965). Some other investigators are of opinion that it forms from the vesicle or vacuole appearing in the cytoplasm (Binford, 1913 and Fasten, 1926). However, electron microscope studies of Yasuzumi *et al.*,. (1960) reveals that certain dense granules appear in interzonal spindle region, as the acrosomal precursor. Pochon-Masson, (1965, 1968) concluded that the acrosome incrustaceans is derived from the fusion of dilated ergastoplasmic cisternae, and the nuclear envelope derivatives may also contribute to the acrosome. From the ultrastructural investigations of Moses, (1961b) in *P. clarkii* and Langreth, (1969) in *Cancer* crabs, it is found that the acrosome forms from the fusion of the cytoplasmic vesicles, which appear to be derived from the rough ER. In *P. monodon* the small granule filled vesicles derived from the cytoplasm and the nuclear blebs forms the precursors of acrosomal materials. Other structures involved mainly in the acrosome formation are smooth ER elements and nuclear envelops (Diwan *et al.*, 2008).

The acrosomal core forms by an in pocketing of the acrosomal vesicle in the regions closest to the nucleus. Fibrous and granular materials are noted within the acrosomal core in the early spermatozoa of *P. monodon* (Diwan *et al.*, 2008). A similar mode of formation of the acrosomal core is described in other decapods (Moses, 1961 a, b; Yasuzumi *et al.*, 1960; Pochon- Masson, 1965 and Langrenth, 1969). These investigators also noted similar type of fibrous tubular materials in connection with the acrosomal complex.

A morphological study by earlier investigators has resulted in the crustacean decapod gametes being divided into two classes, the unistellate sperm of the natantians (shrimps) and multistellate sperm of the reptantians (Lynn and Clark,

1983). Studies on the fine structure of the sperm in *P. monodon* reveals that, it is more or less round and unistellate as in other natantians and superficially it can be distinguished into three regions viz. 1) aposterior main body 2) the central cap region and 3) the anterior spike. Ultrastructure of the sperm in other natantians has been studied by investigators like Clark *et al.*, (1973) in *P. aztecus*; Lu *et al.*,(1973) in *P. setiferus;* Kleve *et al.*,. (1980) in *S. injentis;* Mohammed and Diwan, (1994) in *P. indicus* and they described more or less similar type of spermatozoa in these shrimps.

The nucleus of spermatozoa in *P. monodon* is de-condensed which is a typical feature of the decapod crustacean sperms, and a nuclear envelope is absent as in other natantians (Pochon-Masson, 1969; Clark *et al.*,1973; Lu *et al.*,1973; Kleve *et al.*,1980; Mohammed and Diwan, 1994). The nucleus of the natantian sperm is found confined to the main body (Pochon-Masson, 1969; Clark *et al.*,1973 and Mohammed and Diwan, 1994). The spermnuclei of *P. monodon* confirm this typical natantian pattern. The nucleus extends to the plasma membrane at its outer region with no intervening region of cytoplasm as shown in *P. aztecus* (Clark *et al.*,1973) and in *M. rosenbergii* (Lynn and Clark, 1983). Loosely packed fibrillar materials constituted the majority of the cell body as in other shrimps like *P. aztecus* (Clark *et al.*,1973; Kleve *et al.*,1980); *P. indicus* (Mohammed and Diwan, 1994) and *P. setiferus* (Lu *et al.*,1973). As in other shrimps, the nucleus of the sperm of *P. monodon* is surrounded partially by an amorphous cytoplasmic band that contains small sized vesicles whose functions are unknown (Diwan *et al.*, 2008).

Morphologically complex cap region represents the acrosomal complex in *P. monodon* sperm. This complex cap region of the sperm exhibit structures similar to those described in other penaeids like *P. aztecus* (Clark *et al.*,1973); *P. setiferus* (Lu *et al.*,1973); *S. injenis* (Kleve *et al.*,1980) and *P. indicus* (Mohammed and Diwan, 1994) with slight modifications. The granular central core observed in *P .monodon* is also reported in *S. injentis* (Kleve *et al.*,1980) and in *P. indicus* (Mohammed and Diwan, 1994). Clark *et al.*,(1973) found the central core long with the spike in *P. aztecus*. Thea crosomal materials found in the sperm of *P. monodon* separating the nuclei and the central core of the acrosomal complex are also described in *P. indicus* (Mohammed and Diwan, 1994) and as granule in *P. aztecus* (Clark *et al.*,1973). The single pro-acrosomal granule is located in one side of this sub-acrosomal granule in *P. monodon* whereas in *P. indicus* (Mohammed and Diwan, 1994) noticed the presence of one or two pro-acrosomal granules, either on one or on both the sides.

The spike in the sperm of *P. monodon* is an anterior extension of the acrosomal complex, as both are encompassed in a single membrane. Similarly in *S. injentis*

Kleve *et al.*,(1980) reported that a continuous membrane encloses the acrosomal cap region and its anterior extension, is the spike. In other penaeids also, the spike is considered as an anterior extension of the acrosomal complex (Clark *et al.*,1973; Lu *et al.*,1973; Kleve *et al.*,1980 and Mohammed and Diwan, 1994). Surprisingly no organelles are observed in the fully matured spermtozoa of *P. monodon*, except the small vacuoles and vesicles found in the cytoplasmic band on both the lateral sides. The degeneration of organelles has been reported in many decapods crustaceans (Moses, 1961 a, b; Langrenth, 1969; Clark *et al.*,1973; Lu *et al.*,1973; Kleve, 1980; Hinch, 1980; Jesperson, 1983; McKnight and Hinch, 1986 and Mohammed and Diwan, 1994). It is not only in decapods, among Malacostraca itself, the spermatozoa show considerable modification and loss of organelles (Adiyodi, 1985). From all these characteristics, it shows that the sperm of *P. monodon* belongs to an altered vesicular type, which is different from other flagellate and non-flagellate gametes in other crustaceans (Pochon-Masson, 1983).

The vas deferens of decapod crustaceans function as a site of sperm maturation, encapsulation of sperm into spermatophores, production of seminal fluids and their storage (Hinch and Mc Knight, 1986). Many studies have described the structure of the vas deferens and its role in the formation of the spermatophore in crustaceans through light and electron microscopy (Cronin, 1947; Mathews, 1954; Hinch and Walker, 1974; Malek and Bawab, 1974 a, b; Kooda-Cisco and Talbot, 1986; Hinch and McKnight, 1986; Talbot and Beach, 1989; Chow *et al.*,. 1991; Ro *et al.*,1990; Vasudevappa, 1992; Mohammed and Diwan, 1994). However, such studies in penaeids are limited to *P. kerathurus* (Malek and Bawab 1974 a, b); *P. setiferus* (Ro *et al.*,1990 and Chow *et al.*,1991); *P. vannamei* (Chow *et al.*,. 1991); *Metapenaeus dobsoni* (Vasudevappa, 1992) and *P. indicus* (Mohammed and Diwan, 1994). Ultrastructural investigations in penaeids on the spermatophore formation have been more limited (Ro *et al.*,1990; Chow *et al.*,1991).

The Vas deferens of decapods crustaceans conveys sperm from the testes to the exterior in the form of a spermatophore. The glandular epithelial cells lining the Vas deferens during the passage of sperm layed own the spermatophore. The mechanism of spermatophore formation has been described for several Penaeid shrimps like in *P. monodon, P. kerathurus* (Malek and Bawab, 1974 a, b); *P. setiferus* (Ro *et al.*,1990, Chow *et al.*,1991); *P. vannamei* (Chow *et al.*,1991) *P. indicus* (Mohammed and Diwan, 1994) and *M. dobsoni* (Vasudevappa, 1992). The processes in which the different layers of the spermatophore wall are laid appear o be similar to other penaeids like *P. kerathurus* (Malek and Bawab, 1974 a, b). As in *P. kerathurus,* spermatophore of *P. monodon* is composed of five acellular layers. Different layers are found secreted in different parts of the Vas deferens by glandular epithelial cells lining the corresponding parts (Diwan *et al.*, 2008).

Spermatophore formation starts at the PVD and continues till the terminal ampoule. The PVD secretes an amorphous material, which forms the sperm-supporting matrix. The epithelial cells in the spermduct of MVD in its ascending limb secrete the first two layers of the spermatophore wall. The sperm mass is found completely encompassed in these two layersin *P. monodon* as in *P. kerathurus* (Malek and Bawab, 1974 a, b) and *P. indicus* (Mohammed and Diwan, 1994). The third and the fourth layer are secreted in the wing duct. The fifth layer is secreted in the wingduct and sperm duct at the descending limb. In *M. dobsoni*, Vasudevappa (1992) has noticed two complete layers for its spermatophores. The first three layers are incomplete, while the fourth and fifth completely surround the spermatophore of *M. dobsoni*. Kooda-Cisco and Talbot (1982) observed three complete layers in *H. americanus*. The morphological and histological structure of the wing of the spermatophore in *P. monodon* is found to be similar to that described in *M. dobsoni* by (Vasudevappa, 1992) and in *P. indicus* by (Mohammed and Diwan, 1994).

The ultrastructure of the Vas deferens of *P. monodon* is found to be similar to that of other decapods crustaceans, such as *Libinia* (Hinsch and Walker, 1974); *Homarus* (Kooda-Cisco and Talbot, 1986) and *P. setiferus* (Ro *et al.*,1990; Chow *et al.*,1991). It is found that the entire lumen of the vas deferens in *P. monodon* is lined by secretory epithelium and that inturn is surrounded by a basal lamina. As in other decapods that have been studied so far, the Vas deferens of *P. monodon* is divided into several morphological and functional regions, while the secretory cells throughout the tracts are ultrastructurally similar. These secretory cells function in the formation of the acellular components of the spermatophores. In *P. monodon* secretion of spermatophoric layers starts in the PVD and continues till the TA. In *P. kerathurus* (Malek and Bawab, 1974 a, b) and in *P. setiferus* (Ro *et al.*,. 1990 and Chow *et al.*,1991) reported that the secretion of the primary spermatophoric layers starts only in the spermduct of MVD. The PVD as well as the blind pouch secrete the sperm supporting matrix and the primary spermatophoric layer. In contrast, Mohammed and Diwan (1991) have reported that in *P. indicus* the PVD is devoid of secretory cells and the spermatophore formation starts in MVD.

Columnar epithelial cells as reported in other penaeids line the blind pouch of *P. monodon*. The small, branched typhlosole projecting into the lumen of the blind pouch is characteristic of *P. monodon*. Malek and Bawab (1974 a, b) reported numerous regularly arranged glandular cells but no typhlosole. Highly secretory epithelial cells are reported in blind pouch of *P. setiferus* (Ro *et al.*,1990). These epithelial cells lining the blind pouch as well as the typhlosole in this area secrete an electron-dense material, which gives a cohesive property to the sperm-supporting matrix. "Filamentous materials" in the sperm-supporting matrix, which keep the sperm cells together, has been reported in *P. kerathurus* (Malek and Bawab,1974,a and b). Filaments, which are probably collagen, have been observed in the sperm supporting matrices of other decapods such as lobsters (Talbot *et al.*,1976; Talbot and Chanmanon, 1980) and Cray fishes (Talbot and Beach, 1989). However, such filaments have not been reported in the sperm-supporting matrix of *P.vannamei* (Ro *et al.*,1990). Thus the presence or absence and the structure of filaments when present in the matrix seem to be variable among decapods. In *P. monodon* the first evidence of orientation of the sperm occurs in the blind pouch as in other penaeids like *P. Kerathurus* (Malek and Bawab, 1974 a, b); *P. setiferus* (Ro *et al.*,1990) and *P. indicus* (Mohammed and Diwan, 1994).

The ascending limb of MVD of *P. monodon* contains two ducts. Heldt (1989) first described the double duct structure of the vas deferens in *P. trisulcatus*, and this character has been shown to be common in the genus penaeus (King, 1948; Malek and Bawab, 1974 b; Champion, 1987; Ro *et al.*,1990; Chow *et al.*,1991; Vasudevappa, 1992 and Mohammed and Diwan, 1994). In *P. monodon* the two ducts have also been reported in the ascending limb of the MVD termed, spermatophoric and wing ducts by Malek and Bawab (1974 b) in *P. kerathurus*. Tall epithelial cells with increased secretory activity line the inner lumen of both the spermatophoric and wing duct of *P. monodon*. Out growths of these epithelial cells as typhlosoles are present in the lumen of both of these ducts. The wing duct contains only a single typhlosole whereas the spermatophoric duct contains three to four typhlosoles. Secretion of fine granular materials of the third spermatophoric layer is found in the wingduct, while the spermatophoric ductc ontains the materials of the secondary spermatophoric layer (Diwan *et al.*, 2008). Mohammed and Diwan (1994) reported in *P. indicus*, that the wingduct secretes the materials of the wing. A similar type of secretion is reported in *P. setiferus* (Rao *et al.*, 1990). In *M. dobsoni* Vasudevappa (1992) described the structure as well as formation of wingmaterial in the wingduct. Bell and Lightner, (1988) mistakenly presumed that, the accessory duct deposited the primary spermatophore layer in *P. stylirostris*. In contrary, Chow *et al.*,(1991) did not find any wing materials in the wing duct of *P. vannamei* and so they re-termed the "wing duct" as the "accessory duct". Secretion of the materials of the fourth spermatophoric layer is found in the wingduct, along with the third layer in the distal end of the ascending limb and also in the descending limb in *P. monodon*. The fifth layer is secreted in both the wing duct at its terminal end, as well as in the sperm duct and continues till the terminal ampoule.

In *P. monodon* the primary spermatophoric layer is heterogeneous and fibrillar in appearance. The region closer to the sperm matrix is only moderately electrondense but the region outside is electron-denser. The secondary spermatophoriclayer is comprised of a matrix of low density, irregular zones of high density and small and large electron dense granules. These primary and secondary spermatophoric layers are secreted by both, anapocrine mechanism and by exocytosis (Diwan *et al.*, 2008). Similar type of secretion has been reported in the segment III of the vas deferens of *P. setiferus* (Ro *et al.*,1990) and in crayfish *Cherax* (Talbot and Beach, 1989). This is in contrast to *Homarus* (Kooda-Cisco and Talbot, 1986); and Libinia, (Hinsch and Walker, 1974). From the ultrastructural observations, it is presumed that the numerous polysomes observed in the cytoplasm of the secretory epithelium are generally responsible for synthesizing the secretory products that become sequestered in the apical blebs; while those products released by exocytosis are synthesized on the RER and processed through the Golgi bodies. One remarkable feature of the apocrine blebs is the absence of organelles within the bleb. The secretory epithelium of *P. monodon* is similar to that of *Homarus* (Kooda-Ciscoand Talbot, 1986); crayfish *Cherax* (Talbot and Beach, 1989) and *P. setiferus* (Ro *et al.*,1990). The third spermatophoric layer present in the spermatophore of *P. monodon* is secreted in the wing duct. Similar type of apocrine as well as exocytotic secretory pattern has been noticed both in the spermatophoric duct and the wingduct. The electron lucent fourth, and the highly electron dense fifth layer is secreted by the epithelial cells of the wing duct by apocrine mode of secretion and exocytosis (Diwan *et al.*, 2008).

3.12 CONCLUSIONS

Precise staging of gonadal maturation of female shrimp is essential for the selection of females in breeding, spawning and production of quality seeds. Moreover as gametogenesis is the core of reproduction, a thorough knowledge is unwarranted for evolving biotechnological methods, which may be helpful in acquiring reproduction in captivity. From the perusal of the literature on the subject it is found that an ultrastructural study of gonad, particularly ovary is unavoidable to understand the various dynamic processes involved during the gamete development. Once the morphological development of gonad correlated with its cytological characteristics is well understood, one can easily predict the reproductive quality of the brooder by observing its morphological characters without disturbing the animal more. Understanding of shrimp reproduction requires combined knowledge from all the aspects viz. the changes in the morphological characters of the gonads, and corresponding cytological figures of the normal animal, which are essential for the better management of brooders for seed production.

Refrences

Arcos, F.G, Palacios, E, Ibarra, A.M, Racotta, I.S, 2005. Larval quality in relation to consecutive spawning in white shrimp *Litopenaeus vannamei* Boone. *Aquac. Res.* 36, 890–897.

Arcos F G., Ibarra A M., Racotta L S., 2011. Vitellogenin in hemolymph predicts gonad maturity in adult female *Litopenaeus (Penaeus) vannamei* shrimp. *Aquaculture.* 316: 1–4.

Andries J.C, 2001.Endocrine and environmental control of reproduction in Polychaeta. *Can. J. Zool.* 79, pp. 254–270.

Adiyodi, K.G. 1982. Some thoughts on the evolution of reproductive hormones in invertebrates In: *Progresses in Invertebrate Reproduction and Aquaculture.* 21-34

Adiyodi, R.G. and T. Subramoniam 1983. Arthropoda-Crustacea. In: *Reproductive Biology of Invertebrates. Vol.1. Oogenesis, Oviposition, and Oosorption* (K.G. Adiyodi & R.G. Adiyodi, eds) pp. 443-495. Wiley, Chichester, Endland.

Adiyodi, R.G. 1985. Reproduction and its control. In: *The Biology of Crustacea.* (Ed.) D.E. Bliss and L.H. Montel. Academic Press, New York.

Aiken, D.E. and S.L. Waddy 1980. Reproductive biology. In: *The Biology and Management of Lobsters.* (Ed. J.S. Cobb and B.F. Phillips) 215-276. Academic Press New York.

Alfaro, J. 1993. Reproductive quality evaluation of male *Penaeus stylirostris* from a grow out pond, *J. World. Aquacult. Soc.*, 24,1: 6-11.

Alfaro, J. and X. Lozano 1993 Development and deterioration of spermatophores in pond-reared *Penaeus vannamei.*, 3. *J. World. Aquacult. Soc.*, 24,4: 522-528.

Alfaro, J., Zuniga and G., Komen, J., 2004. Induction of ovarian maturation and spawning by combined treatment of serotonin and a dopamine antagonist, spiperone in *Litopenaeus stylirostris* and *Litopenaeus vannamei. Aquaculture* 236: 511–522.

Alikunhi, K.H., A Poernomo, S. Adisukesno, M. Budiono and S. Busman 1975. Preliminary observations on induction of maturity and spawning in *P. monodon* Fabricius and *P. merguiensis* de Man by eyestalk extirpation. *Bull. Shrimp. Cult. Res. Cen.*, 1: 1-11.

Ana M. Ibarra, Ilie S. Racotta, Fabiola G. Arcos, Elena Palacios, 2007. Progress on the genetics of reproductive performance in penaeid shrimp. *Aquaculture*. 268: 1–4.

Anchordoguy, T., J.H. Crowe, W.H. Clark, Jr. and F.J. Griffin 1987. Cryopreservation of sperm from the penaeid shrimp *Sicyonia ingentis*. Abstract 18th *Ann. Meet. World. Aquacult. Soc. Gvayaguil*, Ecuador. Jan. 1987.

Anchordoguy, T., J.H. Crowe, F.J. Griffin, and W.H. Clark. Jr. 1988. Cryopreservation of sperm from marine shrimp *Sicyonia ingentis*. *Cryobiology*., 25,3: 238-243.

Anderson, S.L., E.S. Chang and W.H. Clark 1984. Timing of postvitellogenic ovarian changes in the ridgeback prawn *Sicyonia ingentis* determined by ovarian biopsy. *Aquaculture*, 42: 257-271.

Andriantahina F, Xiaolin L, Hao H, Jianhai X, Changming Y, 2012. Comparison of reproductive per for mance and offspring quality of domesticated Pacific white shrimp, *Litopenaeus vannamei*. *Aquaculture*. 324–325

Anonymous, 2005. Indian sea food exports move up in 2004-2005. *Fishing Chimes*. 25: 27-28.

Aoto, T. and H. Nishida 1956. Effect of removal of the eyestalks on the growth and maturation of the oocytes in a hermaphroditic prawn, *Pandalus kessleri*. *J. Fac. Sci. Hokkaido Univ. Ser.*, 612: 412-424.

Apichart Ngernsoungnern, Piyada Ngernsoungnern, Wattana Weerachatyanukul, Prasert Sobhon, Prapee Sretarugsa, Jittipan Chavadej, 2008. The existence of gonadotropin- releasing hormone (GnRH) immune reactivity in the ovary and the effects of GnRHs on the ovarian maturation in the black tiger shrimp *Penaeus monodon*. *Aquaculture*. 279:1–4,

Aquacop 1975. Maturation and spawning in captivity of penaeid shrimps Penaeus merguiensis de man, Penaeus japonicus Bate, Penaeus aztecus Ives, Metapenaeus ensis de Haan and Penaeus semisulcatus de Hann. Proc. 6th Ann. Meet. World Maricult. Soc., 123-132

Arnstein, D.R. and T.W. Beard 1975. Induced maturation of prawn *Penaeus orientalisi* Kishinogue in the laboratory by means of eyestalk removal. *Aquaculture*, 5: 411-412.

Arnold Stuart J., Greg J. Coman, Charis Burridge, Min Rao 2012. A novel approach to evaluate the relationship between measures of male fertility and egg fertilization in *Penaeus monodon*. *Aquaculture* 338–341. 181–189.

Avvare J-C, Michelis R, Tietz A, Lubzens E, 2003. Relationship between vitellogenin and vitellin in a marine shrimp (Penaeus semisulcatus) and molecular characyerization of vitellogenin complemantary DNAs, *Biol Reprod*, 69:355-364.

Ayub Z. and Ahmed M. (2002). A description of the ovarian development stages of penaeid shrimps from the coast of Pakistan. *Aquaculture Research* 33: 767-776.

Bauer, R.T. and J.W. Martin 1991. *Crustacean Sexual Biology* Columbia Univ. Press. New York. 351 pp.

Beams, H.W and R.G. Kessel 1962 Intracisternal granules of the endoplasmic reticulum in the crayfish oocyte. *J. Cell. Biol.*, 13: 158-162.

Beams, H.W. and R.G. Kessel 1963 Electron microscope studies on developing crayfish oocytes with special reference to the origin of yolk. *J. Cell. Biol.*, 18: 621-644.

Beams, H.W. and R.G. Kessel 1980 Ultrastructure and vitellogenesis in the oocyte of the crustacean, *Oniscus asellus. J. Submicrosco. Cytol.*, 12: 17-27.

Behlmer, S.D. and G. Brown 1984. Viability of cryopreserved spermatozoa of the crab *Limulus polyphemus. Int. J. Invert. Rep. Dev.*, 7: 193-199

Bell TA, Lightner DV, 1988. Female Reproductive System. In: Bell TA, Lightner DV, editors. A Handbook of Normal Penaeid Shrimp Histology. Baton Rouge, La: World *Aquaculture* Society;. pp. 27–33

Binford, R.H. 1913. The germ cells and the process of fertilization in the crab *Menippe mercenaria. J. Morphol.*, 24: 147-204.

Bradfield, J.Y., R.L. Berlin, S.M. Rankin and L.L. Keeley 1989. Cloned CDNA and antibody for an ovarian cortical granule polypeptide of the shrimp *Penaeus vannamei. Biol. Bull.*, 177: 344-349.

Bray, W.A., Jr. Leung-Trujillo, A.L. Lawrence and S.M. Robertson 1985. Preliminary investigation of the effects of temperature, Bacterial inoculation and EDTA on sperm quality. *J. World. Maricult. Soc.*, 16: 250-257.

Bray, W.A. and A.L. Lawrence, 1990. Reproduction of eyestalk ablated *P. stylirostris* fed various levels of total dietry lipids. *J. World. Aquacult. Soc.*, 21,1: 41-52

Brown, A. Jr., and D. Patlan. 1974. Colour changes in the ovaries of penaeid shrimp as a determinant of their maturity. *Mar. Fish. Res.*, 36,7: 23-26.

Browdy C.L, M. Fainzilber, M. Tom, Y. Loya, E. Lubzens, 1990. Vitellogenin synthesis in relation to oogenesis in in vitro-incubated ovaries of *Penaeus semisulcatus* (Crustacea Decapoda, Penaeidae) *J. Exp. Zool.*, 255, pp. 205–215.

Browdy, C.L., 1998. Recent developments in penaeid broodstock and seed production technologies: improving the outlook for superior captive stocks. *Aquaculture*, 164: 3–21.

Bury, N.R. and P.J.W. Olive 1993. Ultrastructural observations on membrane changes associated with cryopreserved spermatozoa of two polycheate species and subsequent mobility induced by quinacine. *Invertbr. Rep. Dev.*, 23: (2-3): 139-150.

Carlisle, D.B. and F.G.W. Knowles,1959. Endocrine control in crustaceans, 150 pp. Cambridge Univ. Press, London.

Caro, L.G. and G.E. Palade, 1964. Protein synthesis, storage and discharge in the pancreatic exocrine cells. *J. Cell. Biol.*, 20: 473-495.

Callaghan T R, Bernard M. D, Melony J. S, 2010. Expression of Sex and Reproduction-Related Genes in *Marsupenaeus japonicus. Marine Biotechnology*, Volume 12, Issue 6, pp 664-677.

Cavalli R.O., Scardua M.P. and Wasielesky W.J. 1997. Reproductive performance of different sized wild and pond reared *Penaeus paulensis* females. *J. World. Aquacult. Soc.*, 28: 260-267.

Champion, H.F. B. 1987. The functional anatomy of the male reproductive system in *Penaeus indicus. South African, J. Zool.*, 22: 297-307.

Chan S.M, P.L. Gu, K.H. Chu, S.S, 2003. Tobe Crustacean neuropeptide genes of the CHH/MIH/GIH family: implications from molecular Studies. *Gen. Comp. Endocrinol.*, 134, pp. 214–219.

Chandran, M.R. 1968. Studies on the marine crab, *Charybdis variegata*. I. Reproductive and nutritional cycle in relation to breeding periodicities. *Proc. Indian. Acad. Sci.*, 67: 215-223.

Chang, E.S., W.A. Hertz and G.D. Prestwich, 1992. Reproductive endocrinology of the shrimp *Sicyonia ingentis*: Steroid, Peptide, and Terpenoid Hormones, NOAA. Tech, Reports. NMFS. 106: 1-6.

Chang, C.F., F.Y. Lee and Y.S. Huang, 1993. Purifiction and characterization of vitelline from the mature ovaries of prawn *Penaeus monodon. Comp. Biochem. Physiol.*, 105 b: 409-414.

Chang, C.F., Fang-YI Lee, Y.S. Huang and T.H. Hong, 1994. Purification and characterization of the female specific protein in mature female haemolymph of the prawn *Penaeus monodon. Invert. Rep. Dev.*, 25(3): 185-192.

Chang, C.F. and T.W. Shih, 1995. Reproductive cycle of ovarian development and vitellogenin profile in freshwater prawns, *M. rosenbergii. Invert. Rep. Dev.*, 27: 11-20.

Charniaux-Cotton, H, 1978. L'ovogenese, la vitellogenine et leur controle chezle crustace Amphipode *Orchestia gammarellus.* Co, paraison ovele d'autres Malaacostracers- *Archives de Zoologie Experimentale et Generale* 119: 365-397.

Charniaux-Cotton, H, 1985. Vitellogenesis and its control in malacostracan Crustacea. *Amer. Zool.*, 25: 197-206.

Charniaux-Cotton, H. and G. Payen 1988. Crustacean reproduction. In: *Endocrinology of Selected Invertebrate Types.*, 2: 279-303.

Chamberlain, G.W. and A.L. Lawrence 1981 a. Maturation, Reproduction and Growth of *Penaeus vannamei* and *P. stylirostris* fed natural diets. *J. World. Maricult. Soc.*, 12: 209-224.

Chamberlain, G.W. and N.F. Gervais 1984. Comparison of unilateral eyestalk ablation with environmental control for ovarian maturation of *Penaeus stylirostris. J. World. Maricult. Soc.*, 15: 29-30.

Champion, H.F. B. 1987. The functional anatomy of the male reproductive system in *Penaeus indicus. South African, J. Zool.*, 22: 297-307.

Chow S. 1982. Artificial insemination using preserved spermatophores in the palaemonid shrimp *Macrobrachium rosenbergii. Bull. Jap. Soc. Sci. Fish.*, 48(2): 1693-1965.

Chow, S. Y. Ogasawara and Y. Take,1985. Male reproductive system and fertilization of the Palaemonid shrimp *Macrobrachium rosenbergii. Bull. Jap. Soc. Sci. Fish.*, 48: 177-183.

Chow, S.C. 1987. Growth and reproduction of eyestalk ablated *Penaeus canalicualtus* (Olivier 1811). *J. Exp. Mar. Biol. Ecol.*, 112: 93-107.

Chow, S., M.M. Dougherty. W.J. Doughtety and P.A. Sandifer 1991. Spermatophore formation in the white shrimps. *Penaeus setiferus* and *Penaeus vannamei. J. Crust. Biol.*, 11(2): 201-216.

Chow, S., W. J. Dougherty and P.A. Sandifer. 1991. Unusual testicular lobe system in the white shrimps *Penaeus setiferus* and *Penaeus vannamei* B (Decapoda, Penaeidae): A new character for dendrobranchiata. *Crustaceana*, 60 (1-3): 305-317.

Chung Meng-Yuan, Chun-Hung Liub, Ying-Nan Chenb, , Winton Cheng, 2011. Enhancing the reproductive performance of tiger shrimp, *Penaeus monodon*, by incorporating sodium alginate in the broodstock and larval diets. *Aquaculture*. 312, 1–4.

Clark, W.H. Jr., P. Talbot, R.A. Neal, C.R. Mock and B.R. Salser. 1973. Invitro fertilization with non-motile spermatozoa of the brown shrimp *Penaeus aztecus*. *Mar. Biol.*, 22: 353-354

Coman, F.E., Norris, B.J., Pendrey, R.C., Preston, N.P., 2003. A simple spawning detection and alarm system for penaeid shrimp. *Aquac. Res.* 34: 1359–1360.

Coman, G.J., Arnold, S.J., Thompson, P.J. and Crocos, P.J., 2003. Development of a low water-exchange rearing system for biosecure maturation of *Penaeus monodon*. *World Aquaculture Symposium* 2003, Salvador, Brazil. Abstract.

Coman, G.J., Crocos, P.J., Preston, N.P. and Fielder, D., 2004. The effect of density on the growth and survival of different families of juvenile *Penaeus japonicus* Bate. *Aquaculture* 229: 215–223.

Coman, G.J., Crocos, P.J., Arnold, S.J., Keys, S.J., Murphy and B., Preston, N.P., 2005. Growth, survival and reproductive performance of domesticated Australian stocks of the giant tiger prawn, *Penaeus monodon*, reared in tanks and raceways. *J. World Aquac. Soc.*, 36: 464–479.

Coman G.J., S.J. Arnold, S. Peixoto, P.J. Crocos, F.E. Coman and N.P. Preston 2006. Reproductive performance of reciprocally crossed wild-caught and tank-reared *Penaeus monodon* broodstock *Aquaculture*, 252: 372–384.

Coman G.J., S.J. Arnold, M.J. Jones and N.P. Preston 2007 Effect of rearing density on growth, survival and reproductive performance of domesticated *Penaeus monodon*. *Aquaculture*, 263: 75-83.

Coman, G.J., Crocos, P.J., Arnold, S.J., Keys, S.J., Murphy, B. and Preston N.P., 2007. Growth, survival and reproductive performance of domesticated Australian stocks of the giant tiger prawn, *Penaeus monodon*, reared in tanks and raceways. *J. World Aquacult. Soc.* 36: (4).

Crocos, P.J. and Coman, G.J., 1997. Seasonal and age variability in the reproductive performance of *Penaeus semisulcatus*: optimizing broodstock selection. *Aquaculture*, 155: 55–67.

Crocos, P.J., Preston N.P. Smith D.M. and Smith M.R, 1999. Reproductive performance of domesticated *Penaeus monodon*. Book of abstracts, World *Aquaculture* Symposium 1999, Sydney, Australia, p. 185.

Cronine, L.E. 1947. Anatomy and histology of the male reproductive system of *Callinectes sapidus J. Morphol.*, 81: 201-239.

Cummings, W.C. 1961. Maturation and spawning of the pink shrimp, *Penaeus duorarum. Trans. American Fish Soc.*, 90(4): 462-468.

Cuzon, G., Arena, L., Goguenheim, J. and Goyard, E., 2004. Is it possible to raise, offspring of the 25th generation of *Litopenaeus vannamei* (Boone) and 18th generation *Litopenaeus stylirostris* (Simpson) in clear water to 40g. *Aquac. Res.* 35: 1244–1252.

Dall. W. 1965. Studies on the physiology of a shrimp *Metapenaeus* sap. II. Endocrine and control of moulting, *Aust. J. Mar. Fresh. Rs.* 16: 1-12.

Dall, W., B.J. Hill, P.C. Rothlisherg and Staples. 1990. The biology of penaeids. In: *Advances in Marine Biology*, 27: 488.

Damrongphol, P.N. Eangchuan and B. Poolsanguan 1991. Spawning cycle and oocyte maturation in laboratory maintained giant freshwater prawn. *Aquaculture*, 95: 347-357.

Decaraman, M. and T. Subramoniam 1983. Endocrine regulation of ovarian maturation and cement gland activity in a stomatopod crustacean, *Squilla holoschista. Proc. Indian. Acad. Sci. (Ani, Sci.)* 92: 399-408

Demeusy, N. 1967. Croissance relative d'un caractere sezueles externe male chez la Decapode Brachyore *Careinus memaus* Lc. *R. Hebd. Seane. Acad. Sci. Paris.*

Diwan, A.D. and R. Nagabhushanam 1974. Reproductive cycles and biochemical changes in the gonads of the freshwater crab, *Barytelphusa cunicularisn. Indian. J. Fish.* 21(1): 164-176.

Diwan, A.D., 2005. Current progress in shrimp endocrinology-a review. *Indian journal of experimental biology*, 43,3:209.

Diwan A D, Shoji Joseph and S Ayyappan, 2008. Physiology of Reproduction, Breeding and culture of tiger shrimp, *Penaeus monodon* (Fabricius) Publ: Narendra Publishing House, New Delhi: 300.

Dunn, R. S. and J. Mc Lachlin, 1973. Cryopreservation of echinoderm sperm. *Can. J. Zool.*, 51: 666-669.

Durand, J.B. 1956. Neurosecretory cell types, their secretory activity in the crayfish. *Biol. Bull.*, (Woods Hole),111: 62-67.

Duronslet, M.J., A.I. Yidin, R.S. Wheeler and W.H. Clark, Jr. 1975. Light and fine structural studies of natural and artificially induced egg growth of penaeid shrimp. *Proc. World. Maricult. Soc.*, 6: 105-122.

Eastman-Recks, S.B. and M. Fingerman 1984. Effects of neuroendocrine tissue and cyclic AMP on ovarian growth *in vitro* in the fiddler crab, *Uca pugilator. Comp. Biochem. Physiol.*, 79A: 679-684.

Eastman-Recks, S.B. and M. Fingerman 1985. *In vitro* synthesis of vitellin by the ovary of the Fiddler crab, *Uca pugilator. J. Exp. Zool.*, 233: 111-116.

Eldred, B. 1958. Observations on the structural development of the genitalia and the impregntion of the pink shrimp, *Penaeus duorarum. Bull of the Florida State Board of lonserv.* (Technical series) 23: 1-26.

Emmerson, W.D. 1980. Induced maturation of prawn *Penaeus indicus. Mar. Ecol. Prog. Ser.*, 2: 121-132

Eurrenius, L. 1973. An electron-microscopy study on the developing oocytes of the crab *Cancer pagurus* L. with special reference to yolk formation (Crustacea). *Z. Morphol. Tiere.*, 75: 243-254.

Fainzilber, M., M. Tom, S. Shafir, S.W. Applebaum and E. Labzens 1992. Is there extra ovarian synthesis of vitellogenic in penaeid shrimp? *Biol. Bull.*, 183: 233-254.

Farfante, P.I. 1975. Spermatophores and thelyca of the American white shrimps, genus *Penaeus*, sub genus *Litopenaeus. Fish. Bull.*, 73: 463-486

Fasten, N. 1926. Spermatogenesis of the black-clawed crab, *Lophopano peus* bellus (Stinmpson) Rathbun. *Biol. Bull.*, 50: 277-279.

Fingerman, M. 1987. The endocrine mechanisms of crustaceans. *J. Crust Biol.*, 7(1): 1-24.

Fujino Y, T. Nagahama, T. Oumi, K. Ukena, F. Morishita, Y. Furukawa, O. Matsushima, M. Ando, H. Takahama, H. Satake, H. Minakata, K. Nomoto, 1999. Possible functions of oxytocin/vasopressin-superfamily peptides in annelids with special reference to reproduction and osmoregulation *J. Exp. Zool.*, 284, pp. 401–406.

Gitterle, T., Rye, M., Salte, R., Cock, J., Johansen, H., Lozano, C., Suárez, J.A., Gjerde, B., 2005a. Genetic (co)variation in harvest body weight and survival in Penaeus (*Litopenaeus*) *vannamei* under standard commercial conditions. *Aquaculture* 243, 83–92.

Gitterle, T., Salte, R., Gjerde, B., Cock, J., Johansen, H., Salazar, M., Lozano, C., Rye, M., 2005b. Genetic (co)variation in resistance to White Spot Syndrome Virus (WSSV) and harvest weight in Penaeus (*Litopenaeus*) *vannamei. Aquaculture* 246, 139–149.

Goyard, E., Patrois, J., Peignon, J., Vanaa, V., Dufour, R., Viallon, J., Bedier, E., 2002. Selection for better growth of *Penaeus stylirostris* in Tahiti and New Caledonia. *Aquaculture* 204, 461–468.

Gomes, L.A.O. and J.H. Primavera 1994. Reproductive quality of male *Penaeus monodon*. *Aquaculture* 112: 157-164.

Gomez, R. 1965. Acceleration of development of brain in the crab *Paratelphusa hydrodromous*. *Naturewise*, 52: 217-219.

Griffond, B. and L. Gomot,1979. Ultrastructural study of the follicle cells in the freshwater gastropod *Viviparus viviparous*. *Cell Tiss. Res.*, 202: 25-32.

Gupta, P.S.P., Rao, L.H., 2000. Effect of captive rearing on sperm quality in *Penaeus monodon*. *Indian Journal of Animal Science* 70,7:777–779.

Gynanath, G., and R. Sarojini 1985. Correlation between neurosecretory activity and annual reproductive cycle of the freshwater prawn, *Macrobrachium lamerri*. *Proc. First. Nat. Symp. On Endocrinology of Invertebrates*, Aurangabad, India, 46-50.

Haefner, P.A. 1977. Aspects of the biology of the Jonah crab, *Cancer borealis* in the mid Atlantic Bight. *J. Nat. Hist.*, 11: 303-320.

Haley, S.R. 1984. Spermatogenesis and spermatophore production in the Hawaiian red lobster *Enoplometopas oxidentalis*. *J. Morphol.*, 180: 181-193.

Hathairat Kruevaisayawan, Rapeepun Vanichviriyakit, Wattana Weerachatyanukul, Boonsirm Withyachumnarnkul, Jittipan Chavadej , Prasert Sobhon, 2010. Oogenesis and formation of cortical rods in the black tiger shrimp, *Penaeus monodon*. *Aquaculture*. 301: 1–4.

Heitzmann, J.C., A Diter and Aquacop 1993. Spermatophore formation in the white shrimp, *Penaeus vannamei* dependence on the inter moult cycle. *Aquacutlure*, 116: 91-98.

Hill, b. J. 1975. Abundance, breeding and growth of the crab *Scylla serrata* in two South African estuaries. *Mar. Biol.*, 32: 119-126

Hinsch, G.W and M.H. Walker 1974. The vas deferens of the spider crab, *Libinia emarginata*. *J. Morpho.*, 143: 1-120.

Hinsch, G.W. and D.C. Bennett 1979. Vitellogenesis stimulated by thoracic ganglion implants into destalked immature spider crabs, *Libinia emarginata*. *Tissue & Cell*, 11: 345-351.

Hinsch, G.W. 1980. Spermeogenesis in a hermit crab, *Coenobita clypeatus* II. Sertoli cells. *Tissue & cell*, 12,2: 255-262.

Hinsch, G. W. and V. Cone 1969. Ultrastructural observations of vitellogenesis in the spider crab *Libiana emarginata. J. Cell. Biol.*, 40 b: 336-342.

Hinsch, G.W. 1969. Microtubules in the sperm of the spider crab *Libiana emarginata. J.ultrastru. Res.*, 29: 525-534.

Huberman A. 2000 Shrimp endocrinology. A review. *Aquaculture*, 191: 191-208.

Hudinaga, M. 1942. Reproduction, development and rearing of *Penaeus japonicus. Jap. J. Zool. Sci.*, 1: 463-470.

Huq. A. 1980. Reproductive system of 6 species of *Penaeus* (Decapoda: Penaeidae) *Bangladesh. J. Zool.*, 8: 81-88.

Ibarra, A.M., Racotta, I.S., Arcos, F.G., Palacios, E., 2007. Progress on the genetics of reproductive performance in penaeid shrimp. *Aquaculture* 268, 23–43.

Iamsaard Sitthichai , Siriporn Sriurairatana and Boonsirm Withyachumnarnkul, 2012. Changes in oviduct structure in the black tiger shrimp, *Penaeus monodon*, during ovarian maturation. *J Zhejiang Univ Sci* B. 13(10): 846–850.

Ishida, T., P. Talbot and Kooda-Cisco 1986. Technique for the long term storage of lobster spermatophores. *Gamete Res.*, 14: 183-195.

Janine Cuzin-Roudy, J. and Margaret O' Lkeary Amster 1991. Ovarian development and sexual maturity staging in Antartic krill *Euphausia superba* Dana (Euphasiacea). *J. Crust. Biol.*, 11,2: 236-249.

Jayalectumie, C. and T. Subramoniam 1989. Cryopreservation of the spermatophores and seminal plasma of the edible crab *Scylla serrata. Bio. Bull.*, 177: 247-253.

Jesperson, a. 1983. Spermatogenesis in *Anaspides tasmaniae. Acta. Zool.*, 64,1: 39-46.

Jiang, S.-G., Huang, J.-H., Zhou, F.-L., Chen, X., Yang, Q.-B., Wen, W.-G., Ma, Z.-M., 2009. Observations of reproductive development and maturation of male Penaeus monodon reared in tidal and earthen ponds. *Aquaculture* 292, 121–128.

Joshi, P.K. 1980. Reproductive physiology and neurosecretion in some Indian marine prawns. Ph.D. Thesis, Marathwada University, Aurangabad, India.

Joshi, P.C. and S.S. Khanna 1982. Seasonal changes in the ovary of a freshwater crab, *Potamon koolooen. Proc. Indian. Acad. Sci.*, 91-15: 451-462.

Joshi, V.P. and A.D. Diwan 1992. Artificial insemination studies *Macrobrachium idella*. In: Silas E.G. (Ed). Freshwater prawns. Kerala Agricultural University. Trichur. Pp. 110-118.

Junera, H., C. Zerbib, M. Martin and J.J. Meusy 1977. Evidence for control of vitellogenin synthesis by an ovarian hormone in *Orchestia gammarella* (Pallas). *Gen. Comp. Endocrinol.*, 61: 248-249.

Karoonuthaisiri N, Sittikankeaw K, Preechaphol R, Kalachikov S, Wongsurawat T, Uawisetwathana U, Russo JJ, Ju J, Klinbunga S, Kirtikara K. ReproArray (GTS) 2009. A cDNA microarray for identification of reproduction-related genes in the giant tiger shrimp *Penaeus monodon* and characterization of a novel nuclear autoantigenic sperm protein (NASP) gene. *Comp Biochem Physiol Part D Genomics Proteomics*. 4,2: 90-9.

Kennedy, F.S. and D.G. Barber 1981. Spawning and recruitment of pink, shrimp *Penaeus duorarum* off Eastern Florida. *J. Crust. Biol.*, 1:474-485.

Kerr, M.S. 1969. The haemolymph proteins of the blue crab, *Callinectes sapidus* II. A lipoprotein serologicaly identical to oocyte lipovitellin. *Dev. Biol.*, 20: 1-17

Kessel, R.G. 1983. The structure and function of annulate lamellae: Porous cytoplasmic and intra-nuclear membranes. *Int. Rev. Cytol.*, 82: 181-303.

Khayat M, Lubzens E, Tietz A, Funkenstein B, 1994. Are vitellin and vitellogenin coded by one gene in the marine shrimp *Penaeus semisulcatus*? *J Mol Endocrinol*, 12:251-254.

Khayat M, Lubzens E, Tietz A, Funkenstein B, 1994. Cell-free synthesis of vitellin in the shrimp *Penaeus semisulcatus* (de Haan). *Gen Comp Endocrinol*, 93:205-213.

Khayat, M., Yang, W.-J., Aida, K., Nagasawa, H., Tietz, A., Funkenstein, B. and Lubzens, E., 1998. Hyperglycemic hormones inhibit protein and mRNA synthesis in in vitro- incubated ovarian fragments of the marine shrimp *Penaeus semisulcatus*. *Gen. Comp. Endocrinol.* 110, 307–318.

Kim YK, Tsutui N, Kawazoe I, Okamura T, Kaneko T, Aida K, 2005. Locazisation and developmental expression of mRNA for cortical rod protein in kuruma prawn *Marsupenaeus japonicus*. *Zool Sci* 22:675-680

King, J.E. 1948. A study of the reproductive organs of the common marine shrimp, *Penaeus setiferus. Biol. Bull.*, 94: 244-262.

Kleve, M.G., A.I. Yudin and W.H. Clark. Jr. 1980. Fine structure of the unistellate sperm of the shrimp, *Sicyonia ingentis. Tissue and Cell*, 12,1: 29-45.

Kooda-Cisco, M.J. and P. Talbot 1982. A structural analysis of the freshly extruded spermatophore from the lobster, *Homarus americanus. J. Morphol.*, 172: 193-207.

Kooda-Cisco, M.J. and P. Talbot 1986. Ultrastructure and role of the lobster vas deferens in spermatophore formation. The proximal segment. *J. Morphol.*, 188: 91-103.

Koebler, L.D. 1979. Unique case of cytodifferentiation, spermatogenesis in prawn, *Palaemonetes paludoses. J. Ultrastru. Res.*, 69: 109-120.

Komatsu M, Andi S, 1998. A very high density lipo-protein with clotting ability from hemolymph of sand crayfish, *Ibcacus ciliates. Biosci Biotechnol Biochem* 62:459-463.

Komm, B.S. and G.W. Hinch 1985. Oogenesis in the terrestrial hermit crab, *Coenobita clypeatus* I. Pre vitellogenic oocytes. *J. Morphol.*, 183: 219-224.

Komm, B.S. and G.W.Hinch 1987. Oogenesis in the terrestrial hermit crab *Coenobita clypeatus.* II. Vitellogenesis. *J. Morphol.*, 192: 269-277.

Koskela, R.W., J.G. Greewood and P.C. Rothlisherg, 1992. The influence of Prostaglandin E2, and the steroid hormones, 17α-hydroxyprogesterone and 17β-Estradiol in moulting and ovarian development in the tiger prawn *Penaeus esculentus* Haselll-1879. *Comp. Biochem. Physiol.*, 101A, 2: 295-299.

Kulkarni, G.K.R. Nagabhushanam and P.K. Joshi 1979. Effect of progesterone on ovarian maturation in a marine prawn *Parapenaeopsis hardwickii. Indian J. Exp. Biol.*, 17: 986-987

Kulkarni, G.K. and R. Nagabhushanam 1980. Role of ovary inhibiting hormone from eyestalks of marine penaeid prawns *Parapenaeopsis hardwickii* during ovarian development cycle. *Aquaculture*, 19: 13-19.

Kung SY, Chan SM, Hui JH, Tsang WS, Mak A, He JG, 2004. Vitellogenesis in the sand shrimp, *Metapenaeus ensis*: the contribution from the hepatopancreas- specific vitrellogenin gene (MeVg2). *Biol Reprod* 71:863-870.

Kunwadee Palasina, Walaiporn Makkapana, Tawatchai Thongnoic, Wilaiwan Chotigeat, 2014. Stimulation of ovarian development in white shrimp, *Fenneropenaeus merguiensis* De Man, with a recombinant ribosomal protein L10a. *Aquaculture*. 432, 20: 38–45

Langreth S,1969. Spermiogenesis in the *Cancer* crabs. *J Cell Biol* 43: 575603.

Laufer H., A. Sagi, J.S.B. Ahl, E. Homola, 1992. Methyl farnesoate appears to be a crustacean reproductive hormone Invertebr. *Reprod. Dev.*, 22 , pp. 17–20.

Laufer H, J. Ahl, G. Rotllant, B. Baclaski, 2002. Evidence that ecdysteroids and methyl farnesoate control allometric growth and differentiation in a crustacean. *Insect Biochem. Mol. Biol.*, 32 , pp. 205–210

Lawrence, A.L., D. Ward, S. Missler, A. Brown, J. McVey and B.S. Middleditch 1979. Organ indices and biochemical levels of ova from penaeid shrimp maintained in captivity versus those captured in the wild. *Proc. World Maricult. Soc.*, 10: 453-563.

Lawrence and Castle 1991. Reproductive maturity in penaeid shrimps. *Bull. of the Institute of Zool.*, 29: 43-49.

Lee, R.F. and D.L. Puppione 1988. Lipoproteins I and II from the haemolymph of the blue crab *Callinectes sapidus*, lipoprotein II associated with vitellogenesis. *J. Exp. Zool.*, 284: 278-279.

Legrand, J.J., G. Martin and P. Juchault and G. Besse 1982. Controle neuroendocrine de la reproduction chez les Crustaces. *J. Physiol. Paris.*, 78: 543-552.

Leung-Trujillio, and A.L. Lawrence 1985. The effect of eyestalk ablation on spermatophroe and sperm quality in *Penaeus vannamei*. *J. World. Maricult. Soc.*, 16: 258-266.

Leung-Trujillo and A.L. Lawrence 1987. Observations on decline in sperm quality of *P. setiferus* under laboratory conditions. *Aquaculture*, 65: 363-370.

Lu, C.C., W.H. Clark and L.E. Franklin 1973. Spermatogenesis of the decapod *Penaeus setiferus J. Cell Biol.*, 59: 202-212.

Lu JP, Zhang XH, Yu XY, 2006. Structural changes of oviduct of freshwater shrimp, *Macrobrachium nipponense* (Decapoda, Palaemonidae), during spawning. J Zhejiang Univ-Sci B;7,1:64–69.

Lui, C.W., B.A. Sage and J.D. O' Connor 1974. Biosynthesis of lipovitellin by the crustacean ovary. *J. Exp. Zool.*, 188: 289-296.

Lui, C.W. and J.D. O'Connor 1976. Biosynthesis of lipovitellin by the crustacean ovary. II. Characteristisation of and in vitro incorporation of amino acids into the purified subunits. *J. Exp. Zool.*, 195: 41-52.

Lui, C.W. and J.D. O'Connor 1977. Biosynthesis of crustacean lipovitellin III. The incorporation of labeled amino acids into the purified lipovitellin of the crab *Pachygrapsus crassipes*. *J. Exp. Zool.*, 199: 105-108.

Lubzens E, Ravid T, Khayat M, Daube N, Teiz A, 1997. Isolation and characterization of the high density lipoprotein from the hemolymph and ovary of the penaeid shrimp *Penaeus semisulcatus* (de Hann): apoprioteins and lipids. *J Exp Zool*, 278:339-348.

Lumare, F. 1979. Reproduction of *Penaeus kerathurus* using eyestalk ablation. *Aquaculture*, 18: 203-214.

Lynn, J.W. and W.H. Clark 1983. A morphological examination of sperm egg interaction in the freshwater prawn *Macrobrachium rosenbergii*. *Biol. Bull.* 164: 446-458.

Lytle J.S, T.F. Lytle, J.T, 1990. Ogle Polyunsaturated fatty acid profiles as a comparative tool in assessing maturation diets of *Penaeus vannamei*. *Aquaculture*, 89,287–299

Madhystha, M.N. and P.V. Rangnekar 1976. Neurosecretory cells in the central nervous system of the prawn, *Metapenaeus monoceros Rev. Di. Biol.*, IXXIX: 133-140.

Malek, S.R.A. and J.M. Bawab 1974 a. The formation of the spermatophore in *Penaeus kerathurus* I. Initial formation of a sperm mass. *Crustaceana* 27(1): 73-83.

Malek, S.R.A. and J.M. Bawab 1974 b. The formation of spermatophore in *Penaeus kerathurus* II. Deposition of the main layers of the body of the wings. *Crustaceana*, 27.

Menasveta, P., S.P. Vorakul, S. Rungasupa, N. More and A.W. Fast 1993. Gonadal maturation and reproductive performance of giant tiger prawn *Penaeus monodon* from Andaman Sea and pond reared sources in Thailand. *Aquaculture*, 116: 191-198.

Mathews, D.C. 1954. The development of spermatophoric mass of the rock lobster, *Perribacus antorficus*. *Pacif. Sci.*, 8,1: 28-34.

Mc Croan, J.E. 1940. Spermatogenesis of the cray fish *Cambarus virilis*, with special reference to the Golgi material and mitochondria. *Cytologia.*, 11: 136-142.

Mc Knight, G.E. and G.W. Hinch 1986. Sperm maturation and ultrastructure in *Scyllarus chacel. Tissue and Cell.*, 18,2: 257-266.

Mekuchi M, Ohira T, Kawazoe I, Jasmani S, Suitoh K, Kim YK, Jayasankar V, Nagasawa H, Wilder MN, 2008. Characterization and expression of putative ovarian lipoprotein receptor in the kuruma prawn, *Marsupenaeus japonicus.* Zool Sci 25:428-437.

Menasveta, P., S.P. Vorakul, S. Rungasupa, N. More and A.W. Fast 1993. Gonadal maturation and reproductive performance of giant tiger prawn *Penaeus monodon* from Andaman Sea and pond reared sources in Thailand. *Aquaculture*, 116: 191-198.

Meusy, J.J. and G.G. Payen 1988. Female reproduction in Malacostracan Crustacea. *Zool. Soc.*, 5: 217-265.

Meunpol O., E. Duangjai, R. Yoonpun, 2005. Determination of Prostaglandin E2 (PGE2) in polychaetes (*Perinereis* sp.) and its effect on *Penaeus monodon* oocyte development in vitro Proceedings of Larvi'05-Fish & Shellfish Larviculture Symposium European *Aquaculture* Society, Special Publication, Belgium , p. 6.

Meunpola O, Saowaluck Iam-Paic, Wanvipa Suthikraid, Somkiat Piyatiratitivorakulb, 2007. Identification of progesterone and 17α-hydroxyprogesterone in polychaetes (*Perinereis* sp.) and the effects of hormone extracts on penaeid oocyte development in vitro. *Aquaculture.* 270, 1–4, 28.

Mohamed, K.S. 1989. Studies on the reproductive endocrinology of the penaeid prawn *Penaeus indicus* H.Milne Edwards. Ph.D. thesis, Cochin Univ. Sci. Tech., 252 p.

Mohammed, S.K. and A.D. Diwan 1991. Effect of androgenic gland ablation on sexual characters of the male Indian white prawn. *Penaeus indicus* H. Milne Edwards., *Indian J. Exp. Biol.*, 29: 478-480.

Mohammed, S.K., K.K. Vijayan and A.D. Diwan 1993. Histomorphology of the neurosecretory system in the Indian white prawn *Penaeus indicus.* H. Milne. Edwards. *Bull of the Institute of Zool.*, 32(1): 39-53.

Mohammed, S.K. and A.D. Diwan 1994. Spermatogenesis and spermatophore formation in Indian white prawn *Penaeus indicus. J. Mar. Biol. Ass. India.*, 33,1-2: 180-192.

Mohammed, S.K. and A.D. Diwan 1994. Vitellogenesis in the Indian white prawn *Penaeus indicus. J. Aquacult. Trop.*, 9: 157-172

Moses, M.J. 1958. A flagellate spermeogenesis in the cray fish. Anat. Res., 130: 343-351. Moses, M.J. 1961 a. Spermiogenesis in the cray fish Procambarus clarkii. I. Structural characterization of the mature sperm. *J. Biophysic. Biochem. Cytol.*, 9: 222-228.

Moses, M.J. 1961 b. Spermiogenesis in the cray fish *Procambarus clarkii*. 2. Description of stages. *J. Biophysio. Biochem. Cytol.*, 10: 301-333.

Motoh, H. 1978. Preliminary histological study of the ovarian development of the giant tiger prawn, *Penaeus monodon. Quart. Res. Rep.*, 2,4: 4-6

Motoh, H. and P. Buri 1980. Development of external genetalia of the giant prawn, *Penaeus monodon. Bull. Jap. Soc. Sci. Fish.*, 40,2: 149-155.

Motoh, H., 1981. Studies on the fishery biology of the giant tiger prawn, *Penaeus monodon* in the Philippines. Technical Report No.7. *Aquaculture* Department, SEAFDEC, Philippines, pp.128.

Motoh, H. 1985. Biology and Ecology of *Penaeus monodon*. Proc. First Internatl. Confr. Cult. Of Penaeid prawns/shrimps pp. 27-36.

Muthu, M.S., A. Laxminarayana and K.H. Mohammed, 1979. Induced maturation and spawning of *Penaeus indicus* without eyestalk ablation. *Mar. Fish. Inform. Serv. T & R* No. 9: 6

Nakamura, K. 1974. Studies on the neurosecretion of the prawn *Penaeus japonicus* I. Positional relationship of the cells group located on the supra oesophageal and optic ganglion. *Mem. Fac. Fish. Kagoshina Univ.*, 23: 175-184.

Nagabhushanam, R., P.K. Joshi and G.K. Kulkarni 1980. Induced spawning in prawn *Parapenaeopsis stylifera* using a steroid hormone 17Hydroxy-Progesterone. Ind. *J. Mar. Sci.*, 227

Nagabhudshnam, R., P.K. Joshi and G.K. Kulkarni 1982. Induced spawning in *Parapenaeopsis stylifera* using a steroid hormone 17 hydroxy-progesteron. *Proc. Symp. Coastal Aquaculture*, 1: 37-39.

Nagabhushanam R., R. Sarojini and S. Sambasiva 1992. Reproductive endocrinology of the Indian Marine shrimp *Metapenaeus affinis*. In: *Aquaculture* research needs for 200 A.D. pp. 163-180.

Nanda, D.K. and P.K. Ghosh 1985. The eyestalk neurosecretory system in the brackishwater prawn, *Penaeus monodon*, A light microscopical study. *J. Zool. Soc. India.*, 37: 25-38.

Nath, V. 1937. Spermatogenesis of the prawn *Palaemon lamarrei*. *J. morphol.*, 61: 149-153.

Nath, V. 1965. Animal gametes (Male). A morphological and cytochemical account of spermatogenesis. *Asia Publishing House, Bombay*. 78-93.

Ngernsoungnern, P., Ngernsoungnern, A., Kavanaugh, S., Sobhon, P., Sower, S.A. and Sretarugsa, P., 2008. The presence and distribution of gonadotropin-releasing hormone-liked factor in the central nervous system of the black tiger shrimp, Penaeus monodon. *General and comparative endocrinology*, 155,3:613-622.

Nguyen Binh Thanh, Shunsuke Koshiob, Kazutaka Sakiyama, Manabu Ishikawa, Saichiro Yokoyama, Md. Abdul Kader, 2012. Effects of polychaete extracts on reproductive performance of kuruma shrimp, *Marsupenaeus japonicus* Bate. – Part II. Ovarian maturation and tissue lipid compositions.*Aquaculture* 334–337: 65.

O'Connor, C.Y. 1979. Reproductive periodicity of *Penaeus esculentus* population near low inlets, Queensland, Australia, *Aquaqculture*, 16: 153-162.

Okumura T, Kim YK, Kawazoe I, Yamano K, Tsutsui N, Aida K, 2006. Expression of vitellogenin and cortical rod proteins during induced ovarian development by eyestalk ablation in the kuruma prawn, *Marsupenaeus japonicus. Comp Biochem Physiol* 143A:246-253.

Otsu, T. 1960. Precoucious development of the ovaries in the crab, *Potemont dehani*, following implantation of the thoracic ganglion. *Annot. Zool. Jpn.*, 33: 90-96.

Okamura, T., C.H., Han, Y. Suzuki, K. Aida and I. Hanyu, 1992. Changes in hemolymph vitellogenin and ecdysteriod levels during the reproductive and non-reproductive molt cycles in the freshwater prawn *Macrobrachium nipponense. Zool. Sci.*, 9: 37-45.

Oyama, S.N. 1968. Neuroendocrine effect on ovarian development in the crab *Thalamita crenata* Latreille studied *in vitro*. Ph.D. Diss. Univ. Hawaii, Honolulu, Hawaii.

Oyama, Y. and S. Kakuda 1987. Scanning electronmicroscopic observations on the spermatozoa of the prawn *P. japonicus. Nippon Suisan Gakkaishi*, 53: 975-977.

Paulus, J.E. and H. Laufer 1987. Vitellogenocytes in the hepatopancreas of *Carcinus maenas. Inter J. Invertbr. Rep. & Dev.*, 11: 29-44.

Payen, G.G. 1986. Endocrine regulation of male and female genital activity in crustaceans. A retrospect and perspectives. In: *Advances in Invertebr. Rep.*, 4: 125-134.

Penn, J.W. 1980. Spawning and fecundity of the western King prawn, *Penaeus latisulcatus* Kishinouye, in Western Australian waters. *Aust. J. Mar. Fresh.*

Perryman, E.K. 1969. Procambarus simulans – Light induced change in the neurosecretory cells and in ovarian cycle, *Trans. Am. Microsco. Soc.*, 88: 514-524.

Peixoto, S., Cavalli, R.O., D'Incao, F., Wasielesky, W., Aguado, N., 2002. Description of reproductive performance and ovarian maturation of wild Farfantepenaeus paulensis from shallow waters in southern Brazil. Nauplius 10, 149–153.

Peixoto, S.,Wasielesky,W., D'Incao, F., Cavalli, R.O., 2003. Reproductive performance of similarly-sized wild and captive Farfantepenaeus paulensis. J. World Aquac. Soc. 34, 50–56.

Peixoto, S., Cavalli, R.O.,Wasielesky,W., D'Incao, F., Krummenauer, D., Milach, Â., 2004. Effects of age and size on reproductive performance of captive *Farfantepenaeus paulensis* broodstock. *Aquaculture* 238, 173–182.

Peixoto, S., Coman, G.J., Arnold, S.J., Crocos, P.J. and Preston, N.P., 2005. Histological examination of final oocyte maturation and atresia in wild and domesticated *Penaeus monodon* broodstock. *Aquaculture Research* 36: 666–673.

Peixoto S, Wilson Wasielesky Jr., Ricardo C. Martino, Ângela Milach, Roberta Soares Ronaldo O. Cavalli, 2008. Comparison of reproductive output, offspring quality, ovarian histology and fatty acid composition between similarly-sized wild and domesticated *Farfantepenaeus paulensis.*

PeixotoS, Wilson Wasielesky, Ronaldo O. Cavalli, 2011. Broodstock maturation and reproduction of the indigenous pink shrimp *Farfantepenaeus paulensis* in Brazil: An updated review on research and development.

Perez-Velazquez, M., Bray, W.A., Lawrence, A.L., Gatlin III, D.M. and Gonzalez-Felix, M.L., 2001. Effect of temperature on sperm quality of captive *Litopenaeus vannamei* broodstock. *Aquaculture*, 198: 209–218.

Phiriyangkul P, Puengyam P, Jakobsen IB, Utarabhand P, 2007. Dynamics of vitellogenin mRNA expression during vitellogenesis in the banana shrimp Penaeus (*Fenneropenaeus merguiensis*) using real-time PCR. *Mol Reprod Dev* 74:1198-1207.

Pillai, K.K. and N.B. Nair 1970. The reproductive cycle of three decapod crustaceans from the South West coast of India. *Curr. Sci.*, 400: 161-162.

Pillai, K.K. and N.B. Nair 1971. The annual reproductive cycle of *Uca annulipes, Portunus pelagicus* and *Metapenaeus affinis* (Decapoda: Crustacea) from the South West coast of *India. Mar. Biol.*, 11: 152-166.

Pillai, R.S. 1960. Studies on the shrimp *Cardina laevis* (Heller). 11. The reproductive system. *J. Mar. Biol. Ass. India*, 2: 226-236.

Pochon-Masson, J. 1965 b. Schema general du spermatozoid vesiculaire des Decapods.*C.R. Acad. Sci. Paris*, 260: 5093–5103.

Pochon-Masson, J. 1968 a. L'ultrastructure des spermatozoids vesiculares chez les crustaces decapods avant et au cours de leur devagination experimentale. I. Brachyoures et anomeures. *Ann. Sci. Nat. Zool. Biol. Anio. Ser.*, 12: 10: 1-100.

Pochon-Masson, J. 1968 b. L'ultrastructure des spermatozoids vesicularis chez les crustaces decapods avant et au cours de leur devagination experimentale. II. Macroures. Disussion et conclusions. *Ann. Sci. Nat. Zool. Biol. Anio. Ser.*, 12: 10: 367-372.

Pochon-Masson, J. 1969. Ultrastructure du sperm atozoide de *Palaemon elagans* (de Man) Crustacea (decapoda). *Arch. Zool. Exp. Gen.*, 110: 363-372.

Pochon-Masson, J. 1983. Arthropoda-crustacea. In: "Reproductive biology of Invertebrates", Vol. II. Spermatogenesis and sperm function. (K.G. Adiyodi and R.G. Adiyodi (eds.). 407449. Wiley, Chichester, England.

Pochon-Masson, J. 1983. Arthropoda-crustacea. In: "Reproductive biology of Invertebrates", Vol. II. Spermatogenesis and sperm function. (K.G. Adiyodi and R.G. Adiyodi (eds.). 407449. Wiley, Chichester, England.

Poltana P, 2005. Development of the polychaete *Perinereis nuntia brevicirrus* and its prostaglandin F2 alpha content in the atokous stage 10th International Congress on Invertebrate Reproduction and Development 18–23 July 2004, Newcastle upon Tyne, UK Abstract. 8 pp.

Pongtippatee, P., Vanichviriyakit, R., Chavadej, J., Plodpai, P., Pratoomchart, B., Sobhon, P., Withyachumnarnkul, B., 2007. Acrosome reaction in the sperm of the black tiger shrimp *Penaeus monodon* (Decapoda, Penaidae). *Aquaculture* Research 38, 1635–1644.

Pratoomchat, B.S., Piyatiratitivorakul and P. Menasveta 1993. Sperm quality of pond reared and wild-caught *Penaeus monodon* in Thailand. *J. World Aquacult. Soc.*, 24(4): 530-540.

Preechaphol R, Rungnapa L, Kanchana S, Sirawut K, Bavornlak K, Narongsak P and Piamsak M, 2007. Expressed Sequence Tag Analysis for Identification and Characterization of Sex-Related Genes in the Giant Tiger Shrimp *Penaeus monodon. Journal of Biochemistry and Molecular Biology,* 40:4.

Preechaphol R, Sirawut K, Pattareeya P, Piamsak M, 2010. Isolation and characterization of progesterone receptor - related protein p23 (Pm-p23) differentially expressed during ovarian development of the giant tiger shrimp *Penaeus monodon. Aquaculture.* 308.

Primavera, J.H. 1980. Broodstock of sugpo (*P. monodon*) and other penaeid prawns. Extension manual, No. 7. Tigbuan, Iloilo: SEAFDEC *Aquaculture* department. 24 p.

Primavera, J.H. 1982. Studies on broodstock of sugpo *(P. monodon*) and other penaeids at the seafdec *Aquaculture* department. *Proc. Symp. Coastal Aquaculture*, 1: 28-36.

Primavera, J.H. 1985. A review of maturation and reproduction in closed thelycum penaeids. In: Y. Taki, J.H. Primavera and J.A. Lloberera (eds.) *Proc. First Internat. Cont. on the Culture of Penaeid Prawns/Shrimps*, SEAFDEC City, Philippines, 47-64.

Quackenbush, L.S. and L.L. Keeley 1988. Regulation of vitellogenesis in the Fiddler crab *Uca pugilator. Biol. Bull.*, 175: 321-331.

Quackenbush, L.S. 1989. Yolk protein production in the marine shrimp *P. Vannaemi. J. Crustacea Biol.* 9,4: 509-516.

Quackenbush, L.S. 1989 b. Vitellogenesis in the shrimp, *Penaeus vannamei* In vitro studies of the isolated hepatopancreas and ovary. *Comp. Biochem. Physiol.*, 94 B: 253-261.

Quackenbush, L.S. and W.F. Herrkind 1981. Regulation of moult and gonadial development in the spiny lobster, *Panulirus argus. J. Crust. Biol.*, 3: 34-44

Quackenbush, L.S. 1991. Regulation of vitellogenesis in penaeid shrimp. In: Frontiers in Shrimp Research, P.F. Dehoach, W.J. Dougherty and M.A. Davidson (eds.), *Elsevier*, Amsterdam, 1991, pp. 125-140.

Qunitio, E.T., A. Hara, K., Yamauchi, T. Mizushime and A. Fuji 1989. Identification and characterization of vitellin in a hermaphrodite shrimp, *Pandalus kessleri. Comp. Biochem. Physiol.*, 94 B,3: 445.

Qunitio, E.T., A. Hara, K. Yamauchi and A. Fugi 1990. Isolation and characterization of vitellin from the ovary of *Penaeus monodon. Invertbr. Reprod. And Dev.* 17,3: 221-227.

Quinitio, E.T., R.M. Cabellero and L. Gustilo 1993. Ovarian development in relation to changes in the external genitalia in captive *P. monodon*. *Aquaculture*, 114: 71-81.

Quinn and K.J. Barbara 1987. Reproductive biology of *Scylla* spp. (Crustacea; Portunidae) from the Labu estuary in Papua new Guinea. *Bull. Mar. Sci.*, 41,2: 234-241.

Rangneker, P.V. and Deshmukh 1968. Effect of eyestalk removal on the ovarian growth of the marine crab, *Scylla serrata*. *J. Anim. Morph. Phsio.*, 15: 503-511.

Rankin, S.M., J.Y. Bradfield and L.L. Keeley 1989. Ovarian protein synthesis in the South American Shrimp, *Penaeus vannamei* during the reproductive cycle. *Invertbr. Reprod. Dev.* 15: 27-33.

Rao, P.V. 1968. Maturation and spawning of the penaeid prawns of South West coast of India. *FAO Fish. Rep.*, 57: 285-302.

Rao, P.V. 1969. Genus *Parapenaeopsis*. *CMFRI Bulletin No.* 14: 127-158.

Rao, Ch, N.K., K. Shakunthala and S.R. Reddy 1981. Studies on the neurosecretion of the thoracic ganglion in relation to reproduction in female, *Macrobrachium lanchesteri* (de Man), *Proc. Indian. Acad. Sci. (Anim. Sci.)* 90: 503-511.

Rao, G. Sudhakara, 1990. An assessment of the penaeid prawn seed resources of the Godavari estuary and the adjacent backwaters. *Indian J. Fish.*, 37: 99-108.

Ratenu, J.G. and C. Zerbib 1978. Etude ultrastructurale des follicules ovocytaires chezle crustace Amphipode, *Orchestia gammarellus Cr. Acad. Sci. Parid.*, 286: 65-68.

Renard, P. 1992. Cooling and freezing tolerances in embryos of the pacific oyster *Crassostrea gigas* methanol and sucrose effects. *Aquaculture*, 92: 43-57.

Ro, S., P. Talbot, J. Leung-Trujillo, A.L. Lawrence 1990. Structure and function of vas deferens in the shrimp *Penaeus setiferus* Segments 1-3. *J. Crust. Biol.*, 10,3: 455-468.

Ryan, E.P. 1967. Structure and function of reproductive system of the crab *Portunus sanguinolentus* (Herbst). The male system. *Proc. Symp. Crustacea, Ser. III*, Part 2: 506-521.

Sagi, A., Y. Milner and D. Cohen 1988. Spermatogenesis and sperm storage in the tests of the behaviorally distinctive male morphotypes of *M. rosenbergii*. *Biol. Bull.*, 174: 330-336.

Sagi, A., Y. Soroka, E. Snir, O. Chomsky, J. Calderon and Y. Molner 1995. Ovarian protein synthesis in the prawn *Macrobrachium rosenbergii*: Does ovarian vitellin synthesis exist?. *Invertebr. Rep. Dev.*, 27,1: 41-47.

Santiago, A.C., Jr. 1977. Successful spawning of cultured *P. monodon* after eyestalk ablation. *Aquaculture*, 11,1: 185-196.

Sawada, Y. and M.C. Chang 1964. Tolerance of honey bee sperm to deep freezing. *J. Econ. Entomol.*, 57: 891-892.

Shafir, S., M. Ovadia and M. Tom. 1992 a. *In vivo* incorporation of labeled methionine into proteins vitellogenin and vitelline in females of the penaeid shrimp *Penaeus semisulcatus. Biol. Bull.*, 183: 242-247.

Sheikhmahmud, F.S. and U.B. Tambe 1958. Study of prawns: The reproductive organs of *Parapenaeopsis stylifera* (M. Edw.). *J. Univ. Bom.*, 23: 99-110.

Shigekawa, K. and W.H. Clark 1986. Spermatogenesis in the marine shrimp, *Sicyonia ingenitis. Dev. Gro. Dif.*, 28: 95-112.

Shoji Joseph, 1997. Some studies on the reproductive endocrinology of tiger prawn *P. monodon* Fabricius, Ph. D Thesis, Cochin University of Science and Technology.

Shigekawa, K. and W.H. Clark 1986. Spermatogenesis in the marine shrimp, *Sicyonia ingenitis. Dev. Gro. Dif.*, 28: 95-112.

Shirley H.K. Tiu, Jerome H.L. Hui, Abby S.C. Mak, Jian-Guo He, Siu-Ming Chan, 2006. Equal contribution of hepatopancreas and ovary to the production of vitellogenin (PmVg1) transcripts in the tiger shrimp, *Penaeus monodon. Aquaculture.* 254: 1–4.

Schoenmakers, H. J. N., CH. G. Van Bohemen and S. J. Dieleman. 1981. Effects of the Oestradiol-17β on the ovaries of the starfish Asterias rubens. *Develop. Growth Differ.* 23,2:125-135.

Solis, N.B. 1988. Biology and ecology. In: Biology and culture of *Penaeus monodon*. Publ. *Seafdec Aquacult. Dept.*, 1-36.

Subramoniam, C.B. 1965. On the reproductive cycle cflf *Penaeus indicus. J. Mar. Biol. Ass. India*, 7(2): 284-290.

Subramoniam T, 2011. Mechanisms and control of vitellogenesis in crustaceans. Fisheries Science 77: 1–21.

Susan, M.R.. J.Y. Bradfield and L.I. Keeley 1993. Ovarian development in the South American white shrimp, Penaeus vannamei. NOAA, Techn. Rep. N.M.F.S., 106: 27-33.

Swetha C H, Sainath S B, Reddy P R,ReddyP S, 2011. Reproductive Endocrinology of Female Crustaceans: Perspective and Prospective. *J Marine Sci Res Development* S3:001.

Tahara D, Katsuyoshi S, Hisatake H, 2005. Hemolymph vitellogenin levels during final maturation and post- spawning in the female kuruma prawn, *Marsupenaeus japonicus*. *Aquaculture*. 245:1–4.

Tan-Fermin, J.D. and R.A. Pudadera 1989. Ovarian maturation stages of the wild giant tiger prawn *Penaeus monodon*. *Aquaculture*, **77**: 229-242.

Takayanagi, H., Y. Yamamoto and N. Takeda 1986. An ovary stimulating factor in the shrimp *Paratya compressa*. *J. Exp. Zool.*, 240: 203-209

Tartakof, a. and P. Vassalli 1978. Comparative studies of intercellular transport of secretory proteins. *J. Cell. Biol.*, 79: 694-707

Talbot, P., R.G. Summers, B.L. Hylander, E.M. Deough and L.E. Franklin 1976. Role of calcium in the acrosome reaction: An analysis using Ionophore A23187. *J. Exp. Zool.*, 198(3): 383-392.

Talbot, P. and P. Chanmanon 1980. Morphological features of the acrosome reaction of the lobster sperm and the role of the reaction in generating forward sperm movement. *J. Ultrastru. Res.*, 70: 287-297.

Talbot, P. and E. Beach 1989. Role of the vas deference in the formation of the spermatophore of the crayfish. Cherax. (1989). *J. Crust. Biol.*, 9,10: 9-24.

Talbot, P., D. Howard, J. Leung-Trunillo, T.W. Lee, W.Y. LI.H.R.D. and A.L. Lawrence, 1989. Characterisation of the male reproductive tract degenerative syndrome in captive penaeid shrimp (*Penaeus setiferus*). *Aquaculture*, 78: 365-377.

Talbot P, Helluy S. Reproduction and Embryonic Development, 1995. In: Factor JR, editor. Biology of the Lobster *Homarus american*. NY: Academic Press. 177–180.

Thomas, M.M. 1974. Reproduction, Fecundity and sex ratio of the green tiger prawn *Penaeus semisulcatus*. *Indian J. Fish.* 21: 151-163.

Ting Sze Lo, Zhaoxia Cui, Janice L.Y. Mong, Queenie W.L. Wong, Siu-Ming Chan, Hoi Shan Kwan, Ka Hou Chu, 2007. Molecular Coordinated Regulation of Gene Expression during Ovarian Development in the Penaeid Shrimp. Marine Biotechnology. Volume 9, Issue 4, pp 459-468.

Tinikul Y, Jaruwan P, Ruchanok T, Panat A, Charoonroj C, Nipon S, Tanes P, Peter H, Prasert S, 2014. Effects of gonadotropin-releasing hormones and dopamine on bovarian maturation in the Pacific white shrimp, *Litopenaeus vannamei*, and their presence in the ovary during ovarian development. *Aquaculture*. 420–421:15.

Tirmizi, N.M. 1958. A study of the developmental stages of the thelycum and its relation to the spermatophore in the prawn *Penaeus japonicus*. *Proc. Zool. Soc.*, 131: 231-244

Tirmizi N.M. and B. Khan 1970. "A handbook on Pakistani Marine prawn". University of Karachi, Karachi.

Tirmizi N.M. and Javed, W. 1976. Study of juvenile of *Metapenaeus stebbing Nobili* (Decapoda, Penaeidae) with particular reference to the structure and development of the genitalia. *Crustaceana*, 30: 55-67.

Tiu SH, Hui JH, Mak AS, He JG, Chan SM, 2006. Equal contribution of hepatopancreas and ovary to the production of vitelogenin (PmVg1) transcripts in the tiger shrimp, *Penaeus monodon*. *Aquaculture* 254: 666-674.

Tiu SH, Benzie J, Chan SM 2008. From hepatopancreas to ovary: molecular characterization of a shrimp vitellogenin receptor involved in the processing of vitellogenin. *Biol Reprod* 79:66-74.

Tom, M., M. Goren and M. Ovaldia 1987. Localization of the vitelline and its possible precursors in various organs of *Parapenaeus longirostris*. *Inter. J. Invert. Rep. Dev.*, 12: 1-12.

Tom M, M. Fingerman, T.K. Hayes, V. Johnson, B. Kerner, E. Lubzens, 1992. A comparative study of the ovarian proteins from two penaeid shrimps, *Penaeus semisulcatus* de Hann and *Penaeus vannamei* (Boone) *Comp. Biochem. Physiol.*, 102 B, pp. 483–490.

Treerattrakool S, Panyim S, Chan SM, Withyachumnarnkul B, Udomkit A, 2008. Molecular characterization of gonad-inhibiting hormone of *Penaeus monodon* and elucidation of its inhibitory role in vitellogenin expression by RNA interference. *FEBS J* 275:970-980.

Trentini, M. and Scanabissi, F. S. 1978. Ul-trastructural Observations on the Oogenesis of *Triops cancriformis* (Crustacea, Noto-straca), I. Origin and Differentiation of Nurse Cells. *Cell Tiss. Res.*, 194: 71-77

Tsutsui N, Kawazoe I, Ohira T, Jasmani S, Yang W-J, Wilder MN, Aida K 2000. Molecular characterization of a cDNA encoding vitellogenin and its expression in the kuruma prawn, *Penaeus japonicus*. *Zool Sci* 17:651-660.

Tsutsui N, Katayama H, Ohira T, Nagasawa H, Wilder MN, Aida K 2005. The effects of crustacean hyperglycemic hormone- family peptides on vitelogenin gene expression in the kuruma prawn, Marsupenaeus japonicus. *Gen Comp Endocrinol* 144:232-239.

Tsutsui N, Ohira T, Kawazoe I, Takahashi A, Wilder MN, 2007. Purification of sinus gland peptides having vitellogenesis-inhibiting activity from the whiteleg shrimp *Litopenaeus vannamei. Mar Biotech* 9:360-369

Tsukimura, B. and F.I. Kamemoto 1988. Organ culture assay of the effects of putative reproductive hormones on immature *Penaeus vannamei* ovaries. *J. World Aquacult. Soc.*, 19: 288

Tsang WS, Quackenbush L.S, Chow BK, Tiu SH, HE JG, Chan SM 2003. Organization of the shrimp vitellogenin gene. Evidence of multiple genes and tissue specific expression by the ovary and hepatopancreas. Gene 303:99-109.

Tuma, D.J. 1967. A description of the development of primary and secondary sexual characters in the banana prawn, *Penaeus merguiensis* De Mann. *Aust. J. Mar. Freshwater Res.*, 18: 73-88.

Uawisetwathana U , Rungnapa L, Amornpan K, Juthatip P, Sirawut K, and Nitsara K, 2011. Insights into Eyestalk Ablation Mechanism to Induce Ovarian Maturation in the Black Tiger Shrimp. *PLoS One.* 6,9: 24427.

Van-Herp, F. and G.G. Payen 1991. Crustacean neuroendocrionology perspectives for the control of reproduction in aquacultural systems. *Bull. Inst. Zool. Acad. Sinica Monograph.*, 16: 513-539.

Van-Herp, F. 1992. Inhibiting and stimulating neuropeptides controlling reproduction in crustacea. *Invertebr. Rep. Dev.*, 22(1-30): 21-30.

Vasudevappa, C. 1992. Growth and reproduction of the penaeid prawn *Metepenaeus dobsoni* (Miers) in brackishwater environment. Ph.D. Thesis, Cochin Univ. of Sci. & Tech. 276 pp. Wyban, J.A., C.S. Kee, J.N. Sweerney and W.K. Richards, Jr. 1987. Observations on the development of a maturation system for *Penaeus vannamei. J. World Aquacult. Soc.*, 18,3: 198-200.

Villaluz, D.K., Villaluz, a., Ladeera, B., Sheik, M. and A. Gonzaga 1969. Reproduction, larval development and cultivation of sugpo (*Penaeus monodon*). *Phillip. J. Sci.*, 98: 205-236.

Weitzman, M.C. 1966. Oogenesis in the tropical land crab *Gecarcinus lateralis. Z. Zellforsch*, 75: 109-119.

Wouters, R., Piguave, X., Bastidas, L., Calderon, J. and Sorgeloos, P., 2001. Ovarian maturation and haemolymphatic vitellogenin concentration of Pacific white shrimp *Litopenaeus vannamei* (Boone) fed increasing levels of total dietary lipids and HUFA. *Aquac. Res.* 32: 573–582.

Xie Y, Fuhua Li, Bing W, Shihao Li, Dongdong W, Hao J, Chengsong Z, Kuijie Yu , Jianhai X, 2010. Screening of genes related to ovary development in Chinese shrimp *Fenneropenaeus chinensis* by suppression subtractive hybridization. *Comparative Biochemistry and Physiology, Part D* 5. 98–104

Yano, I. 1985. Induced ovarian maturation and spawning in greasy back shrimp, *Metapenaeus ensis* by progesterone. *Aquacult.* 47: 223-229.

Yano, I. 1987. Effect of 17 alpha-Hydroxy-progesteron on vitellogenin secretion in kuruma prawn *Penaeus japonicus*. *Aquaculture*, 61: 49-57.

Yano, I. and Chinzei 1987. Ovary is the site of vitellogenin synthesis in kuruma prawn *Penaeus japonicus*. *Comp. Biochem. Physiol.*, 86 B (2): 213-218.

Yano, I. 1988. Oocyte development in the kuruma prawn *Penaeus japonicus*. *Marine Biology*, 99: 547-553

Yano, I. 1992. Effect of thoracic ganglion on vitellogenin secretion in kuruma prawn, *Penaeus japonicus*. *Bull. Natl. Inst. Aquaculture*, No. 21: 9-14.

Yasuzumi, G. 1960. Spermatogenesis in animals as revealed by electron-microscopy. VII. Spermatid differentiation in the crab, *Ericheir japonicus*. *J. Biophys. Biochem. Cytol.*, 7: 73-88.

Young, N.J., S.G. Webster and H.H. Rees 1993. Ecdysteroid profiles and vitellogenesis in *Penaeus monodon*. *Invert. Rep. & Devp.*, 24(2): 107-118.

Yu XY, Lu JP, 2006. Observation on the microstructure and ultrastructure of the oviduct of *Macrobrachium rosenbergii* (Crustacea, Decapoda) J Zhejiang Univ (Sci Ed); 33(1):85–88.

Zell, S.R., M.H. Bamford and H. Hidu 1979. Cryopreservation of spermatozoa of American Oyster, *Crassostrea virginica*. *Cryobiology*, 16: 448-460.

Zeeck E, T. Harder, M. Beckman, 1998. The sperm release pheromone of the marine polychaete *Platynereis dumerilii*. *J. Chem. Ecol.*, 24 (1998), pp. 13–22

Zerbib, C. 1980. Ultrastructural observations of oogenesis in the crustacean amphipoda *Orchestia gammarellus*. *Tissue Cell.* 12: 47-62.

CHAPTER 4

NEUROENDOCRINE SYSTEM IN SHRIMPS

4.1 INTRODUCTION

In the culture of penaeid shrimps, the production of healthy seed of a particular organism in sufficient quantity has hindered the potential farming of the species. Controlled reproductive maturation is the major problem in the development of commercial aquaculture programmes in penaeid shrimps. Even though some achievements have already been made in the captive breeding of penaeids, the full control of the process is yet to be seen reveals. Similarly due to the lack of understanding of the entire maturation process, constraints of selection breeding in these animals still remains. In this regard, the basic knowledge of female reproduction and its endocrine control is of prime importance in crustacean aquaculture. A better understanding of the mechanism that regulates the reproduction at its ultrastructural levels is fundamental to the successful aquaculture programmes. Controlling reproduction in captivity could help to provide a reliable round the year supply of juveniles, developing selective breeding progr ammes, and for obtaining disease free spawners.

Any attempt in captive seed production calls for a better understanding of the endocrine mechanism involved in the control of reproduction (Fingerman, 1987). During the past two decades our understanding of crustacean reproductive endocrinology, especially that of females, has grown steadily. Neuroendocrine regulation of reproduction has been reviewed by Legrand (1982), Payen (1986), Fingerman (1987), Charniaux-Cotton and Payer (1988), Meusy and Payen (1988); VanHerp and Payen (1991); Gaytan, *et al.*, (1996); Fingerman (1997); Keller and Sedlmeier, (1998); Khayat, *et al.*, (1998); Webster and Chung, (1998); Keller *et al.*, (1999); Huberman, (2000); Alfaro, (1993); Wouters *et al.*, (2000); Cuzon, *et al.*, (2004); Coman *et al.*, (2006); Kanokpan *et al.*, (2006); Meeratana, *et al.*, (2006); Sellars, *et al.*, (2006), Coman *et al.*, (2007) and Shweta *et al.*, (2011). From the literatures it appears that the classical manipulations being practiced are extirpation and re-implantation of the suspected endocrine tissues. Eyestalk ablation has been used to mature female shrimp in captivity in conjunction with the management of water quality parameters such as water temperature, salinity, photoperiod, light intensity, sex ratio, and nutrition status of animals (Caillouet, 1972, Lumare 1979, Yano 1985, Primavera 1985, Crocoss and Keller 1986, Mohammed and Diwan 1992; Cavalli *et al.*, 1997; Crocos, and Coman, 1997; Browdy, 1998 Pratoomchat, 1998; Palacios *et al.*, 1999 a; Palacios *et al.*, 1999 b; Withyachumnarnkul *et al.*,2001; Hoang, *et al.*, 2002; Coman, *et al.*, 2003; Coman, *et al.*, 2004; Cuzon, *et al.*, 2004; and Coman *et al.*, 2006). Although many observations on endocrine systems have been made on the inhibition/suppression of reproductive maturation by eyestalk hormone(s) since the pioneering work of the 1940s (Panouse, 1943) however, recent researches have focused mostly on organs like brain, thoracic ganglion, ovary etc. functioning closely with gonad stimulating factors or hormone(s) in Crustacea (Eastman-Recks and Fingerman, 1984; Takayanagi *et al.*, 1986; Yano, 1988; Mohammed and Diwan, 1991; Yano, 1992; Yano and Wyban, 1992). Hormones are chemical messengers synthesized and secreted by endocrine glands and released to the circulatory system. Blood carries the hormones to a distant target cell(s) where they alter the cell's physiology. Usually, these hormones mediate long term, pervasive physiological alteration, as opposed to the short term, localized events that are usually mediated by the nervous system (Chang, 1992). Hormones that regulate reproduction in crustaceans include, the gonad inhibiting hormone (GIH), which may be the same as that of the vetillogenin inhibiting hormone (VIH) and the gonad stimulating hormone (GSH) both of which have been described mainly in conjunction with female reproduction (Laufer *et al.*, 1992). It has been reported that GIH/VIH is released from the sinus gland in the eyestalk and inhibits ovarian synthesis of yolk protein vetillogenin in vitro (Eastman-Recks and Fingerman, 1984; Quackenbush and Keeley, 1987; Huberman; 2000 and Diwan; 2005). In contrast, it is reported that a hormone (GSH) released from the thoracic ganglion (Eastman-Recks and Fingerman, 1984) and the brain (Gomes, 1965; Hinch and Benner, 1979; Anilkumar and Adiyodi, 1980; Takayanagi, 1986; Yano and Wyban, 1992; Shoji, 1997; Huberman, 2000 and Diwan, 2005) stimulates ovarian growth and maturation. A gonad-stimulating hormone releasing hormone has been reported from the brain of *P. vannamei* by Yano and Wyban (1992). Knowledge of the hormonal control of maturation in crustaceans however, is not well documented.

Among decapod crustaceans the control of diverse and numerous physiological processes, such as reproduction, moulting, haemolymph glucose - content, rate of heart beat and chromatic adaptation, etc. is mediated by neuroendocrine factors synthesized within the specialized neurosecretory cells

(NSCs) located within or associated with the ganglia of the central nervous system (CNS) (See reviews of Kleinholz, 1976, Cooke and Sullivan, 1982; Sandeman, 1982; Fingerman, 1987; Huberman, 2000 and Diwan, 2005). Some of these peptide neurofactors, for example, pigment aggregating (Fernlund and Jossefson, 1972) and dispersing hormone (Fernlund, 1976) have been identified, purified and synthesized. Considerable information is available concerning specific processes, such as the regulatory phenomena involved in the control of chromatic adaptation (Rao and Fingerman, 1983; Fingerman 1985, 1987 and Huberman 2000 and Diwan 2005). In contrast, neuroendocrine control of other phenomena such as reproduction, osmoregulation etc. has received the attention of many investigators during the last two decades, however the neurofactors involved have been only partially characterized and purified. A clear picture of the control of reproduction has not been unraveled till recently. While there is substantial data available concerning the physiological effects of the homogenates of the optic, supra-esophageal and thoracic ganglia on reproduction information regarding the specific sites of the synthesis of neurofactors involved in the reproduction is meager (McNamara1993). The approach of cytophysiology of the endocrine systems has a great deal to offer when it comes to understanding the hormonal regulation of various life processes mainly reproduction and its control.

Young (1959) described the gross morphology of the central nervous system of *Penaeus* and its structure and function in Penaeoidea, including *Penaeus* by Bullock and Horridge (1965). Optic ganglia of penaeids have described by Dall (1965) and Elofsson (1965). Dall *et al.*,(1990) described the layout of the central nervous system of penaeids. The established endocrine elements in the decapod neurosecretory system are the X-organ sinus gland complex of the eyestalks, neurosecretory cells of the brain and central nervous system, post commissural organs, pericardial organs, the epithelial endocrine systems, the Y-organ and the androgenic gland (Gabe 1953, 1956; Carlisle and Knowles, 1959; Bullock and Horridge, 1965; Cooke and Sullivan, 1982; Skinner, 1985; Charniaux-cottom and Payen, 1985; Fingerman, 1985, 1987; Diwan, 2005).

The morphology as well as the histology of the crustacean endocrine organs in the central nervous system have been described by Carlisle and Knowles, (1959); Nagabhushanam *et al.*, (1992) and Mohammed *et al.*,(1993). Similarly the X-organ sinus gland complex has been described by Dall (1965) in *Metapenaeus bennettee* and Nakamura (1974) in *P. japonicus*, Nanda and Ghosh 1985) in *P. monodon* and Mohammed *et al.*,(1993) in *P. indicus*. Histological investigations of the neurosecretory cell types in the *Penaeus* species has been described by Ramadan and Matta (1967), Nanda and Ghosh (1985), Mohammed *et al.*,(1993) and that of *Metapenaeus* by Madhysta and Rangnekar (1976); Nagabhushanam *et al.*,(1992); Shoji (1997); and Diwan, (2005).

It is now well known that physiologically active substances are produced in the neurosecretory cells located in the ganglionic centers of crustaceans (Bliss *et al.*, 1954, Durand, 1956; Dall, 1965; Fingerman and Oguru, 1968; Adiyodi and Adiyodi, 1970; Nakamura, 1974; Chandy and Kolwalker, 1985; Gyananath and Sarojini, 1985; Nanda and Ghosh, 1985; Nagabhushanam *et al.*, 1986 & 1992; Quackenbush, 1989; Mohammed *et al.*, 1993). Nevertheless investigations on penaeid neurosecretion are described by Dall 1965, in the Australian school shrimp *Metapenaeus* species, Nakamura (1974) in *P. japonicus*; Nagabhushanam *et al.*, (1986) in *P. stylifera*; Nanda and Ghosh (1985) in *P. monodon*; Nagabhushanam *et al.*, (1992) in *M. ensis*; Mohammed *et al.*,(1993) in *P. indicus;* Shoji (1997) in *P. monodon*; Huberman, (2000) and Diwan, (2005). Gyananath and Sarojini (1985) delineated the activity of neurosecretory cells in the fresh water shrimp *M. rosenbergii*.

Even though these studies are describing different types of neurosecretory cells in different ganglionic centers of crustaceans, only few studies have been made to correlate the changes in the neurosecretory activity with physiological events like reproduction (Aoto and Nishida, 1956; Durand, 1956; Hanoaka and Otsu, 1957; Kulkarni and Nagabhushanam, 1980; Gyananath and Sarojini, 1986; Mohammed *et al.*, 1993; Shoji (1997); and Diwan, (2005). Similarly these investigators have described the secretory activity of these NSCs but not the mechanism and/or the various stages in the secretory cycle in relation to any other physiological processes. Neurosecretory cells in different stages are mentioned in the reports of Durand (1956) in crayfishes, Peryman (1969) in *P. clarkii*; Diwan and Nagabhushanam (1974) in *B. cunicularis*; Rao *et al.*, (1981) in *M. lanchesteri*; Decaraman and Subramoniam (1983) in *Squillaholochista*; and Joshi (1989) in *P. koolooense*. However, these investigators did not explain the actual mechanism and/or cyclic stages in neuro secretion. But Damassieux and Batesdent (1977) have reported the cyclic variation in certain NSCs in *A. aquaticus*. Mohammed *et al.*, (1993) gave a detailed description of the various phases in the secretory cycle of different neurosecretory cells in *P. indicus* through light microscopy. Such attempts through electron-microscopy has been restricted to the ultrastructural studies in the neurosecretory cells of shrimp *M. olfersii* maintained in different salinities by McNamara (1993) among crustaceans.

Neurosecretory cells are located in different ganglionic centers of the central nervous system and also in the optic ganglion of decapod crustaceans (Dall, 1965). Furthermore in crustaceans, neurosecretory cells are distributed as distinct groups (Bliss, 1951; Enami, 1951). Various investigators as mentioned above studied different ganglionic centers of different decapods through light microscopy. NSCs of the brain and their various functions have studied by Nakamura (1974), Ramadan and Matta (1976), Fahrenbach (1976), Nanda and

Ghosh (1985) and Mohammed *et al.*,(1993). Various investigators like Enami (1951) in the crab *Sesarma*; Matsumoto (1954) in *Exiocheir japonicus*; Maynard (1985) in nine species of brachyuran crabs and Mohammed *et al.*,(1993) in *P. indicus* studied the NSCs in the thoracic ganglia. Abdominal ganglia of crustaceans are not investigated as thoroughly as other ganglia in the central nervous system. Available information on this aspect is restricted to reports of Skinner (1968) and Kendoh and Hsada, (1986) in crayfish and Mohammed *et al.*,(1993) in *P. indicus*. In contrast, the NSC types in the optic ganglia of various crustacean species have been studied in detail by many investigators like Carlisle (1953), Smith and Naylor (1972), Hisano (1974), Smith (1975), Jaros (1978), Nakamura (1980), Nanda and Ghosh (1985) and Mohammed *et al.*, (1993).

The X-organ sinus gland complex is the most thoroughly investigated part in crustacean endocrinology, as it is the center of many of the inhibitory hormonal factors, especially the gonad inhibitory factors (Fingerman, 1987). Similarly sinus gland is the principal neurohaemal organ involved in the storage and release of neurosecretory materials serving several endocrine functions (Gabe, 1956; Kleinholz, 1976; Highnam and Hill, 1977; Fingerman, 1987). The ultrastructures of the neurosecretory granules stored in the sinus gland of various crustaceans are studied by Fingerman and Aoto, (1959); Knowles, (1959); Meusy, (1968); Andrew *et al.*, (1971); Smith, (1974); Brodie and Halerow, (1977) and Martin *et al.*, (1983). However, such studies in penaeids are scanty. Mohammed *et al.*, (1993) described the light microscopy of the sinus gland but there are no such studies in penaeids through electron microscopy. Similarly many of the above-mentioned investigators have studied the various types of neurosecretory granules in the sinus gland but there are no report showing the types of neurosecretory granules during any of the particular physiological activities.

Similarly, endocrine control of reproduction has been investigated in a wide variety of crustaceans. In this respect amphipods and isopods have received considerable attention. However, the actual mechanism and hormones working behind reproduction have not been reveals in any other crustaceans (Fingerman, 1987). It has been attributed to the eyestalk factors by many (Aoto and Nishida, 1956; Demeusy, 1967; Rangnekar and Deshmukh, 1968; Diwan and Nagabhushanam,1974; Laubier, 1978; Arnstein and Beard, 1975; Webb, 1977; Quackenbush and Herrkind, 1981; Eastman-Recks and Fingerman, 1984; Radhakrishnan and Vijayakaumaran, 1984; Kulkarni *et al.*, 1984; Anilkumar and Adiyodi, 1985; Fingerman, 1987; Quackenbush and Keeley, 1987; Mohammed and Diwan, 1991; VanHerp, 1992; Shoji, 1997) after Panouse (1943) in *Leander serratus*.

A second decapod reproductive hormone found in the brain and thoracic ganglia has been attributed the role of gonad stimulation. Investigators like Otsu (1960); Oyama (1968), Hinch and Bennet (1979), Nagabhushanam *et al.*, (1982); Eastman-Recks and Fingerman (1984); Takayanagi *et al.*,(1986); Yano *et al.*,(1988); Yano (1992); Shoji 1997 and Diwan, 2005 have reported that the thoracic ganglion secretes the GSH in decapods. Similarly Gomez (1965); Diwan and Nagabhushanam (1974); Nagabhushanam *et al.*,(1982); Yano and Wyban (1992); Shoji (1997); Huberman (2000) and Diwan (2005) reported that besides the thoracic ganglia, the brain also secretes an ovarian growth accelerating hormone in crustaceans. In this respect cytological and ultrastructural studies of various endocrine organs during the process of gonadal maturation can give some concrete evidences about which organism are active and what type of factors are produced etc. However, unfortunately various investigators over looked such studies and most of them have concentrated in the isolation of neurosecretory factors. From the foregoing literature it is obvious that very fragmentary information is available in this regard for various crustaceans, especially in marine penaeid shrimps. Mohammed and Diwan (1991) figured out the light microscopic features of the various ganglionic centers of a penaeid shrimp *P. indicus* during the process of gonadal maturation. Still there is no published information on such studies describing the characters of various NSCs engaged in the reproduction through electron microscopy. Hence, in the present chapter details regarding the cytological investigations of various neuroendocrine centers during the process of gonadal maturation are described. Different types of NSCs in various ganglionic centers have been identified through light microscopy and its functional details have been given through electron microscopy studies during the process of gonadal maturation. Ultrastructural features of the different phases of secretory cycle of these NSCs have also been mentioned to get a clear picture of the process of neurosecretion. The mapping of neurosecretory cells in different ganglionic masses has also been described. A thorough description of the ultra structural features of the eyestalk neurosecretory system including the sinus gland has been made in relation to reproduction.

4.2 NEUROENDOCRINE SYSTEM

The long double ventral nerve cord and the paired ganglia corresponding to each segment are the characteristic features of the crustacean dendro branchiate central nervous system. Typical of annulata, the central nervous system of *Penaeus monodon* comprises of a dorsal brain connected to the ganglionated ventral nerve cord below the gut by two large tracts so that the gut passes between these tracts. The dorsal brain or supra-oesophageal ganglion is located within the head lobe in the dorsal part of the protocephalon, protected dorsally by the broadening base of the rostrum. It receives nerves from the sense organs of the head and supplies nerves to the adjacent muscles. The supraoesophageal ganglion consists of three primary lobes; the proto cerebrum, deuterocerebrum and tritocerebrum. From the dorsal surface of the brain i.e from the protocerebrum arise a pair of stout optic nerves, which ends in the optic ganglia. The medulla interna and medulla terminalis of the optic ganglia are derived from the proto cerebral part of the brain. Deuter ocerebrum is the middle part of brain and these supplies nerves to the antennular, antennal appendages and statocyst. While the third part, the tritocerebrum is the postero-ventral part of the brain and supplies nerves to the labrum, the preoral stomato-gastric system and the post oral tritocerebral commisure.

The brain or supraoesophageal ganglion is linked to the ventral nerve cord by two connectives, the tritocerebral connectives that pass around the oesophagus. Midway through the tritocerebral connectives are the pair of tritocerebral ganglia. The tritocerebral connectives arising from the brain meet ventral to the oesophagus, in a large suboesophageal ganglion on this condthoracic segment. The first ganglion of the ventral nervecord, the sub-oesophageal ganglion, together with the following metameric ventral ganglia, receive impulses from sensory end organs of the body and appendages and sends motor impulses to the muscles moving these structures. The brain or the sub-oesophageal ganglion of *P. monodon* is a compound of several major ganglia supplying nerves to mouthparts like mandibles, maxillules, and maxillae.

The ventral nerve cord of *P. monodon* is the fusion product of a "ladder" nervous system where the paired ganglia of each segment have come together at the midline. It contains a longitudinal series of ganglia, interconnected by fused pairs of inter segmental nerve tracts. This ventral nervecord continues posteriorly from the sub-oesophageal ganglion to the five thoracic ganglia each corresponding to the five thoracic segments. Even though these thoracic ganglia are fused across the mid-line, they are longitudinally separated from one another by short-paired connectives. The thoracic ganglia are found from the 3rd segment onwards and the 1st thoracic ganglion in the ventral nervecord supply nerves to the three maxillipedes. The next three thoracic ganglions supply nerves to each of the 3 pereiopods. The 5th thoracic ganglia in the seventh and eighth segments fused together and it supplies nerves to the 4th and 5th pereiopods.

The ventral nerve cord narrowed as it enters the abdomen, where there are six abdominal ganglia. Anterior five abdominal ganglia appear identical but the last one is somewhat enlarged. These abdominal ganglia are also fuse across the midline like thoracic ganglia. The connective that joins the abdominal ganglia is comparatively longer than that of the thoracic ganglia. The different ganglionic centers identified in *P. monodon* and their relative position in the nervous system is given in Fig.1

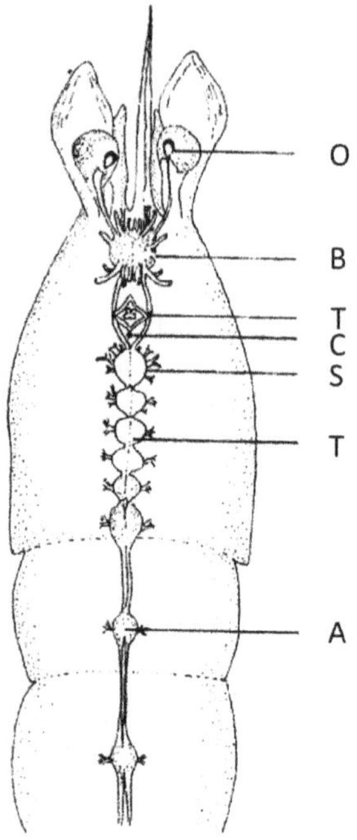

Fig. 1. Diagramatic representation of the neuroendocrine system of *P. monodon.* O – optic ganglia, B – brain (supra-oesophageal ganglion) S – sub-oesophageal ganglion, Tr–tritocerebral ganglion, T–thoracic ganglion, A–abdominal ganglion, C–circumnoesophageal commissure
(**Source:** Diwan *et al.*, 2008)

4.3 NEUROSECRETORY CELL TYPES

In the ventral nerve cord different types of neurosecretory cell are present. These neurosecretory cells are present aggregated form only in the ganglionic masses and not in the nervecord. Nervecord contains only axons of nerve fibers. Neurosecretory cells are characterized by the presence of large, centrally located, weakly stained spherical nucleus, abundant cytoplasm and conspicuously staining granules in their perikarya. Distinct cyclic changes take place in their granular cytoplasm during different phases of maturation of the animal. Along with these NSCs in each ganglion some other non- neurosecretory cells are also present in abundance. They are structurally different from the neurosecretory cells. They are tiny, deep stained cells whereas, neurosecretory cells are larger and not so deeply stained.

Histologically NSCs differ in their size, general shape of the cell body, presence or absence of vacuoles in cytoplasm and in the appearance of secretory product. Based on these characters the NSCs in different ganglions are classified arbitrarily into 4 morphological type's viz. Giant neurons (GN) (Figs.2, 3), 'A' cells (Figs.4, 5), B cells and C cells (Fig.6). Ultrastructure of these NSC cells shows a comparatively large electron dense nucleus and a less electron-dense but active cytoplasm. The cytoplasm contains abundant supplies of cell organelles like stacks of RER, large accumulations of ovoid granules and electron lucent vesicles.

1. Giant Neurosecretory cell (GN-cells)

Giant neurons as indicated by its name are the largest of the NSC present in most of the Penaeid shrimps. Generally cell bodies of these GN cells are large and their shape varies from oval to spherical. Most of the GN cells are unipolar with a short axon (Fig. 2). Pericellular capillary plexus are also present around each GN cells. Glial cell nuclei are distributed in limited numbers in all the ganglia except the optic ganglia. Sizes of these large cell bodies vary from 110 to 154 μ in diameter and possess much cytoplasm and a large nucleus of 28 to 44 μ in diameter. The large nuclei of the GN cells sometimes considerably differ from the round nuclei of other cells. The modified large nuclei appear as flat, oval or crescent shaped in different cells (Fig.3). The nucleus often contains more number of nucleoli, the number varied from 8 to 14. These nucleoli are dispersed in the chromatic material usually towards the peripheral region lying against the nuclear membrane. Even though there are minor differences in the size and shape among GN cells, all of them are rich in cytoplasm with granular materials around the nucleus. Further, not all of the GN cells show the presence of large amount of secretory materials at all times. In certain stages a very thin rim of cytoplasm encompasses the large nucleus in these cells. Occurrence of small vacuoles in the cytoplasm is a remarkable feature of these cells on certain occasions. Round, oval or horse-shoe shaped vacuoles of varying sizes occur in the perikarya of GN cells with reduced amount of cytoplasm. In these cells, cytoplasm is not granular in nature due to the absence of secretory granules.

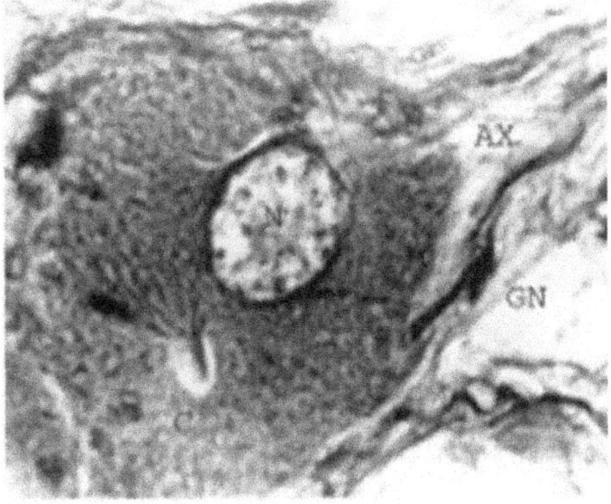

Fig. 2. Light microgpah of the GN cells of *P. Monodon* (N-nucleus, C-cytoplasm, Ax-axon, Nu-nucleolus) X100
(**Source:** Diwan *et al.*, 2008)

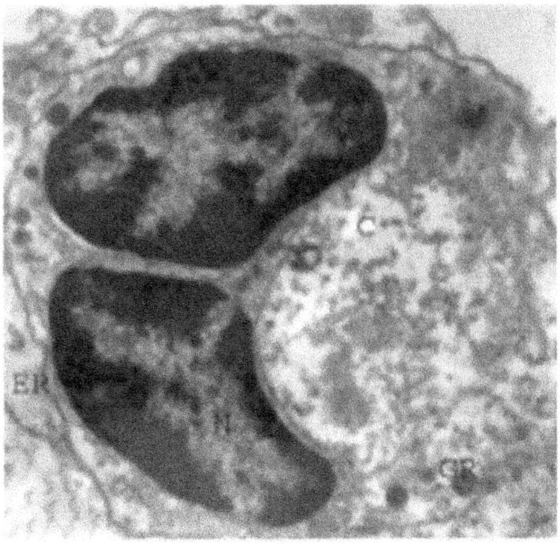

Fig. 3. Electron micrograph of the GN cells *P. Monodon* (N-nucleus,C-cytoplasm, ER-endoplasmic reticulum, Gr-granules) X 4,600
(**Source:** Diwan *et al.*, 2008)

Ultrastructurally the GN cells are richly endowed with abundance of highly active cell organelles and secretory granules in their perikaryon. Variably shaped large central nucleuses are provided with numerous gaps or nuclear pores. The electron dense cytoplasm around the nucleus contains stacks of rough endoplasmic reticulum (RER), Golgi complex's, electron-dense Golgi derived granules and glycogen particles associated with electron-lucent vesicles made of RER (Fig.3). These large and highly electron-dense secretory granules appear as round or spherical, homogenous and bound by a thin limiting membrane. Most of the granules are dispersed throughout the cytoplasm. Free and poly ribosome and abundant oval to spherical mitochondria are present in the cytoplasm. Multi vesicular bodies and inclusion bodies containing flocculent materials are also present.

The RER consists of flattened cisternae running parallel to each other as well as oval to polygonal granule filled electron lucent vesicles. The intra-cisternal space of the parallel RER is slightly electron dense owing to the presence of some flocculent material. Highly active large Golgi complexes are present in the cytoplasm in association with these RER. In the well-developed Golgi complex the cisternae are occasionally enlarged at the ends and contain a moderately dense material, apparently giving rise to the elementary granule. Highly electron-dense large round or spherical secretory granules as well as variably shaped vesicular bodies are present in the perikayon of these cells. The perikarya of these GN cells are delimited by a capsular glial like system of 5- 6 paired membranes. These cells possess glial cell nuclei in between these layers.

2. Neurosecretory cell A type

'A' cells (Fig. 4) are much smaller than GN cells and are present in all the ganglia of shrimp. Even though these neurosecretory cells are located in all ganglia, their occurrence is limited especially in optic and abdominal ganglia. All the other ganglia contain comparatively more number of 'A' cells. These cells possess an average cell diameter of 79.5 ± 7.36 μ The centrally placed nucleus is spherical or oval with a nuclear diameter ranging from 18.5 to 27.8μ.The vesicular nucleus contains a small nucleolus. The cytoplasm is more or less homogeneously distributed with secretory inclusions. Some of these cells occur with axonal process but most of the A cells are without long axons, like GN cells.

The large central nucleus and the characteristic, clumped distribution of the nuclear pores within the nuclear envelope are evident in these cells. The perikarya of most of the A cells are highly active with abundant cell organelles. RER in these cells appear as narrow elongated cisternae aligned parallel in stacks of up to 8 cisternae or as randomly oriented ramified cisternae (Fig. 5). Large Golgi complex also occurs in association with those RER elements and active mitochondria. Mitochontria profiles are rounded or ovoid and the cristae are distinct and rather straight. Free and polyribosomes are apparent in association with these mitochondrial bodies. Multi vesicular bodies and inclusion bodies containing flocculent to dense material are also occasionally present in the cytoplasm. The perikarya of these cells are delimited by the capsular system in the GN cells. Here the limiting membrane is double layered with glial cell nuclei located in between these layers.

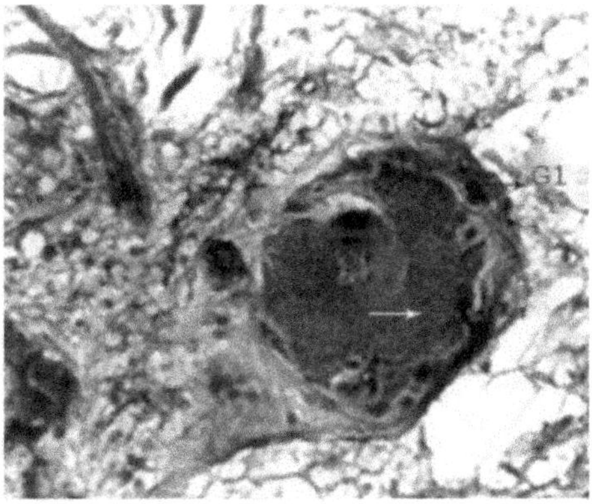

Fig. 4. Light micrograph of the A cells *P. Monodon* (N– nucleus, C – cytoplasm, Nu – nucleolus) X100 (**Source:** Diwan *et al.*, 2008)

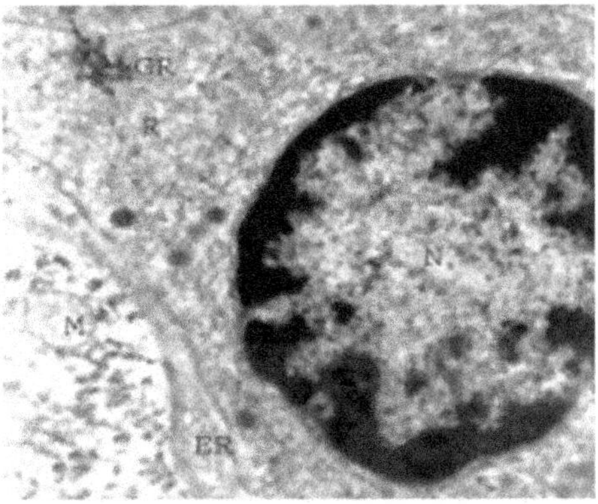

Fig. 5. Electron–micrograph of the A cells *P. Monodon* (N-nucleus, M-mitochondria, ER – endoplasmic reticulum, R –ribosomes, G-granules) X 4,600 (**Source:** Diwan *et al.*,2008)

3. Neurosecretory cell B-type

These medium sized cells are apparent in large numbers in all the ganglia. Their shape varies from oval to polygonal with an average cell diameter of 39.69 ± 6.12μ. These medium sized cells are always present in groups of 10 to 25 cells in ganglionic masses (Fig.6).The nuclear diameter varies from 10 to 20μ. Nucleolus is present near the nuclear membranes. Axonal processes are not distinct in all cells but during certain stages, small axonal processes are feebly visible. Generally B cells showed only little sign of secretory activity differing from the other GN and A cells.

Their cytoplasm is fairly homogenous and of compact appearance. The fine cytoplasmic inclusions are sparsely distributed within the perikarya. Occurrences of peripheral vacuoles lacking secretory inclusions are also seen. Capillary plexus and glial cells are apparent in their pericellular regions during certain times.

Ultrastructural studies show that these ovals to polygonal cells possess a large nucleus and that perikarya of most of these cells are electron lucent and devoid of many cell organelles. The fine structure of these perikarya is appreciably different from that of the larger GN and A cells. Large concentrically arranged membraneous system is present in the perikarya of B cells, apart from the usual RER system on certain occasions. Numerous round/oval mitochondria occur in the perikaryon.

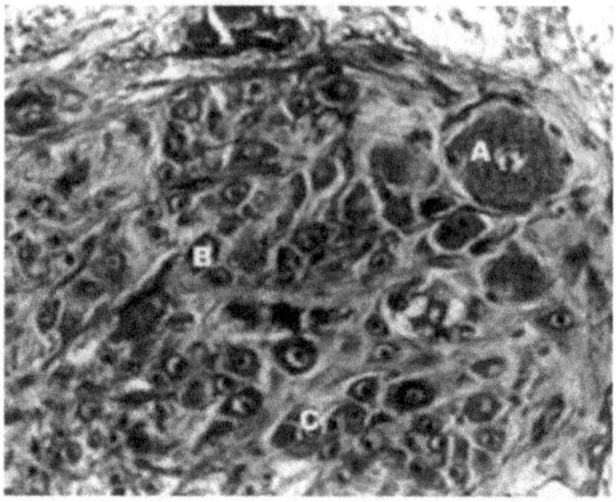

Fig. 6. Light micrograph of the neurosecretory cells (A-A cells, B-B cells, C-cells, N-nucleus, Cy-cytoplasm) X100 (**Source:** Diwan *et al.*, 2008).

4. Neurosecrectory cell C-type

C type cells are the smallest NSCs present with very limited numbers in all the ganglia. These cells are usually present in groups in association with the B cells. Their conical or elliptical shape tapering in narrow axons characterizes these small cells. The size range of these cells varies from 12 to 24 µ with an average diameter of 18.6 ±3.5µ. The centrally placed round nuclei contain conspicuous intra-nucleolar materials. The nucleus cytoplasm ratio is high due to the lesser amount of cytoplasm. Occurrences of peripheral vacuoles are noticed in these C cells.

Electron microscopic studies show that the fine structures of the perikarya of these C cells were appreciably different from that of the other larger cell bodies. The nucleus of these neurosecretory cells is characterized by a conical to polygonal shape (Fig. 6). This centrally placed nucleus contains highly electron-dense chromatin materials. Nuclear pores are not as evident as in the other cells. The electron-lucent perikaryon of these cells is devoid of secretory granules or cell organelle. The relatively abundant RER has an electron-lucent narrow lumen but the slender parallel cisternae are slightly electron- denser than the remaining cytoplasm due to the presence of a flocculent material. Vacuoles of varying size and shape are also present here and there in the cytoplasm. Highly electron-dense small granular materials are also present in certain maturing C cells indicating its secretory activity.

4.4 NON-NEUROSECRETORY CELLS

Existence of a few clusters of unique tiny, deep stained easily identifiable neurons are found freely and in association with afore said cell types mentioned earlier in all the ganglia. Most of these non-neurosecretory cells are round in shape but vary occasionally (Fig.7). The size range of these cells varies from 8.4 to 14.5μ. The disproportionately large nucleus is round and has an average diameter of 10.5μ. Ultrastructural observations indicate that these cells have a large nucleus and a thin rim of cytoplasm. Neurosecretory materials are not present in the cytoplasm of these cells and the limiting membrane is not distinguishable.

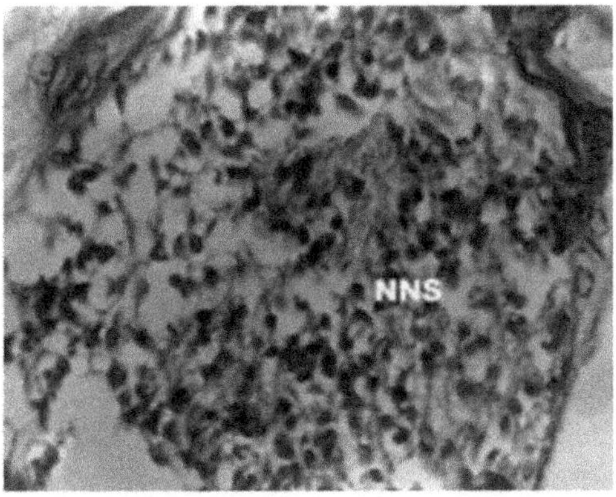

Fig. 7. Light micrograph of the non-neurosecretory cells (NNS) (N-nucleus) X100 (**Source:** Diwan *et al.*, 2008)

Details of the neurosecretory and the non-neurosecretory cells are given in the Table 1.

Table 1. Characteristics of different neurosecretory cell types of shrimp *P. Monodon*

Cell types	Cell diameter (μ)	Nuclear diameter	Nos. of
GN cells	130.12±12.55	35.65±3.95	8-14
A cells	79.50±7.36	22.20±5.17	1
B cells	39.69±6.12	16.90±2.90	1
C cells	18.60±3.50	10.10±2.20	1

(**Source:** Diwan *et al.*, 2008)

4.5 NEUROSECRETORY CELLS AND SECRETORY CYCLE

There are cyclic changes in the perikarya of the NSCs in concurrence with the synthesis of the neurosecretory materials. Different phases are identified during the secretory cycle, on the basis of the cytoplasmic changes and appearance of the secretory products in their perikaryon. The identifying characters of the NSCs are the presence of large amounts of cytoplasm and the abundant supply of secretory granules in their perikaryon. The secretory cycle of these NSCs are classified into 3 different phases viz. quiescent, vacuolar and secretory. These three phases of the secretory cycle are common for all the four types of NSCs. Generally the appearance of the cell body, presence or absence of vacuoles in the cytoplasm and the appearance of secretory products in the cytoplasm etc. are the main criteria in separating different stages of the neurosecretory cycle. Even though there are distinguishable differences in the morphological nature of the secretory granules, there is a basic pattern in their mode of secretion and discharge pattern, in different NSCs.

1. Quiescent Phase (Q-phase)

This forms the in active or resting phase of NSCs. The NSCs in their quiescent phase contain only a thin rim of homogenous lightly staining cytoplasm. But these cells are characterized by a large round, easily distinguishable nucleus filling almost the entire cell volume. The limiting membrane of the nucleus as well as the cytoplasm of these resting cells is very prominent compared to that of the other phases. The nucleus of the GN and A cells contains highly distinguishable scattered nucleoli. While B and C cells possess deeply stained clumped nucleolus in the centre of the karyoplasms. A few glial cells are also present surrounding the outer margin of these resting with capillary plexes occasionally. Axonal filaments of the quiescent cells are devoid of flocculent cytoplasmic materials.

Ultrastructurally, the perikarya of the resting NSCs appear smooth and electron- lucent with minimal cytoplasmic activity. The large electron-dense nucleus occupies almost the entire cellular volume. The characteristic clumped distribution of the nuclear pores within the nuclear envelope is not evident in these resting cells. Here the nuclear envelope is continuous and smooth without any interruptions (Fig.8). The electron- lucent perikaryon of these NSC contains only a minimum number of cell organelles. The most dominant type of cell organelle found in the perikaryon of these quiescent phase NSCs is the rough and smooth endoplasmic reticular elements. Smooth ER is present either as randomly oriented ramified cisternae or as narrow elongated cisternae aligned parallel in stacks of up to 15 cisternae. Small parallel stacks of golgi complex are also present in association with these endoplasmic reticular elements. Some free and polyribosomes and oval to spherical mitochondria are also encounter in the electron- lucent perikarya. A few numbers of electron-lucent multi vesicular and inclusion bodies are present along the outer perikaryon. Electron-lucent and moderately electron–dense vacuoles and/or vesicles are less abundant.

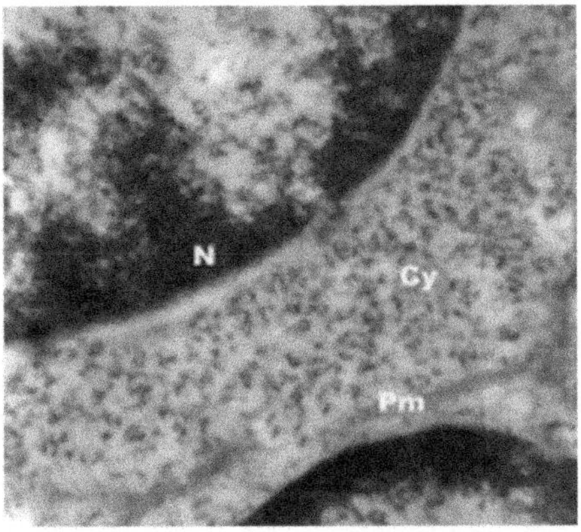

Fig. 8. Electron-micrograph of the Q-phase A cell of *P. monodon* (N-nucleus, Cy-cytoplasm, Pm-plasma membrane) X4, 600 (**Source:** Diwan *et al.*, 2008).

2. Vacuolar Phase (V-phase)

The NSCs of this particular phase are characterized by the presence of numerous vacuoles (Fig. 9). Irrespective of its size and shape all the NSC types possess vacuoles of different size and shape during its V-phase of the secretory cycle. Even though there are minor differences, the vacuoles of all the NSC types possess certain basic characters. These vacuoles originate as small round membrane bound bodies in the cytoplasm and later towards the final stages these vacuoles undergo some structural changes and get transforms into larger vacuoles of various size and shape in different cell types.

Histologically, GN and A cells possess variably shaped larger vacuoles (Fig.9). Most of the times, these expanded vacuoles almost free of inclusions except for a few fuchisinophilic granules. These vacuoles are rather larger and measure upto 42μ in size and most of the time found around the periphery of the cell, although sometimes it is located more centrally in the cytoplasm. These largely expanded ovals to horseshoe shaped vacuoles are usually found empty but occasionally a lightly staining material encounter in some of these vacuoles. The cytoplasm around these vacuoles is flaky in appearance due to the presence of small secretory granules.

The nucleus of these vacuolar phase NSCs contains rather diffused chromatin materials. The limiting membranes of the nucleus as well the cytoplasm is not as clearly detectable abs in the case of resting cells. The glial cell nuclei around these neurosecretory cells exhibit hyper activity during this particular stage. The entire cell volume encompasses by these active glial cells.

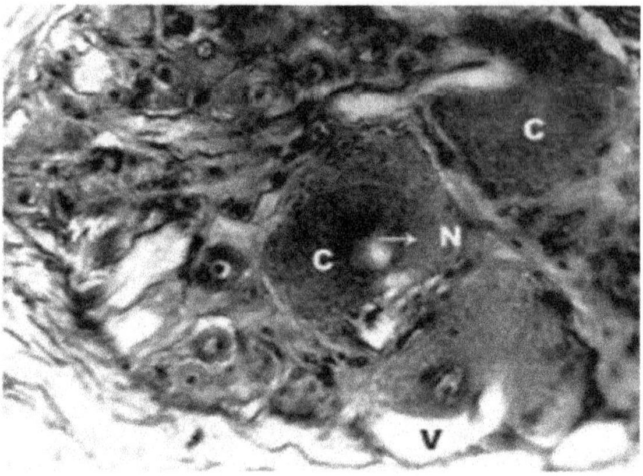

Fig. 9. Light micrograph of the V-phase GN cells of *P. Monodon* (N–nucleus, C–cytoplasm, V–vacuole) X400 (**Source:** Diwan *et al.*, 2008)

Whereas B and C cells possess rather small, uniform round vacuoles and these vacuoles are not sharp lyde limited from the cytoplasm. The diameters of these vacuoles vary from 8 to 12 μ and give a coarse appearance to the cytoplasm. These vacuoles in the cells are not restricted to a particular position but spread throughout the entire cytoplasm. Along with these vacuoles, the cytoplasm also contains some secretory granules. These secretory granules are present apparently at random, throughout the cytoplasm. Here also the limiting membrane of the nucleus as well as the cytoplasm is not distinct as in the case of resting cells. Clumped nucleolar materials occupy at the central part of the nucleus. The hypertrophied glial nuclei along with some capillary plexes are present around the pericellular margin of these cells.

In sharp contrast to the resting condition, ultrastructural features of the NSCs at the vacuolar phase exhibits conspicuous structural changes that are indicative of highly active condition of these cells during this phase. These V-phase NSCs possess a highly active nucleus and cytoplasm. The large nucleus is present with its electron-dense nucleolar materials at the centre of these NSCs. The characteristic distributions of the nuclear pores within the nuclear envelope are particularly evident in these cells. Streaming of small granular materials from the nucleus to the perinucleolar cytoplasm is apparent through these nuclear pores. Numerous vesicles packed with granules of dense nature are also apparent in the cytoplasm around the nucleus. The perikarya around this active nucleus contain numerous highly active organelles like rough endoplasmic reticulum, large Golgi complex round to oval mitochondria with distorted cristae, large numbers of moderately electron-dense multi vesicular bodies with developing secretory granules and various types of lysosomal structures.

Numerous rough and smooth endoplasmic reticular elements are apparent near the nuclear as well as plasma membrane of this vacuolar phase NSCs. The ER is present in different morphological forms like short cisternae of less dilation, randomly oriented ramified cisternae or parallely arranged stacks of cisternae. Parallely arranged stacks of cisternae are present usually around the nuclear membrane, whereas ramified cisternae are more apparent in close apposition to the outer plasma membrane. Active RER in the form of short cisternae are also visible throughout the perikaryon and small electron-lucent vesicles are budded off from these RER elements (Fig. 10).

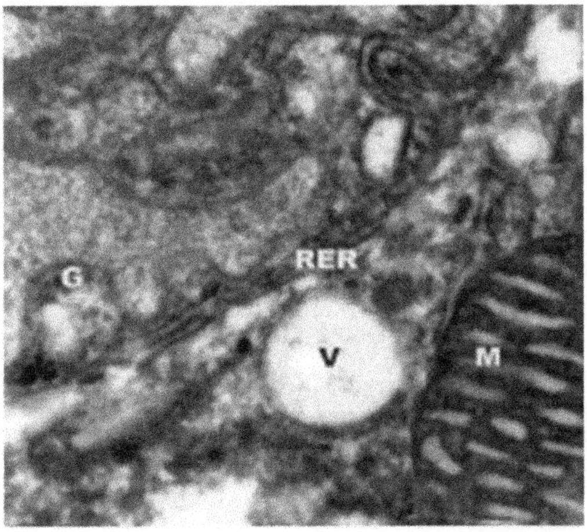

Fig. 10. Electron-micrographs of the V-phase GN cells of *P. Monodon* (N-nucleus, M-mitochondria, G-Golgi body, RER-rough endoplasmic reticulum, V-vacoule, Gr-granules) X15,500 (**Source:** Diwan *et al.*, 2008)

Large golgi complex in association with these stacked ER elements gets distributed randomly throughout the cytoplasm. They are either in the form of 'C-shaped' profiles or in the form of ramified cisternae comprised of 4-5 saccules. Golgi complex produces small vesicles of 2 types, one which is often irregular and electron lucent located around the lateral margins of the Golgi saccules and other vesicles exhibit an electron dense matrix with a diffuse coating, associated with the trans or mature face of golgi complex. These vacuoles are mainly present in association with the Golgi complex with 'C-shaped' profiles.

Large numbers of active mitochondria are apparent throughout the cytoplasm, in the form of groups and sometime scattered in the dense cytoplasm of the vacuolar phase NSCs. Their profiles vary from round or ovoid to irregular and their cristae are distinct and rather straight. Matrix of the mitochondria possesses some small irregularly shaped electron dense deposits. Rough endoplasmic reticulum bound vacuoles of variable sizes are abundant in the perikaryon of these neurosecretory cells. Their profiles are usually rounded or ovoid and these electron lucent bodies are seen empty during most of the times. Apart from these empty vacuoles, single vesicular structures and multi vesicular bodies are apparent in moderate numbers in the perikaryon of these vacuolar phase NSCs (Fig.10). Different types of vesicular bodies encounter in the cytoplasm, round to oval membrane bound vesicles packed with moderately electron-dense granules in close proximity to the nucleus and a moderate number of clear vesicles dispersed throughout the cytoplasm.

Another important cell organelle encounter in these NSCs is the variably shaped lysosomal bodies. Lysosomes containing secretory granules are also present in very few numbers. Free and polyribosomes are also abounding throughout the active dense cytoplasm. The glial cells encompassing these NSCs are also very active during this vacuolar phase and these cells contain active nuclei and highly active cell organelles. Abundant supply of small cisterns of RER is seen in their cytoplasm. Other cell organelles present in the cytoplasm are the vesicular bodies and inclusion bodies with flocculent to dense materials. As a whole, remarkable increase in the synthetic activity takes place during this vacuolar phase of the NSCs compared to the previous resting phase. However, there is a paucity of neurosecretory material or membrane limited neurosecretory granules within the cytoplasm, apparently indicating that the sub-cellular processes have not yet completed.

3. Secretory Phase

Irrespective of the cellular dimensions or cell types all the NSCs in this particular phase are rich in darkly stained granular cytoplasm. NSCs in the secretory phases are apparently ready for the release of its secretory products. In contrast to the previous stages, these cells appear darkly due to the presence of secretory products. All the cell types show an increase in size due to the accumulation of secretory granules in their cytoplasm.

The large vacuoles present in the GN and A cells undergo some morphological changes, when the cell enters the secretory phase. Instead of variably shaped vacuoles of the preceding stages, small round vacuoles are apparent in this stage. Moreover most of these vacuoles contain small secretory granules. These small sized vacuoles gradually shift to the outer cytoplasm from their previous perinuclear position. Due to the extremely granular nature in the outer cytoplasm, the cell membrane shows broken appearance. The cytoplasm in the perinuclear area becomes highly granular.

During the secretory cycle, in addition to the changes apparent in the cytoplasm, the nucleuses also exhibit some concomitant changes in its size and morphological appearance. The clearly round profiles of the nucleus become irregular and broken and small granules make its appearance in the vicinity of the modified nuclei. Streaming of granules from the nucleus is apparent during this S-phase (Fig.11).

At the later stages of the secretory cycle the round nucleus become oval to crescent shaped and then expands, and rapidly regain its round shape during its next resting or quiescent phase. The axonal processes of these cells also become more prominent during this phase. Granules of similar nature also occur to some extent in the axons of these cells. Streaming of secretory granules towards the axonal process is apparent in these cell bodies. Randomly distributed hypertrophied glial cell nuclei (2-3m) are particularly evident in these secretory cells. Blood capillaries are present prominently at the outer margin of these cells (Fig.11).

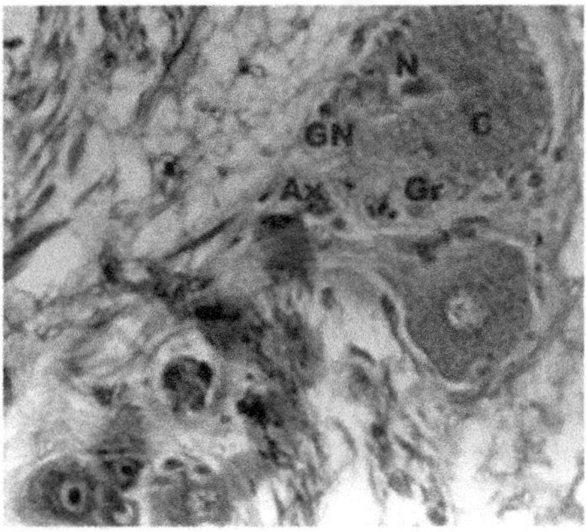

Fig. 11. Light micrograph of the S-phase GN cells (N–nucleus, C–cytoplasm, V– vacuole, Gr–granule Gn–glial cell nuclei) X400 (**Source:** Diwan *et al.*, 2008)

Cytoplasm of B and C cells exhibit similar pattern of secretory activity as in the GN and A cells, but in a lesser extent. Peak secretory activities in these cells are evident from the dark secretory granules spread throughout the cytoplasm. The scattered vacuoles which are apparent throughout the cytoplasm during the preceding vacuolar phase, showed considerably reduced in this stage. These vacuoles get shifted towards the outer cytoplasm and the entire cell volume gets filled with these dark secretory granules (Fig.11). These dark secretory granules are encounter in most of this round to oval vacuoles. Aggregations of small secretory granules are also occurring on the nuclear membranes of the centrally located nucleus and perinucleolar area. The hyper trophied glial cells and capillary net works are not prominent in the B & C type cells as in the case of GN and A cells. Ultrastructural features of these secretory phase NSCs are conspicuously different from that observed in the perikarya of the previous stages. Here the distribution of cellular organelles are limited in the S-phase cell's than the vacuolar phase cells however, greater quantity of various secretory granules are apparent in the perikarya of these secretory phase NSCs (Fig. 12). A marked increase in the number of highly electron dense secretory granules occurs in all the NSC types. Irrespective of the cell types, all the NSCs during their secretory phase possess RER, highly active Golgi complex, mitochondria, lysosomes etc in addition to their electron dense secretory granules and developing vesicles (Fig.12, 13).

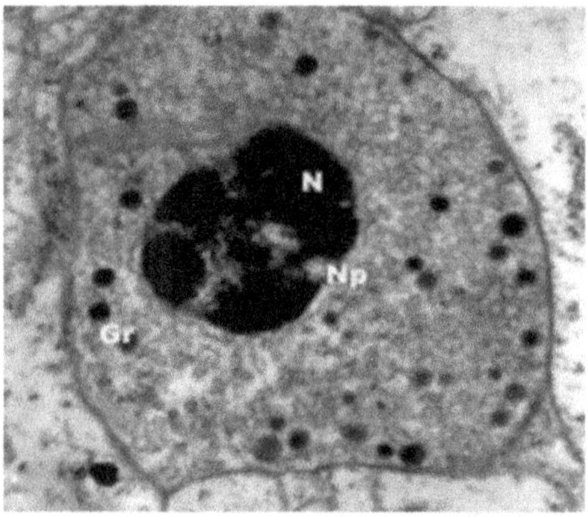

Fig. 12. Electron-micrographs of the S-phase GN cells (N-nucleus, M-mitochondria, G-Golgi bodies, RER – rough endoplasmic reticulum, V – vacoule, Gr-granules, MVB – multi vesicular bodies) X4600 (**Source:** Diwan *et al.*, 2008)

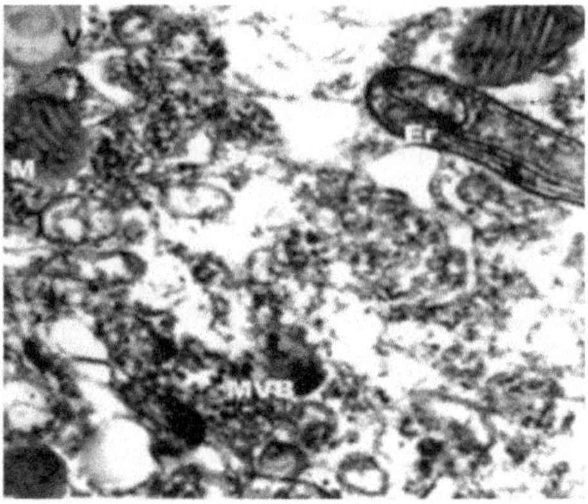

Fig. 13. Electron-micrographs of the S-phase A cells(N–nucleus, M–mitochondria, G–Golgi bodies, RER – rough endoplasmic reticulum, V – vacoule, Gr-granules, MVB – multi vesicular bodies, Np-nuclear pores) X15,500 (**Source:** Diwan *et al.*, 2008)

The rough endoplasmic reticulum in the form of short dilated cisternae dominate over other morphological forms which include flattened, parallely arranged elongated cistenae and concentrically arranged membraneous whorl of elements. The processes of granule synthesis commence in the cisternae of rough endoplasmic reticulum, which contain a flocculent material. Numerous, small sub-spherical to oval vesicular bodies are present in the cytoplasm in association with the RER. Similar types of small secretory granules, which occur in the inner side of the cisternae of RER, are also present in these vesicles as it matures (Fig.13). These moderately electron dense vesicles, packed with secretory granules and derived from RER elements, are abundant in the perikaryon especially around the RER and active golgi elements of these neurosecretory cells. During the next stage these vesicular bodies with moderate electron dense granules near the golgi complex get fused with the outermost golgis accule. The outer golgis accules are generally dilated and contain granules of low and moderate electron density.

Size of these granules decreased towards the maturing or trans-face, however, the electron density of their contents increased. Accumulations of finely granular, electron-dense materials are apparent within the saccules especially at the cis and trans saccules of these golgi complex. After the passage of these precursor granules through the golgi complex, the mature neurosecretory granules with their electron dense contents are budded off from the margins of the trans-most saccules. On occasions, the entire trans-most saccule appears to separate from the golgi complex and subsequently get divided into spherical membrane bounded electron dense neurosecretory granules in the cytoplasm. In addition to these highly electron dense granules, the golgi complex in these activated perikarya produce large numbers of coated vesicles. These coated vesicles contain an electron-lucent material and are most frequently but not exclusively associated with the trans-most saccule of the golgi complex which is not at all incorporated in the production of electron dense neurosecretory granules.

In the process of secretory granule formation three consecutive morphological stages are identified on the basis of their electron density shape and their relative position. The stages include trans, central and peripheral granules. Trans-granules are the newly formed granules and they are still connected to the golgi apparatus. These trans secretory granules show the same high electron density as the dense material in the golgi apparatus. Whereas, the central secretory granules are found budded off from the golgi apparatus and occurred at some distance from the trans-face of the golgi apparatus. These central granules are slightly more electron-dense and smaller than the trans-secretory granules. Thirdly, the peripheral secretory granules those which leaves golgi apparatus completely lies in the perikarya. These secretory granules exhibit the same electron density but are more regular and spherical in morphological appearance.

Moderate numbers of mitochondria that are present particularly close to the plasma membrane their profiles are irregular and cristae often appear indistinct and distorted. Small granular accumulations visible during the vacuolar phase are not visible in these cells of secretory phase. Various types of lysosomal structures containing secretory granules encounter in these secretory cells. The most dominant subcellular structures encountered in these NSCs are the various types of dense cored secretory granules and electron lucent vacuoles of variable size and shape (Fig 14). These neurosecretory granules include highly electron-dense, homogenous and rather large round granules bounded by a thin limiting membrane. Most of these types of granules are dispersed throughout the cytoplasm. These moderately electron dense round granules form the 2nd dominant type of granules. Comparatively smaller & highly electron-dense round to oval granules not bounded by any type of limiting membranes compose the third type of secretory granules. The numbers of these smaller granules are very less in most of the NSC types. Apart from these scattered granules some grouped elementary granules also encounter in some of this NSCs. Lysosome like dense bodies of various sizes and shapes often packed with small elementary granules are present along with these various types of neuro secretory granules. These electron dense granular elements are present in association with the RER elements. In addition to these variably shaped electron dense neurosecretory granules numerous electron-lucent vesicles and vacuolar bodies of different size and shape are also apparent in these NSCs (Fig.14).

4.5 DISTRIBUTION AND MAPPING OF NSCS IN DIFFERENT GANGLIA

I. Supra-Oesophageal Ganglion

In most of the penaeid shrmpis supra-oesophageal ganglion or cerebral ganglion is richly endowed with an abundant supply of NSC groups, forms of different NSCs. These NSC groups are present in the peripheral regions immediately underneath the connective tissue sheath. Totally 14 NSC groups are present in this ganglion and they are not identical. Some of them are larger and contain greater number of secretory cells. In the dorsal side of the cerebral ganglion there are six groups of neurosecretory cells, and these groups are larger with greater number of NSCs. The anterior dorsal groups (ADG) are the first and the largest NSC group present in the cerebral ganglion. This pear shaped NSC group consists of a few GN cells, a large number of A cells and a very few number of B cells. This group lies at the centre of the anterior most regions, in between the junction of the optic nerves and antennary nerves. GN cells are present at the anterior most region of these pear shaped cell group. Most dominated 'A' cells occur in the entire cell group. A pair of median dorsal groups (MDG) forms the second pair of NSC groups, in the middle part of the ganglion in front of the junction of the maxillary nerve. This NSC group contains majority of B cells and a few 'A' cells at the outer margin of the cell group. The third or posterior dorsal group (PDG) is present at the point of origin of the circum oesophageal connectives at the posterior most region of the main body. This group is composed of 4-5 'A' cells and a few B cells. This is the smallest NSC group present in the dorsal side of the cerebral ganglion.

Whereas the ventral side of this ganglion has 8 NSC groups, and these entire cell groups are smaller with fewer number of NSCs. Distribution of the cell groups in ventral is almost similar to that in the dorsal side. The first pair of anterior ventral group (AVG) is the largest in the ventral plane and present in front of the optic nerve. GN cells are totally absent in this cell group but 3-4 A cells are present at the anterior most margin. B cells are the dominant cell type and a few numbers of C cells also encounter in this group (Fig.15). In the middle part there are two NSC groups. The 1st median ventral groups (IMVG) contain A, B and C cell types. B cells are exclusively dominating in this group over 2-3 A cells, and a few C cells are present in the upper and lower margins respectively. Whereas, the second median ventral group is found formed (II MVG) exclusively of B and C type cells. These two median ventral groups are present in front of the maxillary nerve. Apart from all these, a pair of posterior ventral group (PVG) is present in front of the circum-oesophageal connective. This NSC group contains A, B and C cells in almost equal proportions. The GN cells are totally absent in the ventral part of the cerebral ganglion in contrast to the dorsal part where they are in abundant.

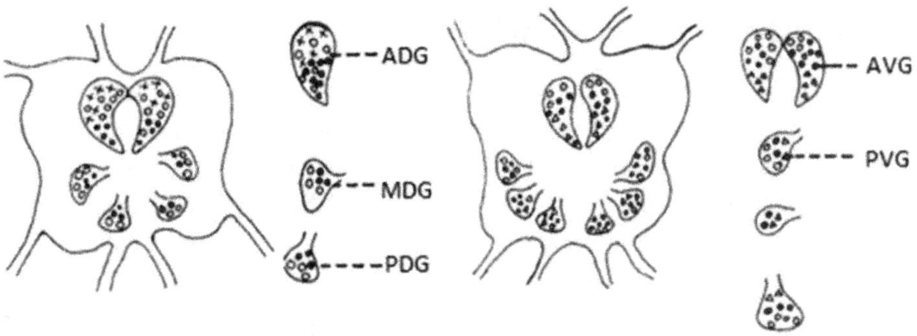

I. Dorsal View **II. Ventral View**

Fig. 14. Diagramatic representation of the distribution and mapping of NSCs in the supra oesophageal ganglion of *P Monodon* I Dorsal view II Ventral view ADG–anterior dorsal group, MDG–median dorsal group, PDG–posterior dorsal group, AVG anterior ventral group, MVG–median ventral group, PVG–posterior ventral group. X–GN cells, O–A cells, O–B cells,–C cells. (**Source:** Diwan *et al.*, 2008)

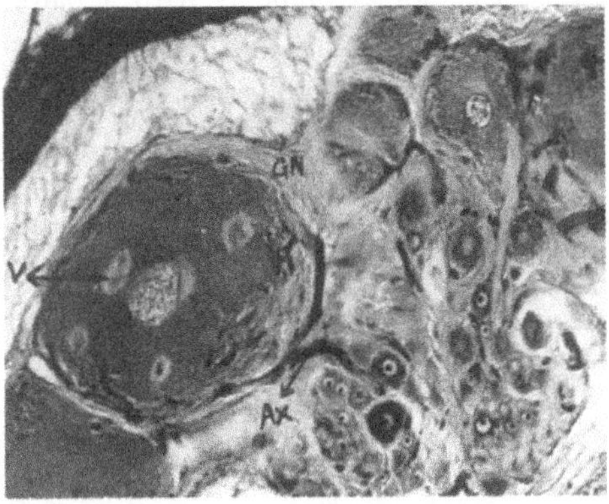

Fig. 15. GN type NSC in the subesophageal ganglion of *F. indicus* with axon (Ax) and vacuoles (V). X 100. (**Source:** Mohamed and Diwan, 1993)

Similarly C cells occur in almost all the NSC groups in the ventral side but they are completely absent in the dorsal side. Whereas A & B cells occur abundantly in dorsal and ventral sides of the cerebral ganglion.

II. Tritocerebral Ganglion

Trito cerebral ganglion is the smallest ganglion in the CNS of *P. monodon*. This paired ganglion is located in the midway of each tritocerebral connectives, connecting the cerebral ganglion to the sub oesophageal ganglion. The ganglion possesses a single NSC group with one or two GN and A cells (Fig.15). 'A' cells are always present in association with those GN cells. GN cells of this ganglion are always larger in size.

III. Sub-Oesophageal Ganglion

Sub oesophageal ganglion is a compound structure supplying nerves to different mouth parts and possesses the highest number of NSC groups. A maximum number of 16 NSC groups are identified in this ganglion on its dorsal, median and ventral planes. Unlike the other ganglions, NSC groups with all the cell types are present in median plane of this ganglia and the total number of the secretory cells are also higher.

In the dorsal side of this ganglion, there are 3 NSC groups. The firstone, the anterior dorsal group (ADG) is not paired but united as single large compound NSC group. This comprises mainly of (Giant neurosecterory cell) GN and A cells in equal proportions and a few B cells. Scatterly distributed GN and 'A' cells are present almost in the entire area of the cell group. Whereas limited numbers of B cells are occur at the interior side of the cell group

The posterior side of the dorsal plane is with a paired neurosecretory cell group, i.e. the posterior dorsal group (PDG). This PDG is endowed exclusively with B cells and a few C cells. In the median plane of this ganglion, a total of 9 NSC groups are present with varying sizes (Fig.16). A single NSC observed in the central region of the median plane, 'the central median' (CM) comprises of large numbers of GN, A and B cells. The remaining 4 pairs of NSC groups, the median lateral group (MLG) are located at the peripheral regions of the ganglia and contains only A and B cells. These medio-lateral groups are smaller compared to the central median groups. Only 4 NSC groups encounter, in the ventral side of the sub esophageal ganglion (Fig. 16). Even though there are only 4 NSC groups in the ventral plane of the sub esophageal ganglion, they contain the highest number of NSCs. The anterior ventral group (AVG) is a single large NSC group observed at the anterior most region of the ventral place, which contains a few GN and A cells, but majority of cell type are of Bcells. The large ventro-median group (VMG) at the centre is the largest NSC group in the ventral plane. It is composed mainly of GN and A cells in equal proportions. A few B cells also encounter in the lower most part of this cell group. At the posterior end of this ganglion a pair of small NSC groups viz. the posterior ventral groups (PVG) is present with a few NSCs, which exclusively consists of oval to polygonal B cells and small C cells. The sub oesophageal ganglion contains the highest number of NSC groups and cell types. There are lots of GN and A cells in the dorsal and ventral planes. While the median plane contain more B and C cells. The larger GN and 'A' type cells are completely absent in the posterior end of the ganglion.

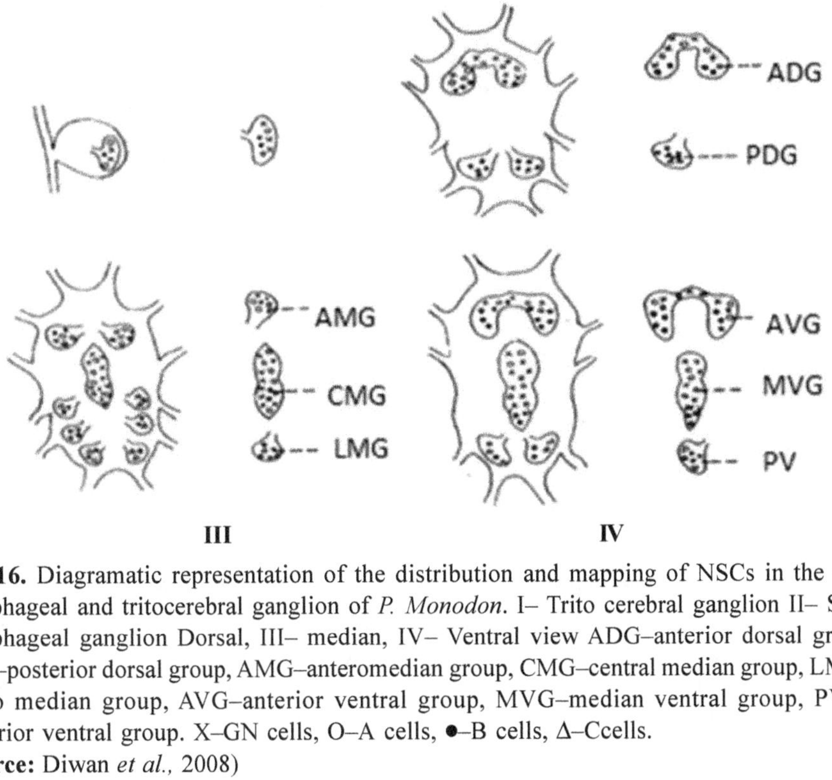

Fig. 16. Diagramatic representation of the distribution and mapping of NSCs in the sub-oesophageal and tritocerebral ganglion of *P. Monodon*. I– Trito cerebral ganglion II– Sub–oesophageal ganglion Dorsal, III– median, IV– Ventral view ADG–anterior dorsal group, PDG–posterior dorsal group, AMG–anteromedian group, CMG–central median group, LMG–latero median group, AVG–anterior ventral group, MVG–median ventral group, PVG–posterior ventral group. X–GN cells, O–A cells, •–B cells, Δ–Ccells.
(**Source:** Diwan *et al.*, 2008)

IV. Thoracic Ganglion

There are five thoracic ganglia innervating the five thoracic segments. These ganglia are connected in their longitudinal axis by short-paired connectives. In all these ganglions there are 3NSC groups with few GN and A cells, and higher B and C cells in their dorsal side. The anterior most regions have a large single NSC group, i.e. the anterior dorsal group (ADG). This ADG contains only few

GN & A cells at their outer margin and here B and C cells dominate in this cell group (Fig.17). Posteriorly there is a pair of small NSC group named posterior dorsal group (PDG) and the same is located in front of the junction of the ventral nerve cord. This cell group contains 2-3'A' cells and comparatively more number of B and C cells.

A single median NSC group encounter in the ventral side is termed as verntro-median group (VMG) (Fig. 17), which is oblong with, flattened anterior and posterior ends. Like the VMG group of the sub esophageal ganglion this also contains mainly GN and 'A' cells. Very few B cells are also encounter at the posterior end of the cell group; however, C cells are absent in this.

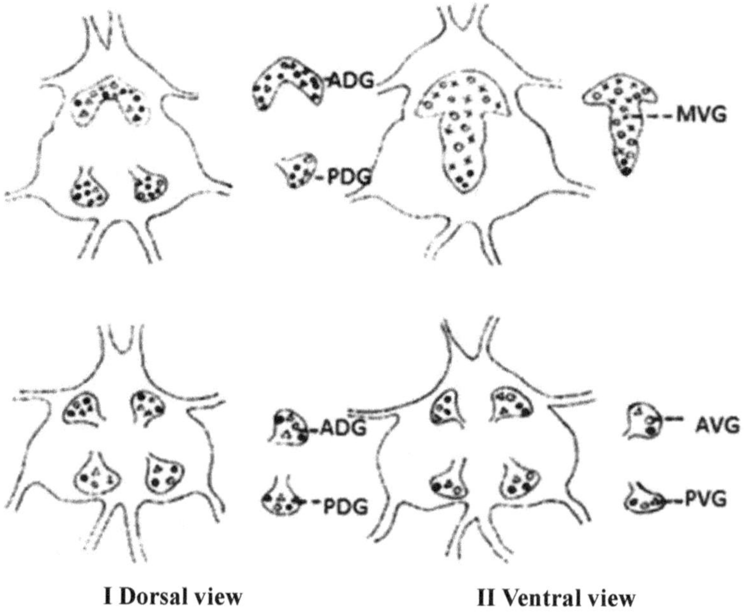

Fig. 17. Diagramatic representation of the distribution and mapping of NSCs in the thoracic and abdominal ganglia *P. Monodon*. A Thoracic and B Abdomina ganglia ADG–anteriordorsal group, PDG–posteriordorsal group, MVG–medianventral group, PVG–posteriorventra lgroup. X–GN cells, O–A cells, •–B cells, Δ–C cells
(**Source:** Diwan *et al.*, 2008)

V. Abdominal Ganglion

There are five abdominal ganglia corresponding to each abdominal appendage, and these are usually smaller in size compared to other ganglia. Histologically this ganglion contains only few NSCs. Even though there are 8 NSC groups together in both dorsal and ventral planes, the total number of neuro secretory cells is less compared to other ganglia. Both the dorsal and ventral sides contain two pairs of NSC groups, one at the anterior most region just in front of the junction where the ventral nerve cord joins the ganglia, the 'anterior dorsals' and the other at the posterior end, the 'posterior dorsals' just in front of the junction from where the VNC leaves the ganglion in the dorsal side. Both of these NSC groups contain the three types of NSCs A, B, and C cells however, their number seems to be very less compared to other ganglia.

4.6 Eyestalk Neurosecretory System

The endocrine system of the eyestalk consists of large prominent, stalked compound eyes attached to the head region by the proximal end of the optic stalk or peduncle. The eyestalk is a complex structure composed of radial units the ommatidia and covered externally by a thin layer of transparent cuticle demarcated into facets. The clear 'V' shaped notch on the dorsal side of the eyestalk is prominent. The outer lying cuticle of the eye is subtended basally with a layer of epicornea genous cells and crystalline cone cells (Fig. 18).

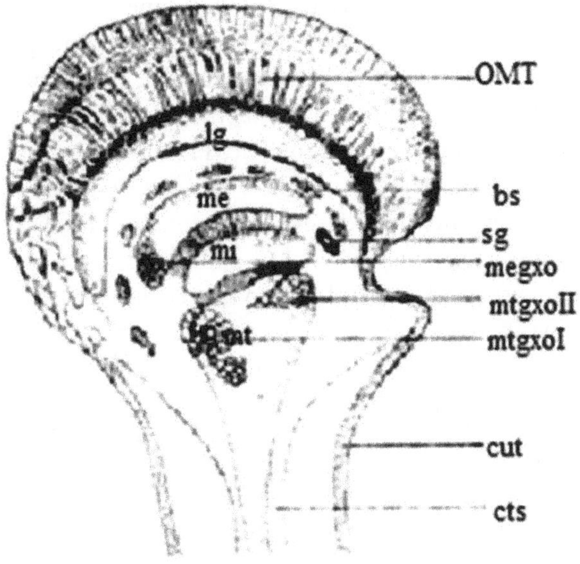

Fig. 18. Schematic representation of the important structures in the right eyestalk of the shrimp as seen from the dorsal side.omt–ommatidia, lg–lamina ganglionaris, bs–blood sinus, medulla externa, megxo –medulla externa, ganglionic X–organ, sg–sinus gland, mi–medulla interna, mt– medulla terminalis, mtgxoI –medulla terminalis ganglionic X-organone, mtgxoII– medulla terminals ganglionic X-organ two, cut–cuticle and cts–connective tissue sheath. (**Source:** Diwan *et al.*, 2008)

Below the crystalline cone cells are the retinular cells. These together form the dioptric portion. The basement membrane divides the distal dioptric portion of the ommatidia from the proximal ganglionic portion. Proximal ganglionic portion of the eye contains three ganglia, the distal one the medulla externa (ME), followed by the medial one the medulla interna (MI) and the most proximal one the medulla terminalis (MT). Between the ommatidia and the medulla externa is the lamina ganglion aris (LG). Lamina ganglion aris is located exactly between the distal fasciculate zone of the dioptric-portion and the proximal medulla externa. LG is devoid of any NSCs. between the LG and the ME, vascular layer and ten to fifteen haemolymph lacunae (blood sinuses) are often visible. The neurohaemal organ, i.e. the sinus gland, is invariably found associated with one such vascular process.

The ME is an inverted cup shaped lobe immediately beneath the LG. The distal surface of the lobe is undulating and the NSCs are present in the dorsal extremities of ME. Fibers of the neuropile and blood capillaries are found in this region. Below the crystalline cone cells are the retinular cells. These together forms the dioptric portion. The basement membrane divides the distal dioptric portion of the ommatidia from the proximal ganglionic portion. Proximal ganglionic portion of the eye contains three ganglia, the distal one the medulla externa (ME), followed by the medial one the medulla interna (MI) and the most proximal one the medulla terminalis (MT). Between the ommatidia and the medulla externa is the lamina ganglion aris (LG). Lamina ganglion aris is located exactly between the distal fasciculated zone of the dioptric-portion and the proximal medulla externa (Fig.19). LG is devoid of any NSCs. Between the LG and the ME, vascular layer and ten to fifteen haemolymph lacunae (blood sinuses) are often visible. The neurohaemal organ, i.e. the sinus gland, is invariably found associated with one such vascular process.

The ME is an inverted cup shaped lobe immediately beneath the LG. The distal surface of the lobe is undulating and the NSCs are present in the dorsal extremities of ME (Fig.19). Fibers of the neuropile and blood capillaries are found in this region.

MI is similar in structure to the ME including the inter-spaced capillaries, though slightly smaller in size and devoid of NSC. The sinus gland situates between this ME and MI (Fig. 20). The most proximal ganglia are the knobs like medulla terminalis on the surface of which, the principal NSC groups are located. Fibers from these cells directly forms the optic nerve. Immediately surrounding the medulla terminalis and the other ganglia is the broad cortical glia, followed by a thin, dense neurilemma. MT, like other ganglia contains matrix of capillaries and lacunae.

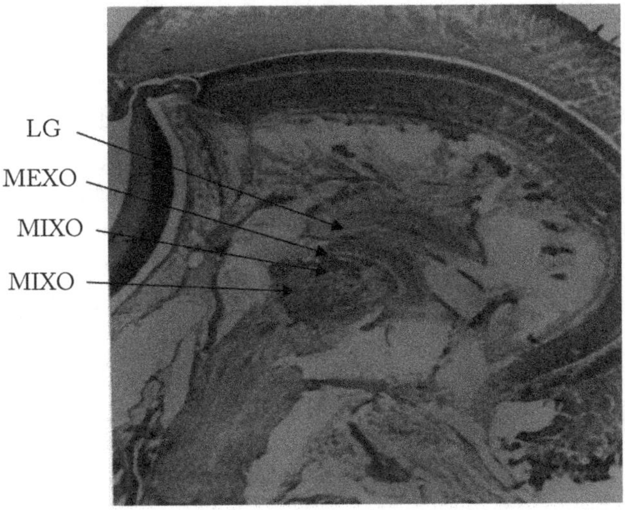

Fig. 19. Cross section of neuroendocrine complex of the eyestalk in Penaeid shrimp.

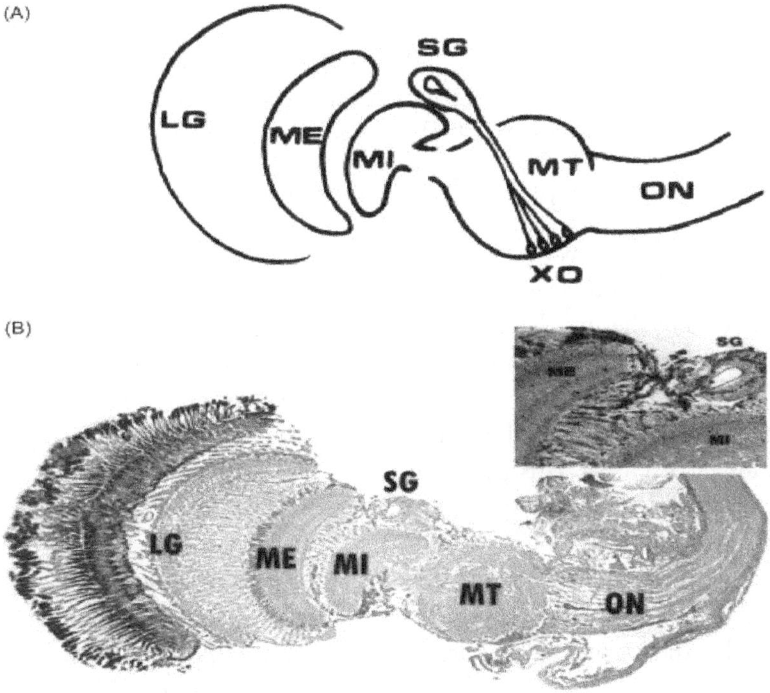

Fig. 20. Neuroendocrine complex of the eyestalk showing different lobes of the optic ganglion.

4.7 DISTRIBUTION OF NSC IN THE EYESTALK

There are mainly three NSC groups in the eyestalks of the shrimp. These cell groups are named on the basis of their location. Among the medullary lobes, neurosecretory cells are found only in the MT and ME (Fig. 21). NSCs groups are absent in the lamina ganglion aries and MI of the eyestalk. The principal NSC groups are located at the apical portion of the MT.

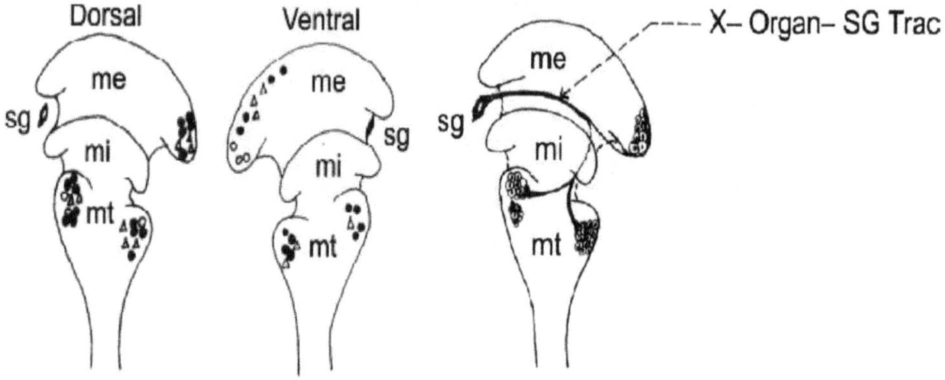

I. Mapping of NSCs in the eyestalk **II.X–Organ sinus gland tract in the eyestalk**

Fig. 21. Diagramatic representation of the distribution and mapping of NSCs in the eyestalk (I) and the X–organ sinus gland tract(II), me–medulla externa, mi–medulla interna, mt–medulla terminalis, sg–sinus gland, O–A cell, •–B cell, Δ–C cell
(**Source:** Diwan *et al.*, 2008)

Dorsally there are two distinct NSC groups, at the top most regions of the lateral sides of MT. However, there is only one NSC group at the ventral side, of the MT. Among the two NSC groups encounter in the dorsal side one is at the right side of the right eye and is termed as the medulla terminalis X organ-I, (MTGXO-I) and the NSC group on the left side is termed the medulla terminal is X organ-II (MTGXO-II). The MTGXO-I is small and contains comparatively lesser number of NSCs. It is pyriform in shape with a tapering end from where the axon terminal leaves the ganglion. The axon terminal is joins the main axon coming from MTGXOII before it leaves MT. The sinus gland is positioned immediately distal to this MTGXO-I group (Fig.20). MTGXO-I is the only NSC group present in the right side of the right eye where the sinus gland is located. All other NSC groups are present at the opposite or left side of the right eye. MTGXO-I contains a few A & C cells, and majority of B cells (Fig.21).

MTGXO-II is also located in the upper most area of the MT on the opposite side of MTGXO I, which is a comparatively larger NSC group containing more secretory cells. Even though a few secretory cells are dispersed in the lower part of MT, majority of the cells are located in the apical part itself. Further a small group of NSCs is located in the close proximity to the MTGXOII. The MEGXO is present on the lateral surface of the ME, on its rostral side. These groups appear as diffuse, with a number of B & C and 2-3 'A'cells on dorsal and ventral surfaces (Fig.22). Neurosecretory axon leaves from this ellipsoid cell group (MTGXO II) at its tapering lower end, which joins the axon coming from the MTGXO-I before leaving the MT. Type B & C cells are the most abundant cell types on both dorsal and ventral aspects of this cell group with few larger 'A' cells (Fig 23 & 24).

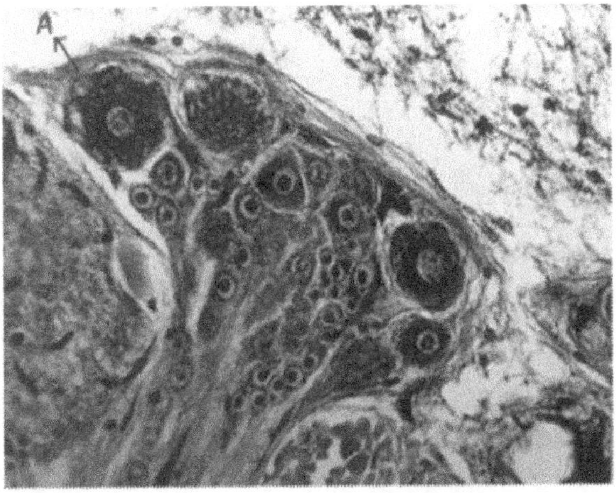

Fig. 22. The pyriform MTGXO I of *F. indicus* with A, B and C cells. Note the combined axons leaving the X organs. X 100. (**Source:** Mohamed and Diwan, 1993).

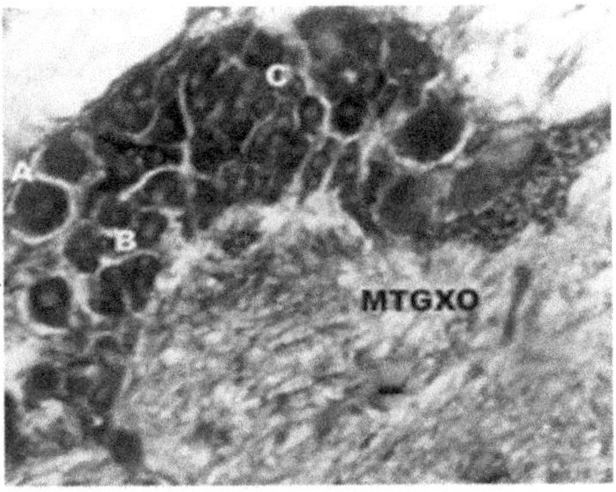

Fig. 24. Light micrograph of the medulla externa X-organ with A, B, C, type cells X400 (**Source:** Diwan *et al.*, 2008)

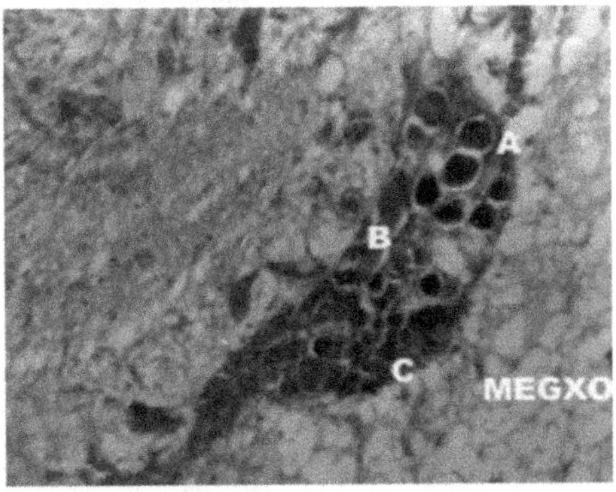

Fig. 23. Light micrograph of the medulla terminalis X-organ with A, B, C cells x400 (**Source:** Diwan *et al.*, 2008)

4.8 SINUS GLAND AND THE X-ORGAN SINUS GLAND TRACT

The sinus gland is the principal neurohaemal organ in the eyestalk. It is located between the medulla externa and medulla interna in the dorso-lateral surface of the eyestalk, connected with some axon bundle coming from the ganglionic part of the optic ganglia. In most of the shrimp the sinus gland is in association with the finger like blood sinuses around the medullary lobes. The sinus gland itself is a blood sinus surrounded by numerous heavily stained axon endings on the dorso-lateral side of the optic ganglia (Fig 25). In the right eyestalk, it is located in the right side between the ME and MI. Light microscopically this gland is roughly elliptical and at its lateral plane is more flattened. At the central part of the gland is the internal blood sinus. In adult animals (above150 mm TL) the sinus gland measures about 250-275μ long and 100-120 μ wide (Fig 26).

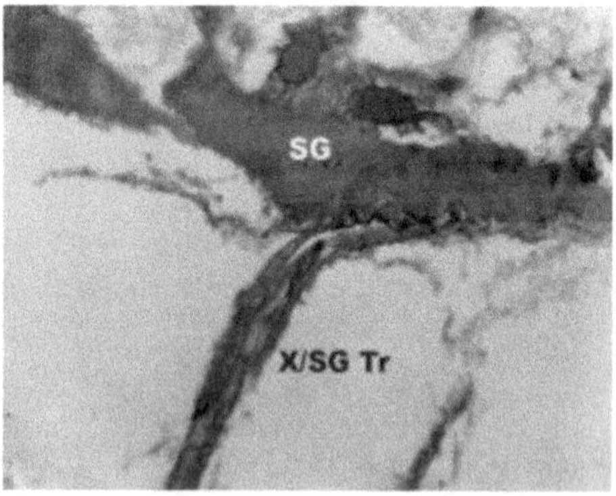

Fig. 25. Light micrograph of the X-organ sinus gland tract entering the sinus gland of *P. Monodon* (SG sinus gland) x100
(**Source:** Diwan *et al.*, 2008)

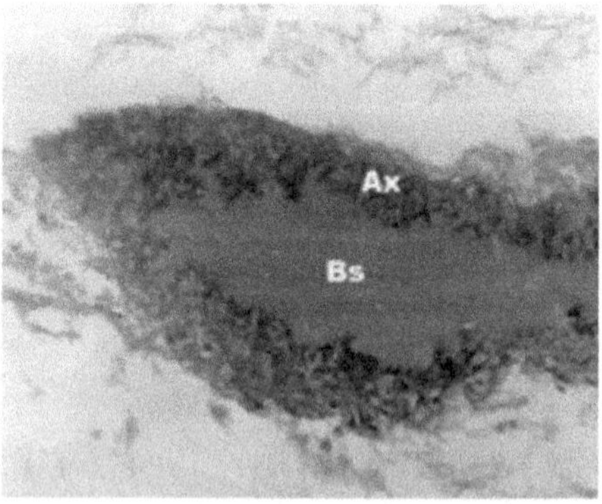

Fig. 26. Light micrograph of the sinusgland (Ax-axon bundles in *P. Monodon* BS-blood sinus) Fig. 172x100, Fig. 173, x 400
(**Source:** Diwan *et al.*, 2008)

Ultrastructurally the sinus gland is composed of axons, which terminate in the bulb shaped enlargements just beneath the basement membrane lining the blood sinus. Biomolecular aspects of the sinus gland shows that the gland is build up of 4 main components, the axons, axon terminals, glial cells and blood sinus. A thin membrane separates the sinus gland with its heavily stained axonal endings and neurosecretory materials from the blood sinus (Fig 27). As a neuro haemal organ, the sinus gland receives the axons from the perikarya of the neurosecretory cells, which are localized in other ganglionic parts of the neuroendocrine systems (Fig. 28).

The profiles of these unmyelinated axons are filled to a greater or lesser extent by bound granules. Transversely cut ends of these axons at their pre-terminal part shows the presence of microtubules at their central part. Preterminally, these narrow axons are fit closely together and contain only neurotubules, running along its entire length. However, the terminal region is characterized by the presence of a greater number of membrane bound granules. This terminal area of the axon is swollen, for increasing its storage volume, and also but on to a basement membrane that separates the axon from the blood sinus. These axon terminals abutting on the blood sinus of the gland contain the neurosecretory granules, electron lucent vesicles and small mitochondria. Golgi complex is also present in certain times. The granules are of varying electron density and generally round to oval in shape. A few granules are dumb bell shaped or have an electron dense head and a tail of decreasing electron-density. Besides these electron-dense granules, electron-lucent vesicles with a diameter of 20-50nm, the so-called synaptic vesicles are also apparent in these axon terminals. Two types of electron-lucent vesicles with a diameter of 500-600A° are present usually scattered among neurosecretory granules in certain axon terminals. The other larger electron-lucent vesicles are irregular in shape and have a diameter of 2μ. Those terminals which possess these vesicles contain only few neurosecretory granules. This suggests the presence of vesicles as an indication of the active neuron.

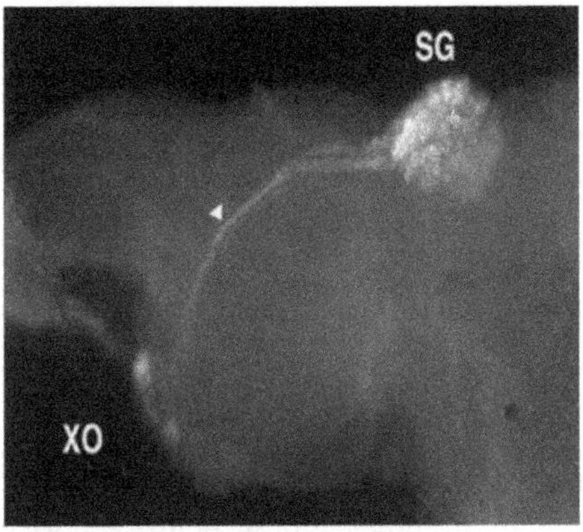

Fig. 27. Sinus gland tract with X- organ of medulla terminalis of Penaeid shrimp.

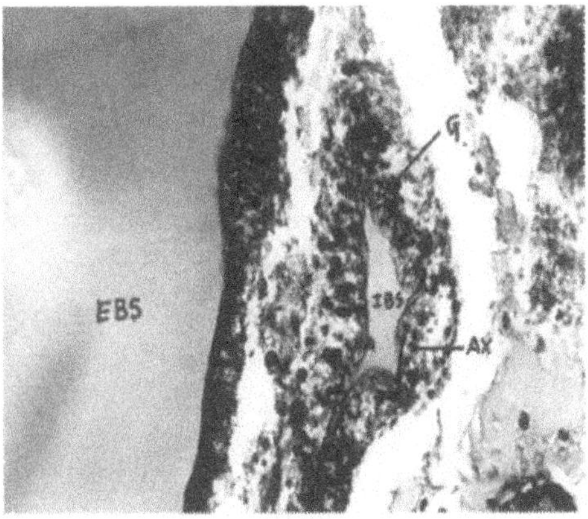

Fig. 28. Detail of the sinus gland of adjacent to the external blood sinus (EBS) (*F. indicus*). Note the abundance of neurosecretory granules (G), bulbous axonal endings (AX) and internal blood sinus (IBS). X 400 (**Source:** Mohamed and Diwan, 1993).

Passing between many of the neurosecretory axons are the glial cells that form the nuclear component of the sinus gland. The greater part of the glial cell is occupied by the nucleus, which is irregular in shape and measured up to 3μ. The cytoplasm contains microtubules, vesicular structures, mitochondria, endoplasmic reticulum and free ribosomes. Rough endoplasmic reticulum is often situated around the mitochondria. golgi complex is encountered rarely within the glial cells (Fig. 29 & 30).

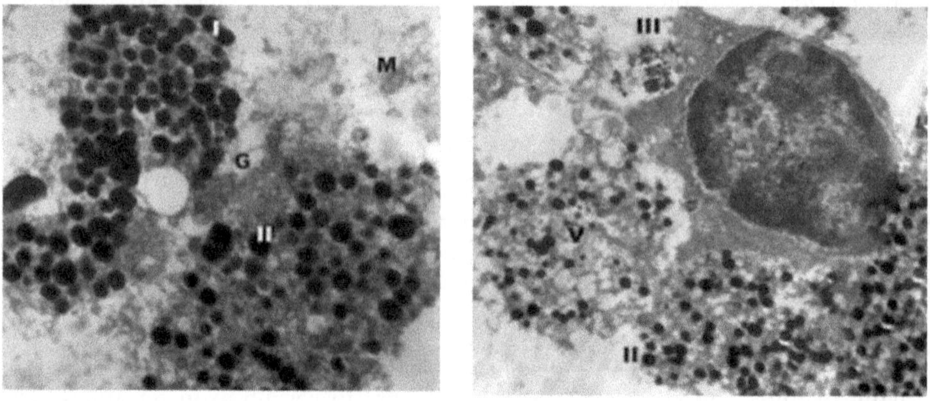

Figs. 29 and 30. Electron –micrograph of the terminal end of the axon bundles around the blood sinus of the sinus gland showing different types of neurosecretory granules of *P. Monodon* (I-type1,II–type2, III–type 3, IV–type 4, V–type 5 granules, M–mitochondria, G–Golgi bodies, V–vesicles, GN–glial cell nuclei, SV–synaptic vesicles and MB–membrane separating the blood sinus and the sinus gland) X21,500 and X31,000
(**Source:** Diwan *et al.*, 2008).

The blood sinus forms the remaining major component of the sinus gland, and the structure less basement membrane forms an acellular boundary between the axons and internal blood sinus, and it varies from 150-450 nm in thickness. Sometimes the basement membrane becomes convoluted in certain areas, and extensions of which occur in between the axon terminals. Axon terminals in the neighborhood of the blood lacunae are characterized by the storage and release of neurosecretory granules. These granules vary in shape, size and electron-density.

In shrimp five different types of axon terminals are tentatively identified in the sinus gland on the basis of characters of its granules. These neurosecretory granules do not mix within the individual axons. The five different types of neurosecretory granules in the axon terminals are described here as type I to V.

The type I terminals (Fig. 29) contain considerably large neurosecretory granules averaging (1700A°) in diameter. These granules are round, elliptical or oval in shape and have very electron dense, homogenous contents, which completely fill the enclosing membranes. These less frequent axon terminals are compactly packed with highly electron-dense granules of almost same size and shape. Nothing other than the granules is apparent in these terminals.

The type II axons are the most common type in the sinus gland of *P. monodon*. These axon terminals contain granules ranging from (1200-2300A°)

in diameter and are oval in shape (Figs.29 & 30). A few numbers of larger granules with high electron density are also present, along with the other moderately larger electron-dense granules. These axon terminals contain loosely packed granules, in contrast with the compactly packed granules of type I terminals. Electron-lucent vesicles, microtubules and mitochondria are found in these axon terminals.

Type III terminals contain the smallest neurosecretory granules in the sinus gland (Fig.31). The diameter of these small granules varies between (700- 2300 A°).These round granules are identical in size and shape and contain homogenous, moderately dense cores. Nothing other than the loosely packed small granules is apparent in these axon terminals. These small axon terminals are not very common in the sinus gland.

The type IV terminals contain the largest neurosecretory granules, averaging (700-1200 A°) in diameter. These granules are round to oval in shape and possess highly electron-dense homogenous contents, which completely fill the enclosing membranes. These axon terminals contain only thickly packed large granules. Other electron lucent vesicles or cell organelle like mitochondria, Golgi complex etc. are not at all apparent.

The type V axon terminals contain granules of varying diameters and shape. The diameters of this round to irregular neurosecretory granules vary between (1800-2600A°). Granules present are rather dispersed than aggregated. Mainly 2 types of granules are found in these axon terminals. One type very few in numbers, very large with a diameter of (1900 A°), highly electron-dense, and the other smaller granules with less electron density but in large numbers. Along with these granules some other microtubules and electron-lucent vesicles are also visible.

The optic ganglion has a prominent axon bundle and this nerve bundle originates from the NSC groups and terminates in the sinus gland (Fig.28). This nervous tract from the X organ to the sinus gland termed as X-organ sinus gland tract. This predominant nerve bundle at the ganglionic part of optic ganglion is clearly visible in the MI (Fig. 21) especially at its terminating end at the sinus gland, and also in the middle part of the medulla interna & medulla terminalis. Further, it is connected with NSC group located at the dorso-lateral surface of the medulla terminal is the so-called MTGXO-I. Additional axons, originating from the small NSC group in the immediate neighborhood and small axon profiles from the NSCs located at the other side of the MT, join together and form a small axon bundle. This bundle also joins the main tract as it passes forwards. Axon bundles from the MEGXO are also found joining this tract during its course through the Mt to the sinus gland. The exact points where these small axon bundles' join the main tract are not clear. The schematic representation of the X organ sinus gland tract is given in (Fig.21).

The X organ sinus gland tract near the sinus gland is approximately 18.2µ min thickness. However its only 10-12 µ thick when it leaves MTGXO-I. When the axon from the MTGXO-II joins with the axons of MTGXO-I above the MGXOII, their thicknesses get increased from 12 to 15 µ. These two together enter the MI where the axon from MEGXO also joins and these three axons together precede dorso-laterally towards the sinus gland.

4.9 NEUROENDOCRINE CONTROL OF REPRODUCTION

The NSCs in different ganglionic masses during different maturity stages indicate concomitant changes in the secretory phases of these cells during the process of maturation. These specific and dominant changes indicate their significant roles in the control of reproduction. The mean percentage occurrence of NSCs in the eyestalk and central nervous system at different stages of their secretory cycle is varying during the various maturity stages of *P. monodon* and the same are presented in the Table 2.

Table 2. Percentage of total neurosecretory cells in eyestalk and central nervous system of immature and maturing *P. monodon*

Secretory	Eyestalk		Central Nervous System	
phase	Immature (%)	Maturing (%)	Immature (%)	Maturing (%)
Q	20-35	45-60	45-50	50-60
V	15-25	20-27	20-35	25-30
S	50-55	15-25	15-25	10-15

Q–Quiescent, S–Secretory, V–Vacuolar (**Source:** Diwan *et al.*, 2008)

4.10 EYESTALK NEUROSECRETORY CELLS DURING MATURATION PROCESS

The eyestalk X-organ complex of immature shrimp has a higher number of active cells with more than 70% of the NSCs in both the MTGXO and MEGXO. Due to the presence of large numbers of active NSCs, both the cell groups (MTGXO and MEGXO) are particularly prominent and clear in the eyestalks of immature animals than in other maturing stages (Figs. 31 & 32). NSC at the S-phase contains highly granular cytoplasm and V-phase is characterized by the presence of vacuoles. Streaming of cytoplasmic granules is apparent in many of the S-phase NSCs. Ultrastructural observations indicate that, the neurosecretory granules of different size, shape and electron-density are apparent in the perikarya of the active NSCs of eyestalk X-organ complex from the immature animals. Increased secretory activity is due its abundant supply of developing vacuoles and vesicular bodies. Perikaryon of most of these cells possesses numerous highly active cell organelles, especially the very active Golgi complex with moderately electron-dense flocculent materials. Free and poly ribosomes are apparent throughout the cytoplasm of these active cells. Multi vesicular bodies also encounter in association with lysosomal structures. Exocytosis of fully developed granules is apparent at the outer membranes of these NSCs. Among different types of NSCs, A type cells exhibited more secretory activity followed by B cells in the eyestalk X-organ complex of immature shrimps. More than 60% of the A cells are at the S-phase with highly granulated cytoplasm. Vacuolar phase A cells are also apparent in equal numbers in these NSC groups. 'A' cells at the resting or quiescent phase are not completely absent. Remaining 10-12% of these A cells are at the resting or Q-phase in all the cell groups of the eyestalk. Such a heightened neuro secretory activity is not there in B & C cell types. However 50% of the B cells exhibit secretory activities by its active coarse cytoplasm due to the presence of secretory granules inter spaced with small active cytoplasmic vacuoles. Among the C cells, 20-35% of them exhibit coarse cytoplasm, which indicates the secretory activity. However, most of C cells are at the Q-phase of its secretory cycle.

In the early maturing stage the number of the S-phase cells in the eyestalk complex decreased considerably. Consequently the percentage of Q-phase cells increase drastically after the initiation of the vitellogenesis. However, the streaming

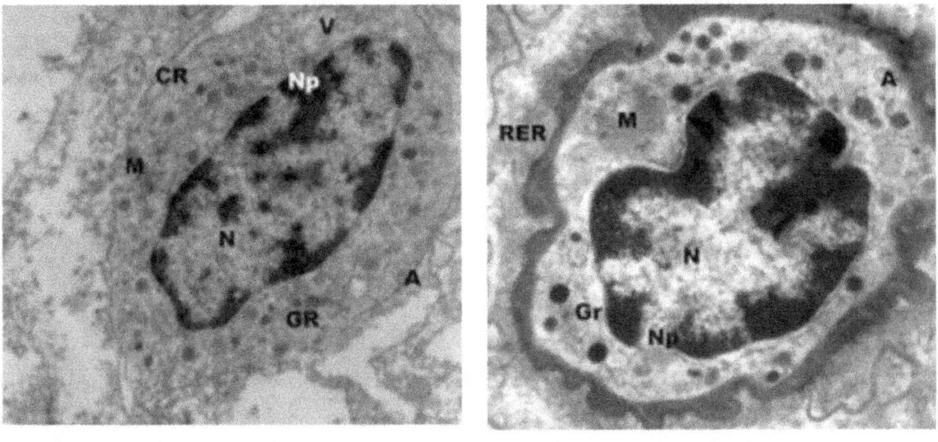

Figs. 31 & 32. Electron–micrograph of the active A and B cells in the eyestalk of immature *P. monodon* (N–nucleus, M–mitochondria, RER–rough endoplasmic reticulum, Np–nuclear pore, Gr–granules and V–vacuole) X4, 600 (**Source:** Diwan *et al.*, 2008)

of the granular cytoplasm and hypertrophied glial cell nuclei encompassing these cells indicate the secretory activity of the existing S-phase. Whereas in the eyestalk X-organ complex of the mature animals, a higher percentage of Quiescent phase NSCs become prominent. About 60% of the NSCs are at the resting or 2 phase of the secretory cycle. Remaining NSCs are still at their vacuolar as well as secretory phases. Even though the cytoplasm of these active cells appears coarse and granular, the cytoplasmic streaming in the previous stages is not apparent in any of these active cells. Increase in the number of Q-phase cells are observed in case of 'A' cells (Figs. 33 & 34). A four-fold increase is seen in the percentage of Q-phase A cells i.e. from 12 to 46% is seen. Other two types of NSCs also exhibit concomitant increase in the percentage of Q-phase cells.

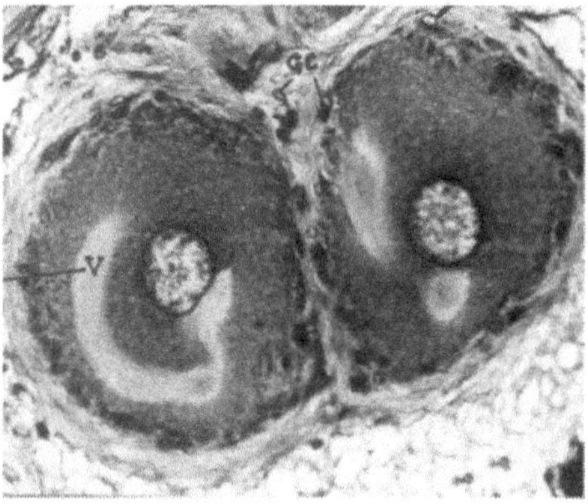

Fig. 33. Vacuolar phase (active) GN cells in subesophageal ganglion of *F. indicus*. Vacuoles (V) originate in the peripheral region of the cell and later coalesce to form a pericellular vacuole (CV). Also note glial cells (GC) and capillary plexus (CP). X 200. (**Source:** Mohamed and Diwan, 1993).

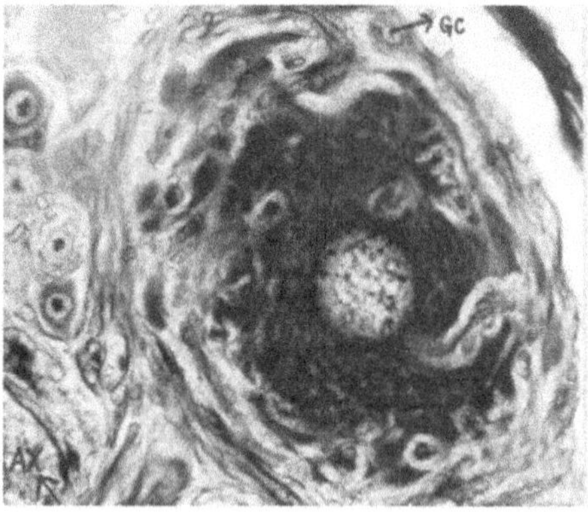

Fig. 34. Secretory phase GN cells in the subesophageal ganglion of *F. indicus*. Note extremely granular and frothy cytoplasm interspersed with vacuoles (V) and hypertrophied glial cells (GC). X200 (**Source:** Mohamed and Diwan, 1993).

4.11 NEUROSECRETORY CELLS AND THEIR CORRELATION WITH OVARIAN MATURATION

In the eyestalk X-organ complex, most of the NSCs in the ganglions of CNS from the immature shrimps are at the Q-phase of their secretory cycle. Nearly 70% of the total NSCs are in the resting Q-phase. Vacuolar cytoplasm is visible in nearly 20-24% of the NSCs, with granular cytoplasm and hypertrophied glial cells present are few in numbers. When the shrimp enters the early maturing stage the events change, and about more than 20% of the total cells are in secretory phase. Large numbers of V-phase cells are present in this early maturing stage. Vacuolated cytoplasm is apparent in nearly 32% of the total cells. However

50% of the NSCs are still at the resting or quiescent phase in the early maturing stage. The active cells at S-phase in this stage of maturity mainly comprise of larger GN and A type cells. Sudden changes in the morphology of cytoplasm occur in these larger cells. A few cells exhibit cytoplasmic streaming to some extent. Hypertrophied glial cell nuclei are apparent around these cells Fig 34.

Variably shaped large cytoplasmic vacuoles are in the V-phase NSCs mainly comprising of GN and A cells. However, B & C cell types at V- phase, do not contain well-defined larger cytoplasm vacuoles. Instead, small round vacuoles interspaced with minute granular cytoplasm give a coarse appearance to those cells. V-phase cells are comprised of 55-65% of GN and A cells, 20-27% of B cells and remaining C cells.

NSCs at their active secretory phase dominate in the vitellogenic animals, followed by vacuolar phase (Figs.35). Most of these cells are actively engage in neurosecretion. More than 48% of S-phase NSCs occur in all the ganglia of CNS during its vitellogenic stage of development (Fig.36). Vacuolated cytoplasm is present in nearly 30% of the cells. Consequently a drastic decline takes place in the percentage of Q- phase cells in all the ganglia. The percentage of these resting cells decreases from 70 to 20 in the ganglia of the late vitellogenic animal. Light and electron microscopy structure of the ganglionic centres of fully matured shrimp just before ovulation show large numbers of almost empty NSCs. Ultra structurally these empty cells possess only their centrally placed nucleus and the outer plasma membrane with a few membrane systems. However, the large numbers of resting NSCs among these empty cells indicate that, they regain their resting phase within short fractions of time. All the cell types in a ganglion do not react in the same manner. Hyperactivity is found in GN and A cells. However, B cells exhibit only moderate activity whereas comparatively less activity encounter in the smallest C type cells.

Highest secretory activity encounter in the GN cells followed by A cells. Almost 90% of the GN cells are either at S or V phase of secretion in these ganglions. Q-phase GN cells are present only in limited numbers. Similarly more than 70% of the A cells are granular and have vacuolar cytoplasm (Fig. 37). Streaming of cytoplasmic granules is observed to some extent in their axonal processes. Whereas the B & C types of cells do not exhibit such a heightened neurosecretory activity, as found in GN and A cells. The cytoplasm of these cells, at their active S & V phases contain small round vacuoles interspaced by granular cytoplasm. The hypertrophied glial cell nuclei are not prominent in these cells, like the other cells. Streaming of the cytoplasm is not apparent in these small cells.

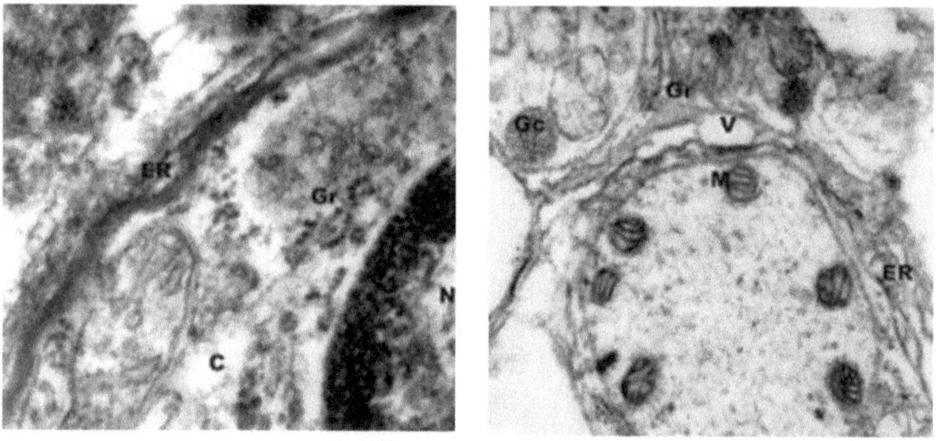

Figs. 35 and 36. Electron micrographs showing the corresponding neurosecretory phases in the ganglionic centers of the central nervous system of *P. Monodon* (N-nucleus, Gr-granules, C-cytoplasm, M-mitochondria, Er-endoplasmic reticulum, V-vacuole, MVB-multi vesicular bodies, GN-giant neurons, A-A type cells) X11,500 (**Source:** Diwan *et al.*, 2008)

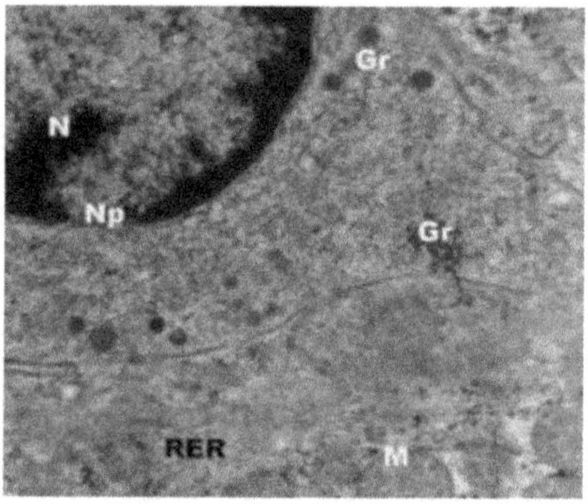

Fig.37. Electron-micrographs of the neurosecretory cells the central nervous system during the corresponding Vitellogenic phases of *P. monodon* (N- nucleus, Np- nuclear pore, RER-rough endoplasmic reticulum, Gr- granules, M- mitochondria) X 11,500
(**Source:** Diwan *et al.*, 2008)

Even though the general pattern of NSC phases are similar in all the ganglionic centres of the CNS, slight differences are noticed between different ganglionic centres. Highest percentage of active neurosecretory cells are occurring in the sub esophageal ganglion, followed by thoracic ganglions. Cerebral ganglion contains comparatively lesser number of active cells in the mature animals. However, higher numbers of S-phase NSCs are present in the cerebral ganglion during the early maturing state of ovary. Neurosecretory cells packed with granules of variable size, shape and electron-density are apparent in various ganglionic centers of mature animals.

Crustaceans utilize neuropeptide hormones for regulating diverse physiological processes including reproduction. The endocrine control of reproduction has been investigated with a wide variety of crustaceans. In this respect, the amphipods and isopods have received considerable attention (Fingerman, 1987). Investigations in decapods are, mainly in the crabs and crayfishes and the *Penaeid* group has not received that much attention in this respect. However, the neuroendocrine control of reproduction has not been elucidated clearly in any group of crustaceans (Rao, 1993).

The central nervous system of the shrimps comprises of a dorsal brain, and a ventral nerve cord with a pair of ganglia in each segment. According to MacLaughl in (1983), the central nervous system (CNS) of crustaceans basically consists of a large supra- oesophageal ganglion, the brain, and a ventral nerve cord with a pair of ganglia corresponding to each embryonic somite. In spite of the considerable variation in the number of segmental ganglia and structure of the ventral nerve cord a more or less similar type of central nervous system (CNS) is described in many in vertebrates by various investigators (Enami, 1951; Matsumoto, 1958; Fingerman and Aoto, 1959; Lake, 1970; Dall *et al.*, 1990; Mohammed *et al.*, 1993). Among the penaeids the neuroendocrine system of *P. indicus* described by Mohammed *et al.*, (1993) closely resembles the pattern seen in *P. monodon* (Diwan *et al.*, 2008).

Four different types of neurosecretory and a non-neurosecrectory cell type are described in the ganglionic centers of *P. monodon* Nanda and Ghosh (1985). Neurosecretory cells (NSC) present in the ganglionic centers of *P. monodon* are distinct from the small non-neurosecretory cells. NSCs described here, are characteristically similar to the NSCs described in other crustaceans by various investigators (Enami, 1951; Durand, 1956; Matsumoto, 1958; Fingerman and Aoto, 1959; Lake, 1970; Nakamura, 1974; Diwan and Nagabhushanam, 1974; VanHerp, *et al.*, 1977; Chandy and Kolwalker, 1985; Nanda and Ghosh, 1985; Mohammed *et al.*, 1993). In *P. indicus*, Mohammed *et al.*, (1993) also reported four different types of NSCs while in *P. japonicus*, Nakamura (1974) described 7 different types of NSCs. However, only three different types of NSCs are identified in the ganglionic centers of *P. japonicus* and *P. kerathurus* (Ramadan and Matta 1976). In the ganglionic centers of *P. stylifera*, Nagabhushanam, *et al.*,(1976) identified 8 different types of NSCs. In the fresh water shrimp *Palaemonpaucidens*, (Hisano, 1974) and *Macrobrachium rosenbergii*, (Deitz, 1982) six different types of NSCs have been identified. Descriptions of the NSC types, found in the optic and other ganglia of different species are varied, probably because of the differences in histological staining procedures, cyclic secretory activity, species differences and human subjectivity. In spite of this, various investigators like Durand (1956), Passano (1960), Adiyodi and Adiyodi (1970),

Quackenbush (1989), Nagabhushanam *et al.*, (1992) and Mohammed *et al.*,(1993) have made attempts to relate the variously distributed NSCs in crustaceans.

Classification of the NSCs in the shrimp is generally in the present chapter is done based on their shape, size, (descending order) and morphological and biomolecular characteristics of these cells and their secretory products. The GN cells are the largest NSCs noted in *P. monodon*. Similar type of GN cells has been described in *P. indicus* in the same terminology (Mohammed *et al.*, 1993). Similarly, the GN cells described in *P. monodon* are comparable to cell type I and II of *P. japonicus* (Nakamura,1974) and to the giant A cells described by Matsumoto, (1954) in the thoracic ganglia of the Japanese freshwater crab *Eriocheir japonicus* in its shape, size and histological features. All the above- mentioned investigators have described only the light microscopical features.

Aperusal of the available literature showed a paucity of information in the ultrastructural features of the NSCs in Crustacean's. Only few published reports are available with regard to electron-microscopic studies of the NSCs in crustaceans (Fingerman and Aoto, 1959; Hisano, 1976; Andrew *et al.*, 1978; Chataigner *et al.*, 1978; Nakamura, 1980; Mc Namara, 1993). Among the various investigators, ultra structural descriptions of the various NSC types have been reported only in fresh water shrimp *P. paucidens* (Hisano 1976); and in the isopod *Spheroma serratum* (Chataigner *et al.*, 1978). In penaeids, Nakamura (1980) studied the PAS positive cells of supra-oesophageal ganglion of *P. japonicus*, and Mohammed *et al.*, (1993) have showed the neurosecretory granules of *P. indicus*. Ultrastructural NSC type of *P. monodon* shows the highly active perikarya as well as the mode of synthesis of neurosecretory materials within their cytoplasm. The perikarya of the GN cells of *P. monodon* during the process of gonadal maturation possess become active with numerous cell organelles and secretory granules. More or less similar features have been described for the A type NSCs of fresh water shrimp *P. paucidens* by Hisano, (1976) and the A type cells of the isopod *S. serratum* by Chataigner *et al.*, (1978).

In its shape, size and histological features A cells of *P. monodon* are comparable with the 'V' cells of *Sesarma dehaani* (Enami, 1951); type II and III cells of *P. paucidens* (Hisano, 1974); type III cells *P. japonicus* (Nakamura, 1974); B cells of *P. Serratus* (Van Herp *et al.*, 1977); and A cells of *Charybdislucifera* (Chandy and Kolwalkar, 1985). The size and histo morphological features of A cells in *P. monodon* show close resemblance to the A cells described by Nanda and Ghosh (1985) in the same species. A cells of *P. monodon* also showed similarity to the cell types I, II, III and IV of *P. stylifera* (Nagabhushanam *et al.*, 1986) and the A cells of *P .indicus* (Mohammed *et al.*, 1993).

B cells in *P. monodon* are smaller than GN and A cells. The B cells are found similar to the B cells described by Nanda and Ghosh in the same species. These cells exhibit more or less similar characters of the type IV cells of *P. paucidens* described by Hisano (1974) and of the cell types V, VI &VII of *P. indicus* (Mohammed *et al.*, 1993). Ultrastructurally perikarya of these cells did not exhibit a high secretory activity. Active cell organelles in the perikarya of certain B cells of *P. monodon* are similar to the D and E type cells of *S. serratum* (Chataigner *et al.*, 1978). However, secretory granules are less in the perikarya of B cells of *P. monodon*. The C cells are smallest NSC type in *P. monodon* and show similar morphological characters to the C cells described by Nanda and Ghosh (1985) in the same species and the C cells of *P. indicus* (Mohammed *et al.*, 1993). D cells described by VanHerp (1977) in *P. serratus* are also comparable to the pyriform C cells of *P. monodon* .

Apart from the NSC types, a single type of non-neuro secretory cells is also reported in the ganglionic centers of *P. monodon* through light microscopy. Almost all the investigators have referred this as the small NSCs presumably based on their mere occurrence midst the NSC group. In *P. indicus* Mohammed *et al.*,(1993) described similar types of non-neuro secretory cells. The non-neurosecretory cells of *P. monodon* do not show any signs of secretory activity similar to such cells described in *P. indicus* by Mohammed *et al.*,(1993). Matsumoto (1954) classified such cells as D-cells and emphatically stated the absence of secretory activity in these cells. These cells are usually present in groups in *P. monodon* and they contain a single conspicuous nucleoli as described by Matsumoto, (1954) in *E. japonicus* and Mohammed *et al.*, (1993) in *P. indicus*. Matsumoto (1954) and Mohammed *et al.*, (1993) also have reported that the cell boundaries of such cells are different to discern.

The non-neurosecretory cells of *P. monodon* during the period of gonadal maturation show a cyclic secretory activity. In crustaceans, very few workers like Enami (1951) in *Sesarmadehaani* and Matsumoto (1954 &1958) in *E. japonicus* have described the cyclic activity of the NSCs in detail. These authors have noted the formation of vacuoles as a significant part of the secretory cycle. In *P. monodon* also the vacuole formation is the first phase in the activation of the cells during its secretory cycle. Vacuoles of various shapes and sizes occur in the NSCs of *P. monodon* especially in GN and A type cells. The diameter of these NSCs is found to be increasing as the vacuole formation continues. This vacuolar stage is followed by a secretory stage during which the entire cytoplasm gets filled with the secretory materials at the expense of the vacuoles. The large and variously sized vacuoles of GN and A cells are present towards the periphery of the cells. Various authors in several other crustaceans have been made identical observations (Matsumoto, 1954; Parameswaran, 1956; Chandy and Kolwalker, 1985).

Peripheral vacuoles forming pericellular vacuolar ring is distinctive feature of the B & C type neurosecretory cells of *P. monodon* (Diwan *et al.*, 2008). Mohammed *et al.*, (1993) reported similar types of observations in *P. indicus*.

The ultrastructure of the perikarya of the NSCs particularly GN and A type cells in *P. monodon*, is almost similar to those observed in *P. paucidens* by Hisano (1976), in the isopod *S. serratum* by Chataigner *et al.*, (1978) and in *M. olfersii* by Mc Namara (1993). In *P. monodon* the perikarya of the NSCs at their quiescent phase contain only electron lucent cytoplasm with few numbers of cell organelles. Even though ultrastructural studies are not available on these aspects, light microscopical studies of Gyananath and Sarojini (1985) in *M. dobsoni*, Nagabhushanam et *al.*, (1992) in *M. affinis* and Mohammed *et al.*, (1993) in *P. indicus* reported that the NSCs are with smooth cytoplasm and less synthetic activity during the in resting phase. The ultrastructural studies of McNamara (1993), in *M. olfersii* showed less cytoplasmic activity in the NSCs during the resting stage of the animal. According to him (McNamara, 1993) the perikarya of NSCs at their resting phase did not contain much sub-cellular organelles or secretory materials.

However, drastic changes occur in the ultrastructure of the vacuolar phase NSCs, especially that of the GN and A cells of *P. monodon*. The characteristic features of the vacuolar phase NSCs are presence of the highly active cell organelles and the abundant supply of variously shaped vacuoles of different sizes. The active sub-cellular machinery of these vacuolar phase NSCs indicate its active role in the neurosecretion (Diwan *et al.*, 2008). Similar types of neurosecretory cells with highly active cell organelles are described in *M. Olfersii* after its introduction to a higher salinity (McNamara, 1993). According to McNamara (1993), the appearance of the highly active sub-cellular organelles and variously sized and shaped vesicles are indications of the activations of sub-cellular machinery for the synthesis of its secretory products. In *P. monodon* also vacuoles later get converted into vesicular bodies packed with neurosecretory granules. Similarly, variously shaped vacuoles are described in association with the cisterns of RER and Golgi bodies in the NSCs of *P. paucidens* (Hisano, 1976) and in *S. serratum* (Chataigner, *et al.*, 1978).

In contrast to the vacuolar phase NSCs, the ultrastructure of the secretory phase NSCs of *P. monodon* contains numerous thick and highly electron-dense secretory granules and granule filled dense vesicles along with the active cell organelles. In these cells no more empty vesicles are encountered, however, the fully packed secretory granules gave the high electron-density to these cells are markable increase in the synthetic activity of the neurosecretory granules is reported in the NSCs of *M. Olfersii* when exposed to seawater (McNamara, 1993).

The NSCs of *P. monodon* shows an increased synthetic activity during its secretory phase followed by the vacuolar phase. Similar a high synthetic activity is noticed in association with the Golgi complex as well as cisternae of RER of these NSCs where abundant vacuoles of varying sizes and shapes are apparent (Mc Namara, 1993). Secretory activity is more pronounced in GN and 'A' type cells of *P. monodon* in contrast to the Mc Namaras observations in *M. Olfersii* where he observed an increased synthetic activity in all the cell types. However, the sub cellular machinery and its synthetic mechanism are found more or less similar in both of these crustaceans. Sub-cellular organelles like Golgi complex as well as cisterns of RER are found actively participating in the synthesis of secretory materials in the NSCs of *P. monodon.*

As in the NSCs of *M. olfersii* described by McNamara (1993), the Golgi complex is the most actively participating sub-cellular organelle in the NSCs of *P. monodon* during the synthesis of neurosecretory materials. Similar to the observations of Hisano (1976) in *P. paucidens* and Chataigner *et al.*, (1978) in *S. serratum* and McNamara (1993) in *M. olfersii,* the cisterns of RER are the most dominating type of cell organelle in the secretory type NSCs of *P. monodon.* Here the commencement of neurosecretion starts in these cisternae of RER and the small vesicles associated with them. Later on, some of these vesicles with moderately electron-dense flocculent materials move to the proximity of Golgi bodies, as noticed in *M. Olfersii* (Mc Namara, 1993), get modified into the secretory vacuoles with some structural modifications. In contrast to *M. olfersii,* synthesis of neurosecretory materials takes place in association with the active mitochondria in the NSCs of *P. monodon.* Abundant supply of small free and poly ribosomes in the cytoplasm as well as attached to the ER elements is a characteristic feature of the secretory phase NSCs of *P. monodon.* Similar observations have been made in the NSCs of other crustaceans (Hisano 1976, Chataigner *et al.*, 1978) and McNamara, 1993). The most dominating feature in the secretory phase of NSCs of *P. monodon* is the presence of five various types of neurosecretory granules. Similarly in the NSCs of the eyestalk of *P. paucidens,* Hisano (1976) has described 6 various types of neuro secretory granules. However, the perikarya of the B & C type NSC do not show an increased synthetic activity like that of GN and A cells in *P. monodon* during the process of gonadal maturation.

The role of neuroglia in the secretory cycle of *P. monodon* is almost similar to that in *P. indicus* described by Mohammed *et al.*, (1993). Some indications of glial activity have been reported by Lake (1970) in the crab *Paragrapsus gaimardii.* The glial cells surrounding the NSCs of *P. monodon* become hypertrophied during vacuolar phase of the NSCs and give a broken appearance to the plasma membrane of the NSCs during its secretory phase. The glio-neuronal relationship is vague in crustacean NSCs. However, Mohammed *et al.* (1993) reported that glial cells may serve as a source of nutritive material during the synthetic phase of the NSCs and they may also serve as a media between the NSC and the surrounding capillary network, thus giving a glio-vacuolar relationship similar to one existing in the vertebrates (Gless and Meller, 1969). Ultrastructure of glial cells surrounding the NSCs of *P. monodon* show a 5 layered membraneous system with numerous RER elements. This is indicative of their role in the transport of materials to the circulatory cells. Some sorts of synthetic activity also occur in the membrane systems surrounding the active nuclei. Nuclear and nucleolar participation in the cellular synthetic activity is well established in all active cells (DeRobertes *et al.*, 1975). Whereas in the light microscopic studies of Mohammed *et al.* (1993) in *P. indicus* the nuclear involvement is found only in the B & C type cells. However, participation of nucleus has also been observed in the synthetic activity of these neurosecretory materials from nuclear pores and aggregations of nuclear materials around these pores in *M.olfersii* (McNamara,1993).

Mapping of the NSC types in different ganglionic centers of decapod crustaceans is thoroughly investigated in pleocyematans by Adiyodi and Adiyodi (1970) and in penaeids by Mohammed *et al.* (1993). The neurosecretory cell groups of various ganglionic centers of other penaeids are meager. The cerebral ganglion of *P. monodon* is richly endowed with an abundant supply of NSC groups of different types of NSCs. A total of 14NSC groups are identified in the cerebral ganglia of *P. monodon*, whereas Nakamura (1974) observed only 8NSC groups in *P. japonicus* and Mohammed *et al.* (1993) observed 11NSC groups in the cerebral ganglia of *P. indicus*. The location of the NSC groups in the cerebral ganglion of *P. monodon* is basically similar to those observed in *P. Japonicus, P. stylifera, P. indicus* and *Cardina laevis* by Nakamura (1974), Nagabhushanam *et al.* (1986) Mohammed *et al.* (1993) and Pillai (1961) respectively. However, no central dorsal group, as found in *P. indicus*, is seen in the cerebral ganglion of *P. monodon*. Here 6 NSC groups are located in the dorsal side and 8 NSC groups in the ventral side. Mohammed *et al.* (1993) observed only 5NSC groups in the dorsal and 6 in the ventral side of the cerebral ganglion.

With regard to the remaining ventral ganglia there is surprisingly only one study in penaeids by Mohammed *et al.*, (1993) in *P. indicus*. The tritocerebral ganglion of *P. monodon* consist one or two large GN and one or two medium sized 'A' cells. In contrast, in crabs *Sesarma dehaani* (Enami, 1951) and *Paragrapsus giamardi* (Lake, 1970) the tritocerebral ganglion contains only medium and small sized NSCs. However, in the tritocerebral ganglia of *P. indicus*, Mohammed *et al.*,(1993) usually found one and occasionally two GN cells and a medium sized

A cell. The sub-oesophageal ganglion of *P. monodon* contains a total of 16NSC groups. By virtue of its size, magnitude and number of NSC groups, the sub esophageal ganglion of *P. monodon* is comparable to the sub esophageal ganglion of *P. indicus* (Mohammed *et al.*,1993) and fused thoracic ganglion of pleocynatans (Adiyodi and Adiyodi, 1970). All the four neurosecretory cell types are found in this ganglion with as light dominance of GN and A cells, whereas 15NSC groups have been reported in the sub-oesophageal ganglion of *P. indicus* and here the GN and A cells are the dominant types, (Mohammed *et al.*,1993). The thoracic ganglion of *P. monodon* contains 4 large NSC groups with a prominent dominance of GN and A cells. In *P. indicus* also Mohammed *et al.*,(1993) reported 4NSC groups.

Similarly 4 NSC groups with A, B and C type cells are identified in the abdominal ganglia of *P. monodon* which are found to be identical in *P. indicus* also (Mohammed *et al.*,1993). The most striking feature noticed in the NSC groups in the optic ganglia of *P. monodon* is the total absence of GN cells. MEGXO of *P. monodon* is a well-defined NSC group with 3 different cell types viz. A, B and C cell. Active NSCs are found in the MEGX-organ of immature animals. However, all these cell types are at their quiescent stage during the process of reproductive maturation. Similarly NSCs in the ME form a well defined group in Natantia such as *Pandalus borealis* (Carlisle, 1959), *P. serratus* (Humbert, 1965; Van Herp, *et al.*, 1977), *P. paucidens* (Hisano, 1974, 1976) and *P. japoncius* (Nakamura, 1974). However, in other Natantia, such as *Typhlatyagarciai* (Juberthie- Juepau, 1976) and *Atyaephyra desmaresti* (Boissou *et al.*, 1976) the NSCs in the MEGXO are more or less scattered in the ME. Only two different cell types are reported in the MEGXO of crabs by Matsumoto, (1958), in *Paragrapsusgiamardii* by Lake, (1970) and in *Palaemon serratus* by Humbert *et al.*,(1981).

MTGXOI in case *P. monodon* is composed of a small NSC group located at the dorsal side of the MT near the SG. It contains 'A', 'B' and 'C' types of NSCs. Similar descriptions have given for the MTGXO of the same species (Nanda and Ghosh, 1985) and for *P. indicus* by Mohammed *et al.*, 1993. The MTGXO II of *P. monodon* is composed of a cluster of NSC groups located at the opposite side of the MTGXOI. Here a few NSC are found to have invaded the dorsal part of the MT region. This cell group may have identified as the MIGXO by Nanda and Ghosh (1985) in *P. monodon*. However, in *P. indicus* Mohammed *et al.*,(1993) described this as MTGXOII which is similar in *P. monodon*.

The sinus gland (SG) of *P. monodon* appears as an elongated finger like body, associated with one of the vascular process in the optic ganglia. Bulbous axonal endings of various shapes and size and the internal blood sinus are the major components of the SG in *P. monodon*. Mohammed *et al.*,(1993) also described a similar type of SG in *P. indicus*. The structure of the SG in *P. monodon* as well as its dorso-lateral position in between the MI and ME, closely resembles the findings of earlier workers in other crustaceans (Dall, 1965; Nakamura, 1974). The 'S' shaped X-organ sinus gland tract in *P. Monodon* almost resembles identical to that of other decapod crustaceans (Adiyodi and Adiyodi, 1970). Similarly the axon bundles from MTGXO MTGXO II and I unite at the level of MT and then moves dorsally through the space in between the MI and ME and on the way it joins with the axon bundle from the MEGXO before entry into the SG. These pathways clearly indicate that the mode of discharge of neurosecretory material in the eyestalk is via the axonal transport. Based on the electro- physiological and cobalt ion ionophoretic studies of Andrew *et al.*, (1978) in *Orconectes virilis* and Jaros (1978) in *O. limosu*, showed a more or less similar pathway for the X-organ sinus gland tract.

Ultrastructural studies of SG of *P. monodon* that it indicates is composed of axons, which terminate in the bulb like enlargements just beneath the basement membrane lining the blood sinus. Earlier workers also described a similar structure for the SG of various crustaceans like *Procambarus clarkii* by Bunt and Ashby (1967), *Uca pugilator* by Silverthorn (1975), *Carcinus maenas* by Nordmann (1977) and *Ligiaoceanica* by Martin *et al.*,(1983). The sinus gland *P. monodon* also show five different types of axon endings based on morphology of the granules occupying the axon. This observation is comparable to the sinus gland of other decapods. However, only three types in *Porcellio dilatatus* (Martin, 1972) and two types in *Gecarcinus lateralis* (Hodge and Chapman, 1958), *Camberallus shufeldti* (Fingerman and Aoto, 1959) and *O. nais* (Shivers, 1969) have been reported.

In Crustacean's unlike the vertebrates and other invertebrates like insects, very few studies have been made to correlate changes in the neurosecretory system with physiological events particularly reproduction, even though neurosecretory elements are known to control this phenomenon (Adiyodi and Adiyodi, 1970). Similarly information regarding the sub- cellular machinery responsible for the synthesis of neurosecretory materials in relation to reproduction is lacking in penaeids. Light and electron microscopical studies on the NSCs of different ganglionic centers of *P. indicus* during the process of maturation indicate concomitant changes in the activity of NSCs. Demassieux and Balesdent, (1977) have reported a cyclic variation in function of certain NSCs in *Ascellus aquatics* in relation to reproduction. Gyananath and Sarojini (1985) delineated the cyclic activity of NSCs in relation to reproduction in a freshwater shrimp, *M. lamerii.* In *P. indicus*, Mohammed *et al.*, (1993) described similar types of cyclic activity of neurosecretary cells in relation to gonadal maturation.

In *P. monodon* the neurosecretory cells of the supra-oesophageal, sub-oesophageal and thoracic ganglion are highly active in the early maturing and late- maturing stage. The NSCs of brain showed its maximum secretory activity during early-vitellogenic phases. However, NSCs of sub esophageal and thoracic ganglion are maximally active in the late maturing *P. monodon*. In the thoracic ganglia of crabs, Matsumoto (1958) reported that the NSCs are maximally active in its early stages of maturation. Similarly in the crayfish *Procambarus simulors*, Perryman (1969) correlated the stages of ovarian development with varying amount of neurosecretory material in cell type III of the cerebral ganglion. Diwan and Nagabhushanam (1974) demonstrated a seasonal activity in the thoracic ganglia of the fresh water crab *Barytelphusa cunicularis*. Such a high secretory activity during ovarian maturation has also been observed in specific NSCs of the thoracic ganglion of *Macrobrachium lanchestrii* (Rao *et al.*, 1981), *M. kistensis* (Mirajkar *et al.*, 1983) and *Potamon koolooense* (Joshi, 1989).

4.13 CONCLUSION

From the above account on the different NSCs located in various ganglionic centers during the process of gonadal maturation, it can be concluded that these neurosecretory cells play active roles in the regulation of reproduction in *P. monodon*. Significant changes occur in the secretory activity of these cells during different stages of gonadal maturation. In the immature animals the NSCs of eyestalk, especially A & B type of cells are found highly active indicating that, these cells are engaged in the secretion of the gonad- inhibiting factors. At the same time most of the NSCs in various ganglionic centers of CNS are at their resting Q-phase. This shows they are not actively participating in the secretion of GIH. In contrast during the early stages of gonadal maturation, the NSCs of eyestalks do not show a heightened secretory activity as that of immature animals. During the early stages of maturation, GN and A cells of brain secrete some factors, which in turn activate the thoracic and sub-esophageal ganglia to secrete the gonad stimulating factors. Concomitantly, an increase in the number of S-phase GN and A type NSCs occur in sub-oesophageal and thoracic ganglia. Heightened secretory activity take place in these cells and the other B & C cells of these ganglia, during the subsequent stages of oogenesis, indicating its active role in the synthesis and secretion of gonad-stimulating factors. In order to correlate the functional aspects of different neurosecretory cells with various physiological changes, it is necessary to carry out the molecular studies of these NSCs using TEM and SEM techniques.

References

Adiyodi, K.G and R.G Adiyodi, 1970. Endocrine control of reproduction in Decapod Crustacea. *Biol. Rev.*, 45: 121-165.

Alfaro, J. 1993. Reproductive quality evaluation of male *Penaeus stylirostris* from agrow out pond, *J. World. Aquacult. Soc.*, 24(1): 6-11.

Anilkumar, G. and K.G. Adiyodi, 1985. The role of eyestal hormones in vitellogenesis during the breeding season in crabs *Paratelphusa hydrodromous. Biol. Bull.*, 169: 689-695.

Aoto, T. and H. Nishida, 1956. Effect of removal of the eyestalks on the growth and maturation of the oocytes in a hermaphroditic prawn, *Pandalus kessleri. J. Fac. Sci. Hokkaido Univ. Ser.*, 612: 412-424.

Andrews, P.M., D.E. Copeland and M. Fingerman, 1971. Ultrastructural study of the neurosecretory granules in the sinus gland of the blue crab *Callinectes sapidus. Z. Zelforsch.*, 113: 461-471.

Andrew, R.D. and R.R. Shivers 1976. Ultrastructure of neurosecetory granule exocytosis by crayfish sinus gland induced with ionic manipulations. *J. Morphol.*, 150: 253-278.

Andrew, R.D., I. Orchard and A.S.M. Saleddunin, 1978. Structural revaluation of the neurosecretory system in the crayfish eyestalk. *Cell. Tiss. Res.*, 190: 235-246.

Anilkumar, G. and K.G. Adiyodi 1985. The role of eyestalk hormones in vitellogenesis during the breeding season in crabs *Paratelphusa hydrodromous. Biol. Bull.*,169: 689-695.

Arnstein, D.R. and T.W. Beard 1975. Induced maturation of prawn *Penaeus orientalisi* Kishinogue in the laboratory by means of eyestalk removal. *Aquaculture*, 5: 411-412.

Bliss, D.W. 1951. Metabolic effects of sinus gland or eyestalk removal in the crab,*Gecarcinus lateralis. Anat. Reod.* 111: 502-503.

Bliss, D.W. and J.R. Beyer 1964. Environmental regulations of growth in decapod crustacean *Geracarcinus lateralis. Gen. Comp. Endocrinol.*, 4: 15-41

Boissou, F., D. Huguet and M. Vincent 1976. Note preliminaire sur la repartition des cellules neurosecretrices cephaliques de la revette d'eau douce *Atyaephyra desmaresti* Millet (Crustacea, Decapoda) *Bull. Soc. Nat. Quest Fr.*, 74: 91-100.

Brodie, D.A. and K. Halerow 1977. The ultrastructure of the sinus gland of *Gammarus oceanicus. Cell. Tiss. Res.*, 182: 557-564.

Browdy, C.L., 1998. Recent developments in penaeid broodstock and seed production technologies: improving the outlook for superior captive stocks. *Aquaculture*, 164: 3–21.

Bullock, G.E. and G.A. Horrdige 1965. Structure and function of the nervous systems of invertebrates. Vol. II. W.H. Freeman and Company, San Francisco.

Bunt, A.H. and E.A. Ashby 1967. Ultrastructure of the sinus gland of the crayfish, *Procambarus clarkii. Gen. Comp. Endocrinol.*, 9: 334-342.

Carlisle, D.B. and F.G.W. Knowles 1959. Endocrine control in crustaceans, 150. Cambridge Univ. Press, London.

Carlisle, D.B. and F.G.W. Knowles 1959. Sexual biology of *Pandalus borealis*. I. Structure of incretory elements. *J. Mar. Biol. Ass. U.K.*, 38: 381-395.

Caillouet, C.W. 1972. Ovarian maturation induced by eyestalk ablation in pink shrimp *Penaeus duorarum* Burkenroad. *Proc. World. Maricult. Soc.* 3: 205-225.

Cavalli R.O., Scardua M.P. and WasieleskyW.J. (1997) Reproductive performance of different sized wild and pond reared *Penaeus paulensis* females. *J. World. Aquacult. Soc.*, 28: 260-267.

Chang, E.S., W.A. Hertz and G.D. Prestwich 1992. Reproductive endocrinology of the shrimp *Sicyonia ingentis*: Steroid, Peptide, and Terpenoid Hormones, NOAA. Tech, Reports. NMFS. 106: 1-6.

Chandy, J.P. and D.G. Kolwalkar 1985. Neurosecretion in the marine crab, *Charybdis variegata. Ind. J. Mar. Sciences*, 14: 31-34.

Charniaux-Cotton, H. and G. Payen 1988. Crustacean reproduction. In: *Endocrinology of Selected Invertebrate Types*, 2: 279-303.

Chataigner, J.P., G. Martin and P. Juchault 1978. Etude Histologique, cytologique, et experimentale des centres neurosecreteurs cephaliques du flabellifere *Shaeroma serratum. Comp. Endocrinol.*, 35: 52-69.

Coman, G.J., Arnold, S.J., Thompson, P.J. and Crocos, P.J., 2003. Development of a low water-exchange rearing system for biosecure maturation of *Penaeus monodon. World Aquaculture Symposium* 2003, Salvador, Brazil. Abstract.

Coman, G.J., Crocos, P.J., Preston, N.P. and Fielder, D., 2004. The effect of density on the growth and survival of different families of juvenile *Penaeus japonicus* Bate. *Aquaculture* 229: 215–223.

Coman G.J., S.J. Arnold, S. Peixoto, P.J. Crocos, F.E. Coman and N.P. Preston 2006. Reproductive performance of reciprocally crossed wild-caught and tank-reared *Penaeus monodon* broodstock *Aquaculture*, 252: 372–384.

Coman, G.J., Crocos, P.J., Arnold, S.J., Keys, S.J., Murphy, B. and Preston N.P., 2007. Growth, survival and reproductive performance of domesticated Australian stocks of the giant tiger prawn, *Penaeus monodon*, reared in tanks and raceways. *J. World Aquacult. Soc.* 36:4.

Cooke, I.M. and R.E. Sullivan 1982. Hormones and neurosecretion. In: *Biology of Crustacea*, 3: 205-290. Academic Press.

Crocos, P.J. and Coman, G.J., 1997. Seasonal and age variability in the reproductiveperformance of *Penaeus semisulcatus*: optimizing broodstock selection. *Aquaculture*, 155: 55–67.

Crocos, P.J. and J.D. Keller 1986. Factors affecting induction of maturation and spawning of the tiger prawn *Penaeus esculentus* under laboratory conditions. *Aquaculture*, 58: 203-214.

Cuzon, G., Arena, L., Goguenheim, J. and Goyard, E., 2004. Is it possible to raise, offspring of the 25th generation of *Litopenaeus vannamei* (Boone) and 18th generation *Litopenaeus stylirostris* (Simpson) in clear water to 40 g? *Aquac. Res.* 35: 1244–1252.

Dall. W. 1965. Studies on the physiology of a shrimp *Metapenaeus* sap. II. Endocrine and control of moulting, *Aust. J. Mar. Fresh. Rs.* 16: 1-12.

Dall, W., B.J. Hill, P.C. Rothlisherg and Staples. 1990. The biology of penaeids. In: *Advances in Marine Biology*, 27: 488 p.

Decaraman, M. and T. Subramonim, 1983. Endocrine regulation of ovarian maturation and cement gland activity in a stomatopod crustacean, *Squilla holoschista. Proc. Indian. Acad. Sci. (Ani., Sci.)* 92: 399-408.

Demassieux, C. and M.L. Balesdent 1977. Les cellules a caractere neurosecreteur des ganglions cerebraux et de la chaine nerveuse chez le crustace' isopode *Asellus aquati* cus Linne variations cycliques des cellules de type B en fonction de la reproduction. *C.R. Acad. Sci. Paris.*, D, 207: 207-210.

Demeusy, N. 1967. Croissance relative d'un caractere sezueles externe male chez la Decapode Brachyore *Carcinus memaus* Lc. *R. Hebd. Seane. Acad. Sci. Paris.*

DeRobertes, E.D.P., W.W. Nowinski and F.A. Saez. 1975. Cell Biology. 6th ed. Saunders, Philadelphia, USA.

Diwan, A.D. and R. Nagabhushanam 1974. Reproductive cycles and biochemical changes in the gonads of the freshwater crab, *Barytelphusa cunicularisn. Indian. J. Fish.* 21(1): 164-176.

Diwan, A.D., 2005. Current progress in shrimp endocrinology-a review. *Indian journal of experimental biology*, 43,3:209.

Diwan, A D and Joseph, Shoji and Ayyappan, S, 2008. *Physiology of reproduction, breeding and culture of tiger shrimp Penaeus monodon (Fabricius).* Narendra Publishing House, New Delhi.

Durand, J.B. 1956. Neurosecretory cell types, their secretory activity in the crayfish. *Biol. Bull.*, (Woods Hole), 111: 62-67.

Eastern-Recks, S.B. and M. Fingerman, 1984. Effects of neuroendocrine tissue and cyclic AMP on ovarian growth *in vitro* in the fiddler crab, *Uca pugilator. Comp. Biochem. Physiol.*, 79A: 679-684.

Elofsson, R. 1969. The development of the compound eyes of *Penaeus duorarum* with remarks on the nervous system, *Z. Zellforsch.*, 97: 323-350.

Enami, M. 1951. The sources and activities of two chromatophorotropic hormones in crabs of the genus *Sesarma* II Histology of incretory elements. *Biol. Bull.*,101: 241-258.

Fahrenbach, W.H. 1976. The brain of the horseshoe crab *Limulus polyphemus*. I. Neuroglia. *Tissue & Cell*, 8,3: 395-410.

Fingerman, M. 1985. The physiology and pharmacology of crustacean chromatophores. *Amer. Zool.*, 25: 233-252.

Fingerman, M. 1987. The endocrine mechanisms of crustaceans. *J. Crust Biol.*, 7,1: 1-24.

Fingerman, M.,1997. Roles of neurotransmitters in regulating reproductive hormone release and gonadal maturation in decapod crustaceans. Invertebr. *Reprod. Dev*, 31,1–3: 47–54.

Fingerman, M. and T. Aoto 1959. The neurosecretory system of the dwarf crayfish *Cambarellus shufeldti*, revealed by electron and light microscopy. *Trans. Amer. Mocrosco. Soc.*, 78: 305-317.

Fingerman, M. and C. oguro 1968. Chromatophore control and neurosecretion in the mud shrimp, *Upogebia affinis. Biol. Bull.*, 124: 24-30.

Ferlund, and L. Jossefson 1972. Crustacean colour change hormone: amino acid sequence and chemical synthesis. *Science*, 177: 173-175.

Ferlund, P. 1976. Structure of a light adapting hormone from the shrimp, acid sequence and chemical sysnthesis. *Science*, 177: 173–175

Gabe, M. 1953. Sur quelques applications de la coloration par las fuchsine paraldehyde. *Bull. Micro. Applqu.*, 3: 153-162.

Gabe, M. 1956. Histologie compares de la glands de mue des Crustaces Malacostraces. Annales des Sciences Naturelles *Zoologie et Biologie Animale* 18: 145-152.

Gless, P. and K. Meller 1969. Morphology of neuroglia, in G.H. Bourne ed.. The structure and function of nervous tissue. 1:301-323.

Gomez., R. 1965. Acceleration of development of brain in the crab *Paratelphusa hydrodromous. Naturewise*, 52: 217-219.

Gyananath, G., and R. Sarojini 1985. Correlation between neurosecretory activity and annual reproductive cycle of the freshwater prawn, *Macrobrachium lamerri. Proc. First. Nat. Symp*. On Endocrinology of Invertebrates, Aurangabad, India. 46-50.

Hanoaka, K.I. and T. Otsu 1957. The source of the ovarian inhibiting hormone in the eyestalks of the crab, *Potamon dehaani. J. Fac. Sci. Hakkaido Univ.* (Ser. VI) 13: 379-383.

Highnam, K.C. and L. Hill 1977. The Comparative Endocrinology of the Invertebrates. 2nd ed. Univ. Park. Press. Baltimore, USA.

Hinsch, G.W. and D.C. Bennett 1979. Vitellogenesis stimulated by thoracic ganglion implants into destalked immature spider crabs, *Libinia emarginata. Tissue & Cell*, 11: 345-351.

Hisano, S. 1974. The eyestalk neurosecretory cell types in the freshwater prawn, *Palaemon paucidens*. I. A. Light microscopical study. *J. Fac. Sci. Hokkaido Univ. Ser.*, 6. 19: 503-514.

Hisano, S, 1976. Neurosecretory cell types in the eyestalk of the freshwater prawn *Palaemon paucidens. Cell Tissue Res.* 166: 511.

Hoang, T., Lee, S.Y., Keenan, C.P. and Marsden, G.E., 2002. Effects of age, size, and light intensity on spawning performance of pond reared *Penaeus merguiensis. Aquaculture* 212: 373–382.

Hodge, M.H. and G.B. Chapman, 1958. Some observations on the fine structure of the sinus gland of a land crab, *Gecarcimus lateralis. J. Biophys. Biochem. Cytol.*, 4: 571-574.

Huberman A. 2000 Shrimp endocrinology. A review. *Aquaculture*, 191: 191-208

Humbert, C. 1965. Etude experimentale de l'ornage X (pars distalis) dans les changements de couleur et la mue de la revetten *Palaemon serratus*. *Trans. Inst. Sci. Cherfien, Ser. Zool. No.*, 32: 86 pp.

Bellon-Humbert CF, van Herp, GE, Strolenbergy CM, Denuce, JM, 1981. Histlogical and physiological aspects of the medulla externa X organ, a neurosecretory cell group in the eyestalk of *Palaemon serratus* (Crustacea, Decapoda). Natantia. Biol. Bull. (Woods Hole), 160:11-30.

Jaros, P.P. 1978. Tracing of Neurosecretory neurons in crayfish optic ganglia by cobalt ion tophoresis. *Cell. Tissue. Res.*, 194: 297-302.

Joshi, P.C. 1989. Neurosecretion of brain and thoracic ganglion and its relation to reproduction in the female crab, *Potamon Koolooense*. *Proc. Indian Acad. Sci. (Anim. Sci)*. 98: 41-49.

Juberthie-Jupeau, L. 1976. Sur le systeme neurosecreteur du pedoncule oculaire d'un Decapode souterrain microphtalme *Typhlatya garcial*. *Ann. Speleol.*, 31: 107-114.

Kanokpan Wongprasert, Somluk Asuvapongpatana, Pisit Poltana, Montip Tiensuwan and Boonsirm Withyachumnarnkul, 2006 Serotonin stimulates ovarian maturation and spawning in the black tiger shrimp *Penaeus monodon* *Aquaculture*, 261: 1447–1454.

Keller, R. and Sedlmeier, D., 1998. A metabolic hormone in crustaceans: the hyperglycemic neuropeptide. In: Laufer, H., Downer, R.G.H. Eds., Endocrinology of selected invertebrate types vol. II A.R. Liss, New York, 315–326.

Keller R., Kegel G., Reichwein B., Sedlmeier D. and Soyez D. 1999. Biological effects of neurohormones of the CHH/MIH/GIH peptide family in crustaceans. In Recent Developments in Comparative Endocrinology and Neurobiology (Roubos, E. W., Wendelaar Bonga, S. E., Vaudry, H. and De Loof, A., eds.), pp. 209±212, Shaker Publishing B.V., Maastricht.

Kendoh, Y. and M. Hisada 1986. Neuroanatomy of the terminal ganglion of the crayfish, *Procambarus clarkii*. *Cell Tissue Res.*, 243: 273-288.

Khayat, M., Yang, W.-J., Aida, K., Nagasawa, H., Tietz, A., Funkenstein, B. and Lubzens, E., 1998. Hyperglycemic hormones inhibit protein and mRNA synthesis in in vitro- incubated ovarian fragments of the marine shrimp *Penaeus semisulcatus*. *Gen. Comp. Endocrinol.* 110, 307–318.

Kleinholz, L.H. 1976. Crustacean neurosecretory hormones and physiological specificity. *Amer. Zool.*, 16: 151-166.

Knowles, F. 1959. The control of pigmentary effectors. *Comp. Endocrinol.*, 147: 223-232.

Kulkarni, G.K. and R. Nagabhushanam 1980. Role of ovary inhibiting hormone from eyestalks of marine penaeid prawns *Parapenaeopsis hardwickii* during ovarian development cycle. *Aquaculture*, 19: 13-19.

Kulkarni, G.K., Nagabhushanam, R. and P.K. Joshi 1984. Neuroendocrine control of reproduction in the male penaeid prawn, *Parapenaeopsis hardwickii. Hydrobiologia*, 108: 281-289.

Lake, P.S. 1970. Histochemical studies of the neurosecretory system of *Chierocehalus diaphanous*. Prevose (Crustacea: Anostraca). *Gen. Comp. Endocrinol.*, 14: 1-15.

Laufer, H., A. Sagi, J.S.B. Ahl and E. Homola 1992. Methylfernasoate appears to be a crustacean reproductive hormone. *Invertbr. Rep. And Dev.*, 22,1-3: 17-20.

Laubier, B.A. 1978. Ecophysiologie de la crevette *Penaeus japonicus* trios annees d'experience on mileu controle. *Oceano. Acta.*, 1: 135-150.

Legrand, J.J., G. Martin and P. Juchault and G. Besse 1982. Controle neuroendocrine de la reproduction chez les Crustaces. *J. Physiol. Paris.*, 78: 543-552.

Madhystha, M.N. and P.V. Rangnekar 1976. Neurosecretory cells in the central nervous system of the prawn, *Metapenaeus monoceros Rev. Di. Biol.*, IXXIX: 133-140.

Martin, G. 1972. Contribution a letude ultrastructurale de la glande du sinus del'Oniscoide *Porcellio dilatatus C.R. Hebd. Seances Acad. Sci. Ser. D. Sci. Nat., Paris* 275: 839-842.

Martin, G., R. Maissiat and P. Girard 1983. Ultrastructure of the sinus and lateral cephalic nerve plexus in the isopod *Ligia oceanica. Gen. Comp. Endocrinol.*, 52: 38-50.

Matsumoto, K. 1954. Neurosecretion in the thoracic ganglion of the crab, *Eriocheir japonicus. Biol. Bull.*, 106: 60-68.

Matsumoto, K. 1958. Morphological studies on the neurosecretion in crabs. *Biol. J. Okayama Univ.*, 4: 103-176.

Maynard, D.M. 1965. Thoracic neurosecretory structures in brachyura. 1. Gross anatomy. *Biol. Bull.*, 121: 316-329.

Mc Namara, J.C. 1993. Exposure to high salinity medium and neurosecretion in the anteromedial cells of the supraoesophageal ganglion of the freshwater shrimp *Macrobrachium olfersii J. Crust. Biol.*, 13,3: 409-422.

Meeratana, P., Withyachamnarnkul, B., Damrongphol, P., Wongprasert, K., Suseangtham, A. and Sobhon, P., 2006. Serotonin induces ovarian maturation in giant freshwater prawn broodstock, *Macrobrachium rosenbergii* de Man. *Aquaculture*, 260,1–4: 315–325.

Meusy, J.J. 1968. Precisions nouvelles sur ;'ultrastructure de la glande du sinus d'un Crustacea *Carcinus maenas* L. *Bull. Soc. Zool. Fr.*, 93: 291-299.

Meusy, J.J. and G.G. Payen 1988. Female reproduction in Malacostracan Crustacea. *Zool. Soc.*, 5: 217-265.

Mirajkar M, R Sarojini, R Nagabhushanam. 1984. Histophysiology of the androgenic gland in freshwater prawn *Macrobrachium kistnensis. J. Anim. Morphol. Physiol.* 31: 211-220.

Mohammed, S.K. and A.D. Diwan 1991. Neuroendocrine regulation of ovarian maturation in the Indian white prawn *Penaeus indicus. Aquaculture*, 98: 381-393.

Mohammed, S.K., K.K. Vijayan and A.D. Diwan 1993. Histomorphology of the neurosecretory system in the Indian white prawn *Penaeus indicus*. H. Milne. Edwards. *Bull of the Institute of Zool.*, 32,1: 39-53.

Nagabhudshnam, R., P.K. Joshi and G.K. Kulkarni 1982. Induced spawning in *Parapenaeopsis stylifera* using a steroid hormone 17 hydroxy-progesteron. *Proc. Symp. Coastal Aquaculture*, 1: 37-39.

Nagabhushanam, R., R. Sarojini and P.K. Joshi, 1986. Observation on the neurosecretory cells of the marine penaeid prawn, *Parapenaeopsis stylifera. J. Adv. Zool.* 7: 63-70.

Nagabhushanam R., R. Sarojini and S. Sambasiva 1992. Reproductive endocrinology of the Indian Marine shrimp *Metapenaeus affinis*. In: *Aquaculture research needs for 2000* A.D,. 163-180.

Nakamura, K. 1974. Studies on the neurosecretion of the prawn *Penaeus japonicus* I. Positional relationship of the cells group located on the supra oesophageal and optic ganglion. *Mem. Fac. Fish. Kagoshina Univ.*, 23: 175-184.

Nakamura, K. 1980. Electron-microscopical observation of the PAS positive cells in the supra-cesophageal ganglion of the prawn, *Penaeus japonicus* Bate. *J. Jap. Soc. Sci. Fish.*, 46 ,1: 1211-1215.

Nanda, D.K. and P.K. Ghosh 1985. The eyestalk neurosecretory system in the brackishwater prawn, *Penaeus monodon*, A light microscopical study. *J. Zool. Soc. India.*, 37: 25-38.

Nordmann, J.J. 1977. Ultrastructural appearance of neurosecretory granules in the sinus gland of the crab after different fixation procedures. *Cell. Tiss. Res.* 185: 557-563.

Otsu, T. 1960. Precoucious development of the ovaries in the crab, *Potemont dehani*, following implantation of the thoracic ganglion. *Annot. Zool. Jpn.*, 33: 90-96.

Oyama, S.N. 1968. Neuroendocrine effect on ovarian development in the crab *Thalamita crenata* Latreille studied *in vitro*. Ph.D. Diss. Univ. Hawaii, Honolulu, Hawaii.

Palacios E., Rodr|£ guez-Jaramillo C. and Racotta I.S. (1999a) Comparison of ovary histology between different-sized wild and pond-reared shrimp *Litopenaeus vannamei* (*Penaeus vannamei*). *Invertebrate Reproduction and Development* 35: 251-259.

Palacios E. & Racotta I.S. and Acuacultores de La Paz (APSA) (1999b) Spawning frequency analysis of wild and pond reared Pacific White Shrimp *Penaeus vannamei* broodstock under large-scale hatchery conditions. *J. World Aquacult. Soc.*, 25: 180-191.

Panouse, J.B. 1943. Influence du peduoncule oculare sur la eroissance de louaire chez la crevette *Leander serratus*. *Ann. Inst. Oceranog*, 23: 65-147.

Parameswaran, R., 1956. Neurosecretory cells of the central nervous system of the crab, *Paratelphusa hydrodromous*. *Journal of Cell Science*, 3,37:75-82.

Passano, L.M. 1960. Moulting and its control. In: Physiology of Crustacea, Vol. I, 473-536.Ed. T.H. Waterman. Academic Press, Newyork.

Payen, G.G. 1986. Endocrine regulation of male and female genital activity in crustaceans.A retrospect and perspectives. In: *Advances in Invertebr. Rep.*, 4: 125-134.

Perryman, E.K. 1969. *Procambarus simulans* – Light induced change in the neurosecretory cells and in ovarian cycle, *Trans. Am. Microsco. Soc.*, 88: 514-524.

Pillay, R.S. 1961. Studies on the shrimp, *Cardina laevis* (Heller) IV. Neurosecretory system. *J. Mar. Biol. Ass. India.* 3: 146-152.

Pratoomchat, B., 1998. Size at first maturity of male *Penaeus monodon Fabricius* from pond-reared stock. *Journal of Aquaculture in the Tropics* 13: 189–194.

Primavera, J. H. 1985. A review of maturation and reproduction in closed thelycum penaeids. In: Y. Taki, J.H. Primavera and J.A. Lloberera (eds.) Proc. First Internat. Cont. on the Culture of Penaeid Prawns/Shrimps, SEAFDEC City, Philippines, pp. 47-64.

Quackenbush, L.S. 1989. Yolk protein production in the marine shrimp *P. vannaemi. J. Crustacea Biol.* 9,4: 509-516

Quackenbush, L.S. and L.K. Keeley 1987. Vitellogenesis in the shrimp *P. vannamei. Amer. Zool.*, 26: 810 A.

Quackenbush, L.S. and W.F. Herrkind 1981. Regulation of moult and gonadial development in the spiny lobster, *Panulirus argus. J. Crust. Biol.*, 3: 34-44.

Radhakrishnan, E.V. and M. Vijayakumaran 1984. Effect of eyestalk ablation in the spiny lobster *Panulirus homarus* 3. On gonadial maturity. *Ind. J. Fish.*, 31: 209-216.

Rangneker, P.V. and Deshmukh 1968. Effect of eyestalk removal on the ovarian growth of the marine crab, *Scylla serrata. J. Anim. Morph. Phsio.*, 15: 503-511.

Ramadan, A.A. and C. Matta 1976. Studies on the central nervous system of the prawn: I. A histological study for mapping the neurosecretory cells. *Folia Morphologie.* 24: 284-288.

Rao, Ch, N.K., K. Shakunthala and S.R. Reddy 1981. Studies on the neurosecretion of the thoracic ganglion in relation to reproduction in female, *Macrobrachium lanchesteri* (de Man),*Proc. Indian. Acad. Sci.* (*Anim. Sci.*) 90: 503-511.

Rao, K.R. and M. Fingerman 1983. Regulation and mode of release and mode of action of crustacean chromatophorotropins. *American Zoologist.* 23: 223-232.

Rao. G S, subramaniam V T, rajamani M, Sampson PE, Maheswarudu G, 1993 Stock assessment of *Penaeus* Spp. Off the east coast of India . *Indian Journal of Fisheries* 40, 1-2: 1-19.

Sellars, M.J., Keys, S.K., Cowley, J.A., McCulloch, R., Preston, N.P., 2006. Effect of rearing environment on Mourilyan Virus load in *Penaeus japonicus* during maturation in tank systems. *Aquaculture*, 252: 240–245.

Shivers, R.R. 1969. Possible sites of release of neurosecretory granules in the sinus gland of the crayfish, *Orconectes nais. Z. Zellforsch.* 97, 38-44.

Shoji Joseph, 1997. Some studies on the reproductive endocrinology of tiger prawn *P. monodon* Fabricius, Ph. D Thesis, Cochin University of Science and Technology.

Swetha C H, Sainath S B, Reddy P R,ReddyP S, 2011. Reproductive Endocrinology of Female Crustaceans: Perspective and Prospective. *J Marine Sci Res Development* S3:001.

Silverthorn, S.U. 1975. Hormonal involvement in thermal acclimation in the fiddler crab *Uca pugilator* (Bosc)- I. Effect of eye-stalk extracts on whole animal respiration. *Comp. Biochem. Physiol.* 50A: 281-283.

Skinner, D.M. 1968. Isolation and characterization of ribosomal ribonucleic acid from the crustacean *Gecarcinus lateralis. J. Exp. Zool.*, 169: 347-356.

Smith, G. 1974. The ultrastructure of the sinus gland of *Carcinus maenas. Cell. Tissue.Res.*, 155: 117-125.

Smith, G. 1975. The neurosecretory cells of the optic lobe in *Carcinus maenas. Cell.Tissue. Res.*, 156: 403-409.

Smith, G. and E. Naylor 1972. The neurosecretory system of the eyestalk of *Carcinus maenas. J. Zool. Lond.*, 166: 313-321.

Takayanagi, H., Y. Yamamoto and N. Takeda 1986. An ovary stimulating factor in the shrimp *Paratya compressa. J. Exp. Zool.*, 240: 203-209.

Van Herp, F., C. Bellon-Humbert, J.T.M. Luub and A. Van Wormhoudt. 1977. A histo physiological study of the eyestalk of *Palaemon serratus* (Pennant) with special reference to the impact of light and darkness. *Arch. Biol.* (Bruxelles). 88:257-278

Van-Herp, F. 1992. Inhibiting and stimulating neuropeptides controlling reproduction in crustacea. *Invertebr. Rep. Dev.*, 22,1-30: 21-30.

Van-Herp, F. and G.G. Payen 1991. Crustacean neuroendocrionology perspectives for the control of reproduction in aquacultural systems. *Bull. Inst. Zool. Acad. Sinica Monograph.*, 16: 513-539.

Webster, S. G. and Chung, J. S. 1999 Roles of moult-inhibiting hormone and crustacean hyperglycemic hormone in controlling moulting in decapod crustaceans. In Recent Developments in Comparative Endocrinology and Neurobiology (Roubos, E. W., Wendelaar Bonga, S. E., Vaudry, H. and De Loof, A., eds.), pp. 213-216, Shaker Publishing B.V., Maastricht.

Yano, I. 1985. Induced ovarian maturation and spawning in greasy back shrimp, *Metapenaeus ensis* by progesterone. *Aquacult.*, 47: 223-229.

Yano, I. 1988. Oocyte development in the kuruma prawn *Penaeus japonicus. Marine Biology*, 99: 547-553.

Yano,I. And J.a. Wyban 1992. Induced ovarian maturation of *Penaeus vannamei* by injection of lobster brain extract. Bull. Natl. Res. Inst. *Aquaucldture*, 21: 1-7.

Young, N.J., S.G. Webster and H.H. Rees 1993. Ecdysteroid profiles and vitellogenesis in *Penaeus monodon. Invert. Rep. & Devp.*, 24,2: 107-118.

CHAPTER 5

NEUROENDOCRINE CONTROL OF MATURATION, BREEDING AND SPAWNING

5.1 INTRODUCTION

Successful domestication of the candidate species is the footstep of any economically viable culture programmes. The key to domestication lies in controlled and enhanced reproduction of the broodstock shrimps. Now, the aquaculture technology of penaeid shrimps has developed remarkably and at present the shrimps are cultured on a large scale in captivity. The evolution of modern shrimp culture demands captive reproduction and seed production through larvi culture in hatcheries at required time. However, the fundamental problem in shrimp mariculture industry is the lack of predictable abundant supplies of off springs of known heritage. The growing demand for quality shrimp seed from the farmers and entrepreneurs, coupled with uncertainty of their availability from nature at the appropriate time in required quantities has prompted research on problems connected with shrimp seed production.

In many decapod crustaceans, control of gonadal maturation is a major problem in developing commercial aquaculture programmes (Yano, 1992). The giant tiger shrimp *Penaeus monodon* is one of the largest and hardiest penaeid species in the world, and it has been identified as the most favourable candidate species for intensive aquaculture by many of the countries. The most successful technique for inducing maturation in captivity has been the removal of the eyestalk, which is known to be the site of production and storage of gonad inhibiting hormones (Fingerman, 1987).

The classical trials of Panouse (1943) on the female shrimp *Leander serratus* demonstrated for the first time that, removal of eyestalk during exualin activity led to the rapid increase in ovarian size and precocious egg deposition. His (Panouse, 1943), studies have been elaborated over last 50 years in various crustaceans, such as *Pandalus kessleri* (Aoto and Nishida, 1956), *Carcinus means* (Demeusy, 1967), *Scylla serrata* (Rangnekar and Deshmukh, 1968) *Cragoncragon* (Bomirski and Klek, 1974), *Barytelphusa cunicularis* (Nagabhushanam and Diwan, 1974), *Panulirus argus*, (Quackenbush and Herrnkind, 1981), *Ucapugilator*, (Quackenbush and Keeley, 1988). However, the importance of this finding to penaeid shrimp culture lay dormant for 30 years until Caillouet, (1972) applied the eyestalk removal/gonadal development principle to *Penaeus duorarum*. Later a flurry of research activity on similar lines i.e. by applying unilateral eyestalk ablation to various penaeid species, under various conditions continued until recent times and some of the worth mentioning contributions made on these lines are by Arnstein and Beard (1975); Alikunhi *et al.* (1975); Aquacope (1975); Lumare (1979); Emmerson (1980); Kulkarni and Nagabhushanam (1980); Primavera (1982); Chamberlain and Gervais (1984); Wyban *et al.* (1987); Mohammed and Diwan (1991) Benzie, (1997); Cavalli *et al.*, (1997); Crocos, and Coman, (1997); Yang, *et al.*, (1997); William *et al.*, (1998); Browdy, (1998); Pratoomchat, (1998); Crocos, *et al.*, (1999); Palacios *et al.*, (1999*a*,1999b); Wouters, *et al.*, (1999, 2001, 2001b); Vaca, and Alfaro, (2000); Withyachumnarnkul, *et al.*, (2001); Hoang, *et al.*, (2002); Coman *et al.*, (2003, 2004, 2005, 2006); Peixoto *et al.*, (2003, 2005); Alfaro, *et al.*, (2004); Arcos, *et al.*, (2004); Cuzon, *et al.*, (2004); Mengqing, *et al.*, (2004); Toledo *et al.*, (2005); Kanokpan *et al.*, (2006); Meeratana, *et al.*, (2006); Sellars, *et al.*, (2006).

The successful application of the eyestalk ablation technique to induce maturation and spawning of *P. monodon* under controlled conditions has been a commercial practice in many of the Indo-Pacific countries where this species is present in large quantities in natural seas. However, the percentage of shrimps getting fully matured and spawned in case of *P. monodon* is still low, and problems are more acute especially in areas where the natural population of this species is comparatively less. Even though there are reports about the achievements of maturation from pond reared *P. monodon*, the farmers still depended on the capture of wild ones (Liao, 1992). So there is still a need to refine eyestalk ablation conceptual technique particularly for commercial shrimp species.

In decapod crustaceans two antagonistic hormones one stimulating, presumed to be from brain and thoracic ganglion and the other inhibiting, from the eyestalk, profoundly influence gonad maturation and function. Various investigations are underway to determine the roles of hormones on ovarian maturation, which may lead to practices of induced maturation in accordance with the need for increased aquaculture practices. Many workers suggested that the gonad-inhibiting hormone from the X-organ sinus gland complex of the eyestalk regulate ovarian maturation, (Bomirski *et al.*1 981; Quackenbush and Herrkind, 1983; Meusy *et al.* 1987, Mohammed and Diwan 1991).

In crustaceans, vitellogenin (Vg) is the precursor of egg protein and is a necessary prerequisite for the ovarian oocytes to reach full maturation (Chang *et al.*, 1993). Hormonal regulation of synthesis of Vg is well documented in oviparous vertebrates, (Tata, 1978). Knowledge of hormonal induction of gonadal development in crustaceans however is fragmentary. Otsu, (1960) reported that repeated implantation of pieces of the thoracic ganglion in immature female crab *Potamon dehaani* stimulated accumulation of yolk granules in the oocytes. Oyama (1968), Hinch and Bennett (1979), Eastman-Reck and Fingerman (1984) and Takayanagi *et al.* (1986) reported similar effects of the thoracic ganglion on ovarian maturation in *In vivo* and *In vitro* trials in various crustaceans. Yano *et al.*, (1988) demonstrated that yolk accumulation of *P. vannamei* could be induced and accelerated by implantation of thoracic ganglion from maturing female lobster *Homarus americanus*. Similarly advancement of ovarian stages is demonstrated in *P. Vannamei* by injection of brain extract (Yano and Wyban, 1992) and in *P. Japonicus* by injection of thoracic ganglion extract (Yano, 1992) from maturing females. Paradoxically, several of the crustacean reproductive hormones are only postulated, but direct chemical evidences and its regulatory activities are lacking (Aiken and Waddy, 1980). An area of research where enticing prospects lay both from the point of view of basic and applied research is the study of role of the steroids in the reproduction of invertebrates. Involvement of ovarian steroids in invertebrates especially in crustacean reproductive biology is still unknown. In crustaceans the steroids are apparently necessary for moulting (Stevenson *et al.*, 1979) and reproduction (Adiyodi, 1978). Little work has been done with regard to the effect of exogenous administration of steroids on gametogenesis in crustaceans (Kulkarni *et al.* 1979, 1992; Nagabhushanam *et al.* 1982 and 1987; and Koskela *et al.* 1992; Nieto, *et al.*, 1998, Striker and Smythe, 2001; Sosa, *et al.*, 2004; Treerattrakool *et al.*, 2005; Meeratana *et al.*, 2006).

Various steroid hormones have been identified in the ovaries of decapod crustaceans including estrogen in marine invertebrates (Hagerman *et al.* 1957), estrone in *P. monodon* (Fairs *et al.* 1990), the estrogenic compound 17B-estradiol in *H. americanus* (Couch *et al.* 1987) and in *Nephropsnorvegicus* (Fairs *et al.*, 1989) and progesterone and its related compounds, in *Panulirus japonicus* (Kanazawa and Teshima, 1971); in *Astacus leptodactylus* (Ollevier *et al.* 1986); in *H. americanus* (Couch *et al.*1987); and in *Nephrops norvegicus* (Fairs *et al.* 1989). Occurrence of precursors of steroid hormones and enzyme systems involved in the steroid metabolism has been reported in crustaceans (Kanazawa and Teshima, 1971 and Teshima and Kanazawa, 1971). Junera *et al.*, (1977) first proposed that, a vitellogenin stimulating ovarian hormone is present in the ovary of the amphipod, *Orchestia gammarella*. Since then progesterone (Kulkarni *et al.*, 1979; Yano, 1985), 17α-hydroxy progesterone (Nagabhushanam *et al.*, 1980; Yano,

1987; Tsukimura and Kamemoto, 1988 and Koskela *et al.* 1992) has all been implicated in this role. D'croz *et al.* (1988) isolated 5 prostaglandins and three related compounds from the polychaete *Americonuphis reesei*, which is used as a dietary supplement to accelerate ovarian maturation of penaeid shrimps in Central America. However, the detailed mechanism of ovarian induction by steroids in crustaceans is still unknown. Involvement of estrogenic compounds in vitellogenesis and development of egg has been suggested by Couch, (1987) and Ghosh and Ray, (1992).

Reproductive processes in most of the shrimps are controlled by hormones. Hormones that influence gonadal development and spawning are produced in the brain and thoracic ganglia. The ovaries also produce hormones. Y-organs, mandibular organ (MO) and androgenic glands also produce hormones which promote gonadal development directly or indirectly. In recent years, the roles of neuro regulators have been appraised in enhancing maturation and the spawning process (Charmantier *et al.*, 1997; Fingerman and Nagabhushanam 1992)

5.2 REPRODUCTION PATTERN IN SHRIMPS

Shrimps are bisexual and development of gonads is a slow process, which takes place along with the moulting process during a large part of the adult life. During reproductive cycles the gonads undergo a sequence of major morphological and physiological transformations which require a large amount of energy. In females, development of the oocyte starts with the process of ogenesis and in males with spermatogonial cells. The various stages in the development of the oocytes as well as spermatozoa have been described for several species (Fingerman and Nagabhushanam 1983; Charniaux- Cotton and Payen, 1988; Mohamed and Diwan 1992, Mohamed and Diwan 1991, Shoji 1997). Determination of sexual maturity of live shrimps in early stages is difficult. However, as a shrimp advances towards the ripening phases of gonadal development, identification of the maturation phase is possible. For example, the developing ovary and developed thalycum of female shrimp for storing spermatozoa can be seen through the cephalo thoracic region. In males, the development of white glitter hening spermatheca or sperm boxes is visible on the bases of the fifth pereiopods. Many have developed immunochemical techniques for determining the scale of maturity. Derelle *et al.*, 1986 developed an ELISA titration assay for measuring vitellogenin synthesis in the freshwater shrimp *Macrobrachium rosenbergii*, using a monoclonal antibody. Similar studies have been done with the lobster *H. Americanus* (Tsukimura *et al.*, 1992) and crab *Callinectes sapidus* (Lee and Watson, 1994). The most accurate method now available is based on a micro–anatomical survey of the reproductive organs using histological procedures. Many shrimps are seasonal breeders and in tropical countries generally the breeding season lasts from October to February. Environmental factors particularly temperature, photoperiod and salinity play an important role in enhancing breeding and spawning activity (Van Herp and Payen, 1991). All the marine shrimps breed and spawn in deep sea waters. The larvae migrate to shallow coastal waters for their growth to attain the adult size. Shrimps undergo breeding and spawning several times and produce millions of larvae during their lifetime (Rao 1968). Currently, technologies are available for domestication of shrimps in hatcheries and laboratory conditions so that induction of breeding and spawning at a desired time is possible. The role of the central nervous system and its coordination with neuroendocrine centres, along with environmental cues, are now well understood so far as regulating the reproductive organs is concerned (Fingerman 1997).

5.3 REGULATION OF REPRODUCTION

The reproductive activity of crustaceans is partially under the control of neuroendocrine factors. Panouse (1943) showed the presence of gonad inhibiting hormone (GIH) in the eyestalk of a female shrimp, *P. serratus*. The increase in ovarian growth after eyestalk removal was primarily due to removal of the inhibitory effect on vitellogenesis (Mohamed and Diwan, 1991). Otsu T (1985) showed the existence of the gonad stimulating hormone (GSH) in the brain and thoracic ganglion. Based on these findings it was deduced that gonadal maturation in shrimp was regulated by two antagonistic neurohormones, GIH and GSH. This model concept has been well reviewed (Adiyodi 1985; Yudin *et al.*, 1980; Van Herp and Payen 1991; Mohamed and Diwan 1991; Fingerman 1997; Charmantier *et al.*, 1997; Fingerman and Nagabhushanam 1997). Khalaila *et al.*, (2002) while studying the role of eyestalk-borne hormones on spermatogenic activity in the testis and androgenic gland of the crayfish, *Cherax quadricarinatus*, found that the sinus gland directly controls the activity of the androgenic gland which suggests an endocrine axis like relationship between the sinus gland and AG and the male reproductive system in decapod crustaceans.

5.4 NEUROENDOCRINE FACTORS CONTROLLING REPRODUCTION

1. Eyestalk X-organ sinus gland complex

The X-organ-sinus gland complex, which is the prime neuroendocrine centre in the eyestalks of shrimps, produces hormonal factors (neuropeptides / neurohormones) that control the physiological process of gonads (Fingerman 1997). GIH was localized by specific antibodies in the X-organ SG complex of the lobster *H. americanus*, (Kallen and Meusy 1989) prominently at the meta nauplius stage (Rotlant , 1995). By using the technique of non radio actively labelled cDNA probes it was possible to detect neuropeptides in X-organ SG complex of this lobster and its larvae, which has the property of GIH (De Kleijn, 1992; Rotlant, 1993).This study revealed that GIH is also present in males. In the shrimp *P. varians* and the crayfish *P. Bouvieri* neuropeptides extracted from the eyestalks have a negative effect on the growth of vitellogenic oocytes (Soyez *et al.*, 1991; Aguilar 1992). Now the eyestalk ablation technique is being practiced commercially for inducement of maturation and spawning of shrimps (Hossain *et al.*, 1990; Mohamed and Diwan 1991; Primavera and Caballero 1992). The presence of GIH in embryos and larvae105 may be an indication of an inhibitory role before adolescence. Chang *et al.*, (1992) isolated and purified the peptide from sinus glands of shrimp *S. ingentis* which is responsible for inhibition of ovarian development and spawning. Quackenbush and Keeley, (1988) have also isolated a factor from eyestalks of the shrimp *P. setiferus*, which has inhibitory effect on vitellin synthesis. Similar, studies have been carried out by Quackenbush, (1989) in *Penaeus vannamei.* Huberman, (2000) recently reviewed the biomolecular aspects regarding the work carried out on GIH in crustaceans in general and shrimps in particular. Edomi *et al.*, (2002) while studying GIH of the Norway lobster (*Nephros norvegicus*) reported that the GIH is actively involved in gonad maturation process and plays a more complex role in control of reproduction and moulting. With a combination of the reverse transcription-polymerase chain reaction (RT-PCR) and rapid amplification of cDNA ends approaches, they determined the cDNA sequence of *N. norvegicus* prepro-GIH. The possibility of the involvement of neurohumoral agents in relation to control of reproduction was debated for quite some time, but in recent years experimental evidences are available that 5-HT that is present in the nervous system including the X-organ sinus gland complex has a stimulatory effect on reproductive activity.

In kuruma shrimp *M. japonicus* Okumura and Sakiyama (2004) have reported the induced ovarian development after eyestalk ablation during non- reproductive periods. Tsutsui *et al.*, (2005) mentioned that bilateral eyestalk ablation caused an increase in gonado-somatic index in immature shrimp *M. japonicus* and further it has been also mentioned that there was increase in the vitellogenin mRNA levels in the ovary. Okumura (2007) while working on immature female shrimp *M. japonicus* reported that bilateral eyestalk ablation induced ovarian development and increased ovarian width, vitellogenin levels in haemolymph and vitellogenin mRNA levels in the ovary significantly. The vitellogenin inhibiting hormone (VIG) has been characterized both structurally and functionally in limited species of crustaceans. The gene for VIH have been cloned from several crustacean species including the shrimp *Rimicaris kairei* (Qian, 2009). Molecular analysis of VIH isolated from females of a few crustaceans species shows that they consist of signal peptides and mature peptides. They also show a considerable degree of sequence similarity with moult inhibiting hormome (MIH) (Swetha *et al.*, 2011). With the advancement in crustacean molecular endocrinology considerable work has been carried out using RNAi technology for gene silencing process in which dsRNA triggers sequence specific separation of its cognate mRNA which can be a powerful tool to reveal the gene function. Treerattrakool *et al.*, (2008) worked in the direction attempting on the silencing of VIH using reverse genomic technique in *P. monodon*. In this study they observed the decreased VIH levels along with increased levels of vitellogenesis in the ovary.

In the past decade a vast amount of research has been directed toward the understanding and manipulation of reproduction in penaeid shrimps. In many decapods including penaeids removal of eyestalks that contain the X-organ/sinus gland complex, which produce and stores gonad-inhibiting hormones has become a well-recognized technique for inducing gonadal maturation. The most commonly accepted theory is that a gonad-inhibiting hormone (GIH) is produced in the neurosecretory complexes in the eyestalks. This hormone apparently occurs in nature in the non-breeding season and is absent or present only in low levels during the breeding season (Bomirski and Kleke, 1974; Kulkarni and Nagabhushanam, 1980). By inference then, the reluctance of most penaeids to develop mature ovary in captivity is a function of elevated levels of GIH, and eyestalk ablation lowers the high haemolymph titer of this GIH. Eyestalk ablation in crustaceans has been reviewed by Fingerman (1970); Kleinholz and Keller (1979) and Charniaux-Cotton (1985); Fingerman (1987); Wyban *et al.* (1987) and Liao (1992); Benzie, (1997); Cavalli *et al.*, (1997); Crocos, and Coman (1997); Yang *et al.*, (1997); William *et al.*, (1998); Browdy (1998) Pratoomchat, (1998); Crocos *et al.*, (1999); Palacios *et al.*, (1999a); Palacios *et al.*, (1999b); Wouters *et al.*, (1999, 2001, 2001b); Vaca, and Alfaro, (2000); Withyachumnarnkul, *et al.*, (2001); Hoang *et al.*, (2002); Coman *et al.*, (2003, 2004, 2005, 2006, 2007); Peixoto *et al.*, (2003, 2005); Alfaro et *al.*, (2004); Arcos, *et al.*, (2004); Cuzon, *et al.*, (2004); Mengqing, *et al.*, (2004); Toledo *et al.*, (2005); Kanokpan *et al.*, (2006); Meeratana *et al.*, (2006); Sellars *et al.*, (2006).

A few instances of natural maturation and spawning of unablated and captive *P. monodon* in seawater ponds and tanks have been reported by Liao (1977) from Taiwan, by Primavera (1978) from Philippines and by Aquacop (1979) from Tahiti. The most successful technique in captivity for inducing maturation of penaeids has been the removal of eyestalks. This method has been tried with varying degrees of success by pioneers workers *viz.*, Idyll (1971) in USA, Anstein and Beard (1975) in UK, Alikunhi *et al.* (1975) in Indonesia, Aquacop, (1975) and (1977) in Polynesia, Wear and Santiago (1976), Santiago, (1977) and Primavera, (1978) in Philippines and Halder, (1978) in India. These results indicated that unilateral eyestalk ablation greatly enhances gonadal maturation in penaeids like *P. indicus, P. stylirostris, P. vannamei* and *P. seriterus* and it is a prerequisite for hardy penaeids like, *P. monodon* of good spawnings, the proportion of regression among females, which have started their ovarian development has been reported to be high (Aquacop, 1979). However the use of eyestalk ablation to induce gonadal maturation in *P. monodon* female has been done with success by many other workers like Alikunhi *et al.* (1975), Wear and Santiago (1976), Aquacop (1977, 1979, 1983), Primavera (1978), Primavera *et al.* (1979), Millamena *et al.* (1985), Lin and Ting (1986); Menasveta *et al.* (1989); Benzie, (1997); Cavalli *et al.*, (1997); Crocos, and Coman, (1997); Yang, *et al.*, (1997); William *et al.*, (1998); *Browdy*, (1998) Pratoomchat, (1998); Crocos, *et al.*, (1999); Palacios *et al.*, (1999*a*,1999b); Wouters, *et al.*, (1999, 2001, 2001b); Vaca, and Alfaro, (2000); Withyachumnarnkul, *et al.*, (2001); Hoang, *et al.*, (2002); Coman *et al.*, (2003, 2004, 2005, 2006); Peixoto *et al.*, (2003, 2005); Alfaro, *et al.*, (2004); Arcos, *et al.*, (2004); Cuzon, *et al.*, (2004); Mengqing, *et al.*, (2004); Toledo *et al.*, (2005); Kanokpan *et al.*, (2006); Meeratana, *et al.*, (2006); Sellars, *et al.*, (2006). It is also reported that the stimulation of gonadal maturation in *P. monodon* is dependent on the relative interaction of environmental factors and the age (Laubier, 1978; Emmerson, 1980; 1983), apart from eyestalk factors.

The shrimp *P. monodon* at a salinity of 34‰ using wild stock it is reported that eyestalk ablation significantly enhances ovarian development and spawning. Similarly, at the salinity SH"34‰ shrimps from filtration ponds also showed that gonadal maturation can be enhanced by eyestalk ablation, though the progress of maturation process in these animals is slower then that reported in wild shrimps collected from the open sea.

Generally *P. monodon* appear to take longer time to mature in captivity than other penaeids (Vide Alikunhi *et al.* 1975; Aquacop, 1975). Wear and Santiago (1976) and Santiago (1977) reported that *P. monodon* took approximately 2 months to attain first maturity. However, Primavera (1978) obtained first spawnings in *P. monodon* 22 days after unilateral eyestalk ablation under reduced light intensity and at a salinity of 34‰, while Halder (1978), reported 38 days for gonadal maturation in the same species at salinity of 25‰. It is reported that all of them used outdoor ponds or cages, presumably with high light intensity to mature ablated *P. monodon*. This is presumably why they took longer periods than reported by Primavera (1978) and Emmerson (1980) who used indoor tanks with reduced light conditions. Beard and Wickins (1980) also matured ablated *P. monodon* under reduced light intensity (40-70 lux). It has been noted that water quality also plays a role in the full achievement of the process. Shoji (1997) reported that the shortest period from ablation to the onset of maturation is 22 days and first spawning occur after 25 days in wild shrimps colleted from the open sea. Where as the shrimps collected from filtration ponds took 30 days to reach the late-vitellogenic stage of ovarian development. Spawning could not be achieved in these shrimps till the termination of the experiment after 30 days. Muthu and Laxminarayana (1982) reported that *P. monodon* collected from filtration ponds took 66 days to mature. They concluded that frequent power failures leading to disruption of the recirculatory system collapsed the water quality, which delayed the maturation process.

In terms of the size of the animals, it has been reported that the rate of achievement of gonadal maturity in larger animals is much greater compared to the smaller *P. monodon* and it is comparable with other penaeid shrimps (Shoji, 1997). Chotikun (1988) and Tansutapanich *et al.* (1989) found that optimal age of maturation of pond reared penaeid shrimps is at least 18 months. In the wild, *P. monodon* attained full maturity and spawning at 10 months (Motoh 1981) and optimum shrimp size reported in these studies is > 100 g. Primavera (1978) reported that five-month-old *P. monodon* could mature and spawn after ablation, but produced poor quality larvae. These results indicated the need for older and larger females for successful production of PL from pond-reared broodstock.

It is mentioned that eyestalk ablation conducted in pond reared *P. monodon* at low salinity it did not induce in gonadal maturation; instead it induced repeated moulting (Shoji, 1997). It is found that, that the penaeid shrimps attained maturity faster when the salinity of the water is 30-33ppt. The fact that juveniles of these species live in brackish waters and are usually migrate to the sea for spawning purposes suggests that salinity is one of the important factors that influence the maturation process (Muthu and Laxminarayana, 1982). This is strongly supported by the observations of Silas *et al.*, (MS) that *M. dobsoni* could attain maturity in brackishwater ponds when the salinity increased to 28-30 ppt (See review Muthu and Laxminarayana, 1982). Even *P. indicus* of stage III maturity have been collected by George, (1974) from the brackishwater ponds during the high salinity months. Vasudevappa (1992) reported that *M. dobsoni* could mature through eyestalk ablation during high saline months in brackish waters. Halder (1978) stated that ablated *P. monodon* attained maturity and spawned viable eggs in a brackish water when the salinity is only 25‰ is the only exception reported.

It is reported that a system incorporating unilateral eyestalk ablation, with high salinity, good water quality and reduced light intensity are sufficient to induce the maturation of female *P. monodon* in captivity (Shoji, 1997). Regarding the effect of broodstock source on maturation, it was mentioned that pond reared and wild-caught broodstock of *P. monodon* performs comparably better if the size and age are adequate (Shoji, 1997). Similar observations have made by Srimukda (1987) and Menasveta *et al.* (1994) in the same species and also by Rangnekar and Deshmukh (1968) in *Scylla serrata*; Bomiriski and Klek, (1974 and 1975) in *Crangon crangon*; and Nagabhushanam and Diwan, (1974) in *Barytelphusa cunculsris.*

Demonstrating hormone's effectiveness in inducing Vg synthesis in decapod crustaceans is difficult because females with immature ovaries are strongly affected by a gonad-inhibiting hormone (GIH) compared to vitellogenic females (Yano and Wyban, 1992). Moreover, the complete physiological process in decapod crustaceans are under the control of different hormones secreted by neuroendocrine cells located in different ganglionic centers (Fingerman, 1970). Evidences have also been presented to show the presence of a gonad- stimulating hormone (GSH), which stimulates ovarian maturation, in the thoracic ganglion of vitellogenic crab, shrimp and lobster (Otsu, 1960, Oyama, 1968, Hinsch and Bennet,1979; Eastman-Recks and Fingerman, 1984; Takayanagi *et al.*, 1986; Yano *et al.*, 1988; Yano, 1992).

Shoji (1997) reported that the thoracic ganglion extract is more effective in the induction of ovarian maturation in adult female *P. monodon.* Even though the hormones from the brain of vitellogenic females also have a role in the induction of gonadal maturation, it is not so prominent as that of the hormone from the thoracic ganglion. However, it has been reported that formation of yolk granules, cortical crypts and germinal vesicle breakdown are found in maturing and mature oocytes of *P. vannamei*, which may be induced by a hormone secreted from brain, and it is similar to the development of oocytes induced by GSH secreted from thoracic ganglion (Yano and Wyban, 1992).

2. Cerebral and thoracic ganglia

Otsu (1963) reported that the accumulation of yolk granules in oocytes was stimulated by repeated implantation of pieces of thoracic ganglion in the immature female crab, *Potamon dehanni*. Later many have reported similar effects of thoracic ganglion on ovarian maturation with in vivo and in vitro experiments. In addition to these findings, Yano and Wyban (1992) have demonstrated that ovarian maturation of *P. vannamei* can be induced and accelerated by implantation of pieces of thoracic ganglion tissue prepared from female lobsters (*H. americanus*) with developing ovaries. These results indicate that ovarian maturation may be induced by a gonad-stimulating hormone (GSH) secreted by the neurosecretory cells of the thoracic ganglion of maturing females and that this GSH is not species specific in activity between this shrimp and lobster. Injection of thoracic ganglion extract prepared from maturing females is effective in increasing serum Vg in *P. japonicus*. This means that GSH also stimulates Vg synthesis and or its secretion into the blood in penaeid shrimp.

Yano (1992) showed that injection of thoracic ganglion extract prepared from vitellogenic females was effective in increasing serum Vg, even in immature females. This indicates that after initiation of vitellogenesis, higher amounts of GSH, which are increased by injection of thoracic ganglion extract, accelerate Vg synthesis and its release into the blood. This suggests that in penaeid shrimp, GSH levels may increase further with the advancement of vitellogenesis, parallel to a decrease in the level of GIH.

Yano (1998) while discussing hormonal control of vitellogenesis in penaeid shrimp has mentioned that vitellogenin (Vg) synthesis in the ovarian tissue gets stimulated by the thoracic ganglion extract from vitellogenic females. Vitellogenin synthesis in the ovary is stimulated by vitellogenesis-stimulating hormone (VSH), a neuropeptide of approximately 10000 Da secreted from the thoracic ganglion in the kuruma shrimp, *P. japonicus*. Further he has mentioned that vitellogenin synthesis in incubated previtellogenic ovarian tissues can be stimulated by estradiol-17 β, probably secreted from the ovary, stimulate Vg synthesis in the ovary as a Vg –stimulating ovarian hormone in penaeid shrimp. He has also stated that unilateral eyestalk ablation is effective in increasing serum Vg in the kuruma shrimp.

Further thoracic ganglion extract prepared from vitellogenic kuruma shrimp females was fractionated by gel filtration high-performance liquid chromatography; high Vg-stimulating activity was detected in the fraction corresponding to a molecular weight of 10,000.This fraction was inactivated by trypsin; therefore, this bioactive factor, GSH, may be characterized as a peptide hormone.

Shoji (1997) reported that the thoracic ganglion extract is more effective in the induction of ovarian maturation in adult female *P. monodon*. Even though the hormones from the brain of vitellogenic females also have a role in the induction of gonadal maturation, it is not so prominent as that of the hormone from the thoracic ganglion. However, it has been reported that formation of yolk granules, cortical crypts and germinal vesicle breakdown are found in maturing and mature oocytes of *P. vannamei*, which may be induced by a hormone secreted from brain, and it is similar to the development of oocytes induced by GSH secreted from thoracic ganglion (Yano and Wyban, 1992).

On the other hand, brain extracts prepared from maturing females induced Vg synthesis in *P. japonicus*. A high Vg-stimulating activity was detected in the fraction corresponding to a molecular weight of 1000 to 2000. This fraction was inactivated by trypsin, clearly showing it to be a peptide hormone. Although a supplement of thoracic ganglion extract to the culture medium was effective in keeping vitellogenic oocytes from degenerating, brain extract was not effective in *P. japonicus* in vitro. These findings indicate that brain hormone is different from thoracic ganglion hormone and that the brain works through the thoracic ganglion in regulating vitellogenesis in release of GSH. This suggests the presence of brain hormone stimulates the release of GSH in penaeid shrimp. Therefore, gonad-stimulating hormone-releasing hormone (GSH-RH) is nominated as a possible hormone type in the brain (Yano and Wyban, 1992).

Vg has been identified electrophoretically and immunologically in the hemolymph of vitellogenic female crustaceans. Therefore, extra-ovarian tissue has been suspected for a long time as the site of Vg synthesis in crustaceans. In fact, evidence has been presented to show that Vg is synthesized by the body fat or adipose tissue in amphipods and isopods. Recently, several workers demonstrated that Vg is synthesized on a large scale by the ovaries of crayfish, fiddler crabs, *Uca pugilator*, kuruma shrimps and white shrimps. Considering these observations, the site of Vg synthesis in decapods is different from that in isopods and amphipods. Vg is secreted into the hemolymph after its synthesis and then accumulated in the developing oocytes as vitellin. Recently, evidence has been presented to show that 17a-hydroxy-prosesteron in the freshwater shrimp *Macrobrachium lanchesteri*, and the black tiger shrimp *P. monodon*. The hormone 17,-hydroxy-progresterone is generally stimulating ovarian hormone (VSOH) which controls Vg synthesis in females of *Orchestia gammarella*. It is probable that 17 a-hydroxy-progesterone stimulate Vg synthesis and release into the hemolymph as a VSOH in penaeid and other shrimp. By immune fluorescence Vg was found to occur for the first time in the follicle cells of the oil globule stage I oocytes in early developing ovaries which actively synthesize Vg. The follicle cells are nominated as a possible cell type responsible for ovarian Vg synthesis in kuruma shrimps. It is suggested that 17a-hydroxy-progesterone, probably secreted from the ovary as a VSOH, stimulates Vg synthesis in follicle cells and Vg release into the hemolymph in penaeid shrimp (Ghosh and Ray, 1993; Charmantier *et al.*, 1997).

A considerable amount of research has been conducted in crustaceans that indicate the presence of gonad stimulating factors in cerebral and thoracic ganglia. Implantation experiments carried out by Otsu, (1963) and Gomez and Nayar, (1965) indicated that the thoracic and supra oesophageal ganglia can induce ovarian growth. Injections of aqueous extract prepared from thoracic ganglia and brain as well as *in vitro* incubation experiments indicated vitellogenesis can be stimulated in several crustaceans (Nagabhushanam and Diwan, 1974; Eastman-Reks and Fingerman 1984; Kulkarni *et al.*, 1991; Mohamed and Diwan 1991; Yano 1992; Shoji 1997). Induction of ovarian maturation in *P. vannamei* has been accomplished by injecting lobster brain extract (Yano and Wyban 1992). By injecting thoracic ganglion extract of *P. semisulcatus,* Shoji (1997) demonstrated the possibility of maturation and spawned *P. monodon.* Yano, 1998, 2000 modelled the mode of action of hormones related to maturation. According to him GnRH is produced in the brain and in response to this the thoracic ganglion produces gonadotropins (GnH) and under the influence of GnH maturation is affected. However, the molecular structure of GSH is yet to be elucidated.

5.4 ROLE OF STEROID HORMONES IN REPRODUCTION

Although very little is known of steroid action on maturation, it has been postulated that steroid hormones can stimulate vitellogenesis in crustaceans. In crustaceans the steroids are apparently necessary for moulting (Stevenson *et al.*, 1979) and reproduction (Adiyodi, 1978). The use of exogenous hormones to induce the gonadal maturation and spawning is not well established. However, various steroid hormones have been identified in the ovaries of decapod crustaceans, including estrogen (Hagerman *et al.*, 1957); estrone (Fairs *et al.*, 1990); the estrogenic compound 17–β-estradiol (Couch *et al.*, 1987; Fairs *et al.*, 1989) Progesterone and related compounds (Kanazawa and Teshima, 1971; Teshima and Kanazawa, 1971; Couch *et al.*, 1987; Quinito *et al.*, 1990; Tsukimura and Kamemoto, 1991; Yano, 1985 and 1987). Apart from these there is some more works in penaeids revealing the role of extraneous hormones on the works like, Nieto, *et al.*, (1998), Stricker and Smythe, 2001; Sosa, *et al.*, (2004); Treerattrakool *et al.*, (2005); Meeratana, *et al.*, (2006). In short, vertebrate steroid hormones seem to be present in crustacean tissues and exogenous applications of these hormones produce consistent effects with a role promoting ovarian maturation. It is likely that in future, the use and manipulation of steroid hormones in the regulation of penaeid maturation is a real possibility (Quackenbush, 1991). Junera *et al.* (1977) first proposed that a vitellogenin stimulating ovarian hormone is present in the ovary of the amphipid, *Orchestia gammarella.* Since then progesterone (Yano, 1985 and Nagabhushanam *et al.*, 1982; Yano, 1987; Tsukimura and Kamemoto, 1988 and Koskela *et al.*, 1992) and various estrogens including 17 Beta-estradiol (Couch *et al,* 1987; Fairs *et al.*, 1990; Ghosh and Ray, 1992; Koskela *et al.*, 1992 and Ghosh and Ray, 1994) have been considered in the maturation process.

Administration of exogenous steroids (17α-hydroxyprogesterone) caused marked ovarian development in *P. monodon*, when compared to that of control animals (Shoji, 1997). Similar type of observations have made by Nieto, *et al.*, (1998), Stricker and Smythe, (2001); Sosa, *et al.*, (2004); Treerattrakool *et al.*, (2005); Meeratana, *et al.*, (2006) in other penaeids. This observation is suggestive that, this hormone has got some potential to elicit active ovarian growth in this penaeid shrimp. Whereas in the estradiol injected shrimps marked ovarian development is not found. Kulkarni *et al.*, (1979); and Kulkarni and Nagabhushanam (1982) are able to induce oogenesis in *P. hardwickii* injected with ions of 10 μg of progesterone/shrimp (9-10 cm body length) on alternate days. In contrast Yano, (1985) reported that mortality is great in *M. ensis* when they are injected with similar doses of (10 μg) progesterone or are subjected to repeated injections of the hormone. In *M. ensis* ovarian development and spawning are induced by a single, low dose (0.1 μg/g body wt.) injection of progesterone (Yano, 1985). Sarojini *et al.*, (1985) also reported that administration of progesterone into immature female *Macrobrachium kistnensis* accelerated oogenesis. Nagabhushanam *et al.* (1980 and 1982) reported that spawning has been induced in *P.stylifera* by injecting a single dose of (50μg /shrimp, 80-85 mm) 17-α-hydroxyprogesterone into the abdominal musculature at a lower temperature (20°C), at which this shrimp does not spawn naturally. Later on Yano, (1987) demonstrated that injection of 0.01 mg. 17-α-hydroxy progesterone/g body wt. is effective in stimulating secretion of large amount of Vg. into haemolymph of *P. japonicus*. Conversion of progesterone into 17-α-hydroxyprogesterone has been demonstrated in the ovaries of the crab, *Portunus trituberculatus*) Teshima and Kanazawa, 1971). Therefore based on these observations, it can be deduced that progesterone may serve for ovarian vitellogenesis as a precursor of 17-α-hydroxy progesterone, which stimulates Vg synthesis (Yano, 1987). In *P. monodon* when 20μg /g body wt of 17-α-hydroxyprogesterone has been administered, caused the induction and acceleration of the ovarian maturation to a marked extent. The lack of full maturation and spawning may be due to the limited time and/ or inappropriate dosage (Shoji, 1997). Optimal progesterone doses and injection sequences to induce ovarian maturation and spawning may vary from species to species (Yano, 1985). It is however unclear whether 17-α-hydroxy progesterone acts directly or as a precursor to stimulate Vg synthesis. Nagabhushanam *et al.*, (1982) suspected that 17- α-hydroxy progesterone and 17-α-hydroxyprogesterone, bypasses the synthesis of steroid mediator. The possible conversion of injected steroid hormone into desired hormone, which may be naturally occurring in crustaceans, is also speculated (Nagabhushanam *et al.*, 1987; Sarojini *et al.*, 1990). In contrast, Tsukimura and Kamemoto (1988) have reported that, neighter 17-β-neither estradiol nor progesterone induced the ovarian maturation in *P. vannamei*. Similarly, Anon (1992) failed to effectively enhance ovary development in pre-adult *P. monodon* using 17-α-hydroxyprogesterone and 17 β-estradiol administered in *P. esculentus.*

5.5 ROLE OF ENDOCRINE ORGAN

1. Mandibular organ (MO)

The possible role of the MO in reproductive activity of crustaceans was reviewed by Laufer *et al.*, (1993) and Waddy and Aiken (2000). Terpenoids like methyl farnesoate (MF) and farnesoic acid (FA), both secreted by MO, influence reproduction of both in male and female crustaceans. Liu and Laufer (1996) found that the activity of MO is regulated by sinus gland neuropeptides. It was mentioned that MO-inhibiting hormonal compounds have the similar type molecular masses and amino acid composition as noticed with other sinus gland neuropeptides. Chang *et al.*, (1992) by using a radio labelled photo affinity analog of MF found in *S. ingentis* the presence of MF binding proteins in the ovaries, testis and accessory glands in addition to the haemolymph. The culturing *in vitro* of ovarian tissue of the shrimp, *P. vannamei* in presence of MF resulted in a significant increase in the size of the oocytes. This can be interpreted as the involvement of MF in the early events related to secondary vitellogenesis. MF was reported to increase fecundity in cultured shrimp *P. Vannamei* (Laufer 1992; Laufer *et al.*, 1997). Gunawardene *et al.*, (2000) while working on functions and cellular localization of farnesoic acid O-methyl transferase (FAMeT) in the shrimp, *Metapenaeus ensis* reported that FAMeT directly or indirectly modulates the reproduction and growth of shrimp through methyl farnesoate (MF) by interacting with the eyestalk neuropeptides as a consequence of its presence in the neurosecretory cells of the X-organ-sinus gland.

Yashiro *et al.*, (1998) while carrying studies on the effect of methyl testosterone and 17α-hydroxyprogesteron on the spermatogenesis in the tiger shrimp, *P. monodon* found that the methyl testosterone has enhanced spermatogenesis yielding full sacs from empty sacs within 22-25 days after the administration.

Nagaraju *et al.*, (2002) reported that ovarian index and oocytes diameter in case of shrimp *P. indicus* increased significantly after injection of MF and the increase of the ovary was four fold than those of controls. They also observed that the ovaries of MF received animals, entered into the late vitellogenic stage. Nagaraju (2007) in his review paper has described role of MF in regulating several physiological functions in crustaceans. It has been mentioned that MF is involved in the regulation of reproduction, molting, larval development, morphogenesis, behaviour and general protein synthesis in crustaceans. Further it is also mentioned that MF synthesis and secretion is negatively regulated by an eyestalk peptide, called mandibular organ-inhibiting hormone (MO-IH).

2. Androgenic gland

In decapod crustaceans, the androgenic glands are generally found associated with the terminal portion of the male gamete duct. Earlier investigators have noted the differences and / or uniformities in the function of the androgenic gland among various group of crustaceans in general and hermaphroditic and non hermaphroditic decapods in particular. Mohamed and

Diwan (1991) described the structure of the androgenic gland in *F. indicus* and also demonstrated the impact of bilateral andrectomy on sex reversal of the shrimp. It was reported that andrectomized male shrimps have lost their secondary sexual characters and exhibited absence of sperm in the lumen of their testicular acini. Decapod androgenic glands are not necessary for completion of spermatogenesis and their absence results only in a reduction of spermatogenesis intensity. Mohamed and Diwan, 1991 reported that the lack of androgenic glands in *F. indicus* appeared to inhibit spermatogonial differentiation. Many studies have been devoted to determine the chemical nature of the androgenic hormone. Laufer and Landau (1991) have mentioned that these glands are capable of producing several compounds including proteins and the terpenes, hexahydroxy farnesyl acetone and farnesyl acetone, and the exact role of these compounds is not yet known. Charmantier *et al.*, (1994) reported that the androgenic gland hormone regulates the spermatogenic activity in the testis and is responsible for the development and maintenance of the secondary sexual characteristics in male crustaceans.

3. Y-organs

The role of the Y-organs in crustaceans in the regulation of the moulting process and growth is well known (Echalier 1955, 1959; Passano and Jyssum 1963, Lachaise *et al.*, 1993; Sun 1994; Carlisle 1957; Maissiat 1970 (a, b); Bourguet *et al.*, 1977; Charmantier 1997). As stated earlier, the Y-organ is the source of ecdysteroids and correlations between vitellogenesis and ecdysteroids levels in haemolymph have been reported in some species Okumura *et al.*, 1992; Yudin *et al.*, 1980. The role of ecdysteroids in the regulation of vitellogenesis in female crustaceans is not known. Ecdysteroids may directly or indirectly participate in the regulation of spermatogenesis (Sagi *et al.*, 1991) and induction of gonadal growth in males (Laufer *et al.*, 1993).

4. Ovaries

The ovarian tissue in most crustaceans, particularly in decapods produces vitellogenin-stimulating ovarian-hormone (VSOH) and under the influence of this, the growth of oocytes takes place (Legrand *et al.*, 1982; Payen 1986; Khalaila *et al.*, 2002; Charniaux and Payen 1988; Van Herp and Payen 1991). Besides VSOH, many ecdysteroids have been identified in the follicle cells and oocytes of the ovarian tissue (Chang 1993). Yano, (2000) described in detail the control of the process of vitellogenesis by the endocrine system in penaeid shrimp. He reported that estradiol-17β is effective in increasing serum Vg in the Kuruma shrimp, *M. japonicus.* The hormone estradiol-17 β is generally distributed in the ovary of crustaceans and it was suggested that estradiol-17 β secreted from ovarian follicle cells induces Vg synthesis in the ovary as a Vg-stimulating ovarian hormone in penaeid shrimp Yano, 2000. A positive relationship between Vg levels in haemolymph and circulatory levels of both progesterone and 17 β estradiol has been observed in shrimps and other crustacean animals (Quinitio *et al.*, 1994). Fluctuating levels have also been reported between estradiol and progesterone in the ovary and haemolymph at different vitellogenic stages. In contrast, Okumura and Sakiyama (2004) reported negative relationship between ovarian maturation and haemolymph steroids levels in *M. Japonicus.* Injection of progesterone induced, ovarian development in the shrimp *P. hardwickii* (Kulkarni *et al.*, 1979). Progesterone and estradiol have been reported to stimulate Vg gene expression in both hepatopancreas and ovary explants of *M. ensis* (Tiu *et al.*, 2006). Yano (1987) has reported that administration of 17α- hydroxy progesterone stimulated ovarian development and vitellogenesis process in the kuruma shrimp *M. japonicus.* It is also known that injection of progesterone and 17α- hydroxy progesterone are able to induce ovarian maturation in the shrimp *M. ensis* (Yano I, 1985). Vertebrate types of steroids are also found in the shrimp *P. monodon* and the levels of some of these steroids (progesterone and 17 β estradiol) are found to be high in the ovaries at the terminal phase of maturation suggesting the role of steroids in shrimp reproduction (Quinitio *et al.*, 1994) (Fig.1)

Though the relationship between steroid levels and gonadal maturation process have been established, however still it is not known whether shrimps obtain this type of steroids from the environment that is via feed or synthesized internally. Therefore future studies are required to examine the mode and action of steroid hormone, their binding capacity with nuclear receptors in order to generate physiological actions. Though classical vertebrate steroid hormone and their receptors have been identified for limited number of crustaceans, the exact mechanism by which they exert their effect on crustacean reproduction is less clear than steroid induced responses in mammals (Swetha *et al.*, 2011).

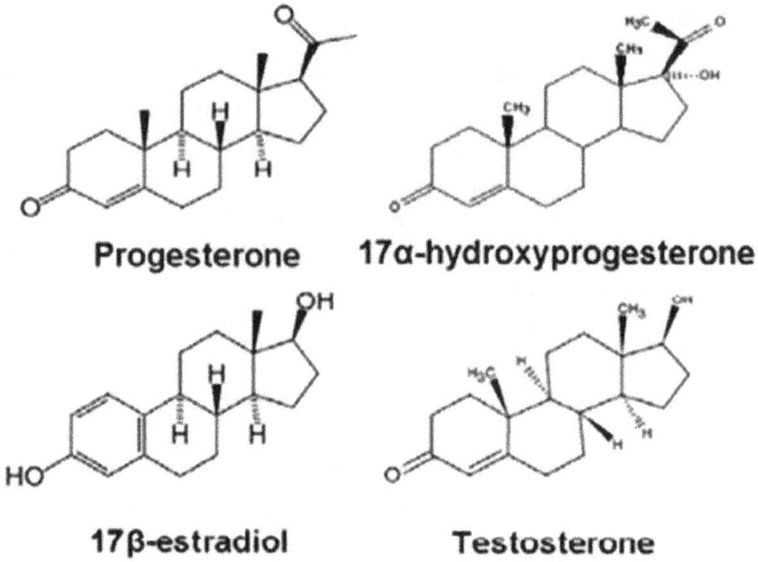

Fig. 1. Structure of some vertebrate-type steroids

5.6 CHEMISTRY OF HORMONES

1. Ecdysteroids

Ecdysteroid or ecdysone (Fig. 2), a moulting hormone, similar to the hormone found in insect prothoracic glands is produced by the Y-organs in crustaceans (Chang and O'Connor 1997). It was also demonstrated that 3-dehydroecdysone is produced by the Y-organ in several crustaceans including the shrimp *P. Vannamei* (Blais *et al.*, 1994). Besides, 3-dehydro ecdysone, the Y-organ also produces 20-hydroxyecdysone (Fig. 3), (Charmantier-Daures and Charmantier 1994), deoxyecdysone (Watson *et al.*, 1989) and several other ecdysteroids (Grieneisen M L, 1994). However, the biosynthetic pathway for the production of ecdysone is not known. Cholesterol is the precusor for ecdysone production. Cholesterol cannot be synthesized by crustaceans (Law *et al.*, 1992) it must be obtained from dietary sources. A number of studies indicated the involvement of ecdysteroids not only in the moulting process but also in the reproduction of crustaceans (Spindler 1989).

2. Neuropeptides

Considerable research on chemical aspects of crustacean neuropeptides has been conducted. Chang *et al.*, (1990) reported an amino acid sequence of lobster MIH. It has been observed that the amino acid sequence of lobster MIH was

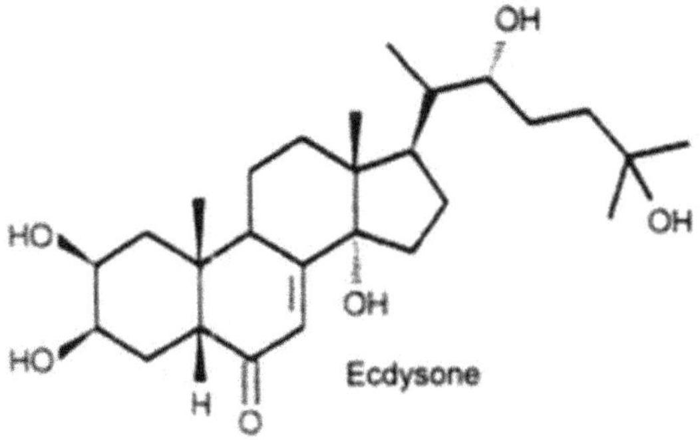

Fig. 2. Ecdysone

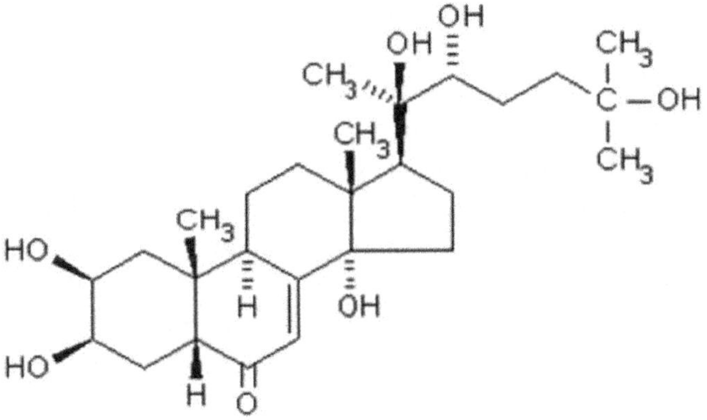

Fig. 3. 20-hydroxyecdysone

almost identical to that of crustacean hyper glycemic hormone A (CHHA) discovered by Tensen *et al.*, (1999) and Aguilar *et al.*,(1996) isolated MIH from crayfish and compared its sequence with other four known peptides of crustaceans. Their lengths vary between 72 and 78 residues and their molecular masses between 8 and 9 k Da. All have six cysteines that form three disulfide bonds. Sun, 1994 by using molecular techniques investigated the chemistry of MIH like neuropeptides of the shrimp *P. vannamei.* She reported that this MIH consists of a 72 residue mature peptide and a 30-residue region of a propeptide. By using similar techniques Aguilar *et al.*, 1997 isolated the MIH and found that it consists of a 72-residue peptide with amino and carboxyl termini and six cysteines forming three disulfide bonds. Similar investigations have been made by Yang *et al.*, (1996) and it was reported that *M. japonicus* MIH consists of a 77-residue peptide with both free amino and carboxyl termini. Sefiani *et al.*, (1996) isolated an MIH like peptide from *P. vannamei* and by mass spectrometry its molecular mass was estimated to be 8627 Da and consists of only 38 residues. Gu and Chan 1998 have isolated MIH from *M. ensis* and detected the amino acid sequence of 77 residues preceded by a signal peptide of 28 residues. Kawakami *et al.*, (2000) chemically synthesized a moult inhibiting hormone (pre-MIH) from the American crayfish, *Procambarus clarkii,* which consist of 75 amino acid residues. This product they found was almost similar to natural pre-MIH chemically. Chang (1997) reported that activity of growth hormone peptide has not been reported in either CHH or MIH assays.

Further it is described that the CHH/MIH/VIH peptide family appears to be a novel group of crustacean neuropeptides (Chang 1997). Much amount of work has been done on the expression of the CHH gene(s). Soyez *et al.*, (1987) isolated a 7500 Da peptide from the sinus gland of *H. americanus* and assayed its GIH activity *in vitro* in a shrimp by measurement of oocytes diameter. They (Soyez 1991) further investigated the structure of GIH and found that this peptide consists of 77 residues and its molecular weight is 9135 k Da. Dircksen *et al.*, (2001) while studying the crustacean hyper glycemic hormone (CHH) and CHH-precusor related peptides from pericardial organ (PO) neurosecretory cells in the shore crab, *Carcinusmaenas,* observed PO-CHH is a 73 amino acid peptide with a free C-terminus. PO-CHH and sinus gland CHH have been found to share an identical N-terminal sequence at positions 1-40 but the remaining sequence, positions 41-73 or 41-72, differs considerably. They have also reported that PO-CHH may have different precursors and CHH genes coding for precusor products are presumably modified at the post-transcriptional or post-translational level. It has been also mentioned that PO-CHH unlike SG-CHH, has neither a hyperglycemic effect nor it is active in inhibition of ecdysteroid production by crab Y-organs. Katayama *et al.*, (2003) studied the structure of moult inhibiting hormone (MIH) from the Kuruma shrimp, *Marsupenaeus japonicus.* It has been reported that the amino acid sequence of MIH is similar to that of crustacean hyperglycemic hormone (CHH). Further, it was shown that the homology- modeled structure of *M. japonicus* CHH was similar to the structure of MIH with the exception of the absence of the N-terminal α-helix and the C-terminal tail. This is the first report on the tertiary structure analysis of a crustacean neuropeptide. This kind of elucidation is expected to provide new insights not only on structure-activity relationship but on the molecular evolution of the CHH family peptides. MF is a secretory product of the MO in crustaceans and is related to insect juvenile hormone as esquiterpene that has a significant effect on growth and reproduction in insects. Laufer *et al.*, (1987a, b) isolated MF from the haemolymph of crabs. Later a number of workers have isolated MF from a number of other crustaceans (Chang, 1997). Wainwright *et al.*, 1996 and Liu *et al.*, 1997 have characterised mandibular organ inhibiting hormones (MOIH) from crabs and found that the hormones have the molecular masses of approximately 8400 Da and the amino acid composition of this neuropeptide is similar to those of MIH and CHH. There is another compound called farnesoic acid secreted by cultured MO from *P. clarkii* (Cusson *et al.*, 1991) and *Cancer magister* (Tamone and Chang 1993). Chang (1997) has mentioned that MF is converted rapidly to farnesoic acid by esterases that are present in tissues.

Swetha *et al.*, (2011) while dealing with reproductive endocrinology of female crustaceans reported the recent developments in molecular advances taken place in the reproduction regulation of different crustacean species via hormones, opioids, neurotransmitters and other molecules. In their paper they have given brief description of female reproductive system, Identification of reproductive Stages, and regulation of reproduction using external factors and endogenous factors. The main emphasis was given on the role of eyestalks in regulation of reproduction. It is reported that various neurohormones released by eyestalk neuroendocrine complex affect several physiological events, including the activity of peripheral endocrine glands, which results in the production of vitellogenin, and gonad development. Besides, eyestalk peptides, it is also possible that other members of hormones belonging to crustacean hyperglycaemic hormones (CHH-family) may have influence on the reproduction. Work has been carried out on CHH receptors on the oocyte membranes of several crustaceans (Webster, 1993). CHH mediated reproduction was studied in penaeid shrimps *P. semisulcatus* (Khayat *et al.*, 1998) and *M. ensis* (Gu *et al.*, 2000). The reproductive inhibition by the eyestalk peptide, mandibular organ inhibiting hormone (MOIH) has also been reported (Wainwright *et al.*, 1996) which affects the reproduction by inhibiting the synthesis and secretion of methylfarneasoate by mandibular organs (Reddy and Reddy, 2006). In addition to CHH-family peptides, red pigment concentrating hormone (RPCH) released by the eyestalk has also stimulatory effect on ovary by releasing the gonad stimulating hormone (GSH) from the brain and thoracic ganglion (Sarojini *et al.*, 1995). GSH which possesses antagonistic activity to VIH is involved in the stimulation of reproduction (Sarojini *et al.*, 1997, Yano and Wyban, 1993) and it is not yet characterized. In addition to the above hormones, methyl farnesoate (Laufer *et al.*, 1987, Reddy *et al.*, 2004) and the steroid moulting hormone, 20-hydroxyecdysone (Chang 1997, Subramoniam, 2000, Gunamalai *et al.*, 2004) have been suggested to be involved in the expression of vitellogenin.

3. Androgenic hormone

Once the physiological function of androgenic hormone was known, efforts were made to purify and characterise the active factors present in this hormone. Hasegawa *et al.*,(1987); Hasegawa *et al.*,(1993) purified two proteins, androgenic gland hormone I (AGHI) and AGH II consisting of 157 and 160 amino acids respectively with molecular weights of 17.0 and 18.3 kDa. The other biological active factors isolated were identified as the terpenes farnesyl acetone and hexhydroxy farnesyl acetone (Ferezou 1977). However, the relative role of these proteins and terpenes is unknown.

5.7 ROLES OF NEUROPEPTIDES / NEUROTRANSMITTERS IN REPRODUCTION

The known neurotransmitters in crustaceans are acetylcholine (Ach), gamma-amino butyric acid (GABA), glutamate, octopamine (OA), dopamine (DA), and 5-hydroxy tryptamine/ serotonin (5-HT). Among the neurotransmitters tested for possible roles in crustacean reproduction and molting, 5-HT, OA, and DA seem to be attracting more attention than the others. Each of these is present in the crustacean nervous system (Fingerman M and Nagabhushanam R, 1997).

1. 5-Hydroxytryptamine (5-HT, Serotonin)

5-HT is a ubiquitous substance found in plants and animals. The Falck-Hillarp fluorescence histochemical technique for monoamines was used by several workers to study the distribution of 5-HT in crustacean tissues. Osborne and Dando (Osborne and Dando, 1970) detected fluorescence characteristic of 5-HT in some cell bodies and in most of the neuropile in the stomatogastric ganglion of the lobster *Homarus vulgaris*. In recent years, immunocytochemical techniques have largely been supplanted the Falck-Hillarp fluorescence technique for localization studies of biogenic amines (Sarojini *et al.*, 2000). In the eyestalk, immunoreactive neurons are present in three of the four optic ganglia (medulla externa, medulla interna, and medulla terminalis). Some of the immunoreactive cell bodies in the medulla terminalis are sent out fibres through the optic peduncle that terminate in the protocerebrum. Immunoreactive cell bodies are also present in the brain. The brain and every ganglion in the ventral nerve cord displayed at least one immunoreactive cell body. In addition, several axons in the pericardial organs are immunoreactive, but the origin and termination of these axons could not be determined. In addition to identification and localization of 5-HT by histochemical methods, precise measurements of tissue concentration of biogenic amines are now possible. With the HPCL technique, 5-HT was detected in the central nervous system and hemolymph of *P. leniusculus* (Elofsson 1983). A series of experiments done elsewhere revealed that 5-HT stimulates ovarian development when injected into the fiddler crab and red swamp cray fish. These crabs showed increased, dose dependent ovarian development (Fingerman and Nagabhushanam,

1997; Sarojini *et al.*, 2000). Supporting evidence for this hypothesis was obtained by determining the effects of two 5-HT agonists on ovarian development in the crab. For this purpose, the 5-HT releaser fenfluramine and the 5-HTpotentiator fluoxetine were used. The ovaries of crabs that received fenfluramine, fluoxetine, 5-HT alone, 5-HT plus fenfluramine, or 5-HT plus fluoxetine exhibited significant increase in ovarian index and oocyte size compared to the ovaries of untreated initial control crabs and saline-injected concurrent control specimens (Sarojini *et al.*, 2000). Supporting evidence for a neurotransmitter role of 5-HT in stimulating GSH release in *P. clarkii* was reported by Sarojini *et al.*, 2000. In *M. japonicus*, Vg synthesis in ovarian pieces incubated with thoracic ganglion pieces prepared from vitellogenic females can be stimulated by5-HT (Yano 2000).

In addition, 5-HT stimulates testicular maturation in *P. clarkii* as shown by the use of a series of 5-HTagonists and antagonists. 5-HT and its agonists induce testicular maturation and help in the development of the androgenic glands. In contrast, 5-HT antagonists had no stimulatory effects on the testes or androgenic glands. This stimulatory action of 5-HT on the testes and androgenic glands was hypothesized to be indirect, i.e. 5-HT stimulates the release of GSH, which in turn activates the androgenic glands to synthesize and release androgenic gland hormone, and the androgenic gland hormone then triggers testicular maturation (Fingerman and Nagabhushanam 1992). While discussing 5-HT modulation of crustacean hyperglycerine hormone (CHH) secretion by isolated cells of the crayfish retina and optic lobe, Escamilla–Chemal *et al.*, (2002) found that retinal CHH secreting cells correspond to a population of retinal tapetal cells and optic lobe CHH-secreting cells correspond to two sub populations of CHH of medulla terminalis–X organ cells. Further, they mentioned that CHH secretion generally increases as a function of 5–HT concentration. Gu *et al.*, 2000 reported the cloning and characterization of the cDNA and the gene encoding the hyperglycemic hormone (CHH-B) of the shrimp *M. ensis.* It was shown that the amino acid sequence of the hyperglycemic hormone of *M. ensis* is identical to that of the CHH-like neuropeptide (CHH-A) of *M. ensis.* They further observed the presence of CHH-B in the eyestalk of vitellogenic females. At the middle stage of gonadal maturation, a minimum level of CHH-B transcript and a maximum level of CHH-A transcript were detected which indicated that both CHH-related neuropeptides may play an important role during the female gonad maturation cycle in shrimp.

Wongprasert *et al.*, (2006) reported that serotonin (5-hydroxytryptamine, 5HT) induces ovarian maturation and spawning in white pacific shrimp *Litopenaeus vannamei.* Based on this work Wongprasert *et al.*, (2006) also carried out studies on effect of serotonin on ovarian maturation and spawning in *Penaeus monodon* and found that the 5-HT injected *P. monodon* developed ovarian maturation and spawning rate at the level comparable to that of unilateral eyestalk ablated shrimp. It was also reported that hatching rate and the amount of nauplii produced per spawner were also significantly higher in 5-HT injected shrimp, compared to the eyestalk ablated shrimp. These results suggest a positive role of 5-HT, possibly directly on the ovary and oviduct, on the reproductive function of female *P. monodon.*

Earlier also Alfaro *et al.,* (2004) reported that injection of combined 5HT and dopamine antagonist, spiperone, in *L. Stylirostris* and *L. vannamei,* stimulated ovarian maturation, spawning and the release of maturation promoting pheromones. Several alternatives have been tried to induce the ovarian maturation, such as injections of GIH antibody, hormones, neurotransmitters and double stranded RNA to knock off GIH mRNA; and results from some of these trials have been proved promising (Poltana *et al.,* 2005, Treerattrakool *et al.,*2005). The effect of serotonin injection on the ovarian maturation and vitellogenin levels of *Fenneropenaeus indicus* was investigated by Santhoshi *et al.,* (2009) and it was revealed that serotonin stimulated ovarian maturation process of *F. indicus* by increasing the vitellogenic levels.

Makkapan *et al.,* (2011) while working on molecular mechanism of serotonin via methyl farnesoate in the ovarian development of white shrimp *F. merguiensis* observed that injection of 5-HT releases methyl farnesoate in to the haemolymph and further it stimulates ovarian development in shrimp. There was a significant correlation observed between the ovarian development stimulated by 5-HT and the haemolymph MF levels. Marsden *et al.,* (2008) mentioned that MF plays important role in regulating reproductive process in crustaceans in general and in *P. monodon* in particular. However, it is further mentioned that factors that determine the extent of its effect whether the MF has stimulatory or inhibitory effect, remain unknown. Until these factors are identified, the application of MF as means of predictably manipulating egg production in captive shrimps remains problematic.

Recently Babu and Reddy (2014) worked on stimulation of ovarian maturation using serotonin (5-HT) hormone on *P. monodon.* It was observed that after using serotonin ovarian maturation was induced when compared to normal shrimps. In their studies, 25 days after treatment with serotonin they found that the mean size of oocytes was significantly increased, which suggest that 5-HT can induce faster maturation in this particular shrimp. Vaca and Alfaro (2000) and Kumlu (2005) while working on *P. semisulcatus* found that 5 HT can induce ovarian maturation in the shrimp and further they also mentioned that the action of serotonin on ovarian maturation and spawning is significant but less effective than eyestalk ablation.

2. Dopamine (DA)

Dopamine (DA) (Fig 4) is present in crustacean nervous systems. Histochemical studies employing the fluorescence method of Falck and Hillarp indicated that catechol amines, including DA, are present and catechol amines exhibit a green fluorescence (Sarojini *et al.*, 2000). However, both DA and norephinephrine (NE) evoke this green fluorescence, both being catecholamines. Much green fluorescence was apparent in the medulla externa and medulla interna of the eyestalk, and in the brain and ventral nerve cord of *Astacus astacus* (Elofsson *et al.*, 1966). As with 5-HT, immune cytochemical procedures have been used to demonstrate the presence of DA in crustaceans. By use of an anti-DA antibody, the presence of DA-like neurons in the terminal abdominal ganglion, intestinal nerve, and axons in the hindgut musculature of *Orconectes limosus* was demonstrated. Mercier *et al.*, 1991 identified two immune reactive neurons in the abdominal nerve cord of *P. clarkii* that contribute axons through the intestinal nerve to the plexus that surrounds the hindgut. HPLC analys is confirmed the presence of DA in this nerve (Fingerman and Nagabhushanam, 1997).

The hormonal control of reproduction in crustaceans is mediated by neurotransmitters in the central nervous system. Serotonin (5-hydroxytryptamine; 5-HT) and dopamine (DA) are two biogenic amines that modulate the synthesis and release of reproductive hormones during gonadal maturation (Fingerman, 1997). Our understanding on neurotransmitters and their receptors that modulate reproductive endocrine system in shrimps is important for the development of alternate ovarian induction strategies. As dopamine may inhibit the release of gonad stimulating factor from the central nervous system, it is possible that blocking dopamine receptor also helps to stimulate ovarian maturation. Sukthaworn *et al.*, (2013) worked on molecular and functional characterization of a dopamine receptor type 1 from *P. monodon*. Dopamine exerts its effect through a membrane bound G protein–coupled receptor. In their study they obtained a cDNA encoding DA receptor of *P. monodon*, so called PemDopR1 by rapid amplification of cDNA Ends strategy. It was reported that the expression of PemDopR1 in the brain throughout ovarian maturation cycle reduced significantly suggesting the inhibitory effect of DA during the early stages of ovarian maturation is mediated via PemDopR1.

Role of dopamine in reproduction

Dopamine when injected into female *P. clarkii*, inhibited ovarian maturation. The Dopamine injected individuals had a smaller mean ovarian index than the control. The crayfish given DA plus 5-HT had a significantly smaller mean ovarian index than the crayfish that received 5-HT alone, but larger than the crayfish that received DA alone. 5-HT exerts its effect on the ovary indirectly, but stimulating GSH release, the inhibitory action of DA on ovarian maturation induced by 5-HT could have been due to (1) inhibition of GSH release, thereby directly counteracting the action of 5-HT; (2) stimulation of release of the GSH antagonist, GIH; or (3) both (1) and (2) (Sarojini *et al.*, 1995, 2000).

3. Octopamine (OA)

The presence of OA in central nervous organs of *H. americanus* was shown by using thin layer chromatography. With the use of radiolabeled compounds and HPLC it was shown that OA is synthesized in the brain of *Orconectes*. By HPLC, OA is detected in the eyestalks, brain and hemolymph of *P. leniusculus*. Additionally, by HPLC, Butler and Fingerman found OA in the central nervous system of *U. panacea* and Luschen *et al.*, (1988) found OA in Y-organs of *C. maenas*. Wallace *et al.*, (1974) by using various staining techniques such as treatment with neutral red and osmium tetroxide and performing OA analysis by use of a radiolabeled OA precursor reported the presence of OA in the specific neurons o the lobster nervous system. Many chromatographic and electrophoretic techniques were able to identify neurons in the second thoracic ganglion of *H. americanus* that appeared to be the ones that contain this biogenic amine. The role of OA in reproduction appears to be at least in part stimulation of contraction of the ovarian walls, liberating the oocytes. (Fig. 4)

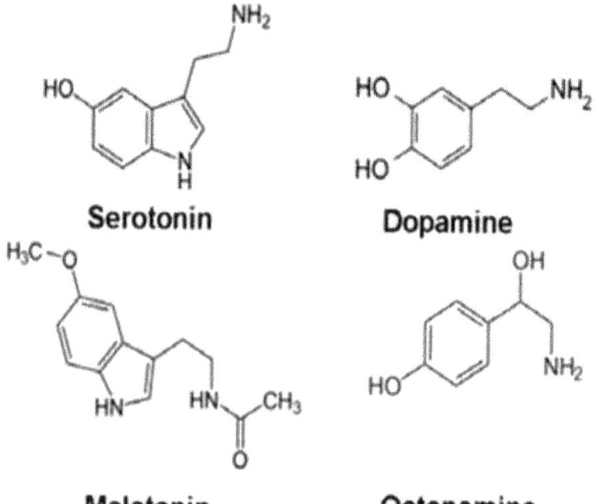

Fig. 4. Neuropeptides and neurotransmitters in reproduction

4. Red pigment-concentrating hormone (RPCH)

RPCH was isolated from eyestalks of the shrimp, *Pandalus borealis*. It was later found to be an octapeptide. Extracts of all portions of the central nervous system, including the brain and thoracic ganglia of the crayfishes *Cambarellus shufeldi* and *Faxonella clypeata* have concentrated red pigment hormone in the integumentary chromatophores, thereby revealing that RPCH is widely distributed throughout the central nervous system of the crayfish (Mancillas *et al.*, 1981).

Role of RPCH in reproduction

RPCH, in addition to its hormonal role in regulating pigmentation, appears to have another hormonal role, stimulation of MF synthesis in the mandibular organ of *P. clarkii*. MF may function as a hormone involved in the regulation of crustacean reproduction to complement the two peptidergic neuro hormones, GIH and GSH that have well documented roles in controlling gonadal maturation in crustaceans (Law 1992).

5. Opioid peptides

The first report of the presence of an opioid-like substance in crustaceans was published by Mancillas *et al.*, (1981). Through immune cytochemistry, they found Leu-Enk-like immune reactivity in all the retinular cells of the spiny lobster, *Panulirus interruptus*, and *P. clarkii*. In addition, such immunoreactivity was also apparent in nerve fibers in chiasma 3 that run from the medulla interna to the medulla terminalis. Later, other crustaceans were similarly studied. Immunoreactivity was apparent in the retinular cells, lamina ganglionaris, sinus gland, optic peduncle, the three chiasma, and medulla terminalis. Immunoreactivity was present in ovarian follicle cells and avitellogenic and early vitellogenic oocytes but not in fully ripe oocytes. In the testes, the immunoreactivity was present in spermatogonia, spermatocytes and spermatids but not in mature spermatozoa. This was the first demonstration of opioid-like immunoreactivity in a crustacean gonad.

Role of opioids in reproduction

The potential involvement of an endogenous opioid system in the regulation of ovarian development in *U. pugilator* has been investigated *in vivo* (Fingerman M *et al.*, 1994). Injection of synthetic Met-Enk into female crabs significantly slowed ovarian maturation. The inhibition was dose dependent.

The mean ovarian index and mean oocytes diameter of the crabs that received the opioid were significantly smaller than the corresponding values for the saline-injected concurrent control specimens. In contrast, injection of the opioid antagonist, naloxone, produced dose-dependent ovarian maturation. It was hypothesized that the opioid (1) stimulates GIH release, (2) inhibits GSH release or (3) does both (1) and (2).The role of an opioid in testicular maturation of the fiddler crab was also investigated. Injection of synthetic Met-Enk into male crabs significantly slowed maturation of the testes. The inhibition was dose-dependent. The mean testicular index, testicular weight, testicular lobe diameter and the number of mature spermatocytes in each testicular lobe were significantly less than the corresponding values for the saline-injected concurrent control crabs. Injection of naloxone produced dose-related testicular maturation (Fingerman *et al.*, 1994). Reddy (2000) while working shrimp *P. indicus* reported the possible involvement of endogenous opioids in the regulation of reproduction where injection of leucine-enkephalin caused significant increase in ovarian index.

5.8 ROLE OF RNA INTERFERENCE IN REPRODUCTION

Number of attempts has been made for advance in maturation in Penaied shrimp in captive condition using several techniques as mentioned above. Currently unilateral eyestalk ablation technique is still popular and is being used in the aquaculture. Unilateral eyestalk ablation technique creates physiological complications in the brooder due to removal of neuroendocrine structures of the eyestalk and creating imbalance in the hormonal levels. This kind of operations of the technique may lead to mortality among mother brooders in the farming system. Therefore, there is an urgent need of evolving an innovative and powerful technology using the knowledge of molecular biology. In this context, recently attempts have been made by Das *et al.*, (2015) wherein they studied captive maturation in the shrimp *P. monodon* by silencing gene responsible for producing gonad inhibiting hormone using RNA interference technique. It has been reported that there was 3-5 fold increase in the expression of androgenic gland hormone transcript in the male but as far as the female is concerned the vitellogenin expression did not occur in the experimental female shrimps. They also mentioned that after silencing GIH gene through RNA interference, moulting process was also accelerated. However, GIH silencing could be a potential alternative technique for eyestalk ablation for acceleration of maturation process in the shrimp and further research in this particular area is the need of the hour. Earlier Treerattrakool *et al.*, (2011) also reported induced maturation in wild and captive *P. monodon* using the technique of RNA interference in silencing of GIH gene. They also carried out gene silencing in wild and farm bred pre-vitellogenic *P. monodon* females by injecting anti GIH double stranded (ds) RNA produced in vitro. Further they observed that there was enhancement in the maturation process in the female shrimps where gene silencing by RNA interference technique was performed.

5.9 CONCLUSION

The shrimp industry today all over the world is developing rapidly, particularly in Southeast Asian countries due to the high demand of shrimp in the international market and also due to its high export value. The success of the industry often depends on the availability of the shrimp resource on a continuous basis. Today the shrimp industry is mostly dependent for this resource from the wild and to meet the demand, shrimps are being exploited continuously and in the process over exploitation takes place, and as a result shrimp catches have dwindled and in some places even the fishery has declined to a large extent. In order to reverse this trend and manage this resource on a sustainable basis the FAO has recently introduced the Code of Conduct for Responsible Fisheries (CCRF) for strict adoption by the all the maritime countries. Even if this code becomes effective, it will take a long time to rejuvenate the depleted stock. Therefore, in order to fulfil the demand of shrimps by the industries on a continuous basis, one of the alternate ways is to produce them in captive conditions. In the recent past several efforts have been made to evolve methods for the induction of maturation and spawning of shrimps in captive conditions. Though eyestalk ablation has been accepted as the most successful method for induction of maturation and spawning, this method does not work for many shrimps of commercial value. Moreover, there are other inherent and social problems in adopting this technique. Hence, the need of the hour is to explore /devise methods involving other possible neuro–endocrine manipulations. The use of neuro humours or neuro transmitters, or extracts from the brain and thoracic ganglia containing reproductive hormones, has to be explored to initiate gonadal maturation. With the advent of technology advancement in molecular biology, we need to explore gene editing (CRISPR-Cas) technology for accelerating maturation process.

References

Adiyodi, R.G. 1978. Endocrine control of ovarian function in crustaceans. *Comp. Endocrinol.*, 45: 25-28.

Adiyodi R G, 1985. Reproduction and its control, in *The biology of crustacea* Vol. 9 edited by D E Bliss and L H Mantel (Academic Press, New York), 147.

Adiyodi R G and T Subramonian, 1983. Arthropoda-Crustacea, in *Reproductive biology of invertebrates* Vol. 1 edited by K G Adiyodi and R G Adiyodi (Wiley, Chichester), 443.

Aguilar-Gayten R, Cerbon M A, Cevallow M A., Lizano M and Huberman A, 1997. Sequence of a cDNA encoding the molt inhibiting hormone from the Mexican crayfish *Procambarus bouvieri* (Crustacea, Decapoda). *Asia Pacific J. Mol. Biol. Biotechnol,* 5:51.

Aguilar M B, Quakenbush L S, Hunt D T, Shabanowitz J and Houbrman A, 1992. Identification, purification and initial charecterization of the vitellogenesis-inhibiting hormone from the Mexican crayfish *Procambarus bouvieri* (Ortmann), presence of a D-amino acid, *Peptides,* 16:1375.

Aguilar M B., Falchetto R., Shabanowitz J, Hunt D F and Huberman, A, 1996. Complete primary structure of the molt inhibiting hormone (MIH) of the Mexican crayfish *Procambarus bouvieri* (Ortmann), *Peptides,* 17:367.

Alfaro, J., Zuniga and G., Komen, J., 2004. Induction of ovarian maturation and spawning by combined treatment of serotonin and a dopamine antagonist, spiperone in *Litopenaeus stylirostris* and *Litopenaeus vannamei. Aquaculture* 236: 511–522.

Alikunhi, K.H., A Poernomo, S. Adisukesno, M. Budiono and S. Busman 1975. Preliminary observations on induction of maturity and spawning in *P. monodon* Fabricius and *P. merguiensis* de Man by eyestalk extirpation. *Bull. Shrimp. Cult. Res. Cen.*, 1: 1-11.

Aiken D E, 1980. Molting and growth, in T*he biology and management of lobsters* Vol. 1 edited by J.S Cobb and B.F Phillips (Academic Press, New York) 91.

Aquacop, 1979. Penaeid reared brood stock: closing the cycle of *Penaeus monodon, Penaeus stylirostris* and *Penaeus vannamei. Proc. World Maricult. Soc.*, 10: 445-452.

Aiken, D.E. and S.L. Waddy 1980. Reproductive biology. In: *The Biology and Management of Lobsters.* (Ed. J.S. Cobb and B.F. Phillips) pp. 215-276. Academic Press New York.

Anon. 1992. Annual Report, Central Institute of Brackishwater Aquaculture, Chennai 600 008, 40.

Aquacop, 1975. Maturation and spawning in captivity of penaeid shrimps *Penaeus merguiensis de man, Penaeus japonicus Bate, Penaeus aztecus Ives, Metapenaeus ensis de Haan* and *Penaeus semisulcatus de Hann. Journal of the World Aquaculture Society* 6,1 4:123-132.

Aquacop, 1979. Penaeid reared brood stock: closing the cycle of *Penaeus monodon, Penaeus stylirostris* and *Penaeus vannamei. Proc. World Maricult. Soc.*, 10: 445-452.

Aquacop, 1982. Reared broodstock of *Penaeus monodon. Proc. Symp. Coastal Aquaculture*, 1: 55-62.

Arcos, F.G., Racotta, I.S. and Ibarra, A.M., 2004. Genetic parameter estimates for reproductive traits and egg composition in Pacific white shrimp *Penaeus (Litopenaeus) vannamei. Aquaculture* 236: 151–165.

Arnstein, D.R. and T.W. Beard, 1975. Induced maturation of prawn *Penaeus orientalise.* Kishinogue in the laboratory by means of eyestalk removal. *Aquaculture*, 5: 411-412.

Aoto, T. and H. Nishida 1956. Effect of removal of the eyestalks on the growth and maturation of the oocytes in a hermaphroditic prawn, *Pandalus kessleri. J. Fac. Sci. Hokkaido Univ. Ser.*, 612: 412-424.

Atwood and D.C. Sandeman 1982 (Academic Press, Orlando) b, 205.

Bazing F, 1976. Mise en evidence des caracteres cytologiques desglandes steroidogenes dans les glanes mandibularies et les glandes Y du crabe *Carcinus maenas* (L) normal etepedoncule, *C.R. Acad. Sci. Paris,* 282: 739.

Beltz B S, 1988. Crustacean neurohormones in *Endocrinology of selected invertebrate Types,* Vol. 2, edited by H. Laufer and R.G.H. Downer (Alan R. Liss, Inc., New York) 235.

Benzie J.A.H, 1997. A review of the effect of genetics and environment on the maturation and larval quality of the giant tiger prawn *Penaeus monodon. Aquaculture* 155: 69-85.

Blanchet M F, 1974. Etude du controle hormonal du cycle d' intermue et de I' exuviation chez *Orchestia gammarella* par microcauterisation des organs Y suivie d'introduction d'ecdysterone, *C.R. Acad. Sci. Paris,* 278: 509.

Blais C, Sefiani M., Toullec J.Y. and Soyez D., 1994. *In vitro* production of ecdysteroids by Y-organs of *Penaeus vannamei* (Crustacea, Decapoda), Correlation with haemolymph titers, *Invert. Reprod. Dev,* 26: 3.

Bomirski, A. and E. Kelk 1974. Action of eyestalk on the ovary in *Rhithropanopeus harrissi* and *Crangon crangon* (Crustacea, Decapoda), *Mar. Biol.* 24: 329-337.

Bourguet J.P, Exbrayat J.M, Trilles J P and Vernet, G. 1977. Mise enevidence et description de l'organe Y chez *Penaeus japonicus* (Bate, 1881) (Crustacea Decapoda, Natantia). *C.R. Acad. Sci. Paris,* 285: 977.

Browdy, C.L., 1998. Recent developments in penaeid broodstock and seed production technologies: improving the outlook for superior captive stocks. *Aquaculture*, 164: 3–21.

Caillouet, C.W. 1972. Ovarian maturation induced by eyestalk ablation in pink shrimp *Penaeus duorarum* Burkenroad. *Proc. World. Maricult. Soc.* 3: 205-225.

Carlisle D B and Knowles F, 1959.*Endocrine control in crustaceans,* (Cambridge University Press, New York).

Carlisle D B, 1957. On the hormonal inhibition of moulting in decapod crustacea. II. The terminal anecdysis in crabs, *J. Mar. Biol. Assoc. U.K,* 36: 291.

Cavalli R.O., Scardua M.P. and WasieleskyW.J, 1997. Reproductive performance of different sized wild and pond reared *Penaeus paulensis* females. *J. World. Aquacult. Soc.*, 28: 260-267.

Chang E S, 1985. Hormonal control of molting in decapod crustacea, *Am. Zool,* 25: 179.

Chang E S, 1989. Endocrine regulation of molting in Crustacea. *Rev. Aquat. Sciences* 1:131

Chang E.S, Prestwich, G.D, and Bruce M.J, 1990. Amino acid sequence of a peptide with both molt-inhibiting and hyper glycemic activities in the lobster, *Homarus americanus, Biochem. Biophys. Res. Commun,* 171: 818.

Chang E S, Hertz W A and Prestwich G D, 1992. Reproductive endocrinology of the shrimp *Sicyonia ingentis,* steroid, peptide and terpenoid hormones, *NOAA Tech. Rep. NMFS,* 106: 1.

Chang E S, 1993. Comparative endocrinology of molting and reproduction, Insects and crustaceans, *Annu. Rev. Entomol,* 38: 161.

Chang, C.F., F.Y. Lee and Y.S. Huang 1993. Purification and characterization of vitelline from the mature ovaries of prawn *Penaeus monodon. Comp. Biochem. Physiol.*,105 b: 409-414

Chang E S and O'Connor J D, 1997. Secretion of a-ecdysone by crabY-organs *in vitro. Proc. Natl. Acad. Sci. USA,* 74: 615.

Chang E S, 1997. Chemistry of crustacean hormones that regulate growth and reproduction, in *Recent advances in marine biotechnology, endocrinology and reproduction,* Vol. 1edited by M Fingerman, R Nagabhushanam, M F Thompson (Oxford and IBH, New Delhi), 163.

Charmantier G, Charmantier-Daures M and Van Harp F, 1997. Hormonal regulation of growth and reproduction in crustaceans, in *Recent advances in marine biotechnolog,*Vol.1 edited by M. Fingerman, R. Nagabhushanam, and Mary Frances Thompson (Oxford and IBH, New Delhi), 109

Charmantier-Daures M Charmantier G, 1994. Les organs endocrines. In: *Traite de Zoologie, Crustaces,*Vol. VII edited by I. J. Forest (Masson, Paris) 595.

Charniaux-Cotton, H. 1985. Vitellogenesis and its control in malacostracan Crustacea. *Amer. Zool.*, 25: 197-206

Charniaux-Cotton H and Payen G G, 1988. Crustacean reproduction, in *Endocrinology of selected invertebrate types,* Vol. 2 edited by H. Laufer and R.G.H. Downer (Alan R. Liss. Inc., NewYork) 279.

Chotikun, S. 1988. Seed production of eyestalk ablated giant tiger prawn (*Penaeus monodon*). Thai Fisheries Gazette.

Coman, F.E., Norris, B.J., Pendrey, R.C., Preston, N.P., 2003. A simple spawning detection and alarm system for penaeid shrimp. *Aquac. Res.* 34: 1359–1360.

Coman, G.J., Crocos, P.J., Preston, N.P. and Fielder, D., 2004. The effect of density on the growth and survival of different families of juvenile *Penaeus japonicus* Bate. *Aquaculture* 229: 215–223.

Coman, G.J., Crocos, P.J., Arnold, S.J., Keys, S.J., Murphy and B., Preston, N.P., 2005.

Growth, survival and reproductive performance of domesticated Australian stocks of the giant tiger prawn, *Penaeus monodon*, reared in tanks and raceways. *J. World Aquac. Soc.*, 36: 464–479.

Coman G.J., S.J. Arnold, S. Peixoto, P.J. Crocos, F.E. Coman and N.P. Preston 2006

Reproductive performance of reciprocally crossed wild-caught and tank-reared *Penaeus monodon* broodstock *Aquaculture*, 252: 372–384.

Cooke, I.M. and R.E. Sullivan 1982. Hormones and neurosecretion. In: *Biology of Crustacea*, 3: 205-290. Academic Press.

Couch, E.R., N. Hagino and J.W. Lee 1987. Changes in estradiol and progesterone immuno-reactivity in tissues of the lobster, *Homarus americanus* with developing and immature ovaries. *Comp. Biochem. Physiol.*, 85A: 765-770.

Crocos, P.J. and Coman, G.J., 1997. Seasonal and age variability in the reproductive performance of *Penaeus semisulcatus*: optimizing broodstock selection. *Aquaculture*, 155: 55–67.

Crocos, P.J., Preston N.P. Smith D.M. and Smith M.R., 1999. Reproductive performance of domesticated *Penaeus monodon*. Book of abstracts, World Aquaculture Symposium 1999, Sydney,Australia, p. 185.

Cusson M, Yagi K J, Ding Q, Duve H, Thorpe A, Mc Neil JN and Tobe S S, 1991. Biosynthesis and release of juvenile hormone and its precursors in insects and crustaceans, The search for a unifying arthropod endocrinology, *Insect. Biochem* 21:1.

Cuzon, G., Arena, L., Goguenheim, J. and Goyard, E., 2004. Is it possible to raise, offspring of the 25th generation of *Litopenaeus vannamei* (Boone) and 18th generation *Litopenaeus stylirostris* (Simpson) in clear water to 40 g? *Aquac. Res.* 35: 1244–1252.

D'Croz, L., V.L. Wong, G. Justine, G. and M. Gupta 1988. Prostaglandins and related compounds from the polycheate worm *Americonuphis reesi* as possible inducers of gonad maturation in Penaeid shrimps. *Rev. de Biol. Trop.* 36(2a): 331-332.

Das R, Gopal Krishna, Himanshu Priyadarshi, Gireesh-Babu P, A. Pavan-Kumar, K.V. Rajendran A.K. Reddy, M. Makesh, Aparna Chaudhari, 2015. Captive maturation studies in *Penaeus monodon* by GIH silencing using constitutively expressed long hairpin RNA. Aquaculture. 448, 512–520.

Dall W, 1965. Studies on the physiology of a shrimp, *Metapenaeus sp.* (Crustacea, Decapoda, Penaeidae) II. Endocrines and control of molting. *Aust. J. Mar. Freshwater Res*, 16: 1.

Derelle E, Grosclaude J, Meusy J J, Unera H J and Martin M, 1986. ELISA titration of vitellogenin and vitellin in the freshwater shrimp *Macrobrachium rosenbergii*, with monoclonal antibody, *Comp. Biochem. Physiol*, 85: 1.

De Kleijn D P V, Coenen T, Laverdure A M, Tensen C P andVan Herp F, 1992. Localization of mRNAs encoding the Crustacean Hyperglycemic Hormone (CHH) and Gonad-Inhibiting Hormone (GIH) in the X-organ sinus gland complex of the lobster *Homarus americanus*, *Neuroscience*,51: 121.

Demeusy, N. 1967. Croissance relative d'un caractere sezueles externe male chez la. Decapode Brachyore *Careinus memaus* Lc. *R. Hebd. Seane. Acad. Sci. Paris.*

Diwan A D and Nagabhushanam R, 1975. The neurosecretory cells of the central nervous system of the freshwater crab *Barytelphusa cunicularis* (Westwood 1836). *Riv Biol* 68: 79.

Diwan A D and Usha T, 1985. Characterization of moult stages of *Penaeus indicus* based on developing uropod setae and some closely allied structures. *Indian J. Fish,* 32: 275.

Dircksen H, Boecking D, Heyn N U, Mandel C, Chung J S, Baggerman G, Verhaert P, Daufeldt S, Ploesch T, Jaros P P, Waelkens E, Keller R and Webster S G, 2001. Crustacean hyperglycaemic hormone (CHH)-like peptides and CHH precursor- related peptides from pericardial organ neurosecretory cells in the shore crab, *Carcinus maenas*, are putatively spliced and modified products of multiple genes, *Biochem. J,* 356:159.

Durica D S, Wu X, Anilkumar G, Hopkins P.M, and Chung A C K, 2002. Characterization of crab EcR and RXR homologs and expression during limb regeneration and oocytes maturation, *Mol Cell Endocrinol,* 189: 59.

Eastern-Recks, S.B. and M. Fingerman 1984. Effects of neuroendocrine tissue and cyclic AMP on ovarian growth *in vitro* in the fiddler crab, *Uca pugilator. Comp. Biochem. Physiol.*, 79A: 679-684.

Echalier G, 1955. Role de l'organe Y dans le determinisme de lamue de *Carcinides (Carcinus) maenas L.* (Crustace Decapode); experiences d'implantation, *C.R. Acad. Sci. Paris,* 240: 1581.

Echalier G, 1956. Effets de l'ablation et de la greffe de I'organe Ysur la mue de *Carcinus maenas L, Ann. Sci. Nat. Zool,* 11: 153.

Echalier G, L'organe Y 1959. Et le determinisme de la croissance etde la mue chez *Carcinus maenas* (L) Crustace Decapode, *Ann. Sci. Nat. Zool,* 12 :1.

Edomi, P., Azzon, E. , Mettulio, R., Pandolfelli, N., Ferrero, E. A. and Giulianini, P. G. 2002. Gonad-inhibiting hormone of the Norway lobster (*Nephrops norvegicus*): cDNA cloning, expression, recombinant protein production and immunolocalisation. Gene, 284: 93-102

Elofsson R, 1983. 5-HT-like immunoreactivity in the central nervous system of the crayfish, *Pacifastacus leniusculus, Cell Tissue Res,* 232:221.

Elofsson R, Kauri T, Nielsen S O and Stromberg J O, 1966. Localization of mono aminergi neurons in the central nervous system of *Astacus astacus. Zeit. Zellforschung,* 74: 464.

Emmerson, W.D. 1980. Induced maturation of prawn *Penaeus indicus. Mar. Ecol. Prog. Ser.*, 2: 121-132.

Emmerson, W.D. 1983. Maturation and growth of ablated and unablated *Penaeus monodon. Aquaculture*, 32: 235-271.

Escamilla-Chimal E G, Hiriart M, Sanchez-Soto M C and Fanjul-Moles M L, 2002. Serotonin modulation of CHH secretion by isolated cells of the crayfish retina and optic lobe, *GenComp Endocrinol,* 125: 283.

FAO, 2000. *The state of world fisheries and aquaculture* (UnitedNations, Rome, Italy) 1.

Fairs, N.J., R.P. Evershed, P.T. Quinlan and L.J. Goad. 1989. Detection of unconjugated and conjugated steroids in the ovary, eggs and haemolymph of decopod crustaceans. *Nephrops norvegicus. Gen. Comp. Endocrinol.*, 74: 199-208.

Fairs, N.J., R.P. Evershed, P.T. Quinlan and L.J. Goad 1990. Changes in ovarian unconjugated and conjugated steroid titer during vitellogenesis in *Penaeus monodon. Aquacult.*, 89: 83-99.

Ferezou J P, Berreur-Bonnenfant, Meusy J J, Barbier M, Suchy M and Wipf H K, 1977. 6, 10, 14-Trimethylpentadecan-2-one and 6, 10, 14-trimethyl-5-trans, 13-pentadecatrien-2-one from the androgenic glands of the male crab *Carcinus maenas, Experientia,* 33: 290.

Fernlund P and Josefsson L, 1972. Crustacean color change hormone, amino acid sequence and chemical synthesis, *Science,* 177: 173.

Fingerman, M. 1970. Perspective in Crustacan endocrinology. *Scientia.*105: 422-444.

Fingerman M, 1987. The endocrine mechanisms of crustaceans, *J Crust Biol,* 7: 1.

Fingerman M, Sarojini R and Nagabhushanam R, 1994. Evidence for the involvement of endogenous opioids in the regulation of ovarian development in the regulation of ovarian development in the fiddler crab, *Uca pugilator, Neuro endocrinology,* 60: 74.

Fingerman M, 1997. Roles of neurotransmitters in regulating reproductive hormone release and gonadal maturation in decapod crustaceans, *Invertebr Reprod Dev,* 31:47.

Fingerman M, 1997. Crustacean endocrinology, a retrospective, prospective and introspective analysis, *Physiol. Zool,* 70: 257.

Fingerman M and Nagabhushanam R, 1997. Role of neuroregulators in controlling the release of the growth and reproductive hormones in crustacea, in *Recent advances in marine biotechnology,* Vol. 1 edited by M, Fingerman, R. Nagabhushanam and Mary Frances Thompson (Oxford and IBH, New Delhi). 109.

Fingerman M. and Nagabhushanam R, 1992. Control of the release of crustacean hormones by neuroregulators, *Comp. Biochem. Physiol,* 102: 343.

Fingerman M, Sarojini R and Nagabhushanam R, 1994. Evidence for the involvement of endogenous opioids in the regulation of ovarian development in the regulation of ovarian development in the fiddler crab, *Uca pugilator, Neuro endocrinology,* 60: 74.

Flegel 1998 (National Centre for Genetic Engineering and Biotechnology, Bangkok), 29.

Gabe M, 1967. *Neurosecretion* (Gauthier-Villars, Paris).

Gabe M, 1953. Sur l'existence chez quelques Crustaces Malacostraces d'un organe comparable a la glande de mue des Insectes,. *C.R. Acad. Sci. Paris,* 237: 1111.

Gabe M, 1956. Histologie comparee de la glande de mue (organeY) des Crustaces Malacostraces, *Ann Sci. Nat. Zool,* 11: 145.

George, K.V. 1974. Some aspects of prawn culture in the seasonal and perennial fields of Vypeen Island. *Ind. J. Fish.*, 21: 1-19.

Ghosh, D.B. and A.K. Ray, 1992. Evidence for physiologic responses to estrogen in freshwater prawn, *Macrobrachium rosenbergii. J. Inland. Fish. Soc. India.*, 24,1: 15-21.

Ghosh, D.B. and A.K. Ray. 1994. Estrogen stimulated lipogenic activity in the ovary of the freshwater prawn, *Macrobrachium rosenbergii. Invert. Repr. Dev.*, 25,1: 41-47.

Gomez R and Nayar K K, 1965. Certain endocrine influences in the reproduction of the crab *Parathelphusa hydrodromous, Zool.Jahrb, Abt. Allg. Zool.Physiol Tiere,* 71: 694.

Grieneisen, M.L., 1994. Recent advances in our knowledge of ecdysteroid biosynthesis in insects and crustaceans. *Insect Biochemistry and Molecular Biology,* 24,2:115-132.

Gu P L and Chan S M, 1998. Cloning of a shrimp *(Metapenaeusensis)* cDNA encoding a nuclear receptor super family member, an insect homologue of E75 gene, *FEBS Lett,* 436: 395.

Gu P.L, Yu K.L and Chan S M, 2000. Molecular characterization of an additional shrimp hyperglycemic hormone : cDNA cloning, gene organization expression and biological assay of recombinant protein,. *FEBS Lett*, 472:122.

Gunamalai V, Kirubagaran R, Subramoniam T, 2004. Hormonal coordination of molting and female reproduction by ecdysteroids in the mole crab *Emerita asiatica* (Milne Edwards). *Gen Comp Endocrinol.* 138: 128-138.

Hagerman, D.D., F.M. Wellington and C.A. Ville 1957. Estrogens in Marine invertebrates. *Biol. Bull.*, 112: 180-183.

Halder, D.D. 1978. Induced maturation and breeding of *P. monodon* under brackishwater pond conditions by eyestalk ablation. *Aquaculture*, 25: 171-274.

Hasegawa Y, Haino-Fukushima K and Katakura Y, 1987. Isolation and properties of androgenic gland hormone from the terrestrial isopod, *Armadillidium vulgure, Gen. Comp.Endocrinol,* 67:101.

Hasegawa Y, Hirose E and Katakura Y, 1993. Hormonal control of sexual differentiation and reproduction in crustacea, *Am. Zool,* 33:403.

Hinsch G W, 1977. Fine structural changes in the mandibular gland of the male spider crab, *Libinia emarginata* (L.) following eyestalk ablation, *J. Morph.* 154: 307.

Hinsch, G.W. and D.C. Bennett, 1979. Vitellogenesis stimulated by thoracic ganglion implants into destalked immature spider crabs, *Libinia emarginata. Tissue and Cell,* 11: 345-351.

Hoang, T., Lee, S.Y., Keenan, C.P. and Marsden, G.E., 2002. Effects of age, size, and light intensity on spawning performance of pond reared *Penaeus merguiensis. Aquaculture* 212: 373–382.

Hossain M N, Kadar M Aand Islam K S, 1990. Induced maturation and spawning of Indian white shrimp *Penaeus indicus* H. Milne Edwards, 1837, *Bangladesh J. Zool,* 18: 245

Huberman A and Aguilar M B, 1989. A neuropeptide with moult inhibiting hormone activity from the sinus gland of the Mexican crayfish *Procambarus bouvieri* (Ortmann), *Comp.Biochem. Physiol,* 93:299.

Huberman A, 2000. Shrimp endocrinology: A review, *Aquaculture,* 191:191.

Idyll, C.P. 1971. Induced maturation of ovaries and ova in pink shrimp. *Comp. Fish Res.*33: 20.

Jorge Alfaro, Gerardo Zúñiga, J Komen, 2004. Induction of ovarian maturation and spawning by combined treatment of serotonin and a dopamine antagonist, spiperone in *Litopenaeus stylirostris* and *Litopenaeus vannamei. Aquaculture.* 236: 1–4.

Junera, H., C. Zerbib, M. Martin and J.J. Meusy 1977. Evidence for control of vitellogenin synthesis by an ovarian hormone in *Orchestia gammarella* (Pallas). *Gen. Comp. Endocrinol.*, 61: 248-249.

Kaeuser G., Koolman J. and Karlson P, 1990. Mode of action of molting hormones in Insects, in *Morphogenetic Hormones of Arthropods*, Vol. 1 edited by A.P. Gupta (Rutgers University Press, New Brunswick), 361.

Kallen J. L and Meusy J J, 1989. Do the neurohormones VIH (vitellogenesis inhibiting hormone) and CHH (crustacean hyperglycemic hormone) of crustaceans have a common precursor, Immuno localization of VIH and CHH in the X organ sinus gland complex of the lobster *Homarus americanus*, *Inver. Reprod. Develop*, 16: 43.

Kanokpan Wongprasert, Somluk Asuvapongpatana, Pisit Poltana, Montip Tiensuwan and Boonsirm Withyachumnarnkul, 2006. Serotonin stimulates ovarian maturation and spawning in the black tiger shrimp *Penaeus monodon Aquaculture*, 261: 1447–1454.

Kanokpan Wongprasert, Somluk Asuvapongpatana, Pisit Poltana, Montip Tiensuwan, Boonsirm Withyachumnarnkul, 2011. Serotonin stimulates ovarian maturation and spawning in the black tiger shrimp *Penaeus monodon. Aquaculture*. 261: 4.

Kanazawa, A. and S.I. Teshima 1971. *In vivo* conversion of cholesterol to steroid hormones in the spiny lobster, *Panulirus japonicus. Bull. Jpn. Soc. Sci. Fish.*, 37: 891-898.

Katayama H, Nagat K, Ohira T, Yumoto F, tanokura M and Nagasawa H, 2003. The solution structure of molt inhibiting hormone from the kuruma shrimp *Marsupenaeus japonicus*, *J. Biol. Chem*, 278: 9620.

Kawakami T, Toda C, Akaji K, Nishimura T, Nakatsuji T, Ueno K, Sonobe M, Sonobe H and Aimoto S, 2000. Synthesis of a molt inhibiting hormone of the american crayfish, *Procambarus clarkii* and determination of the location of its disulfide linkages, *J. Biochem., Tokyo*, 128: 455.

Keller R, 1992. Crustacean neuropeptides, structure, functions and comparative aspects, *Experientia*, 48: 439.

Khalaila I, Manor R, Weil S, Granot Y, Keller R, and Sagi A, 2002. The eyestalk androgenic gland testis endocrine axis in the crayfish *Cherax quadricarinatus*, *Gen. Comp. Endocrinol*, 127: 147.

Khayat M, Yang W, Aida K, Nagasawa H, Tietz A, Funkenstein B, Lubzens E, 1998. Hyperglycemic hormone inhibits protein and m-RNA synthesis *in vitro* - incubated ovarian fragments of the marine shrimp, *Penaeus semisulcatus*. *Gen Comp Endocrinol.* 110: 307-318.

Kleinholz, L.H. and R. Keller, 1979. Endocrine manipulations in crustaceans. In: *Hormones and ovolution*. (ed.) Barrington, E.J.W., Academic Press, New York. 1: 159-213.

Koskela, R.W., J.G. Greewood and P.C. Rothlisherg 1992. The influence of Prostaglandin E2, and the steroid hormones, 17α-hydroxyprogesterone and 17β-Estradiol in moulting and ovarian development in the tiger prawn *Penaeus esculentus* Haselll-1879. *Comp. Biochem. Physiol.*, 101A,2: 295-299.

Kulkarni G, Nagabhushanam R, Joshi P, 1979. Effect of progesterone on ovarian maturation in a marine penaeid shrimp *Parapenaeopsis hardwickii* (Miers, 1878). *Ind J Exp Biol* 17: 986-987.

Kulkarni, G.K. and R. Nagabhushanam 1980. Role of ovary inhibiting hormone from eyestalks of marine penaeid prawns *Parapenaeopsis hardwickii* during ovarian development cycle. *Aquaculture*, 19:13-19.

Kulkarni G K, Glade L and Fingerman M, 1991. Oogenesis and effects of neuroendocrine tissues on *in vitro* synthesis of protein by the ovary of the red swamp crayfish *Procambarusclarkii* (Girard), *J. Crust. Biol*, 11: 513.

Kulkarni, G.K. and R. Nagabhushanam, G. Amaldoss, R.G. Jaiswal and M. Fingerman 1992. In vivo stimulation of ovarian development in the red swamp crayfish, *Procambarus clarkii* by 5-hydroxytryptamine. *Invertbr. Rep. Dev.* 21,3: 231-240.

Kumlu M, 2005. Gonadal maturation and spawning in *Penaeus semisulcatus* de Hann, 1844 by hormone injection. *Turk J Zool* 29: 193-199.

Lachaise F, Le Roux A, Hubert M and Lafont R, 1993. The molting gl and of crustaceans, localization, activity and endocrine control (a review), *J. Crust. Biol*, 13: 198

Laubier, B.A. 1978. Ecophysiologie de la crevette *Penaeus japonicus* trios annees d'experience on mileu controle. *Oceano. Acta.*1:135-150.

Laufer H, Landau M, Borst D and Homola E, 1986. The synthesis and regulation of methyl farnesoate, a new juvenile hormone for crustacean reproduction in *Advances in invertebrate reproduction* 4, edited by M. Porchet, J.C. Andries, A. Dhianaut (Elsevier Science, Amsterdam), 135.

Laufer H, Landau M, Borst D.W and Homola E, 1987. Methylfarnesoate, its site of synthesis and regulation of secretion in a juvenile crustacean, *Insect Biochem*, 17: 1123.

Laufer H, Homola E and Landau M, 1987. Control of methyl farnesoate synthesis in crustacean mandibular organs, *Am. Zool*, 27: 69.

Laufer H, Borst D, Baker FC, Reuter CC, Tsai LW, Schooley DA, Carrasco C, Sinkus M, 1987. Identification of Juvenile hormone-like compound in a crustacean. Science 235: 202-205.

Laufer H, Ahl J S B and Sagi A, 1993. The role of juvenile hormones in crustacean reproduction, *Am. Zool,* 33: 365

Laufer H, and Landau M, 1991. Endocrine control of reproduction in shrimp and other Crustacea, in *Frotiers in shrimp research,*edited by P Deloach, W J Dougherty and M A Davidson(Elsevier Science Publishers, Amsterdam), 65

Laufer H, 1992. Method for increasing crustacean larval production, *United States Patent,* 5:161.

Laufer H, Paddon J and Paddon M, 1997. A hormone enhancing larval production in the Pacific white shrimp *Penaeus vannamei,* in *IV Symposium on Aquaculture in central America, Focusing on shrimp and tilapia* edited by D.E.Alston, B.W. Green, H.C. Clifford (The Latin American Chapter of the World Aquaculture Society, Tegucigalpa, Honduras), 161.

Lee K J, Elton T S, Bei A K, Watts S A and Watson R D, 1995. Molecular cloning of a cDNA encoding putative molt inhibiting hormone from the blue crab, *Callinectes sapidus, Biochem. Biophys. Res. Commun,* 209: 1126.

Lee C Y and Watson R D, 1994. Development of a quantitative enzyme-linked immune sorbent assay for vitellin and vitellogenin of the blue crab *Callinectes sapidus, J. Crust.Biol,* 14: 617.

Liao, I.C. 1977. A culture study on grass prawn *Penaeus monodon* in Taiwan, the patterns, the problems and the prospects. *J. Fish. Soc. Taiwan.* 5,2: 11-29.

Liao, I.C. 1992. Charting the course of aquaculture for the year 2000 and beyond. In: *Aquacutlure research needs for 2000 AD.* 61-72.

Lin, M. N, and Y, Y. Ting. 1986. Spermalophore transplantation and artificial fertilization in grass shrimp.*Bull. Jpn. Soc. Sci. Fish.*52:585-589.

Liu L and Laufer H, 1996. Isolation and characterization of sinus gland neuropeptides with both mandibular organ inhibiting and hyperglycemic effects from the spider crab *Libiniaemarginata, Arch. Insect Biochem Physiol,* 32: 375.

Liu L, Laufer H, Wang Y and Hayes T, 1997.A neurohormone regulating both methyl farnesoate synthesis and glucose metabolism in a crustacean, *Biochem. Biophys. Res.Commun,* 237: 694.

Legrand J J, Martin G, Juchault P. and Besse G, 1982. Controle neuroendocrine de la reproduction chez les Crustaces, *J Physiol,* 78: 543.

Law J H, Ribeiro J M C and Wells M A, 1992. Biochemical insights derived from insect diversity, *Annu Rev Biochem,* 61L: 87.

Lumare, F. 1979. Reproduction of *Penaeus kerathurus* using eyestalk ablation. *Aquaculture,* 18: 203-214.

Luschen W, Jaros P P and Wilig W, 1988. Detection and quantification of epinephrine in the moulting gland of the shore crab, *Hoppe-Seyler's Zeit. Physiol Chem,* 369:1204.

Maissiat J, 1970. Etude experimentale du role de "l'organeY" dansle determinisme endocrine de la mue chez l'Isopode Oniscoide *Porcellio dilatatus* Brandt, *C.R. Acad. Sci. Paris,* 270: 2573.

Maissiat J, 1970. Anecdysis experimentale provoquee chezl'Oniscoide *Ligia oceanica* L. et retablissement de la mue par injection d'ecdysone ou reimplantation de glande maxillaire,*C.R.. Soc. Biol,* 164:1607.

Mancillas J R, McGinty, J F, Severston A I, Karten H and Bloom F E, 1981. Immuno cytochemical localization of enkephalin and substance P in retina and eyestalk neurons of lobster, *Nature.* 293:576.

Mattson M P and Spaziani E, 1985. Characterization of molt inhibiting hormone (MIH) action on crustacean Y-organ segments and dispersed cells in culture and a bioassay for MIH activity, *J. Exp. Zool,* 236:93.

Mattson M.P and Spaziani E, 1986. Regulation of crabs Y-organ steroidogenesis *in vitro,* evidence that ecdysteroid production increased through activation of cAMP-phopho diesterase by calcium-calmodulin, *Mol. Cell. Endocrinol,* 48: 135.

Mattson M P and Spaziani E, 1987. Demonstration of protein kinase C activity in crustacean Y-organs and partial definition of its role in regulation of steroidogenesis, *Molec. Cell. Endocr,* 49: 159.

Meeratana, P., Withyachamnarnkul, B., Damrongphol, P., Wongprasert, K., Suseangtham, A. and Sobhon, P., 2006. Serotonin induces ovarian maturation in giant freshwater prawn broodstock, *Macrobrachium rosenbergii* de Man. *Aquaculture,* 260 (1–4), 315–325.

Mellon DeF Jr and Greer E, 1987. Induction of precocious molting and claw transformation in alpheid shrimps by exogenous 20-hydorxyecdysone, *Biol Bull,* 172: 350.

Mengqing, L.,Wenjuan, J., Qing, C.and Jialin,W., 2004. The effect of vitamin A supplemen-tation in broodstock feed on reproductive performance and larval quality in *Penaeus chinensis. Aquac. Nutr.* 10: 295–300.

Mercier A J, Orchard I and Schmoeckel A, 1991. Catecholaminergic neurons supplying the hindgut of the crayfish, *Procambarusclarkii, Can. J. Zool,* 69: 2778.

Meusy, J.J., G. Martin, D. Soyez, D. Van Deijnen, and J.M. Gallo, 1987. Immunochemical and immuoocytchemical studies of the crustacean vitellogenesis inhibiting hormone (VIH). *Gen. Comp. Endocrinol.*, 67: 333-341.

Millamena, O.M., Primavera, J.H., Pudadera, R.A. and Caballero, R.V., 1986. The effect of diet on the reproductive performance of pond reared *Penaeus monodon* Fabricius broodstock. In: Maclean, J.L., Dizon, L.B., Hosillos, L.V. (Eds.), the first Asian fisheries forum, pp. 593–596.

Mohamed K S and Diwan A D, 1991. Neuroendocrine regulation of ovarian maturation of the Indian White Shrimp *Penaeus indicus* H. Milne Edwards *Aquaculture,* 98: 381.

Mohamed K S and Diwan A D, 1991. Effect of androgenic glandablation on sexual characters of the Indian white shrimp *Penaeus indicus* H. Milne Edwards. *Indian J Exp. Biol,* 29: 478.

Mohamed K S and Diwan A D, 1993.Neurosecretory cell types, their distribution and mapping in the central nervous system of the penaeid shrimp *Penaeus indicus* H. Milane Edwards, *Bull Inst. Acad Sin* 38.

Mohamed K S and Diwan A D, 1992. Biochemical changes indifferent tissues during yolk synthesis in marine shrimp *Penaeus indicus* H. Milne Edwards, *Indian J. Mar. Sci,* 21:30.

Motoh, H., 1981. Studies on the fishery biology of the giant tiger prawn, *Penaeus monodon* in the Philippines.Technical Report No.7. Aquaculture Department, SEAFDEC, Philippines.128.

Muthu, M.S. and A. Laxminarayana 1982. Induced maturation and spawning of Indian penaeid prawns. *Indian J. Fish.*, 24,1 and 2: 172-180.

Nagabhushanam R and Diwan A D, 1974. Neuroendocrine control of reproduction of female crab *Barytelphusa cunicularis, Marathwada University. J.Sci* 6:59.

Nagabhushanam, R., P.K. Joshi and G.K. Kulkarni 1980. Induced spawning in prawn *Parapenaeopsis stylifera* using a steroid hormone 17 Hydroxy-Progesterone. *Ind. J. Mar. Sci.*, 227.

Nagabhudshnam, R., P.K. Joshi and G.K. Kulkarni 1982. Induced spawning in *Parapenaeopsis stylifera* using a steroid hormone 17 hydroxy-progesteron. *Proc. Symp. Coastal Aquaculture*, 1: 37-39.

Nagabhushanam, R., P.K. Joshi and R. Sarojini 1987. Effect of exogenous steroids on reproduction in *Parapenaeopsis stylifera. Advances in aquatic biology and fisheries*: 201-206.

Nagaraju GP, Ramamurthi R, Reddy PS, 2002. Methyl farnesoate stimulates ovarian growth in *Penaeus indicus.* In: Harikumar VS (Ed.*), Recent Trends in Biotechnology. Agrobios,* India. 1: 85-89.

Nagaraju G P, 2007. Is methyl farnesoate a crustacean hormone? *Aquaculture.* 272: 1–4.

Nagur Babu K., Reddy D.C., 2014. Stimulation of ovarian maturation using serotonin (5-hydroxytryptamine) hormone on tiger shrimp, *Penaeus monodon* (*Fabricius*). *World Journal of Pharmacy and Pharmaceutical Sciences.* 3- 4.

Nieto J., D. Veelaert, R. Derua, E. Waelkens, A. Cerstiaens, G. Coast, B. Devreese, J. Van Beeumen, J. Calderon, A. De Loof, L. Schoofs, 1998. Identification of one tachykinin- and two kinin-related peptides in the brain of the white shrimp, *Penaeus vannamei, Biochem. Biophy. Res. Commun.* 248: 406– 411.

Okamura, T., C.H., Han, Y. Suzuki, K. Aida and I. Hanyu 1992. Changes in hemolymph vitellogenin and ecdysteriod levels during the reproductive and non-reproductive molt cycles in the freshwater prawn *Macrobrachium nipponense. Zool. Sci.*, 9: 37-45.

Okumura T and Sakiyama K, 2004. Hemolymph levels of vertebrate-type steroid hormones in female kuruma shrimp *Marsupenaeus japonicus* (Crustacea: Decapoda: Penaeida) during natural reproductive cycle and induced ovarian development by eyestalk ablation. *Fish Sci* 70: 372-380.

Okumura T, 2007. Effects of bilateral and unilateral eyestalk ablation on vitellogenin synthesis in immature female kuruma shrimps, *Marsupenaeus japonicus. Zool Sci.* 24: 233-240.

Ollevier, F.D., Declerk, H. Diederick and A. Deloof, 1986. Identification of non-ecdysteroids in the haemolymph of both male and female *Astacus leptodactylus* (crustacea) by gas chromatography/mass spectrometry. *Gen. Comp. Endocrinol.*, 61: 214-228.

Osborne N N and Dando M R, 1970. Monoamines in the stomato gastric ganglion of the lobster, *Homarus vulgaris, Comp. Biochem. Physiol,* 32: 327.

Otsu, T. 1960. Precoucious development of the ovaries in the crab, *Potemont dehani*, following implantation of the thoracic ganglion. *Annot. Zool. Jpn.* 33: 90-96.

Otsu T, 1963. Bi hormonal control of sexual cycle in the fresh water crab *Potamon dehaani. Embyologi*a, 8:1.

Oyama, S.N. 1968. Neuroendocrine effect on ovarian development in the crab *Thalamita crenata* Latreille studied *in vitro.* Ph.D. Diss. Univ. Hawaii, Honolulu, Hawaii.

Palacios E., Rodr|£ guez-Jaramillo C. and Racotta I.S.,1999a. Comparison of ovary histology between different-sized wild and pond-reared shrimp *Litopenaeus vannamei* (*Penaeus vannamei*). *Invertebrate Reproduction and Development* 35: 251-259.

Palacios E. and Racotta I.S. and Acuacultores de La Paz (APSA),1999b. Spawning frequency analysis of wild and pond reared Pacific White Shrimp *Penaeus vannamei* broodstock under large-scale hatchery conditions. *J. World Aquacult. Soc.*, 25: 180-191.

Panouse J B. 1943. Influence de l'ablation de pedoncle ocularie surla croissance de l'ovaire chez la crevette *Leander serratus, C R Acad. Sci. Parris,* 217: 535.

Passano L M and Jyssum S, 1963. The role of the Y-organ in crabs pro ecdysis and limb regeneration, *Comp Biochem Physiol,* 9:195.

Passano L M, 1960. Molting and its control, in: *The physiology of crustacea,* Vol. 1 edited by T.H. Waterman (Academic Press, New York) 473.

Payen G G, 1986. Endocrine regulation of male and female genital activity, a retrospect and perspectives in *Advances in invertebrate reproduction* Vol. 4 edited by M. Porchet, J.C. Andries and A. Dhainaut (Elsevier Science Publishers, Amsterdam), 125

Peixoto, S., Wasielesky, W., D'Incao, F. and Cavalli, R.O., 2003. Reproductive performance of similarly-sized wild and captive *Farfantepenaeus paulensis. J. World Aquacult. Soc.* 34: 50–56.

Peixoto, S., Coman, G.J., Arnold, S.J., Crocos, P.J. and Preston, N.P., 2005. Histological examination of final oocyte maturation and atresia in wild and domesticated *Penaeus monodon* broodstock. *Aquaculture Research* 36: 666–673.

Poltana, P., Lerkitkul, T., Pongtippatee-Taweepreda, P., Asuvapongpatana S., Wongprasert, K., Sriurairatana, S., Chavadej, J., Sobhon, P., Olive, J.W. 2005. Culture of polychaete *Perinereis cf. nuntia* and its development. In: Biothailand Proceedings, the International Conference on Shrimp Biotechnology: New Challenges through Thai Shrimp Industry. 4–5 November 2005, Queen Sirikit National Convention Center, Bangkok, Thailand. National Center for Genetic Engineering and Biotechnology, National Science and Technology Development Agency, Ministry of Science and Technology, 41.

Pratoomchat, B., 1998. Size at first maturity of male *Penaeus monodon* Fabricius from pond-reared stock. *Journal of Aquaculture in the Tropics* 13: 189–194.

Primavera, J.H. 1978. Induced maturation and spawning in 5 month old *Penaeus monodon* Fabricius by eye stalk ablation. *Aquaculture,* 13: 355-359.

Primaverand J.H., Lim, C. and Borlongan, E., I979. Effect of different feeding regimes on repro- duction and survival of ablated Penaeus monodon Fabricius. Southeast Asian Fish. *Dev. Cent. Aquacult. Depart, Q. Res. Rep.,* 3: 12-14.

Primavera, J.H. 1982. Studies on broodstock of sugpo *(P. monodon)* and other penaeids at the seafdec aquaculture department. *Proc. Symp. Coastal Aquaculture,* 1: 28-36.

Primavera J H and Caballero R M H, 1992. Light color and ovarian maturation in unablated and ablated giant tiger shrimp *Penaus monodon, Aquaculture,* 108: 247.

Quackenbush, L.S. and W.F. Herrkind 1981. Regulation of moult and gonadial development in the spiny lobster, *Panulirus argus. J. Crust. Biol.*3: 34-44.

Qunitio, E.T., A. Hara, K. Yamauchi and A. Fugi 1990. Isolation and characterization of vitellin from the ovary of *Penaeus monodon. Invertbr. Reprod. And Dev.* 17(3): 221-227.

Quinitio ET, Hara A, Yamauchi K, Nakao S, 1994. Changes in the steroid hormone and vitellogenin levels during the gametogenic cycle of the giant tiger shrimp, *Penaeus monodon. Comp Biochem Physiol* 109: 21-26.

Quackenbush L S, 1986. Crustacean endocrinology a review, *Can.J. Fish. Aquat. Sci,* 43:2271.

Quackenbush L S and Keeley L L, 1988. Regulation of vitellogenesis in the fiddler crab, *Ucapugilator, Biol. Bull,* 175: 321.

Quackenbush L S, 1989. Vitellogenesis in the shrimp *Penaeus vannamei*, in vitro studies of the isolated hepatopancreas and ovary, *Comp. Biochem. Physiol*, 94: 253.

Quackenbush, L.S. 1991. Regulation of vitellogenesis in penaeid shrimp. In: Frontiers in Shrimp Research, P.F. Dehoach, W.J. Dougherty and M.A. Davidson (eds.), Elsevier, Amsterdam. 125-140.

Qian YQ, Dai L, Yang JS, Yang F, Chen DF, 2009. CHH family peptides from an 'eyeless' deep-sea hydrothermal vent shrimp, *Rimicaris kairei*: characterization and sequence analysis. *Comp Biochem Physiol B Biochem Mol Bio*154: 37-47.

Rangneker, P.V. and Deshmukh 1968. Effect of eyestalk removal on the ovarian growth of the marine crab, *Scylla serrata*. *J. Anim. Morph. Phsio.*15: 503-511.

Rao K R, 1965. Isolation and partial characterization of the moult inhibiting hormone of the crustacean eyestalk, *Experientia*, 21:593.

Rao P V, 1968. Maturation and spawning of the penaeid shrimp of southwest coast of India, *FAO Fish Rep*, 57:285.

Reddy P S, 2000. Involvement of opioid peptides in the regulation of reproduction in the shrimp *Penaeus indicus*. *Naturwissenschaften* 87: 535-538.

Reddy PR, Nagaraju GPC, Reddy PS, 2004. Involvement of methyl farnesoate in the regulation of molting and reproduction in the freshwater crab *Oziotelphusa senex senex*. *J Crust Biol* 24: 511-515.

Reddy PR and Reddy PS, 2006. Isolation of peptide hormones with pleiotropic activities in the freshwater crab, *Oziotelphusa senex senex*. *Aquaculture* 259: 424-431.

Rotlant G, De Kleijn D P V, Charmantier-Daures M, Charmantier G and Van Herp F, 1993. Localization of crustacean hyperglycemic hormone (CHH) and gonad-inhibiting hormone (GIH) in the eyestalk of *Homarus gammarus* larvae by immunocytochemistry and in situ hybridization, *Cell Tissue Res*, 271: 507.

Rotlant G, Charmantier-Daures M, De Kleijn D P V, Charmantier G and Van Herp F, 1995. Ontogeny of neuroendocrine centers in the eyestalk of *Homarus americanus* embryos, ananatomical and hormonal approach, *Invert. Reprod. Develop*, 27: 233.

Santhoshi S., V. Sugumar, N. Munuswamy, 2009. Serotonergic stimulation of ovarian maturation and hemolymph vitel logenin in the Indian white shrimp, *Fenneropenaeus indicus. Aquaculture*, 291:3–4.

Santiago, A.C., Jr. 1977. Successful spawning of cultured *P. monodon* after eyestalk ablation. *Aquaculture*, 11,1: 185-196.

Sarojini, R., M.S. Mirajkar and R. Nagabhushanam 1985. Effects of steroids on oogenesis and spermatogenesis of the freshwater prawn, *M. kistensis. Comp. Physiol. Ecol.*,10,1: 7-11.

Sarojini, R., S.S. Rao and K. Jayalakshmi 1990. Effects of steroids (estradiol and esterone)

on the ovaries of the marine crab, *Scylla serrata. Comp. Physiol. Ecol.*, 15(1): 21-36.

Sarojini R, Nagabhushanam R and Fingerman M, 1995. Mode of action of the neurotransmitter 5-hydroxytryptamine in stimulating ovarian maturation in the red swamp cray fish, *Procambarus clarkii,* An *in vivo* and *in vitro* study. *J. Exp. Zool,* 271:395.

Sarojini R, Nagabhushanam R, Fingerman M, 1997. An *in vitro* study of the inhibitory action of methionine enkephalin on ovarian maturation in the red swamp crayfish, *Procambarus clarkii.* Comp Biochem Physiol 115: 149-153.

Sarojini R, Nagabhushanam R and Fingerman M, 2000. New technology for enhancing reproductive maturation in economically important crustacea for aquaculture, in *Recent advances in marine biotechnology,* Vol. 4 edited by M. Fingerman and R. Nagabhushanam (Oxford and IBH publishing Co. Pvt. Ltd. New Delhi) 177.

Sagi A, Karp L, Milner Y, Cohen D, Kuris A M and Chang E S, 1991. Testicular thymidine in corporation in the shrimp *Macrobrachium rosenbergii.* Molt cycle variations and ecdysteroid effects *in vitro, J. Exp. Zool,* 259:229.

Sefiani M., Le Caer J.P, and Soyez D, 1996. Characterization of hyperglycemic and molt-inhibiting activity from sinus glands of the penaeid shrimp *Penaeus vannamei, Gen.Comp.Endocrinol,* 103: 41.

Sellars, M.J., Keys, S.K., Cowley, J.A., McCulloch, R., Preston, N.P., 2006. Effect of rearing environment on Mourilyan Virus load in *Penaeus japonicus* during maturation in tank systems. *Aquaculture*, 252: 240–245.

Shoji Joseph, 1997. Some studies on the reproductive endocrinology of tiger prawn *P. monodon* Fabricius, Ph. D Thesis, Cochin University of Science and Technology.

Swetha C H, Sainath S B, Reddy P R,ReddyP S, 2011. Reproductive Endocrinology of Female Crustaceans: Perspective and Prospective. *J Marine Sci Res Development* S3:001.

Silva Gunawardene YIN, Chow BKC, He JG, Chan SM, 2001. The shrimp FAMeT cDNA is encoded for a putative enzyme involved in the methylfarnesoate (MF) biosynthetic pathway and istemporally expressed in the eyestalk of different sexes. *Insect Biochem Mol Biol.*;31:1115–1124

Skinner D M, 1985. Molting and regeneration, in *The Biology of crustacea,* Vol. 9 edited by D.E. Bliss and L.H. Mantel (Academic Press, New York) 43.

Skinner D M Graham D E Holland C A Mykles D L Soumoff C and Yamaoka L H, 1985. Control of molting in Crustacea, in *Crustacean issues, Factors in adult growth* Vol3, edited by A W Wenner (A A Balkema, Rotterdam) 3.

Snyder M J and Chang E S, 1991. Ecdysteroids in relation to the molt cycle of the American lobster *Homarus americanus.* I Hemolymph titers and metabolites, *Gen. Comp. Endocrinol,* 81: 133.

Sonobe H, Kamba M, Ohta K, Ikeda M and Naya Y, 1991. *In vitro*secretion of ecdysteroids by Y-organs of the crayfish, *Procambarus clarkii, Experientia.* 47:948.

Sosa, M.A.A., Spitzer, N., Edwards, D.H., Baro, D.J., 2004. A crustacean serotonin receptor: cloning and distribution in the thoracic ganglia of crayfish and freshwater prawn. *J. Comp. Neurol.* 473: 526–537.

Soyez D, Le Caer, J P, Noel P Y and Rossier J, 1991. Primary structure of two isoforms of the vitellogenesis inhibiting hormone from the lobster *Homarus americanus, Neuropeptide,* 20: 25.

Soyez D, Van Deijnen J E and Martin M, 1987. Isolation and characterization of a vitellogenesis- inhibiting factor from sinus gland of the lobster, *Homarus americanus, J. Exp.Zool,* 244: 479.

Spaziani E, Mattson M.P and Rudolph P H, 1994. Regulation of crustacean molt-inhibiting hormone, *Perspect Comp Endocrinol,* 243.

Spindler K D, Van Wormhoudt A, Sellos D and Spindler-Barth, M, 1987. Ecdysteroid levels during embryogenesis in the shrimp. *Palaemon serratus* (Crustacea, Decapoda), quantitative and qualitative changes, *Gen Comp Endocrinol,* 66: 116.

Spindler K D, 1989. Hormonal role of ecdysteroids in Crustacea, Chelicerata and other arthropods, in *Ecdysone from chemistry to mode of action,* edited by J. Koolman. Georg (Thieme Verlag, Stuttgart).

Srimukda, B. 1987. Hatching and nursery operations of giant tiger prawn *P. monodon.* Technical Report no. SCS/GEN/82/40. SCSP, Manila. 125.

Stevenson, J.R., P.W. Armstrong, E.S. Cheng and J.D. O'Conner 1979. Ecdysome titers during moult cycle of the crayfish *Orconectes samborni. Gen. Comp. Endocrinol.,* 34: 20-25.

Stricker, S.A., Smythe, T.L., 2001. 5-HT causes increase in cAMP that stimulates, rather inhibits, oocyte maturation in marine memertean worms. *Development* 128: 1415–1427.

Subramoniam T, 2000. Crustacean ecdysteroids in reproduction and embryogenesis. *Comp Biochem Physiol.* 125: 135-156

Suchitraporn Sukthaworn, Sakol Panyim, Apinunt Udomkit, 2013. Molecular and functional characterization of a dopamine receptor type 1 from *Penaeus monodon. Aquaculture* 380.

Sun P S, 1994. Molecular cloning and sequence analysis of a cDNA encoding a molt-inhibiting hormone-like shrimp *Penaeus vannamei, Mol. Mar. Biol. Biotechnol,* 3:1.

Sun P S, 1995. Expression of the molt-inhibiting hormone-like gene in the eyestalk and brain of the white shrimp *Penaeus vannamei, Mol. Mar. Biol. Biotechnol,* 4: 262.

Swetha C. H., S.B. Sainath, P. Ramachandra Reddy and P. Sreenivasula Reddy, 2011. Reproductive Endocrinology of Female Crustaceans: Perspective and Prospective. *J Marine Sci Res Development,* S3:001

Takayanagi, H., Y. Yamamoto and N. Takeda, 1986. An ovary stimulating factor in the shrimp *Paratya compressa. J. Exp. Zool.,* 240: 203-209.

Tamone S L and Chang E S, 1993. Methyl farnesoate stimulates ecdysteroid secretion from crab Y-organs *in vitro, Gen. Comp. Endocrinol,* 89: 425.

Tansutapanich, a., N. Puvapanit, T. Sangkornthankig, S. Huakhum and P. Wonglek 1989. Integrated culture of marine shrimp brood stocks with *Artemia* culture in earthen ponds. *Thai Fish.Gazette.* 42: 281-290.

Tata, J.R. 1978. Induction and regulation of vitellogenin synthesis by estrogen. In: *Biochemical actions of hormones* 5: 397-431. Acad. Press., N.Y.

Teshima, S.I. and A. Kanazawa 1971. Bioconversion of progesterone by the ovaries of crab, *Portunus trituberculatus. Ge. Comp. Endocrinol.*, 17: 152-157.

Tensen C P, Coenen T and Van Herp F, 1999. Detection of mRNA encoding Crustacean Hyperglycemic Hormone (CHH) in the eyestalk of the crayfish *Orconectes limosus* using nonradio active *in situ* hybridization, *Neuroscience Lett,* 124:178.

Tiu SH, Hui JH, He JG, Tobe SS, Chan SM (2006) Characterization of vitellogenin in the shrimp *Metapenaeus ensis*: expression studies and hormonal regulation of *Me*Vg1 transcription *in vitro. Mol Reprod Dev* 73: 424436.

Toledo, J.D. and Kurokura, H. 2005, Aquaculture Research, 36: 666-673.

Treerattrakool, S., Panyim, S., Udomkit, A. 2005. A DNA cloning and functional study of gonad-inhibiting hormone (GIH) from the eyestalk of *Penaeus monodon.* In: Biothailand Proceedings, the International Conference on Shrimp Biotechnology: New Challenges through Thai Shrimp Industry. 4–5.

Treerattrakool S, Panyim S, Chan SM, Withyachumnarnkul B, Udomkit A, 2008. Molecular characterization of gonad-inhibiting hormone of *Penaeus monodon* and elucidation of its inhibitory role in vitellogenin expression by RNA interference. *FEBS* J 275: 970-980.

Treerattrakool S, Panyim S, Udomkit A, 2011. Induction of ovarian maturation and spawning in *Penaeus monodon* broodstock by double-stranded RNA. *Mar. Biotechnol.*13, 163-169.

Tsukimura, B. and F.I. Kamemoto 1988. Organ culture assay of the effects of putative reproductive hormones on immature *Penaeus vannamei* ovaries. *J. World Aquacult. Soc.*, 19: 288.

Tsukimura, B. and F.I. Kamemoto, 1991. *In vitro* stimulation of oocytes by presumptive mandibular organ secretions in the shrimp, *Penaeus vannamei. Aquaculture*, 92: 59-66

Tsukimura B and Borst D W, 1992. Regulation of methyl farnesoate in the hemolymph and mandibular organ of the lobster, *Homarus americanus, Gen. Comp. Endocrinol,* 86: 597.

Tsukimura B, Waddy S, Burow C W and Borst D W, 1992. The regulation of vitellogenesis in the lobster, *Homarus americanus, Amer. Zool,* 32: 28.

Tsutsui, N., Kim, Y.K., Jasmani, S., Ohira, T., Wilder, M.N. and Aida, K., 2005. The dynamics of vitellogenin gene expression differs between intact and eyestalk ablated kuruma prawn *Penaeus* (*Marsupenaeus*) *japonicus. Fisheries science*, 71,2:249-256.

Vaca AA, Alfaro J, 2000. Ovarian maturation and spawning in the white shrimp, *Penaeus vannamei*, by serotonin Injection. *Aquaculture* 182: 373-385.

Van Herp F and Payen G G, 1991. Crustacean neuro endocrinology, perspectives for the control of reproduction in aquacultural system, *Bull. Inst. Zool., Acad Sin*, Monograph, 16:513.

Vernet G and Charmantier-Daures M, Mue, 1994. Autotomie et Regeneration, in *TraITE DE Zoologie, crustaces,* Vol. VII edited by J. Forest (Masson, Paris) 107.

Vijayan K K Mohamed K S and Diwan A D, 1993. On the structure and moult controlling function of the Y-organ in the shrimp *Penaeus indicus* H. Milne Edwards, *J. World Aquaculture. Society,* 24: 516.

Vijayan K K and Diwan A D, 1993 Studies on the physiology of moulting in the penaeid shrimp *Penaeus indicus* H. Milne Edwards. *CMFRI spl. Publ.* 56: 6.

Vijayan K K and Diwan A D, 1994. The mandibular organ of theshrimp *Penaeus indicus* H. Milne Edwards and its in consequential role in moulting process, *J. Aqua. Biol,* 9:45.

Vijayan K. K, Mohamed K S and Diwan A D, 1997. Studies on moult staging, moulting duration and moulting behaviour in Indian white shrimp *Penaeus indicus.* H. Milne Edwards. (Decapoda, Penaeidae), *J Aqua. Trop,* 12: 53.

Waddy S L and Aiken D E, 2000. Endocrionology and the culture of homarid lobsters, Vol. 4 in *Recent advances in marine biotechnology* edited by M Fingerman, R. Nagabhushanam (Science Publishers, USA) 195.

Wainwright G, Webster S G, Wilkinson M C and Chung J S, 1996. Structure and significance of mandibular organ-inhibiting hormone in the crab, *Cancer pagurus.* Involvement in multihormonal regulation of growth and reproduction, *J. Biol. Chem,* 271: 12749.

Walaiporn Makkapan, Lamai Maikaeo, Teruo Miyazaki, Wilaiwan Chotigeat, 2011. Molecular 0mechanism of serotonin via methyl farnesoate in ovarian development of white shrimp: Fenner openaeus merguiensis de Man. *Aquaculture.* 321: 1–2.

Wallace B G, Talamo B R, Evans P D and Kravity E A, 1974. Octapamine selective association with specific neurons in the lobster nervous system, *Brain Res,* 76: 349.

Watson R D, Spaziani E and Bollenbacher W E, 1989. Regulation of ecdysone biosynthesis in insects and crustaceans, a comparison in *Ecdysone, from chemistry to mode of action,* edited by J. Koolman (Theime Med. Publ., Stuttgart),188.

Wear, R.G. and A. Santiago, Jr. 1976. Induction of maturity and spawning in *Penaeus monodon* Fabricius by unilateral eyestalk ablation (Decapoda, Natantia). *Crustaceana,*31: 218-220.

Webster S G, 1986. Neurohormonal control of ecdysteroid biosynthesis by *Carcinus maenas* Y-organ *in vitro* and preliminary characterization of the putative molt-inhibiting hormone (MIH), *Gen. Comp. Endocrinol,* 61: 237.

Webster S G and Keller R, 1986. Purification, characterization and amino acid composition of the putative molt-inhibiting hormone (MIH) of *Carcinus maenas* (Crustacea, Decapoda), *J. Comp. Physiol.* 156: 617.

Webster S.G and Keller R, 1989. Molt inhibiting hormone, in *Ecdysone, from chemistry to mode of action,* edited by J. Koolman (Theime Med. Publ., Stuttgart) 211.

Webster S G and Dircksen H, 1991. Putative molt inhibiting hormone in larvae of the shore crab *Carcinus maenas* L, an immune cytochemical approach, *Biol. Bull,* 180:65.

Webster SG, 1993. High-affinity binding of putative moult-inhibiting hormone MIH and crustacean hyperglycaemic hormone CHH to membrane bound receptors on the Y-organ of the shore crab *Carcinus maenas. Proc R Soc London* 251: 53-59

William A. Bray, Addison L. and Lawrence 1998 Successful reproduction of *Penaeus monodon* following hyper saline culture. *Aquaculture*, 159: 275 – 282.

Withyachumnarnkul, B., Plodphai, P., Nash, G., and Fegan, D., 2001. Growth rate and reproductive performance of F4 domesticated *Penaeus monodon* broodstock. The 3rd National Symposium of Marine Shrimp, Queen Sirikit National Convention Center, Bangkok, Thailand. 33–40.

Wouters, R., Gomez, L., Lavens, P. and Calderon, J., 1999. Feeding enriched Artemia biomass to *Penaeus vannamei* broodstock: its effect on reproductive performance and larval quality. *J. Shellfish. Res.* 18: 651–656.

Wouters R., Lavens P., Nieto J. and Sorgeloos P, 2001a. Penaeid shrimp broodstock nutrition: an updated review on research and development. *Aquaculture* 202: 1-21.

Wouters, R., Piguave, X., Bastidas, L., Calderon, J. and Sorgeloos, P., 2001b. Ovarian maturation and haemolymphatic vitellogenin concentration of Pacific white shrimp *Litopenaeus vannamei* (Boone) fed increasing levels of total dietary lipids and HUFA. *Aquac. Res.* 32: 573–582.

Wyban, J.A., C.S. Kee, J.N. Sweerney and W.K. Richards, Jr. 1987. Observations on the development of a maturation system for *Penaeus vannamei. J. World Aquacult. Soc.*, 18,3: 198-200.

Yang W J, Aida K., Terauchi A., Sonobe H. and Nagasawa H, 1996. Amino acid sequence of a peptide with molt-inhibiting activity from the kuruma shrimp. *Penaeus japonicus, Peptide,* 17: 197.

Yang, W.-J., Aida, K. and Nagasawa, H., 1997. Amino acid sequences and activities of multiple hyperglycemic hormones from the kuruma prawn, *Penaeus japonicus. Peptides* 18: 479–485.

Yano I, 1985. Induced ovarian maturation and spawning in greasy back shrimp *Metapenaeus ensis,* by progesterone. *Aquaculture* 47: 223-229.

Yano I, 1987. Maturation of kuruma shrimps *Penaeus japonicus* cultured in earthen ponds. NOAA Tech Rep NMFS 47: 3-7.

Yano, I. 1988. Oocyte development in the kuruma prawn *Penaeus japonicus. Marine Biology,* 99: 547-553.

Yano I, 1992. Effect of thoracic ganglion on vitellogenin secretionin kumura shrimp *Penaeus japonicus,. Bull. Nat. Res. Inst. Aquac,* 21: 9.

Yano I. and Wyban J A, 1992. Induced ovarian maturation of *Penaeus vannamei* by injection of lobster brain extract, *Bull.Nat. Res. Inst. Aquac,* 21:1.

Yano I, Wyban JA, 1993. Induced ovarian maturation of *Penaeus vannamei* by injection of lobster brain extract. *Bull Natl Res Inst Aquacult* 21: 1-7.

Yano I, 1998. Hormonal control of vitellogenesis in penaeid shrimp, in *Advances in shrimp biotechnology* edited by T.W.

Yano I, 2000. Endocrine control of reproductive maturation in penaeid shrimp, in *Recent advances in marine biotechnology* Vol. 4 edited by M. Fingerman and R. Nagabhushanam (Oxford and IBH publishing Co. Pvt. Ltd. New Delhi) 161.

Yashiro R, Na-anant P, Dumchum V, 1998. Effect of methyltestosterone and 17a-hydroxy-progesterone on spermatogenesis in the black tiger shrimp, *Penaeus monodon* Fab. In Flegel TW (ed) Advances in shrimp biotechnology, 33.

Yudin A I, Diener R A, Clark W H and Chang E S, 1980. Mandibular gland of the blue crab *Callinectes sapidus, Biol. Bull,* 159: 760.

Zeleny C, 1905. Compensatory regulation, *J Exp Zool,* 2: 1.

CHAPTER 6

GAMETE PHYSIOLOGY AND CRYOPRESERVATION OF GAMETES, EMBRYOS AND LARVAE

6.1 INTRODUCTION

Among crustaceans the prospects of shrimp, lobster and crab culture have improved noticeably during the past decade, as the biology and husbandry of these crustaceans have been more thoroughly investigated. Presently shrimp culture is progressing to the point where genetic programs can begin to help increase the production target (Tave and Brown, 1981). Technological improvements in the areas of reproductive biology, however, are needed for the further development of the shrimp Aquaculture industry in India. In sufficiently sound breeding programs, it is often necessary to produce and identify full sib-families and/ or half sib-families and identify all parents. The current spawning technique of placing many shrimp in large tanks is not adequate for such programs. By establishing modern techniques for preservation and utilization of male genetic materials of known heritage, as in higher animals, for future use, the additional efforts and costs for the maintenance of male brood stock in the hatcheries can be avoided. Cryopreservation of viable gametes from known heritage can solve these problems.

Although shrimp can be in bred captivity, there are many problems associated with maturation and insemination that have not yet been solved. The major problem often arising during breeding programs of penaeid prawns is the absence or loss of the stored spermatophores from the wild caught and laboratory matured shrimp. Another problem is the failure of shrimp to copulate in captivity. In order to solve these problems and simplify selection studies, Persyn (1997) developed a method of artificial insemination in penaeid prawns. Artificial insemination involves manual intrusion and attachment of the spermatophores inside the thelycum of female. This technique will allow the production or/ and identification of the parent strains and progeny, and is necessary during the production of intra and inter- specific hybrids because behavioral and structural differences of thelyca among strains and species can serve as a pre fertilization barrier (Aiken *et al.*, 1984). Manual spermatophore transfer can circumvent these problems and enable the breeders to produce valuable hybrids.

Artificial insemination results in high yield, in terms of nauplii per spawn (Alfaro 1993). Artificial insemination has been successfully done in many decapods crustaceans by researchers like Sandifer and Lynn (1980) in *Penaeus monodon*, Chow (1982) in *Macrobrachium rosenbergii*, Bray *et al.* (1983) in *P. monodon*, Talbot *et al.*, (1986) n *Homarus americanus* and Joshi and Diwan (1992) in *M. idella* in laboratories. However this technique is not practiced in most of the commercial hatcheries due to unawareness of its high potentialities. The technique involves the extraction of spermatophores, and then their placement in the thelycum, which is a complicated process and needs utmost care. A small mistake in the technique may lead to the failure of the whole attempt. One of the biggest problems in spermatophore transfer is the amount of time needed to position the spermatophore properly. Shrimp are easily stressed by short periods out of water and the stressed condition for a longer periods leads to an unsuccessful attempt. Strong healthy undamaged spermatophores enable quick artificial insemination. Usually the spermatophore extrusion method determines the condition of the spermatophores. Sandifer *et al.* (1984) suggested that the use of electrical stimulation for obtaining spermatophores is an adequate method for penaeid shrimp. This technique has been widely used by many scientists for the ejaculation of spermatophores from various decapods crustaceans like lobsters (Kooda-Cisco and Talbot 1983) and shrimp (Sandifer *et al.*, 1984, Harris and Sandifer 1986, Alfaro and Lozano 1993).

Successful cryopreservation of gametes would open new perspectives in any culture operations. The ability to cryopreserve gametes and embryos is an important insurance for research and breeding programs. With aquatic animals, the identification of parental strains and progeny, management of better quality and/or wild stocks of farming etc. are hurdles which are difficult to overcome. The ability to freeze the gametes of aquatic organisms and to store them for longer periods without deterioration would be of considerable value for the genetic improvement of Aquaculture.

In recent years, due to cognizance of cryopreservation of fertile gametes to augment animal production elaborate studies are being made to determine the sperm-egg characteristics of several economically important shrimp species. In most cases, female and male shrimp release their gametes into aquatic environment and egg –sperm interaction is expedited by a closely timed release of gametes.

Once contact between an egg and sperm is established, sperm of most species are induced to undergo an acrosome reaction as a prerequisite to fertilization. Penetration of the eggs by fertilizing sperm stimulates the egg to develop (Clark *et al.,* 1974).

The gametes of marine animals are quite unique in many ways and show remarkable differences from terrestrial animals. Research pertinent to gamete structure (spermatozoa and egg) and induction of gamete activity in marine animals, especially of invertebrates, is limited and not much information is available.

6.2 STRUCTURE OF EGGS

The fertilized lobster egg is surrounded by two coats, outer and inner, and the inner coat originates from the ovary. This kind of structure has reported in *H. americanus* (Herrick 1966), *H. vulgaris* (Yonge 1937), and *Crangon vulgaris* (Yonge 1955). Cheung (1966) has described the outermost coat of the fertilized oocyte of the crab, *Carcinus maenas,* and noted that the ovary gives rise to the new additional coats which surround the oocytes. Adding to the above information, Talbot and Goudeau (1988) have also reported that the lobster (*H. americanus*) egg coat undergoes a swelling process during spawning and the formation of the inner coat takes place through a complex cortical reaction (Fig 1).

Penaeid oocytes show extracellular crypts with feathery cortical rods in them (Drouslet *et al.,* 1975, Clark *et al.,* 1974, 1980). These cortical rods are said to be of oocytes origin, as reported by Lynn and Clark (1975) and Clark and Lynn (1977). The extracellular crypts with cortical rods are delimited from external environment by forming a thin vitelline membrane (Clark *et al.,* 1980, Pillai and Clark 1987). In contrast, oocytes of the caridean shrimp *M. rosenbergii* do not show any extracellular crypts. These oocytes surrounded by a thick investment coat and even the germinal vesicle has been reported to be absent in the oocytes (Lyann and Clark 1983a).

6.3 INDUCTION OF EGG ACTIVITY

Information pertaining to egg activation among crustaceans is very limited. Clark and his team have done detailed studies on egg activation in penaeid prawns, viz., *Penaeus japonicus, Penaeus aztecusans, Sicyonia ingentis.* In fact, egg activation means such events as release of eggs from meiotic arrest immediately after the ovulation process, formation of the hatching envelope around a newly created zygote and switching on the bio-synthetic machinery necessary for embryonic development. Fertilization is not requisite to bring out the activational changes in eggs. The eggs can even be activated upon contact with seawater at the time of spawning. Therefore, an egg viability assay, similar to that of sperm viability, becomes an important tool in cryopreservation technology.

For egg activation the following procedure is generally described in the literature. Live spawner shrimps are collected from wild and after carefully assessing the maturity phase of the spawners they are dissected out to remove the eggs and the spermatophores separately. The eggs and sperm are then transferred to a 500 mL beaker containing 200 mL artificial seawater (S 35%). The water is swirled gently and intermittently for about 5-10 minutes. This process allows activation and fertilization of eggs. Generally, inactive and freshly ovulated eggs lack an enveloping jelly layer but possess jelly precursor material within extracellular crypts of the eggs and a very thin vitelline envelope. Upon activation, first the jelly precursor is expelled from the egg surface crypts, surrounds the egg, and is transferred into the jelly layer. Later in the second step, egg resumes its first meiotic division by releasing the first polar body, which appears very distinctly in the activated eggs. This is again followed by the release of a second polar body after sometime. Finally, the hatching envelope (protective extracellular matrix) forms around the egg. Up to this stage of activation, the fertilization process is not required, but is needed for the activated eggs to execute embryonic cleavaeges.

Upon activation with seawater, penaeid shrimp oocytes lose their vitelline membrane and the cortical rods present inside the extracellular crypts disperse in seawater, as reported by Lynn and Clark (1975). *P. aztecus* eggs also undergo a similar reaction and cortical jelly precursor has been reported to contain mostly proteins and carbohydrates (Lynn and Clark 1987). Cortical rod release of the penaeid oocytes is described not as cortical reaction, but due to exocytotic release of cortical vesicles (Talbot and Goudeau 1988). The cortical jelly contains a trypsin like substance that initiates acrosomal filament formation in the sperm of *S. ingentis* (Lindsay and Clark 1992).

Prior to spawning, penaeid shrimp (*P. aztecus, P. setiferus, and S. ingentis*) eggs are arrested in the first metaphase plate of meiosis (Lynn *et al.*, 1991). The resumption of meiotic maturation and formation of the hatching envelope in the egg have been reported to be triggered by exposing the eggs to seawater (Pillai and Clark 1987). Mention of the need of external Mg++ and Ca++ ions for such reactions as prerequisites has been made by Pillai and Clark (1987) and Lynn *et al.* (1991). Both fertilized and unfertilized ova in *S. ingentis* have been observed to undergo the same sequence of reactions, but unfertilized ova need more time to complete the process and undergo unequal cleavages (Pillai and Clark 1987). Detailed similar descriptions of egg activation for *P. indicus* and *Parapenaeopsis stylifera* have been by Diwan and Joseph (1994) and Suresh and Diwan (1999) respectively.

Pongtippatee *et al.*, (2004) carried out egg activation studies in the black tiger shrimp *P. monodon*. They have reported morphological changes in the eggs upon contact with sea water. As soon as the egg was released into sea water, the cortical rods began to emerge from the crypts on the periphery of the egg, and elevated the thin investment coat that covered the surface of the egg. Further it is mentioned that sperm in the first phase of acrosome reaction were seen on both the egg and surface of the investment coat.The rods protruded fron the surface and were completely expelled out within 45 seconds. Later cortical rods began to break up and formed the jelly layer around the egg. By this time, the interaction between the sperm at the second phase of the acrosome reaction and egg is also noted. Formation of hatching envelope occurs at this stage and the first and second polar bodies are extruded (Fig 1).

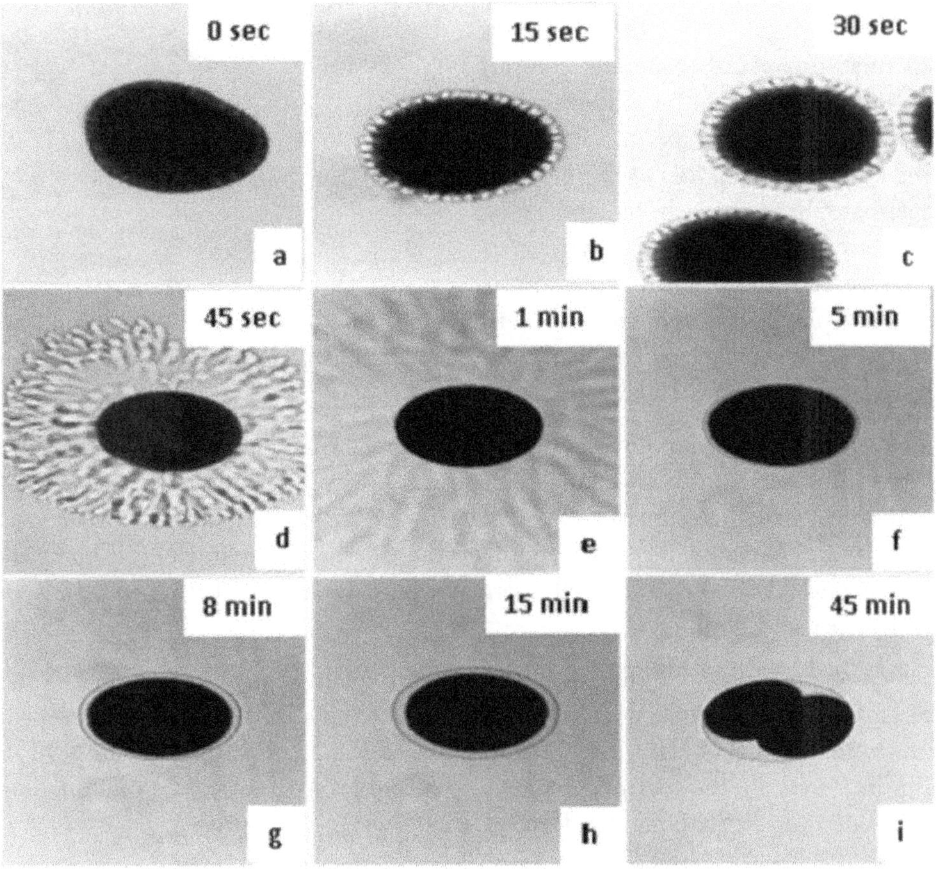

Fig. 1. Light micrographs showing the cortical reaction of *P. monodon* eggs from the unreacted stage, through the release of polar body and formation of a hatching envelope , to the first mitotic stage. (**Source:** Pongtippatee *et al.*, 2004)

6.4 STRUCTURE OF SPERMATOPHORES

The use of spermatophores as the main mode of sperm transfer by crustaceans is important because most crustaceans spermatozoa, and particularly all decapods spermatozoa, are aflagellate and non-motile (Felgrnhauer and Abele 1991), and hence need a vehicle for their transfer to females in the absence of a copulatory organ. This review examines the morphological diversity of spermatophores among natantians and retantians.

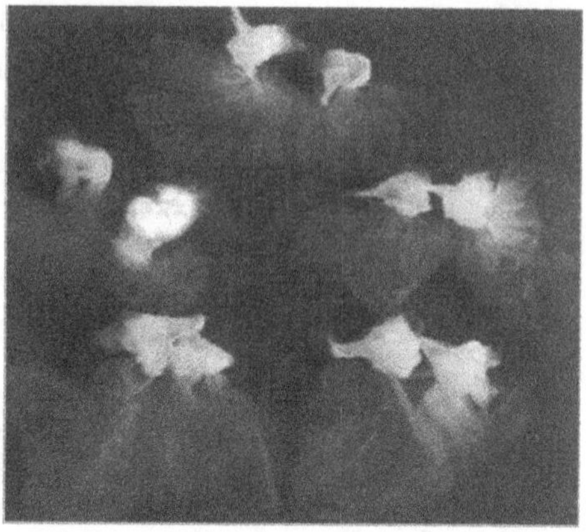

Fig. 2. Fully developed spermatophore with well developed wings.

In lobsters and crayfish like *H. americanus* (Kooda-Cisco and Talbot 1982), *Pacifastacus leniusculus* (Dudenhausen and Talbot 1983), *Enoplonetupus occidentalis* (Haley 1984), *Panulirus homarus* (Radha and Subramoniam 1985) the spermatophores are generally complex masses and consist of several investment layers to protect the sperm cells. But in the brachyuran, *Libinia emarginata,* the spermatophores have been reported to be simple spherical structures transferred in a fluid medium of seminal plasm (Hinsch and Walker 1974). Added to this, in the anomuran crabs *Emerita asiatica* and *Albunea sumnista* (Subramoniam 1984) the spermatophores are pedunculate.

Penaeid shrimp produce spermatophores of varying complexities and deposit them into thelycum of the females (Calman 1909). Depending on the type of thelycum, open thelycum (*Litopenaeus sp.* and *P.setiferus*) or closed thelycum (*P. aztecus*) the characteristics of the spermatophore varies (Calman 1909). The figure 2 illustrates the fully developed spermatophore with well developed wings in typical penaeid shrimp.

In sharp contrast to all other penaeid shrimp, the species belonging to the *Sicyonidae* do not produce a spermatophore, but transfer the sperm in the form of seminal plasm (Burkemroad 1934, Clark *et al.,* 1993, Subramoniam 1993). But in *P. stylifera,* the sperm are contained within transparent, oval and numerous spermatophores (Shaikmahmud and Tambe 1985).

6.5 STRUCTURE OF SPERMATOZOA

In decapods, spermatozoa are immotile, aflagellate cells that display an uncondensed nucleus and one or several appendages. These appandages have been described for many crustaceans, such as, *Callinectus sapidus* (Brown 1966), *L. emarginata* (Hinsch 1969, 1971, 1974, 1986), *Pinaisia so.* (Reger 1970), *Panulirus sp.* (Talbot and Summers 1978), *H. americanus* (Talbot and Chanmanon 1980), *Portunus pelagicus* (Jamieson 1989a), xanthid crabs (Jamieson 1989b), *Petalomna lateralis* (Jamieson 1990) and *Uca tangeri* (Medina and Rodriguez 1992). It has been reported that these appendages bring the sperm into contact with the oocytes (Medina 1992)

In contrast to reptantian, the natantian sperm are unistellate, non flagellated, non-motile and conical shaped, as reported for *P. aztecus* (Clark *et al.*, 1973), *S. ingentis* (Kleve *et al.*. 1980; Shigekawa and Clark 1986), *Rhyncocinetus typus* (Barros *et al.*,1986), *P. monodon* and *P. indicus* (Diwan *et al.*, 1994). But the sperm of *P. stylifera* is rod shaped with a short spike (Shaikmahumad and Tambe 1958). In sharp contrast to all the above mentioned sperm, the sperm of the caridean prawn *M. rosenbergii* has been reported as "umbrella"-shaped with short spike (Lynn and Clark 1983b) (Fig 3).

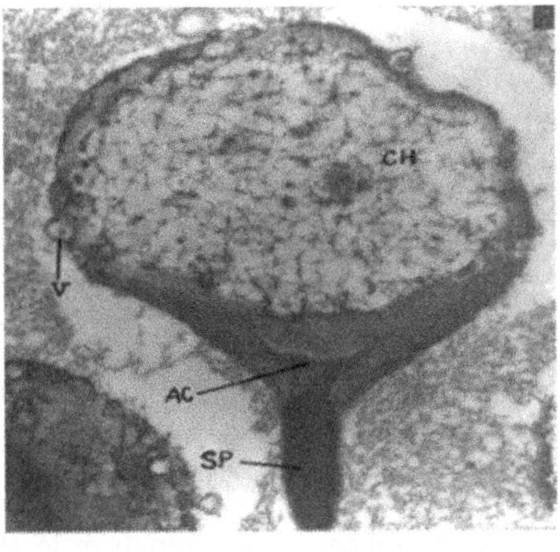

Fig. 3. Electron micrograph of the unistellate sperm of *P. indicus* showing spike (SP), acrosome (AC) vesicle (V) and uncondensed chromatin (CH). X 9000.
(**Source:** Mohamed and Diwan, 1994).

6.6 INDUCTION OF ACTIVITY IN SPERMATOZOA

The morphological and ultrastructural features of non-motile (unreacted) spermatozoa of *P. indicus* and *P. monodon* have been reported by Diwan and Joseph (1994). They mentioned that superficially the spermatozoa appear to have three regions, viz. 1. a posterior main body, 2. a central cap region and 3.

anterior spike. The posterior main body is an elongate sphere housing an uncondensed nucleus followed by the central cap region which includes acrosomal vesciles. The nucleus is not membrane bound and extremely fibrillar. Prominent membrane bound vesciles are commonly seen near the margin of the cell body. In the cap region two portions can be distinguished by differences in electron density. Toward the anterior side a distinct spike is seen. The ultrastructure of spike shows that it contains an amorphous electron dense material with some cross-striations in between (Fig 4 & 5).

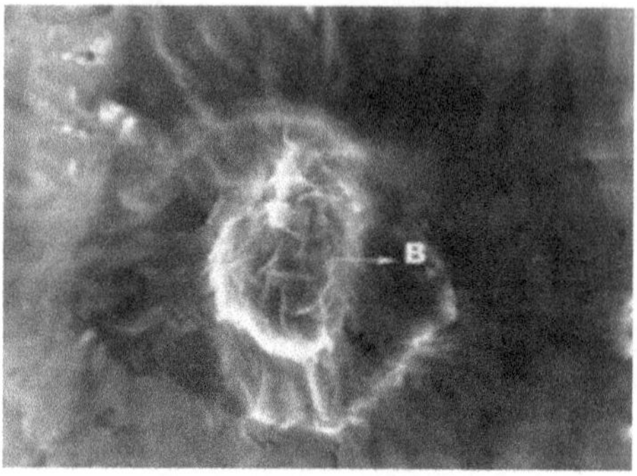

Fig. 4. SEM of acrosome reaction of the sperm X 5000. B-body explosion
(**Source:** Diwan and Joseph 1994)

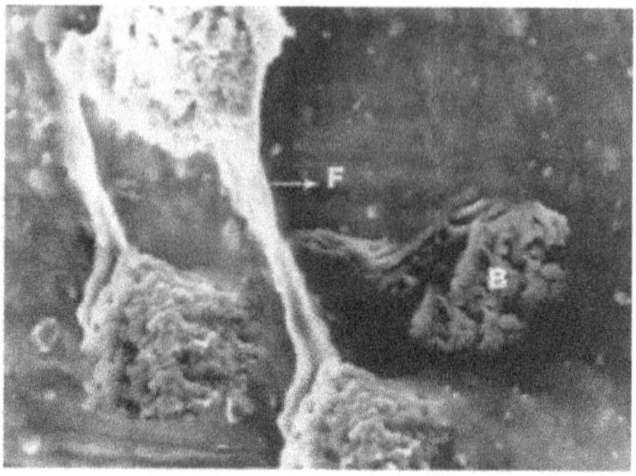

Fig. 5. SEM of acrosome reacted sperm. F-Filament, B-body
(**Source:** Diwan and Joseph 1994)

Several attempts have been made to determine the structural details of crustacean spermatozoa through Scanning electron microscope (SEM) and Transmission electron microscope (TEM) studies. Lynn and Clark (1983a) have done SEM studies on sperm-egg interaction in the freshwater prawn *M. rosenbergii* whereas for the penaeid prawn *S. ingentis*, Clark *et al.* (1981) made an in depth investigation on sperm-egg activational changes through ultrasturctural means. In *P. stylifera* attempts have been made to study the structural details of gametes during fertilization process (Suresh and Diwan, 2000).

Though crustacean sperm appear to be non-motile they become active just prior to fertilization. The activated or reacted sperm exhibit a total change in their morphological anatomy. It has been found that in the wild when a sperm contacts an egg at the time of spawning, it binds with the tip of the spike of the egg's vitelline envelope and in no time gets activated. The first manifestation of activated sperm in *P. indicus* and *P. monodon* as reported by Diwan and Joseph (1994) is the loss of anterior spike, which immediately results in externalization of the acrosomal vesicle contents and excytosis of acrosomal vesicles. With the exocytosis of acrosomal vesicles sperm activation is completed by the formation of a long acrosomal filament.

Clark and others at the Bodega Marine Laboratory in California have done extensive studies and described the detailed events of physiological control and mechanisms of sperm-egg activation, fertilization and early development for the penaeid prawns *S. ingentis*. In fact, the non-motile sperm of decapod crustaceans are a richly diverse group in terms of both morphology and types of activational changes that precede fertilization. For example un-reacted reptantian sperm possess numerous appendages that emanate from the nucleus. During the acrosome reaction the sperm undergo an "eversion" process by forming an amorphous material. On the other hand, sperm of caridean natantian have been described as unistellate, possessing only one appendage and do not undergo an acrosome reaction as a prerequisite fertilization. But the sperm of penaeid natantians though they are unistellate, undergo an acrosome reaction during sperm-egg interaction.

What is known to be the acrosome reaction has been variously described earlier as sperm "explosion" or sperm "eversion" or "degeneration". Baker and Austin in 1963 correctly identified the explosive phenomenon as an event analogous to the acrosome reaction of flagellated sperm. The fine structural events of the acrosome reaction which occur during sperm-egg interaction have been described by Brown (1966) for the blue crab, *C. sapidus*. While numerous studies of sperm development and mature sperm structure exist in the literature, knowledge about the acrosome reaction of decapods sperm is comparatively scarce. Talbot and Chanmanon (1980) have mentioned that the acrosomal reaction of the sperm is a must and is necessary for generating the forward movement of these otherwise immotile spermatozoa.

In recent times it was discovered that the acrosomal reaction can be artificially induced by using a solution hypotonic to seawater. Some workers have activated the acrosomal reaction in the laboratory by using either eggwater or divalent ionophores. Egg water is generally collected from freshly spawned eggs. To collect eggwater, a female prawn is spawned in a beaker containing 500 mL of seawater (S-30%). After spawning the spawner is removed and the eggs are allowed to settle. The whole volume of water is slowly reduced to one third by decanting and the eggs are then gently swirled in the remaining water for 10-15 minutes. After allowing the eggs to settle the supernatant fliud (eggwater) is centrifuged and later it is frozen and stored in liquid nitrogen until needed. For induction of spermatozoa apart from eggwater, other compounds like the bromo-calcium ionophore, valinomycin and nigricin have been reported as inducing agents for the acrosome reaction (Griffin *et al.,* 1987).

Induction of the acrosome reaction in decapods sperm can be done through various ways, as reported by Fasten (1921). The acrosomal exocytosis undergone by decapod sperm has been reviewed by Dan (1952, 1967) and Austin (1968).

The acrosome reaction undergone by sperm of the reptantian *L. emarginata* is distinct with eversion of subasceosome region (Hinsch 1971) and this eversion process is due to the presence of enzyme hydrolases in the acrosome of sperm. It has also been reported that an egg-derived component activates the sperm to undergo the acrosome reaction in decapods (Shapiro and Eddy 1980, Lopo 1983). In addition to natural induction, it has been reported that reptantian (*H. americanus*) and natantian (*S. ingentis*) sperm can be made to undergo the acrosome reaction by using the divalent calcium ionophore A23187 (Talbot and Chanmanon 1980, Clark *et al.,* 1981).

It has also been mentioned that with the Chinese mitten crab *Eriochier sinensis*, the acrosome reaction can be artificially induced using eggwater or seawater or a $CaCl_2$ solution (Nanshan and Luzheng 1987). Moreover, the acrosome reaction in *S. ingentis* depends on external Ca^{++} (Clark and Griffin 1988), and an eggwater component, and trypsin (Griffin and Clark 1990). In the sturgeon *Acipenser transmontanus*, it has been described that a glycoprotein is an important ingredient involved in the acrosome reaction of the sperm (Cherr and Clark 1985).

The acrosome reaction in *H. americanus* has been found to be distinct and involves eversion of the sub-acrosome region and reorientation of the nuclear arms. This results in a net forward movement of the sperm (Talbot and Chanmanon, 1980, Tsai and Talbot 1993). A similar acrosome reaction has been reported in *Uca tangeri*, with slight modifications in nuclear arms orientation, but there is no forward movement (Medina 1992).

The acrosome reaction of *S. ingentis* differs vastly from that of reptantians (Clark *et al.,* 1981). For the acrosome reaction to take place the sperm of *S. ingentis*, have to undergo a capacitation process within the female seminal receptacle (Wikramanayake *et al.,* 1992). In *S. ingentis*, the acrosome reaction occurs by an exocytosis process followed by extension of an acrosomal filament which consists of a distal petal region (Clark *et al.,* 1981, Anderson *et al.,* 1984, Griffin and Clark 1988). Similar reactions have been reported in *P. monodon* and *P. indicus* (Diwan and Joseph 1994) (Fig 6). The reaction has been reported to be accompanied by a decrease in external pH during the *S. ingentis* sperm acrosome reaction (Griffin *et al.,* 1987). This pH variation has also been reported in the acrosome reaction of sea urchin sperm (Shackmann 1989). In sharp contrast to penaeiodean sperm, the caridean *M. Rosenbergii* sperm does not undergo an acrosome reaction prior to fertilization (Lynn and Clark 1983a). In the bivalve *Laternula limicola* (Hosakaw and Noda 1922) and the sea urchin (Hino and Kato 1990), the acrosome reaction takes place while the sperm penetrates the egg.

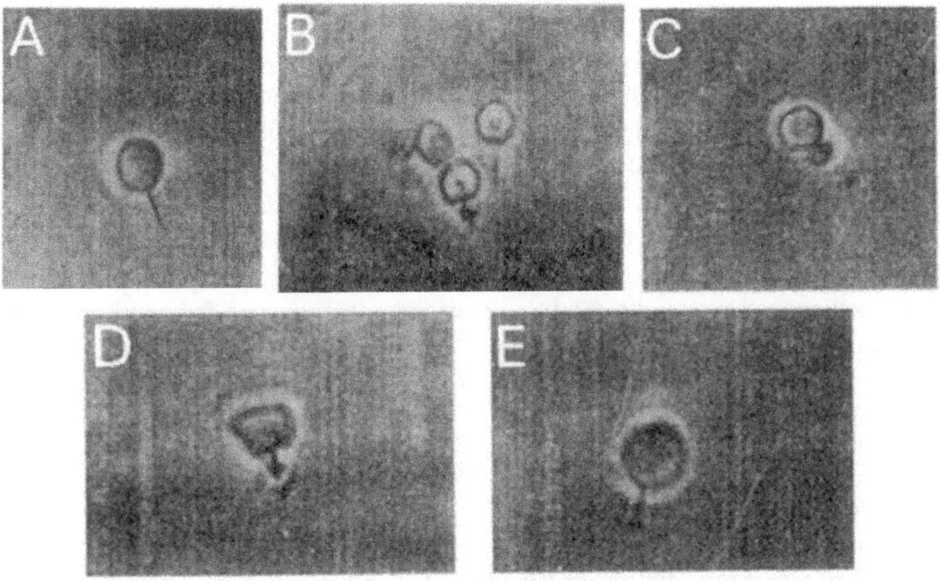

Fig. 6. Phase contrast micrographs of the five activational stages of *Penaeus indicus* spermatozoa X, 400. (A) a unreacted spermatozoa possessing an anterior spike, (B) a spermatozoa that has undergone acrosomal exocytosis and has lost the spike, (C) initiation of acrosomal filament formation, (D) acrosomal filament formation in progress and (E) a spermatozoa that has completed the acrosome reaction by forming an acrosomal filament. (**Source:** Diwan & Joseph, 1998)

Braga *et al.,* (2014) evaluated spermatozoal capacitation of pink shrimp, *Farfantepenaeus paulensis.* This process has direct applications in Aquaculture, and it consists of the ionic, biochemical and morphological changes during the period that the spermatophore is stored or adhered to the thelycum. These changes make the spermatozoa capable of fertilization. While evaluating it was observed that spermatozoa underwent morphological changes related to capacitation. It is mentioned that stored spermatozoa have less condensed chromatin and the acrosomal region is more electron dense and less concave. The most remarkable morphological change that is reported in the spermatozoa of *F. paulensis* was the loss of the flocculent appearance of the sub-acrosomal region which became much denser in capacitated spermatozoa. Similar observations have been earlier reported by Alfaro et al., (2003, 2007) and Aungsuchawan *et al.,*(2011) in the open thelycum of shrimp *Litopenaeus* sps.

Pongtippatee *et al.,* (2007) studied acrosome reaction in the sperm of the black tiger shrimp *P. monodon.* They studied the acrosome reaction of the sperm incubated with egg water and compared the same with actual spawning time. They reported that the sperm taken from female thelycum is composed of a posterior main body, a central cap and anterior single spike. Further they observed that upon contact with egg water, the sperm underwent two phases of acrosome reaction i.e. acrosomal exocytosis and spherical mass formation. During acrosomal exocytosis there is degeneration of the spike of the sperm, swelling of the cap region and rupture of the acrosomal pouch. During spherical mass formation there is polymerization of materials within the subacrosomal region and ended with re-configuration of the subacrosomal region into an electron-dense spherical mass. The acrosome reaction of the sperm observed during spawning revealed similar morphological events, with degeneration of the spike upon contact with the vitelline envelope and formation of the spherical mass while penetrating into jelly material produced by protruding cortical rods. From these findings it is concluded that the acrosome reaction inducers are derived from the vitelline envelope and cortical rods of the egg.

Kruevaisayawan *et al.*, (2008) carried out work on the induction of the acrosome reaction of the sperm of the shrimp, *P. monodon* using trypsin-like Enzyme. They used peptidyl fluorogenic substrates to show the presence of trypsin-like enzymes in *P. monodon* egg water and sperm, but minimal activities of chymotrypsin-like enzymes. In sperm, these trypsin-like enzymes existed both on the sperm surface and in the acrosome. By using this flurogenic substrate, acrosome reaction of the sperm could be initiated and the whole process of acrosome reaction was completed within 45–60 sec. The acrosome reaction comprised of only the acrosomal exocytosis and depolymerization of the sperm head anterior spike. Further they have reported that sperm borne trypsin like enzyme (on the surface and/or in the acrosome) is involved in the acrosome reaction process. This has been proved by using irreversible trypsin inhibitor to abolish the trypsin-like activities in the egg water.

Several methods have been used to assess sperm quality in the male shrimp, mainly sperm count and production of viable sperm (Ceballos-Vazquez *et al.*,2003), and sperm activation (Wang *et al.*, 1995). In flagellate spermatozoa, the energy status of spermatozoids, inferred from ATP levels, is also related to sperm viability in terms of motility (Ellington. 2001, Rurangwa *et al.*, 2002 and Zietara *et al.*, 2004). However, penaeid spermatozoa are non-motile and do not contain mitochondria (Pongtippatee *et al.*, 2007). During fertilization they are trapped in expelled cortical rods or corona and then the acrosome reaction allows a spermatozoon to fuse and penetrate the oocyte membrane (Kruevaisayawan *et al.*, 2008, Li *et al.*, 2010) a process that could use energy from ATP. Energy required for sperm mobility has not been previously investigated. Vazquez-Islas *et al.*, (2013) investigated the energy levels of spermatophores and sperm viability during the moult cycle in intact and bilaterally eyestalk ablated male pacific white shrimp *Litopenaeus vannamei*. It was observed that sexual organs and spermatophores, expressed either as absolute weight or as somatic index and sperm counts were significantly higher in the ablated shrimps. Similarly a continuous increase in the weight of the vas deferens throughout the moult cycle was also observed in the ablated group but not in the control one. Highest ATP level and adenylate energy charge has been reported in the spermatophores visible in the terminal ampoule after ecdysis at the late post moult stage of the first moult cycle. From the detailed studies it has been concluded that there is a cyclic moult regulated energy balance, where new spermatophores are energy recharged in the intact shrimp, while this process is apparently disrupted in ablated shrimp.

Spermatophore Regeneration Time

To determine the time required for regeneration of spermatophore in non-ablated sexually mature males, existing spermatophores are removed from 30 adults at the onset of the trial. Baseline (control) spermatophore's weights, sperm count, morphology and viability are determined from a spermatophore sample (N=8). Each day, terminal ampoule from two to three randomly selected males is dissected, to assess the extent of development if they do not ejaculate the spermatophores upon electro-ejaculation. The sperm quality and quantity (if sperm are present) are assessed using the techniques of Leung-Trujillo and Lawrence, (1985). Spermatophore maturity is assessed based on its morphology and colour (Persyn, 1977).

Effect of Eyestalk Ablation on Sperm Quality

The second series of trials are planned to investigate the effect of unilateral eyestalk ablation on the sperm quality. Eyestalks of 30 shrimps are ablated using the method of Muthu & Laxminarayana, (1977) with an electro-cautery apparatus. After eyestalk (4 to 5 days), the spermatophores are removed from shrimps and the extent of the development of new spermatophore is monitored on a daily basis. The quality of the sperm cells obtained before and after regeneration is evaluated through spermatophore weight, sperm cell counts, percentage of normal and abnormal sperm as well as the viability of the sperm. Percentage of abnormal sperm is obtained by recording, over transect of a microscope slide, number of normal sperm (spherical body and elongate spike) and number of abnormal sperm (mal-forms bodies and bent spikes) (Talbot *et al.*, 1989). At least 100 cells for one of the categories are recorded and from this percentage of normal abnormal cells are calculated.

Viability of Spermatozoa

Acrosome reactions, induced by gamete interaction are observed in sperm associated with eggs spawned from inseminated females. The morphological changes that occur during the *in vitro* acrosome reaction have been studied with the aid of three chemical compounds, namely Bromocalcium ionophore A 23187, Valinomycin and Nigricin (three from Sigma) (Clark *et al* 1981) and a biological agent the egg-water (Griffin *et al.* 1987). The structural changes that took place in the sperm cells are compared with the structural changes that are found in the *in vitro* studies made during the natural spawning. From these observations, the best inductors are identified for the sperm of *P. monodon* and selected for the following trial. In the present study "egg-water" described by Griffin *et al.* (1987), as well as the ionophore A 23187 (Talbot *et al.* 1980), has been identified as the best solutions for the induction of acrosomal filament during the *in vitro* acrosome reaction. The egg-water is extracted as follows. Fully gravid female shrimps ere collected from wild and transported to the laboratory in live condition. They are transferred to glass aquarium tanks containing clean filtered seawater provided with aeration, and observed for spawning. When a female shrimp showed spawning behaviour it is removed from the glass tank and held in 1 l glass beaker with half (500 ml) filtered seawater (S H" 34‰) until eggs are released. Eggs are allowed to settle and after the negatively buoyant eggs settled to the beaker bottom, about 80% of the seawater is drawn off and the eggs are swirled for five minutes with enough force to keep them in suspension. The eggs are again allowed to settle, and after settling the egg, the remaining supernatant fluid now called 'egg-water' is removed by pipette and cleaned by centrifugation for 15 min. at 15,000 rpm (4° C, 15 min.). This egg-water is divided into 5 ml aliquots and stored in liquid nitrogen, if not used immediately.

Extruded spermatophores are individually and gently homogenized to free the sperm cells, using a glass tissue grinder and filtered seawater (S H" ‰). Tissue fragments are separated from the sperm supernatant by hand centrifugation. Assays are conducted in culture tubes of 3 ml capacity. Prior to the addition of sperm, 900 ml of egg water is added to each assay tube. Then immediately the supernatant is removed using a pipette and 100µl of sperm cells are pipetted into the egg water, and mixed thoroughly. After 5 minutes of incubation, 100µl of the sample is removed to another tube and fixed with a drop of 70% ethanol in seawater. The reminder of the sample is allowed to incubate for another 55 minutes. After 1 hr incubation, sperm are examined to determine the number of reacted sperm cells, which had undergone acrosomal exocytosis as described by Clarck *et al.* (1981). Sperm cells are examined using a haemocytometer and phasecontrast microscope, with > 100 sperm cells observed per count and three counts per replication. Sperm are scored for acrosomal status by viewing under a power of 40 X 10 using a phasecontrast microscope. In addition, a portion of the sperm suspension is examined microscopically for normal and abnormal sperm cells. Abnormal cells are distinguished from normal by mal-forms bodies or by bent, short or missing spikes. Percentage of viable sperm is calculated as a percentage of normal sperm. Nimrat et al., (2006) studied the viability of chilled spermatophores using antibiotic treatment and without antibiotics in *L. vannamei*. They used four extenders *viz* mineral oil, ringer's solution, phosphate buffer and 0.85% NaCl and stored the spermatophores at low temperature (2-4^0 C). After 35 days preservation they observed that spermatophores samples preserved without antibiotics showed low viability due to high bacterial growth in the samples and one with antibiotics showed high percentage of viability of sperm. From these results they suggested that chilled storage of spermatophores is a feasible approach for the management and spawning of white shrimp broodstock. Studies on assessment of viability of sperm cells of *L. vannamei* on cryopreservation have been carried out by Uberti *et al.*, (2014). They have used dimethylsulfoxide and ethylene glycol as croprotectants for cropresevation of sperm cells. In both the cryoprotectants they found that there was decrease in cell viability within a longer cryopreservation time. The reason for such decreased viability they assumed is generating potentially toxic metabolites as time of cryopreservation is extended. Suresh Kumar and Diwan (2000) while working on induction of gamete activity and mechanism of fertilization in the shrimp *Parapenaeopsis stylifera* described the ova and spermatozoa characteristics and their mechanism of fertilization in the shrimp. They observed that the inactivated egg is surrounded by a vitelline membrane, which delimited the peripheral cortical crypts. The release of the jelly precursors, the resumption of meiotic maturation and the formation of the hatching envelope in the ova were triggered by exposing the ova to seawater. Further they reported that the spermatophores collected from the female seminal receptacles are found to be oval in shape and transparent with spermatozoa arranged in horizontal rows inside them. The spermatophores measured about 200-250 µm in length and 80-100µm in breadth. The spermatozoa were rod shaped and measured about 26-28 µm in length and 7-8 µm in breadth. Upon activation with egg-water, the spermatozoa underwent activation-related changes (acrosome reaction) both anteriorly and posteriorly. Posteriorly, due to exocytosis, a long filament of 35-40 µm in length was formed. Anteriorly two nuclear arms developed which appeared tentacle-like with globular tips. During sperm-egg interaction *in vitro*, the spermatozoa underwent similar changes and penetrated the ova along with the filament, leaving behind a luminescent extracellular elevation on the surface of the ova.

6.6 CRYOPRESERVATION OF GAMETES

The earliest cells to be cryopreserved successfully are the bovine sperm (Polge *et al.*, 1949). This breakthrough allowed for storage of viable broodstock sperm and led to increased production. Sperm have been chosen for use in cryopreservation studies because of the ease in assessing viability either by motility or fertilization and their commercial importance. Studies on cryopreservation of vertebrate semen especially that of mammals have grown considerably since 1960, but unfortunately similar studies in invertebrate semen have been sparse (Behlmer and Brown 1984). Methods to cryopreserve gametes of aquatic animals are less developed, (Alderson and Mc Neil 1984, Bougrier and Rabenomanana 1986) even though the sperm cryopreservation has been done successfully for a number of commercially important aquatic vertebrate and invertebrate species in the past few years. However, the reproducibility of sperm cryopreservation for invertebrates and aquatic vertebrates needs to be improved and sperm cryopreservation on a large, commercial scale is yet to be developed.

Many attempts have been made to cryopreserve the spermatozoa of aquatic vertebrates, mainly fishes. However, less attention has been paid to invertebrates. The various attempts of cryopreservation in invertebrates include the reports of Sawada and Chang (1964) for the honey bee; Dunn and Mc Lachlin (1973) for echinoderm sperm; and Behlmer and Brown (1984) for *Limulus polyphemus*. Chow (1982) for the first time reported the successful preservation of spermatophores of the fresh water shrimp *M. Rosenbergii*. Ishida *et al.* (1986) later developed a technique for long term storage of lobster (*Homarus*) spermatophores. Spermatozoa of the penaeid prawn *S. ingentis* have been successfully preserved for a period of two months in liquid nitrogen by Anchordouguy *et al.* (1988). Jeyalectumie and Subramoniam (1989) and Joshi and Diwan (1992) have developed a method to cryopreserve viable spermatophores of the mud crab *Scylla serrata* and shrimp *M. idella* respectively. Recently Subramoniam (1993, 1994) has reviewed cryopreservsation of gametes and embryos of a few cultivable crustacean species. Diwan and Joseph (1998) and Joseph (1996) did elaborate studies on cryopreservation of spermatozoa of the penaeid shrimp *P. indicus* and *P. monodon*. They could successfully preserve the viable sperm of both shrimp up to 60 days at -196°C using a combination of DMSO + glycerol (5%) and DMSO + trehalose (0.25 M). the best viability rate of 75-80 % has been reported by them with freeze thawed spermatozoa for both shrimp (Table 1). Joseph (1996) could cryopreserve the spermatozoa of *P. monodon* up to six months at -196°C. The limited number of attempts in the cryopreservation of gametes of decapods crustaceans may partially be due to the fact that, sperm of decapods crustaceans are exception from those o the other animals, as they are non motile.

Cryopreservation of spermatophores

Among the various cryoprotectants tested so far DMSO (dimethyl sulphoxide) and glycerol have been reported to be the best protective media at ultra low temperatures. Trehalose has been reported to be the least protective compound. Glycerol was first reported as effective in protecting sperm freeze thaw damage by Rostand (1946) and Polge *et al.*, (1949). Polge *et al.* (1949) first showed that spermatozoa could be frozen and thawed without losing motility if glycerol was introduced in their suspending medium. Glycerol has been used widely in the cryopreservation of viable fish milt (Horton and Ott 1976). Similarly, glycerol has been used as cryoprotectant in most of the sperm preservation experiments with invertebrates. Successful preservation of spermatophores of shrimp in glycerol at -196°C has also been reported by Chow *et al.*, 1985. Similar observations have been made by Behlmer and Brown (1984) for Limulus spermatozoa. When DMSO was used for spermatophore preservation in *M. Rosenbergii* the post thawing survival was reported to be nil (Chow *et al.*, 1985). While on contrary, Anchordouguy *et al.*, (1988) reported that sperm survival was higher in *S. ingentis* when glycerol was used as a cryoprotectant for the preservation of sperm even though DMSO gave the highest percentage of survival. In *S. serrata* glycerol and DMSO + trehalose gave the highest sperm survival (Jeyalectumie and Subramoniam 1989). Table 1 shows the actual count and perecentage of spermatozoa of *P.indicus* cryopreserved at -196°C for different duration of time.

In the case of *P. monodon* Joseph (1996) has mentioned that the percentage of viability was higher in samples preserved 5% DMSO even after 6 months. The maximum effect was in 5% DMSO; higher and lower concentrations yielded inconsistent results. Similarly, higher percentage viability has been observed by Joseph (1997) when the samples were preserved for 6 months in DMSO (5%) + trehalose (0.25 M). Anchordouguy *et al,*. (1988) reported that highest percentage viability was obtained from samples preserved in 5% DMSO. However it was reported that combining 5% DMSO with other cryprotectants like trehalose, proline sucrose and glycerol in the freezing medium did not alter sperm viability in the marine shrimp *S.ingentis* (Anchordouguy *et al.*, 1988). Whereas in *S. serrata* DMSO + trehalose showed high sperm survival but trehalose when it is used alone did not show good results (Jeyalectumie and Subramonium 1989). Similarly, trehalose offered the least protection in the preservation of *S. ingentis* sperm (Anchordouguy *et al.*, 1987) (Table 2).

In nature during mating and insemination of aquatic organisms there is no control over the progeny. On the contrary, by adopting artificial insemination technique, selective breeding can be achieved. One of the basic requirements for this is the ready availability of gametes for the manipulations. Artificial insemination and cryopreservation of the viable gametes are very much important in controlled breeding of the quality parents. In decapod crustaceans, all these above said biotechnological processes require spermatophores; hence acquisition of the spermatophores is very much important. Different spermatophore extraction

Table 1. Actual counts and percentages of activated spermatozoa of *Penaeus indicus* cryopreserved at -196°C for different durations of time.

	Activated spermatozoa				
Duration of cryopreservation	Spermatozoa with spike	Exocytosis	Filament formation	Percentage of activated spermatozoa	Cryopro tectants used
	47 ± 4	82 ± 5	56 ± 5	74.22	Glycerol
	50 ± 2	64 ± 5	53 ± 4	70.07	DMSO
	26 ± 3	56 ± 5	39 ± 2	78.57	DMSO ± Glycerol
	25 ± 3	59 ± 5	56 ± 5	81.92	DMSO ± Trehalose
7 days	48 ± 1	55 ± 5	49± 3	68.46	Trehalose
	67 ± 2	10 ± 3	–	–	Control
	49 ± 3	64 ± 3	54 ± 2	70.7	Glycerol
	48 ± 1	59 ± 2	47 ± 3	68.90	DMSO
	35 ± 2	71 ± 3	55 ± 3	78.19	Glycerol ± DMSO
15 days	26 ± 2	57 ± 3	55 ± 4	80.39	DMSO ± Trehalose
	39 ± 4	35 ± 4	29 ± 3	62.72	Trehalose
	28 ± 2	9 ± 2	-	–	Control
	42 ± 4	47 ± 3	40 ± 3	67.47	Glycerol
	42 ± 3	47 ± 4	42 ± 3	67.99	DMSO
30 days	30 ± 2	57 ± 5	40 ± 3	75.52	Glycerol ± DMSO
	27 ± 4	55 ± 4	40 ± 3	78.10	DMSO ± Trehalose
	31 ± 3	22 ± 3	24 ± 3	59.89	Trehalose
	17 ± 1	7 ± 2	–	–	Control
	50 ± 5	35 ± 5	38 ± 4	58.70	Glycerol
60 days	46 ± 4	24 ± 4	38 ± 3	56.67	DMSO
	42 ± 4	38 ± 3	49 ± 4	67.16	Glycerol ± DMSO
	37 ± 4	27 ± 4	58 ± 5	69.97	DMSO ± Trehalose
	53 ± 5	27 ± 3	35 ± 3	53.73	Trehalose
	42 ± 5	15 ± 3	–	–	Control

Each value is the mean of three determinations. (**Source:** Diwan and Joseph, 1998)

Table 2. Cryopreservation of sperm of Decapod crustaceans.

Species	Cryoprotectants used	Temperature	Preservation period	Percentage of survival of sperm	Method of testing viability	References
Macrobrachium rosenbergii	Glycerol	-196°C	31 days	53	Fertility	Chow *et al.*, 1985
S. ingentis	Trehalose + DMSO	-196°C	2 months	60-70	Acrosome reaction	Anchordouguy *et al.*, 1987
S. ingentis	Trehalose, sucrose, Proline, Glycerol & DMSO	-196°C	1 month	56	Acrosome reaction	Anchordouguy *et al.*, 1988
Scylla serrata	Glycerol, Trehalose, DMSO, DMSO + Trehalose	-196°C -79°C -4°C	30 days	95 & 89	Eosin dye extrusion	Jeyalectumie and Subramonium (1989)
Macrobrachium idella	Ringer's solution	6°C	96 hours	80	Fertility and Larval production	Joshi and Diwan (1992)
P. Indicus	Glycerol + DMSO, Trehalose + DMSO,	-196°C -35°C	60 days	6770	Acrosome reaction	Diwan and Joseph (1998)
P. Monodon	DMSO, DMSO + Trehalose, DMSO + glycerol	-196°C	6 months	65	Acrosome reaction	Joseph (1996)

(**Source:** Diwan and Joseph, 1998)

methods have been studied by many investigators in various decapod crustaceans, which include; dissection, squeezing, extraction, cannulation and electroejaculation techniques (Sandifer and Lynn, 1980 and Lin and Ting, 1986). Three different techniques and their effects, both on spermatophores as well as its donors, are discussed here. It is found that various methods have their own merits and demerits. Dissection and extraction is the earliest technique used for the removal of spermatophore (King, 1964 and Lin and Ting, 1986). The main disadvantage of this method is that the animal has to be sacrificed for the spermatophore extrusion. In *P. monodon* as described in this chapter, a modified microsurgery method, is tried by using a gill irrigator and micro scissor. It is found that though it does not kill the animal the surgery wounded the animal. It is observed that for the regeneration of spermatophores in these shrimps, long period is required, as the healing of the wound has to take place first. The second method often employed for the spermatophore extraction is the squeezing method; which does not require the sacrificing of the animal; however it requires skill for the complete extraction of the whole spermatophore without damages (Lin and Ting, 1986; Alfaro, 1993 and Heitzmann *et al.*, 1993). The electro-ejaculation (Sandifer and Lynn, 1980) of spermatophore has been reported as the most effective and simple method of spermatophoreretrieval in penaeid shrimps. Later Talbot *et al.*, (1983) refined the techniques of electro- ejaculation in *Homarus americanus* and Sandifer *et al.*, (1984) in *Penaeus* sp. The gill irrigator of Tave and Brown (1981) ensured the continuous supply of water to branchial cavity during extrusion, so that the stress of animal due to electric stimulation could be reduced. Many investigators have employed similar method of spermatophore extrusion after Sandifer and Lynn, (1980); for the acquisition of spermatophores for artificial insemination, evaluation of reproductive quality of males and also for sperm preservation experiments in various decapod crustaceans. Various attempts on artificial insemination by different researchers like Sandifer and Lynn, (1980) in *M. rosenbergii*; Bray *et al.* (1983) in *P. setiferus*; Kooda-cisco and Talbot, (1983) in *Homarus americanus*; Talbot *et al.*, (1986) in *H. orientalis*; Joshi and Diwan, (1992) in *M. idella*; Perez, *et al.*, (2001) in *L. vannamei* and Coman *et al.*, (2006) in *P. monodon* showed that the discovery of electro-ejaculation technique to extrude spermatophores in males has simplified the artificial insemination to a greater extent.

Spermatophores are extruded for assessing the reproductive quality of males in various decapods like lobsters (Talbot *et al.*, 1983); Penaeids like *Penaeus vannamei* (Leung- Trujillo and Lawrence, 1985, Alfaro and Lozano, 1993, Heitzmann, *et al.*, 1993) *P. setiferus* (Bray *et al.*, 1983); *P. setiferus, P. vannamei* and *P. stylirostris* (Leung- Trujillo and Lawrence, 1991); *P. monodon* (Pratoomchat *et al.*, 1993): *P. monodon* (Gomes and Primavera, 1994); *P. stylirostris* (Alfaro,

1993) and *P. setiferus* and *P. vannamei* (Rosas *et al.*, 1993) and in other shrimps like *M. rosenbergii* (Harris and Sandifer, 1986) by electro-ejaculation techniques.

There has been little published work on the time required for spermatophore development in various decapods. Lin and Ting (1984) reported requirement of 7 to 11 days for spermatophore generation in *P. monodon* after applying various artificial spermatophore extraction methods. However, the authors do not comment whether the extraction techniques used, affected the spermatophore development time as well as its quality. In *P. monodon* nearly 60% of the electro-stimulated individuals displayed spermatophore regeneration in an average of 10 days, however it took 11 to 12 days in the rest of the 40% shrimps. Leung-Trujillo and Lawrence (1991) recorded a period for full regeneration of spermatophore from 5 to 7 days in *P. setiferus*. Whereas Roasa *et al.*, (1993) reported that 20% of the shrimps displayed spermatophore regeneration in 3 days after electro ejaculation, even though 80% of the shrimps do the same only in 7 days. They further reported that the slight discrepancy between the spermatophore regeneration time recorded by them and those given by Leung-Trujillo and Lawrence (1991) for *P. setiferus* might be due to the population differences. Leung-Trujillo and Lawrence (1985) first reported the evaluation of reproductive quality in male penaeid shrimps; who determined spermatophore weight, sperm count, and percentages of live and abnormal sperm for evaluating the effect of eyestalk ablation in *P. vannamei*. Using the same parameters, Leung-Trujillo and Lawrence, (1987a) reported a decline in sperm quality in *P. setiferus* held in captivity. Bray *et al.*, (1983) used the same approach to evaluate the effect of water temperature, EDTA and vibrio bacterin on sperm quality in captive *P. setiferus*. Alfaro, (1993) followed the same method for reproductive quality evaluation of male *P. stylirostris*.

Rapid deterioration of the spermatophore weight and sperm quality is observed in the regenerated spermatophores of *P. monodon* after electro ejaculations during the 50 days (Diwan and Shoji, 1998). Leung-Trujillo and Lawrence, (1987) and Talbot *et al.* (1988 & 1989) in *P. setiferus* reported similar trends during 35 days of captivity. This deterioration is attributed to degeneration of the reproductive tract in males, associated with bacterial infection (Talbot *et al*, 1988). Melanization has been reported as a problem for *P. stylirostris* (Alfaro, 1993) whereas Chamberlain *et al.*, (1983) found a higher incidence of melanization in manually ejaculated shrimp. Harris and Sandifer (1986) reported that although electro- ejaculation procedure permitted repeated spermatophore expulsion, the frequent application of electrical stimulation induced melanization of the reproductive tissues in *M. rosenbergii*. Alfaro (1993) reported an increase in sperm count and reduction in abnormalities in regenerated spermatophores in *P. stylirostris*. However, further he mentioned that melanization has not been observed in any of the experimental shrimps. In *P. monodon* melanization is found only in males where repeated electro-stimulation has been given for the ejaculation (Diwan and Shoji, 1998).

Eyestalk ablation has been used to induce precocious spermatophore production in sexually immature male penaeids (Alikunhi *et al.*, 1975). Eyestalk ablation has also induced increase in gonad size in *P. setiferus* (Lawrence *et al.*, 1979) and mating frequency in *P. Vannamei* (Chamberlain and Lawrence, 1981 b). However, in *P. monodon* it is found that eyestalk ablation considerably reduced the spermatophore regeneration time from 9 to 12 days (non-ablated) to 6 to 8 days in ablated males (Diwan and Shoji, 1998). Leung-Trujillo and Lawrence, (1991) reported that spermatophore production seem to be controlled by the X-organ sinus gland complex, as ablated *P. setiferus* produced new spermatophore almost twice as fast as unablated males.

In *P. monodon* it is reported that unilateral eyestalk ablation improved sperm quantity and sperm quality (larger spermatophore, increased sperm count, less abnormal sperm) (Shoji, 1998). These findings are similar to the findings of Leung-Trujillo and Lawrence, (1985) who observed that ablation increased sperm count without affecting sperm quality in *P. setiferus*. Similarly, Gomes and Primavera (1993) reported that unilateral eyestalk ablation improved the sperm quality in terms of less abnormal sperm, bigger head and longer spike in *P. monodon*. In contrast to this, Pratoomchat *et al.*, (1993) reported that eyestalk ablation do not increase the spermatophore size or sperm quality, although re-ejaculation significantly increased mortality of ablated males in *P. monodon* (Diwan and Shoji, 1998).

The percentage of viable sperm is determined through the acrosome reaction described by Griffin *et al.*, (1988). Functionally *P. monodon* sperm can be divided into three regions; Nucleus, the acrosomal vesicle and the sub-acrosome. Griffin *et al.* (1988) in *S. injentis* and in by Mohammed and Diwan (1994) in *P. indicus* have made similar observations. Kleve *et al.* (1980) reported that the acrosomal vesicle is a complex membrane bound entity, that encircles the central cap region including the anterior spike and the sub-acrosomal components are found within the core of the central cap region. The acrosome reaction in *P. monodon* is composed of both the exocytosis of the acrosomal vesicle and the generation of the acrosomal filament. Induction of the two phases of the AR in *P. monodon* is temporally separated and sequential in *in-vivo* and *in-vitro*. *In-vivo* upon binding to the ova, sperm undergo acrosomal exocytosis and 20 to 30 minutes later only, the formation of the acrosomal filament takes place (Diwan and Shoji, 1998). Clark *et al.* (1984) in *S. ingentis* described similar observations. Griffin *et al.* (1988) reported that, the two phases of AR in sperm of *S. injentis*, removed from female seminal receptacle are also temporally separated and sequential.

The viability assays of the sperm are made through the acrosome reaction as described by Clark *et al.*, (1981). Three different chemical compounds viz. Bromocalcium ionophore A 23187, Valinomycin and Nigricin are found inducing the *in-vitro* acrosome reaction in the sperm of *P. monodon*, in addition to the biological solution *viz.* egg water collected according to the method and described by Griffin *et al.*, (1987). Among the different above-mentioned inducers "Egg water" gave the best result followed by ionophore A 23187 as in *in-vivo* conditions during natural spawning. The structural details are found similar to those observed in the sperm of *P. monodon* during the AR, closely resembling with that of the previous observations made in the natantians by different investigators like Clark *et al.* (1984) and Anchordouguy *et al.* 1988 in *S. ingentis*. The ultrastructure of the acrosome reaction of natantian sperm during the sperm-egg interactions is described in sperm of *P. monodon* (Yudin *et al.*, 1979). Clark *et al.*, (1981) reported that, the acrosome reaction of the natantian's sperm could be induced by the ionophore A 23187 and egg-water respectively to the same extent as *in vivo* conditions. Clark *et al.*, (1981) summarized the results of various known acrosome inducers *S. injentis* sperm and found that most of the artificial medium used by earlier researchers as acrosome reaction inducers are ineffective in *S. injentis* sperm, and mentioned further that the divalent cation A 23187 is the only substance that induced the acrosome reaction among the different solutions tested. Griffin *et al.*, (1987) reported that, egg water could induce the complete acrosome reaction including both acrosomal exocytosis and formation of the acrosomal filament equally to that of the natural condition. Many investigators achieved *in vitro* induction of acrosomal filament in the natantian sperm with egg water (Griffin *et al.*, 1988; Griffin and Clark, 1990; Anchordouguy *et al*, 1988 and Clark and Griffin 1993 and in *P. monodon* by Pratoomchat *et al.*, 1993).

Cryopreservtion of gametes is widely practiced in animal husbandry and controlled breeding programmes. However, there have been very few reports on the cryopreservation of the crustacean sperm. The temperatures at which the sperm samples are stored, as well as the cryoprotectants used for the dilution of sperm, have definite and significant roles on the achievement of viability of preserved sperm in *P. monodon*. The screening trials carried out for short periods using different cryoprotectants at 3 different temperatures showed wide variation in sperm viability. It is found that –196°C is the best temperature for the storage of intact viable sperm for long periods than the other two temperatures of –30°C and 0°C. In *P. monodon*, the viability of the sperm preserved at –196°C varied from 73% to 20% in samples preserved in different cryoprotectants. The decrease in viability is less in certain cryoprotectants and higher in others. Leung and Lawrence(1991) reported that the storage temperature should be –130°C or below for long term preservations of live cells because the glass transformation temperature for pure water is found to be –130°C. Chow *et al.*, (1985) reported successful preservation of spermatophores of shrimps in glycerol at –196°C. Anchordouguy *et al.*, (1988) reported that, sperm samples of *S. injentis* stores at –196°C for 1 month showed no decrease in viability upon thawing. In *Scylla serrata* the viability of the sperm in 30 days of storage is fairly high ie. 95% in samples stored at 195°C (Jayalectumie and Subramoniam 1989).

In *S. injentis*, Anchordouguy *et al.*, (1988) tested different cooling rates and it is reported that a cooling rate of 1°C/min. is the best of the rates tested. In *P. monodon* is observed that the viability of the preserved sperm cells vary greatly in different cryoprotectants even at –196°C. The decrease in viability was higher in other two temperatures tried (–30°C and 0°C) for all the cryoprotectants used. Temperature is a critical factor in cryopreservation. This is indicated by the wide variation in the percentage viability i.e. from 78% to 20% at the same temperature of –196°C. Trehalose offered the least protection of the compounds tested in *P. monodon*. Glycerol has been first reported as effective in protecting sperm from freeze thaw damage (Rostand 1946 and Polge *et al.*, 1949). Polge *et al.*, (1949) first showed that, the spermatozoa could be frozen and thawed without losing motility if glycerol is introduced in their suspending medium. Glycerol has been used widely in the cryopreservation of viable fish milt (Horton and Ott, 1976). Similarly, glycerol has been used as a cryoprotectant in most of the sperm preservation experiments in invertebrates. Successful preservation of spermatophores of shrimp in glycerol at –196°C has also been reported by Chow *et al.*, 1985. Similar results are obtained for Limulus spermatozoa (Behlmer and Brown, 1984). When DMSO is used for the spermatophore preservation in *M. rosenbergii* the post thawing survival is nil (Chow *et al.*, 1985). While on contrary, Anchordouguy *et al.*, (1988) reported that sperm survival is higher in *S. injentis* when glycerol is used as cryoprotectant for the preservation of sperm even though DMSO gave the highest percentage of survival. In *Scylla serrata* glycerol and DMSO + trehalose gave highest sperm survival (Jayalectumie and Subramoniam, 1989).

In *P. monodon* it is found that, the percentage viability is higher in samples preserved in 5% DMSO even after 6 months. The maximum effect is in 5% DMSO; higher and lower concentrations of this yielded inconsistent results. Similarly higher percentage viability is obtained when the samples are preserved for 6 months in DMSO (5%) + trehalose (0.25M). Anchordouguy *et al.*, (1988) reported that the highest percentage viability is obtained from samples preserved in 5% DMSO. However, it is reported in the same animal that, combining 5% DMSO with other cryoprotectants like trehalose, proline, sucrose and glycerol in the freezing medium do not alter the sperm viability in the marine shrimp *S.*

injentis. Whereas in *S. serrata* DMSO + trehalose gave high sperm survival but trehalose is not an efficient cryoprotectant when it is used alone (Jayalectumie and Subramoniam 1989). Similarly trehalose offered the least protection in the preservation of *S. injentis* sperm. During the present investigation also the lowest sperm survival is found in samples preserved in trehalose.

6.6 CRYOPRESERVATION OF CRUSTACEAN EGGS, EMBRYOS AND NAUPLII

The science of preservation is quite new to the Aquaculture industry. It commenced with an attempt to cryopreserve sperm mainly with the concept of "gene banks". Methods to cryopreserve aquatic animals are less well developed (Stoss and Holtz 1983, Alderson and Mc Neil 1984). Although sperm cryopreservation has been done successfully in a number of commercially important aquatic animals, particularly teleost fishes (Muir and Roberts 1993, and also shellfishes (Subramoniam 1993, Diwan *et al.,* 1994), the technology has not yet reached an advanced level suitable for commercial application.

Unlike those involving spermatozoa, attempts to cryopreserve fish eggs and embryos have been unsuccessful (Hortan and Ott 1976, Erdahl and Graham 1980). The fundamental problems of sufficient dehydration during cooling due to the relative large size of fish eggs and the [presence of membranes of different water permeability (Loeffler and Lovtrup 1970) have not yet been overcome. Studies on salmonids by Harvey and Ashwood-Smith (1982) have shown the permeation of cryoprotectants to be low and dependent on their molecular size. Zell (1978) and Erdahl and Graham (1980) have reported preliminary attempts to freeze the eggs of rainbow trout (*salmo gairdneri*) and concluded that eggs protected with 8-14% DMSO and frozen to -20% have proven viable and further Zell (1978) has also reported the survival of zygotes and eyed embryos following cooling to -55°C. Jensen and Alderdice (1983) studied the changes in mechanical shock sensitivity of coho sensitivity salmon eggs during incubation. Short term storage of *Sarotherodon mossambicus* ova has been reported by Harvey and Kelly (1984) where they used unfertilized ova and stored them in coelomic fluid in closed humidified containers. Optimal temperature for storage was found to be 20°C for 19 hours with 35% ferlitily and they have mentioned that the post storage fertility was increased to 55% by oxygenation of sample containers and the same declined after $1^{1/2}$ hours exposure to temperatures below 18-20°C. Studies have been conducted to cryopreserve embryos of the Japanese medaka (Arri *et al.,* 1987) where it was mentioned that DMSO qualifies as a good cryoprotectant. Survival rates after freezing have been reported to be greatly affected by the concentration of DMSO, duration of treatment the final temperatures of freezing and the developmental stages of the embryos.

Attempts to cryopreserve eggs and embryos of the carp *Labeo rohita* (Ponnaih *et al.*, 1989) were described as unsuccessful. Toxicity tests were carried out with DMSO and methanol, and methanol has been mentioned to be least toxic with a hatching rate of 18% but post-thaw survival has been reported to be nil. In another attempt to cryopreserve the embryos of common carp and rohu, John *et al.*, (1993) have reported that the tail bud stage embryo of the common carp showed more tolerance to cryoprotectants than the morula stage. Nilsson and Cloud (1992) attempted to cryopreserve isolated blastomeres from rainbow trout blastulae. Individual blastomeres were frozen in a medium containing 12% DMSO. Three rates of coiling (3, 1 and 0.3°C/min) through the critical period were compared using a Cryomedi controlled Rate Freezer. All cells were thawed at the same rate in a 4°C water bath for 3 minutes. Blastomeres frozen at the highest cooling rate (3°C/min) had the lowest percentage (6.6%) of viable cells upon thawing. In contrast, the blastomeres frozen at other two coiling rates had been observed to have 31 and 51% viability respectively.

Lui *et al.*, (1992) attempted to cryopreserve goldfish embryos by passing fertilized eggs between optical vesicle and heart formation through a series of DMSO dilutions, viz. 5%, 8%, 10% and 12%, for 5 minutes in each dilution for equilibrium. The embryos were frozen at -15°C and -25°C for 60 minutes. Survival rates were 20% at -15°C and 5% at -25°C respectively. The hatching rates reported ere 15% and 3% for -15°C and -25°C respectively. The first successful attempt at cryopreservation of sea urchin embryos was made by Ahasina and Takahashi (1978). It was found that late embryos of sea urchin survived freeing, at least for a short period of time, at -196°C in the presence of cryoprotectant. The freezing tolerance was reported to be greater in advanced developmental stages. Both ethylene glycol and DMSO exerted significant protection.

Clam and oyster eggs were cryopreserved (Utting *et al.*, 1990) to -196°C and thawed after 30-60 min. and their viability and growth compared with non-frozen control larvae from the same parents. Growth of Manila clam larvae was monitored for 35 days and there was no significant difference between the shell length of the frozen and control larvae. Survival of frozen has been reported to be high, at 84% of the control. In an early trial, Pacific oyster larvae, which had been frozen, were reared successfully to settlement, but survival was reported to be poor. However, after modification of the freezing process, survival of larvae was much higher. Cryopreservation of eggs and embryos of the oyster, *Crassostrea gigas*, by Rana *et al.* (1992) proved to be unsuccessful. Varying degrees of success, however, have been reported using early trocophore (7h) larvae. In all trials, DMSO at concentrations of 1.0 to 1.5M gave consistently better thaw viability than glycerol. At least, 40-50% of the larvae were recovered after thawing. The post-thaw viability was found to increase by 15% by using larger embryos.

Lubzens *et al.* (1992), while attempting cryopreservation of rotifers, proposed that the success in cryopreservation was influenced by factors like the type and concentration of cryoprotectants, rate of cooling, equilibration time, dilution rate following thawing, and the age of the eggs. Survival rates of cryopreserved amictic eggs at -196°C ranged between 5 and 50%.

Taylor (1990) froze *Penaeus vannamei* nauplii to -30°C, which then recovered with successful metamorphosis to the protozoeal stage. Cryopreservation of cirripede naupii was attempted by Kurokura *et al.* (1992) who examined the prospects of this technique for other crustaceans like prawns and crabs. The larvae were transferred to seawater containing 2M DMSO and 0.1M saccharose as cryoprotectants and frozen to -196°C. However, the nauplii failed to survive.

Preston and Coman (1998) studied the effects of cryoprotectants, chilling and freezing on the embryos and nauplii of shrimp, *P. esculentus*. It was concluded that there are several barriers to the successful cryopreservation of penaeid shrimp embryos. The embryos are found to be very sensitive to chilling and osmotic change and its tolerance to -1<"C was seldom more than 20 minutes. Rapid exposure to hyperosmotic or hypo-osmotic conditions has been reported to be lethal to the developing embryos. Tolerance to cryoprotectants varied according to the molecular weight and concentration of the cryoprotectants.

A breakthrough in the field of cryopreservation of shrimp embryos and nauplii has been made by Subramoniam (1993, 1994). For the first time, the nauplii of *P. indicus* were frozen to -30°C and -40°C using liquid nitrogen vapor. The high percentage survival, 82%, of nauplii frozen to -30°C and 63% of nauplii frozen to -40°C using 15% v/v ethylene glycol suggests that rapid thawing and slow dilution can be incorporated in the protocol for preservation of penaeid prawn nauplii. However, initial attempts to vitrify embryos were not successful (Subramoniam and Newton 1993). Simon *et al.* (1994) attempted to determine a suitable freezing medium for cryopreservation of *P. indicus* embryos. Among the five developmental stages studied, gastrulae and 5-hr embryos were found to be resistant to a prolonged contact with cryoprotectants. Of the 11 cryoprotectants used, the combination of methanol with ethylene glycol or DMSO gave hatching rates equal to those of the control. Attempts to cryopreserve embryos and nauplii of the marine shrimp, *P. semisulcatus* (Diwan and Kandasami 1997), have been reported to be successful. It was mentioned that when DMSO and Glycerol prepared were used independently as cryoprotectants, the viability of freeze- thaw embryos was nil. But 40 to 50% of the embryos hatched out successfully to nauplii when preserved in mixtures of DMSO and glycerol prepared in grades of 5 to 20%. Similar success has been reported by the same authors with regard to cryopreservation of nauplii. Screening of stages of embryonic development for resistance to chilling temperatures (+7°C and 0°C) by McLellan *et al.* (1992) indicated that the fifth nauplius (N_5) and first zoea stage of *P. vannamei* are able to withstand chilling stress.

In spite of the foregoing, cryopreservation of eggs and embryos of aquatic animals has really not received an appreciable amount of attention. Attempts to cryopreserve penaeid prawn embryos, so far, have not been successful although standard protocols for several mammalian embryos are available. The limited success achieved in the cryopreservation of viable eggs, embryo and larvae of higher animals was attributed to the large size of the eggs and embryos which interferes in the penetration of cryoprotectants and uniform cooling during the cryopreservation process. Sometimes, the large volume of yolk present in the eggs and embryos tends to develop crystals while freezing and damages the internal parts (Seymour 1994). In shrimp, though the size of their eggs and embryos is small, the eggs have the tendency to absorb water soon after their release, swell and become activated for fertilization. After fertilization, a strong hatching envelope forms around the egg (Clark and Griffin 1993). Therefore, the presence of water and the thick protective envelope surrounding the eggs are some of the disadvantages for successful freezing of viable eggs and embryos (Simon *et al.*, 1994; Diwan, 2000).

6.7 FUTURE RESEARCH AND APPLICATIONS

For developing a successful marine shrimp industry through Aquaculture technology, one of the major constraints is non-availability of sufficient seed and spawners to produce seed at the desired time. Even in the event of the availability of spawners, their maintenance and management become difficult and expensive. Therefore, to ease this problem there is an urgent need to devise a suitable technology for cryopreservation and cryobanking of viable gametes (sperm and eggs) so that production of shrimp can be made sustainable as per the need.

Cryopreservation of gametes of aquatic animals in contrast to the situation in terrestrial vertebrates, particularly mammals, has met with very limited success. Sperm cryopreservation has been successful with a number of commercially important aquatic species, particularly some teleost fishes. However, the reproductive success with cryopreserved sperm in general is still poor and the technology involved requires further refinement. Sperm preservation in aquatic animals is not at the stage of advanced commercial application as seen with domestic mammals. This may partly due to the problems related to the need for a relatively large volume of sperm to fertilize the larger number of eggs produced by aquatic animals.

In gamete preservation, eggs are fundamentally more difficult to freeze successfully than sperm. The main reason is that due to the large size of eggs there will be some interference with penetration of cryoprotectants and uniform cooling during the cryopreservation process. Sometimes eggs with a large yolk sac tend to develop crystals which damage the egg as it freezes. It has been also stated that the chromosomes in the egg are more vulnerable to damage those in sperm. Also the loss of membrane integrity both in sperm and eggs is a critical damaging factor incurred during the freeze/thaw process. Recent evidence has shown that certain key enzymes in the cells get altered/broken down on freezing. Very few attempts have been made on cryopreservation of sperm in decapods crustaceans in general and marine shrimp in particular.

Further efforts are very much needed particularly for improving this technology not only for cryopreservation of spermatozoa, but also eggs, embryos and larvae of economically important shellfish. If proper breakthroughs are made in this sector, then Aquaculture would acquire a prestigious status.

References

Ahasina, E and Takahashi T, 1978. Freezing tolerance in embryos and spermatozoa of the sea urchin. *Cryobiology* 15: 122-14.

Aiken, D.E., S.L Waddy, K Moneland, and Polar S.M, 1984. Electrically induced ejaculation and artificial insemination of the American lobster *Homarus americanus*. *J. Crust. Biol.*4: 519-527.

Alderson, R and Mc Neil A.J 1984. Preliminary investigations of cryopreservation of milt of Atlantic salmon (*Salmo salar*) and its application to commercial farming. *Aquaculture* 43: 351-354.

Alfaro, J. 1993. Reproductive quality evaluation of male *Penaeus stylirostris* from a grow out pond., *J. World. Aquaculture. Soc.* 24: 6-11.

Alfaro J.M., A.L. Lawrence, D. Lewis 1993. Interaction of bacteria and male reproductive system blackening disease of captive *Penaeus setiferus*. *Aquaculture*, 117.

Alfaro, J. and Lozano X. 1993. Development and deterioration of spermatophores in pond reared *Penaeus vannamei*. *J. World Aquacult. Soc.* 24: 522-528.

Alfaro J.M. Alfaro, N. Muñoz, M. Vargas, J. Komen 2003. Induction of sperm activation in open and closed thelycum penaeoid shrimps. *Aquaculture*, 216, pp. 371–381.

Alfaro J.M, K. Ulate, M. 2007. Vargas Sperm maturation and capacitation in the open thelycum shrimp Litopenaeus (Crustacea: Decapoda: Penaeoidea) *Aquaculture*, 270, 436–442.

Alikunhi, K.H., A Poernomo, S. Adisukesno, M. Budiono and S. Busman 1975. Preliminary observations on induction of maturity and spawning in *P. monodon* Fabricius and *P. merguiensis* de Man by eyestalk extirpation. *Bull. Shrimp. Cult. Res. Cen.*, 1: 1-11.

Anchordoguy, T, J.H. Crow, W.H. Clark, Jr., and F.J. Griffin. 1987. Cryopreservation of sperm from the penaeid shrimp *Sicyonia ingentis.* Abstract 18th Ann. Meet. World. *Aquacult. Soc. Gvayaquil, Equador.*

Anchordoguy, T., J.H. Crowe, F.j. Griffin and W.H. Clark, Jr., 1988. Cryopreservation of sperm from the marine shrimp *Sicyonia ingentis. Cryobiology,* 25: 238-243.

Anderson, S.L., E.S. Chang, and W.H. Clark Jr. 1984. Timing post vitellogenic ovarian changes in the ridgeback prawn *Sicyonia ingentis* determined by ovarian biopsy. *Aquaculture* 42: 257-271.

André Braga , Luiz A. Suita de Castro, Luís H. Poersch, Wilson Wasielesky 2014. Spermatozoal capacitation of pink shrimp *Farfantepenaeus paulensis. Aquaculture* 430.

André Braga, C.L. Nakayama, L. Suita de Castro, W. Wasielesky, 2013. Spermatozoa ultrast ructure of the pink shrimp Far fantepenaeus paulensis (Decapoda: Dendrobranchiata) *Acta Zool* , 94, 119–124.

Arii N., K. Namai, Gomi, and T. Nakazawa. 1987. Cryoprotection of medaka embryos during development . *Zool. Sci.* 4: 813-818.

Aungsuchawan S., C.L. Browdy, B. Withyachumnarnkul 2011. Sperm capacitation of the shrimp *Litopenaeus vannamei.* Aquaculture. *Research.*, 42, 188–195.

Austin, C.R. 1968. *Ultrastructure of Fertilization*, Holt, Rinehart and Winston, New York.

Barros, C., E. dupre, and L. viveros. 1986. Sperm-egg interaction in the shrimp *Rhynchocinetes typus. Gamete Res.* 14: 171-180.

Behlmer, S.D and G. Brown. 1984. Viability of cryopreserved spermatozoa of the crab *Limulus polyphemus. Int. J. Invert. Rep. Dev.* 7: 193-199.

Bougrier, S. and L.D. Rabenomana. 1986. Cryopreservation of spermatozoa of the Japanese oyster, *Crassostrea virginica. Aquaculture,* 58: 277-280.

Bray, W.A., G.W. Chamberlain, and A.L. Lawrence. 1983. Increased larval production of *Penaeus setiferus* by artificial insemination during sourcing cruises. *J. World. Maricult. Soc.* 16: 250-257.

Brown, G.G. 1966. Ultrastructural studies of sperm morphology and sperm-egg interaction in decapods *Callinectes sapidus. J. Ultrastruc. Res.* 14: 425-440.

Burkenroad, M.D. 1934. The Penariodea of Lousiana with a discussion on their world relationships. *Bull. Am. Mar. Nat. Hist.* 68: 68-143.

Calman, W.T. 1909. Crustacean. In: *A Treatise on Zoology*, Vol. 8. E.R. Lankester, ed., Adam and Charles Black, London.

Ceballos-Vázquez B.P, C. Rosas, I.S. Racotta, 2003. Sperm quality in relation to age and weight of white shrimp *Litopenaeus vannamei. Aquaculture*, 228 , pp. 141–151.

Cherr, G.N and W.H. Clark Jr. 1985. An egg envelope component induces the acrosome reaction in sturgeon sperm. *J. Exp. Zool.* 234: 75-85.

Cheung, T.S. 1966. The development of the egg membrane and egg attachment in the shore crab *Carcinus maenas*, and some related decapods. *J. Mar. Biol. Assoc. U.K.* 46: 373-400.

Chamberlain, G.W., S.K. Johnson and D.L. Lewis 1983. Swelling and melanization of the male reproductive system of captive adult penaeid shrimp. *J. World. Maricult. Soc.*, 14: 135-136.

Chamberlain, G.W. and A.L. Lawrence 1981a. Maturation, Reproduction and Growth of *Penaeus vannamei* and *P. stylirostris* fed natural diets. *J. World. Maricult. Soc.*, 12: 209 224.

Chow, S. 1982. Artificial insemination using preserved spermatophores in the palaemonid shrimp, *Macrobrachium rosenbergii, Bull. Japan Soc. Scientific Fisheries*, 48:L 177-183.

Chow, S., T. TAki, and T. Ogassawara. 1985. Cryopreservation of spermatophore of the freshwater shrimp *Macrobrachium rosenbergii. Biol. Bull.* 168: 471-475.

Clark, W.H., Jr. and F.J. Griffin. 1988. The morphology and physiology of the acrosome reaction in the sperm of the decapods, *Sicyonia ingentis. Dev. Growth Differ.* 30: 451-462.

Clark, W.H., Jr. and F.J. Griffin. 1993. Acquisition and manipulation of penaeiodean gametes. Pages 133-151 In: CRC *Handbook of Mariculture*, 2nd Edition Vol.1, J.P. McVey,ed., CRC Press, Boca Raton, Forida.

Clark, W.H., Jr., P. HO, and A.I. Y Udin. 1974. Amicroscopic analysis of the cortical specialization and their activation in the egg of the prawn, *Penaeus sp. Am. Zool.* 14: 45.

Clark, W.H., Jr. and J.W. Lynn. 1977. A Mg^{++} dependant cortical rection in the eggs of penaeid shrimp. *J. Exp. Zool.* 200: 177-183.

Clark, W.H., Jr., J.W. Lynn, A.I. Yudin, and P. HO. 1980. Morphology of the cortical reaction in the eggs of penaeid shrimp. *Biol. Bull.* 158: 175-186.

Clark, W.H., Jr., M.G. Kleve and A.I., Yudin. 1981. An acrosome reaction in natantian sperm. *J. Exp. Zool.* 218: 279-291.

Clark, W.H., Jr., P. Talbot, R.A. Neal, C.R. Mock, and B.R. Salser. 1973. *In vitro* fertilization with non-motile spermatozoa of the brown shrimp *Penaeus aztecus. Mar. Biol.* 22: 353-354.

Coman G.J., S.J. Arnold, S. Peixoto, P.J. Crocos, F.E. Coman and N.P. Preston 2006

Reproductive performance of reciprocally crossed wild-caught and tank-reared *Penaeus monodon* broodstock Aquaculture, 252: 372–384.

Dan, J.C. 1952. Studies on the acrosome reaction 1. Reaction to egg water and other stimuli. *Biol. Bull.* 107: 54-66.

Dan, J.C. 1967. Acrosome reaction and lysins. Pages 237-293 in: *Fertilization,* Vol. I, G.B. Metz, and A. Monray, eds., Academic Press, New York.

Diwan, A.D. and S. Joseph. 1994. Induction of gamete activity in the Indian penaeid prawns. *Mar. Fish. Infor. Serv. T & E Ser.,* 133: 15-19.

Diwan, A.D. and S. Joseph. 1998. Cryopreservation of spermatophores of the marine shrimp *Penaeus indicus* H. Milne Edwards. *Fish Gen. Biodiversity Conserve. Nacton Pub.* 5: 243-252.

Diwan, A.D., S. Joseph, and A. Nandkumar. 1994. Cryobanking potentials of marine shrimp gametes. *Mar. Fish. Infor. Serv. T & E Ser.,* 131: 16-18.

Diwan, A.D. and K. Kandansami. 1997. Freezing of viable embryos and larvae of marine shrimp, *Penaeus semisulcatus de Haan. Aquaculture Res.* 28: 947-950.

Diwan, A.D. 2000. Shrimp embryo cryobanking is now possible. *Current Science,* 78(11): 1282-1284

Drouslet, J.J., A.I.Yudin, R.S. Wheeler, and W.H. Clark, Jr. 1975. Light and fine structural studies of natural and artificially induced egg growth of penaeid shrimp. *6th Ann. Meet World Mariculture Society,* Seattle.

Dudenhausen, E.E and P. Talbot. 1983. An ultrastructural comparison of soft and hard spermatophores from the crayfish *Pacifastacus leniuscula* Danal. *Can J. Zool.* 61: 182-194.

Dunn, R.S. and J. Mc Lachlin. 1973. Cryopreservation of echinoderm sperm. *Can. J. Zool.* 51: 666-669.

Ellington W.R.2001. Evolution and physiological roles of phosphagen systems Annual Review of Physiology, 63 (2001), pp. 289–325.

Erdahl, D.A. and E.F. Graham. 1980. Preservation of gametes of freshwater fish. pages 317-326. In: *Proceedings of the 9th international Conference on Animal Reproduction and Artificial Insemination,* Madrid, RT-H2.

Fasten, N. 1921. The explosion of the spermatozoa of zrab, *Lophopanopeus bellus* (Stimpson) Rathbun, *Biol, Bull.* 41: 288-301.

Felgenhauer, B.E., and L.G. Abele. 1991. Morphological diversity of decapods spermatozoa. Pages 322-341 In: *Crustacean Sexual Biology,* R.T. Bauer, and J.W. Martin,eds., *Columbia University* press, New York.Gomes, L.A.O. and J.H.

Gomes, L.A.O. and J.H. Primavera 1994. Reproductive quality of male *Penaeus monodon. Aquaculture* 112: 157-164.

Griffin, F.J., and W.H. Clark, Jr. 1988. Formation and structure of the acrosomal filament in the sperm of *Sicyonia ingentis. J. Exp. Zool.* 246: 94-102.

Griffin, F.J.,W.H. Clark, Jr. 1990. Induction of acrosomal filament formation in the sperm of *Sicyonia ingentis. J. Exp. Zool.* 254: 296-304.

Griffin, F.J.,W.H. Clark, Jr., J.H. Crowe, and L.H. Crowe. 1987. Intracellular pH decreases duringthe *in vitro* induction of the acrosome reaction in the sperm of *Sicyonia ingentis. Biol. Bull.* 173: 311-323.

Haley, S.R. 1984. Spermatogenesis and spermatophore production in the Hawaiian red lobster *Enoplometopus occidentalis* (Randall) (Crustacea, Nephropidae). *J. Morphol.*, 180: 181-193.

Harris, S.E.G. and P.A. Sandifer. 1986. Sperm production and the effects of electrically induced spermatophore expulsion in the prawn *Macrobrachium rosenbergii. J. Crust. Biol.* 6: 633-647.

Harvey, B. and M.J. Ashwood-Smith. 1982. Cryopreservation penetration and supercooling in the eggs of salmonid fishes. *Cryobiology*. 19: 29-40.

Harvey, B and R.N. Kelly. 1984. Short term storage of *Sarotherodon mossambicus ova, Aquaculture.* 37: 391-395.

Herrick, F.H. 1909. Natural history of the American lobster. *Bull. U.S. Bur. Fish*. 229: 749-908.

Hino, A., and K.H. Kato. 1990. Morphological changes in sperm at fertilization. *Proceedings of the 61st Meeting of the Zoological Society of Japan*, 7: 1074.

Hinsch, G.W.1969. Microtubules in sperm of the spider crab, *Libinia emarginata* L. *J. Utlrastruct. Res*. 29: 525-534.

Hinsch, G.W.1971. penetration of oocytes envelope by spermatozoa in the spider crab. *J. Ultrastruct. Res*. 35: 86-97.

Hinsch, G.W. 1973. Sperm structure of Oxyryncha. *Canadian Journal of Zoology*, 51: 421-426.

Hinsch, G.W.1986. A comparison of sperm morphogenesis, transfer and sperm mass storage between two species of crabs, *Ovalipes oculatus* and *Libinia emarginata. Internatioanal J. Invert. Reprod. Develop*. 10: 79-87.

Hinsch, G.W. and M.H. Walker. 1974. The vas deferens of the spider crab, *Libinia emarginata. J. Morphol.* 143: 1-19.

Horton, H.F. and A.G. Ott. 1976. Cryopreservation of fish spermatozoa and ova. *J. Fish. Res. Bd. Can.* 33: 995-1000.

Hosakaw, K. and Y,D. Noida. 1992. Acrosome reaction and sperm penetration in the bivalve, *Laternula limnicola. Zool. Sci.*, 9: 1179.

Ishida, T., P. Talbot, and M. Kooda-Cisco. 1986. Technique for long term storage of Lobster (*Homarus*) spermatophores. *Gametes Res*. 14: 183- 195.

Jamieson, B.G.M. 1989a. Ultrastructural of the spermatozoa of *Ramna ramna* (Oxystomata) and of other exemplified by *Portunus pelagicus* (Brachygnatha). *Zoomorphology* 109: 102-111.

Jamieson, B.G.M. 1989b. the ultrastructure of spermatozoa of four species of xanthid crabs (Crustacea, Brachyura, Dromidacea). J. *Submicroscopic Cytol. Pathol*. 21: 579-584.

Jamieson, B.G.M. 1990. The ultrastructure of the spermatozoa of *Petalomna lateralis* Grey (Crustacea, Brachyura, Dromidacea) and its phylogentic significance, *Internat.J. Invert. Reprod. Develop*. 7: 39-45.

Jensen, J.O.T., and Alderdice. 1983. Changes in mechanical shock sensitivity of coho salmon eggs during incubation. *Aquaculture* 32: 303-312.

Jeyalectumie, C. and T. Subramoniam. 1989. Cryopreservation of spermatophores of thr edible crab, *Scylla serrata. Biol. Bull.* 177: 247-253.

John, G., A.G. Ponniah, W.S. Lakra, A. Gopalkrishna, and A. Barat. 1993. Prelimaniry attempts to cryopreserve embryos of *Cryprinus carpio* (L) and *Labeo rohita (*Hans). *J. Inland Fish . Soc . India* 25: 55-57.

Joseph, S. 1997. Some studies in the reproductive endocrinology of tiger prawn *Penaeus monodon* Fabricius Ph.D. thesis: Cochin University of Science and Technology, Cochin.pp. 291.

Joshi, V.P. and A.D. Diwan. 1992. Artificial insemination studies in *Macrobrachium idella* (Hilgendorf, 1898). Proceedings of Freshwater Prawns. (Silas, E.G. ed.) Kerala Agricultural University Publ., Trichur, 110-118.

Kleve, M.G., A.I. Yudin, and W.H. Clark Jr. 1980. Fine structure of the unistellate sperm of the shrimp , *Sicyonia ingentis*, Natantia. *Tiss.* and *Cell*, 12: 29-45.

Kooda-Cisco, M.J., and P.T. Talbot. 1982. Astructural analysis of the freshly extruded spermatophore from the lobster *Homarus americanus. J. Morphol.* 172: 193-207.

Kooda-Cisco, M.J., and P. Talbot. 1983. Technique for electriacally stimulating extrusion of spermatophores from the lobster, *Homarus americanus. Aquaculture* 30: 221-227.

Kruevaisayawan H, R. Vanichviriyakit, W. Weerachatyanukul, S. Iamsaard, B. Withyachumnarnkul, A. Basak, N. Tanphaichitr, P. Sobhon 2008. Induction of the acrosome reaction in black tiger shrimp (*Penaeus monodon*) requires sperm trypsin- like enzyme activity. *Biology of Reproduction*, 79 pp. 134–141.

Kurokura, H., J.G. Wu, A.C. Anil, and R. KAdo. 1992. Preliminary studies on the cryopreservation of the cirripede nauplius. Ab.211. In: *Workshop on gametes and Embryos Storage and Preservation of Aquatic organism*, Mary Le Roi/France.

Kumar, R. S.; Diwan, A. D, 2000. Remove from marked Records Induction of gamete activity and mechanism of fertilization in the shrimp *Parapenaeopsis stylifera* M. Edwards. *Journal of Aquaculture in the Tropics*. 15. 4. 351-364

Lawrence, A.L., D. Ward, S. Missler, A. Brown, J. McVey and B.S. Middleditch 1979. Organ indices and biochemical levels of ova from penaeid shrimp maintained in captivity versus those captured in the wild. *Proc. World Maricult. Soc.*, 10: 453-563.

Lawrence and Castle 1991. Reproductive maturity in penaeid shrimps. *Bull. of the Institute of Zool.*, 29: 43-49.

Leung-Trujillio, and A.L. Lawrence 1985. The effect of eyestalk ablation on spermatophroe and sperm quality in *Penaeus vannamei. J. World. Maricult. Soc.*, 16: 258-266.

Li G, X. Kang, S. 2010. Mu Induction of acrosome reaction of spermatozoa in Eriocheir sinensis by low temperature. *Cytotechnology*, 62, pp. 101–107.

Lindsay, L.L. and W. H. Clark, Jr. 1992. Protease- induced formation of the sperm acrosomal filament. 34: 189-197.

Lin, M. N, & Y, Y. Ting. 1986. Spermalophore transplantation and artificial fertilization in grass shrimp. Bull. Jpn. Soc. Sci. Fish. 52:585-589.

Loeffler, C.A. and S. Lovtrup. 1970. Water balance in the salmon egg. *J. Exp. Biol.* 52: 291-298.

Lopo, A.C. 1983. Sperm-egg interaction in invertebrates. Pages 269-324. In: *Mechanism and control of fertilization. J*.F. Hartman, ed. Academic Press, New York.

Lubzens, E., A. Hadani, and S. Bedding. 1992. Problems associated with the development of a technique for cryopreservation of rotifers (*Brachionus Plicatilis*) Ab.17. In: *Workshop on Gamete and Embryo Storage and cryopreservation of Aquatic Organisms.* Marly Le Roi/France.

Lui, K.C., T.C. Chu, and H.D. Lin. 1992. Cryosurvival of goldfish embryos between the stages optical vesicle and heart formation under subzero freezing. Ab: 17. In: *Workshop on Gamete and Embryo Storage and cryopreservation of Aquatic Organisms.* Marly Le Roi/France.

Lynn, J.W., and W.H. Clark Jr. 1975. A Mg^{++} dependent cortical reaction in the egg of penaeid shrimp, *J. Cell. Biol.* 251 (Abst).

Lynn, J.W., and W.H. Clark Jr. 1983a. A morphological examination of sperm egg interaction in the freshwater prawn *Macrobrachium rosenbergii. Biol. Bull.* 164: 446-458.

Lynn, J.W., and W.H. Clark Jr. 1983b. The fine structure of the mature sperm of the freshwater prawn *Macrobrachium rosenbergii. Biol. Bull.* 164: 461-470.

Lynn, J.W., and W.H. Clark Jr. 1987. Physiological and biochemical investigations of the egg jelly release in *Penaeus aztecus. Biol. Bull.* 173: 451-460.

Lynn, J.W., M.C. Pillai, P.S. Glas, and J.D. Green.1991. Comparative morphology and physiology of egg activation selected Penoeoidea. Pages 45-

64. In: *Frontiers of Shrimp Research. Development in* Aquaculture *and Fisheries Science,* D.F. Deloach, W.J. Dougherty, and M.A. Davidson, eds., Elsevier-Kodansha Ltd. Amsterdam.

McLellan, M.R., I.R.B. Mac Fadzen, G.J. Morris, G. Hartinez, and J.A. Wyban. 1992. Recovery of shrimp embryos (Penaeus vannamei and Penaeus monodon) from -30°C. Ab.29. In: *Workshop on Gamete and Embryo Storage and cryopreservation of Aquatic Organisms.* 1992. Marly Le Roi/France.

Medina, A. 1992. Structural modifications of sperm from the fiddler crab *Uca tangeri* (Decapod) during early stages of fertilization. *J. Crust. Biol.,* 12: 610-614.

Medina, A. and A. Rodriguez. 1992. Spermiogenesis and sperm structure in the crab *Uca tangeri* (Crustacea, Brachyura), with special reference to the acrosome differentiation. *Zoomorphology* 111: 161-165.

Muir, J.E. and R.J. Roberts. 19933. Conservation and preservation of genetic variation. Pages 305-307. In: *Recent Advances in* Aquaculture, *IV,* (eds., James F. Muir R.J. Roberts.) Blackwell Scientific Publication, London.

Nanshan, D. and Y. Luzheng. 1987. Induction of acrosome reaction of spermatozoa in the *decapod Eriocheir sinensis. Chi. J. Oceanol. Limnol.* 5: 118-123.

Nilsson, E. and J.G. Cloud. 1992. Cryopreservation of isolated blastomeres from rainbow trout blastulae.Ab.15. In: *Workshop on Gamete and Embryo Storage and cryopreservation of Aquatic Organisms.* Marly Le Roi/France.

Nimrat, S.; siriboonlamom, S; zhang, S.; Xu, Y.; vuthiphandchai, V. 2006 Chilled Storage Of White Shrimp (*L. Vannamei*) Spermatophores. *Aquaculture,* 261:944–951.

Persyn, H.O. 1997. Artificial insemination of shrimp. U.S. Patent4, 031, 855, June 28, 4 pp.

Perez, J. A. A.; Pezzuto, P. R.; Ro drigues, L. F.; Valentini, H. & Vooren, C. M. 2001. Relatório da reunião técnica de ordenamento da pesca de arrasto nas regiões Sudeste e Sul do Brasil. Not. Téc. FACIMAR, 5:3-34.

Pillai, M.C., and W.H. clark Jr. 1987. Oocytes activation in th marine shrimp, *Sicyonia ingentis. J. Exp. Zool.* 244: 325-329.

Polge, C., A.U. Smith, and A.S. Parkes. 1949. Revival of spermatozoa after vitrification and dehydration at low temperatures. *Nature* (London), 164: 666.

Ponniah, A.G., A. Gopalkrishna, P.K. Sahoo, A. Barat, and K. Kumar LAl. 1989. Methodology development for cryopreservation of eggs and embryos of Indian major carps and endangered species of *ex situ* conservation. Pp. 16-18. In: *NBFGR Annual Report,* 1988-89: National Bureau of Fish Genetic Resources, Allahabad, India.

Pongtippatee Pattira -Taweepredaa,b, Jittipan Chavadeja, Pornthep Plodpaic, Boonyarath Pratoomchartd, Prasert Sobhona, Wattana Weerachatyanukula, Boonsirm Withyachumnarnkula, 2004. Egg activation in the black tiger shrimp *Penaeus monodon. Aquaculture* 234, 183–198.

Pongtippatee P, R. Vanichviriyakit, J. Chavadej, P. Plodpai, B. Pratoomchart, P. Sobhon, B. Withyachumnarnkul, 2007. Acrosome reaction in the sperm of the black tiger shrimp *Penaeus monodon* (Decapoda, Penaeidae) *Aquaculture Research,* 38, pp. 1635–1644.

Pratoomchat, B.S., Piyatiratitivorakul and P. Menasveta 1993. Sperm quality of pond reared and wild-caught *Penaeus monodon* in Thailand. *J. World Aquacult. Soc.*, 24(4): 530-540.

Preston NP, Coman FE, 1998. The effects of cryoprotectants, chilling and freezing on *Penaeus esculentus* embryos and nauplii. *In* Flegel TW (ed) Advances in shrimp biotechnology. National Center for Genetic Engineering and Biotechnology.

Radha, T. and T. Subramoniam. 1985. Origin and nature of spermaophoric mass of the spiny lobster *Panulirus Homarus. Mar. Biol.* 86: 13-19.

Rana, K.J., B.J. McAndrew, and A. Musa. 1992. Cryopreservation of oyster, Crassostrea gigas eggs and embryos. Ab. 25. In: *Workshop on Gamete and Embryo Storage and cryopreservation of Aquatic Organisms.* Marly Le Roi/ France.

Reger, J.F. 1970. Studies on the fine structure of spermatids and spermatozoa of the crab, *Pinaisia* sp. *J.Morphol.* 132: 89-100.

Rosas, C., Sanchez, A., Chimal, M.E., Saldana, G., Ramos, L. and Soto, L.A., 1993. The effect of electrical stimulation on spermatophore regeneration in white shrimp Penaeus setiferus. *Aquatic Living Resources*, *6*(2), pp.139-144.

Rostand, J. 1946. Glycerine resistance du sperme aux basses temperatures C.R. *Acad. Sci. Paris,* 222: 1524-1525.

Rurangwa E, A. Biegniewska, E. Slominska, E.F. Skorkowski, F. Ollevier, 2002. Effect of tri butylt ion adenylate content and enzyme activities of teleost sperm: a biochemical approach to study the mechanisms of toxicant reduced spermatozoa motility. *Comparative Biochemistry and Physiology*, 131, pp. 335–344.

Sandifer, P.A. and J.W. Lynn. 1980. Artificial insemination of Caridian Shrimp. Pages 271-288. In: Advances Invertebrate Reproduction. W.H. Clark Jr.and S.T. Adams, eds., *Elsevier* New York.

Sandifer, P.A., A.L. Lawrence, S.E.G. harris, G.W. Chamberlain, A.D. Stocks and W.A. Bray. 1984. Electrical stimulation of spermatophore expulsion in marine shrimp *Penaeus* spp. *Aquaculture* 41: 141-187.

Sawada, Y. and M.C. Chamg. 1964. To;erance of honey bee sperm to deep freezing. *J. Econ.Entomol.* 57: 891-892.

Seymour, J. 1994. Freezing time at the zoo. *New Scientist* 1: 21-23.

Shackmann, R.W., R. Christen, and B.M. Shapiro. 1989. Membrane potential depolarization and increased intracellular pH accompany the acrosome reaction of sea urchin sperm. *Proc. Natl. Acad. Sci. USA.* 78: 6060-6070.

Shaikmahumad, F.S. and V.B. Tambe. 1958. Study of Bombay prawns: The reproductive organs of Parapenaeopsis stylifera (M.Edw.). *Journal of the University of Bombay*, 27: 99-11.

Shigekawa, K., and W.H. Clark, Jr. 1986. Spermiogenesis in marine shrimp, *Sicyonia ingentis. Dev. Growth Differ.* 28: 95-112.

Shiporo, B.M. and E.M. Eddy. 1980. When sperm meets eggs: biochemical mechanisms of gamete interaction. *Int. Rev. Cytol.* 66: 257-302.

Simon, C., P. Dumont, F. Cuende, and A. Dilter. 1994. Determination of suitable fre4ezing media for cryopreservation of *Penaeus indicus* embryos. *Cryobiology*, 31: 245-253.

Stoss, J. and W. Holtz. 1983. Successful storage of chilled rainbow trout (*Salmo gairdenri*) spermatoxzoa for upto 34 days. *Aquaculture* 31: 261-274.

Subramoniam, T. 1984. Spermatophore formation in two intertidal anomuran crabs, *Emerita asiatica* and *Albunea sumnista* (Decapod, Anomura). *Biol. Bull.* 11: 78-95.

Subramoniam, T. 1993. Spermatophores and sperm transfer in marine crustaceans. Pages 129-214. In: *Advances in marine biology*. J.H.S. Blaxter, and A.J. Southward, Eds., Academic Press, New York.

Subramoniam, T. and S.S. Newton. 1993. Cryopreservation of penaeid prawn embryos. *Curr, Sci.* 65: 176-178.

Subramoniam, T. 1994. Cryopreservation of crustacean gametes and embryos. *Proc. Indian. Natn. Sci. Acad.* B 60: 229-236.

Suresh, R. and Diwan, A. D., 2000. Induction of gamete activity and mechanism of fertilization in the shrimp Parapenaeopsis stylifera M. Edwards. *Journal of* Aquaculture *in the Tropics* 15(4): 351-364

Talbot, P. and R.G. Summers. 1978. The structure of the sperm from *Panulirus*, the spiny lobster, with special regard to the acrosome. *J. Ultrastruct. Res.* 64: 2456-2460.

Talbot, P. and P. Chanmanon. 1980. Morphological features of the acrosome reaction of lobster (*Homarus*) sperm and role of the reaction in generating forward sperm movements. *J. Ultrastruct. Res.* 70: 287-297.

Talbot, P., C. Thaler and P. Wilson. 1986. Artificial insemination of the America lobster. *Gamete. Res.* 14: 25-31.

Talbot, P. and M. Goudeau. 1988. A complex cortical reaction leads to formation of the fertilization envelope in the lobster *Homarus. Gamete Res.* 19: 1-18.

Tave, D. and A.J. Brown. 1981. A new device to help and facilitate manual spermatophore transfer in penaeid shrimp. *Aquaculture* 25: 299- 301.

Taylor, A. 1990. A double bonus for shrimp farmers. *Fish Farmer* 13: 39-40.

Tsai, K.L. and P. Talbot. 1993. Videomicroscopic analysis of ionophore induced acrosome reactions of lobster (*Homarus americanus*) sperm. *Mol. Reprod. Dev.* 36: 454-461.

Uberti, M.F., Vieira, F.D.N., Salência, H.R., Vieira, G.D.S. and Vinatea, L.A., 2014. Assessment of viability of sperm cells of Litopenaeus vannamei on cryopreservation. *Brazilian Archives of Biology and Technology*, 57,3:374-380.

Utting, S., Comoy, and I. McFadzen. 1990. Frozen larvae, bivalvae-hatcheries could soon reap benefits. *Fish Farmer* 13: 45-47.

Vázquez-Islas Grecia, Ilie S. Racotta, Arlett Robles-Romo, Rafael Campos-Ramos, 2013. Energy balance of spermatophores and sperm viability during the molt cycle in intact and bi laterally eyestalk ablated male Pacific white shrimp *Litopenaeus vannamei. Aquaculture*, 414–415, 1–8

Wang Q, M. Misamore, C.Q. Jiang, C.L. Browdy, 2003. Egg water induced reaction and biostain assay of sperm from marine shrimp *Penaeus vannamei* : dietary effects on sperm quality. *Journal of the World* Aquaculture Society, 26 ,pp. 261–271

Wikremanayake, A.H., K.R Uhlinger, F.J. Griffin, and W.H. Clark Jr., 1992. Sperm of the shrimp *Sicyonia ingentis* undergo a bi-phasic capacitation accompanied by morphological changes. *Dev. Growth Differ.* 34: 341-355.

Yonge, C.M. 1937. The nature and significance of the membrane surrounding the developing eggs of *Homarus vulgaris* and other decapods. *Proc. Zool. Soc. London. Ser. A*, 107: 499-517.

Yonge, C.M. 1955. Egg attachment in *Crangon vulgaris* and other Caridea. *Proc. Roy. Soc. Edinburgh* 65: 369-400.

Zietara M.S, E. Slominska, J. Swierczynski, E. Rurangwa, F. Ollevier, E.F. Skorkowski, 2004. A T P content and adenine nucleotide catabolism in African cat fish spermatozoa stored in various energetic substrates. *Fish Physiology and Biochemistry*, 30 pp. 119–127

Zell, S.R. 1978. Cryopreservation of gametes and embryos of salmonid fishes. *Ann. Biol. Anim. Biochim. Biophys.* 18: 1089-1099.

CHAPTER 7

ARTIFICIAL INSEMINATION TECHNIQUE FOR INDUCEMENT OF LARVAL PRODUCTION

7.1 INTRODUCTION

Crustacean genetic engineering techniques are of a recent origin with most of the work being carried out on the applied Aquaculture aspects, including brood stock development, heritability, and hybridization, disease diagnostics and also disease therapeutics. The primary techniques for stock improvement are stock selection, selective breeding hybridization and sex control. Artificial Insemination basically is a tool employed for improvements in the phenotype and genotype characters, growth rates, disease resistance, and a number of other related aspects. Problems such as fragmentary nature of knowledge, and difficulty in synchronization of mating behaviour of some commercially important decapods can be successfully tackled through artificial insemination. Artificial insemination can thus be useful in increasing production from a maturation of shrimp in captive condition and thereby assist in closing the life cycle of such species. This is believed to be helpful in the domestication efforts for shrimp stocks and also in propagating them to new areas by controlled reproduction. The technique of artificial insemination first patented by Persyn (1977) in open-thelycum of crustacean species has undergone several modifications (Sandifer & Smith, 1979; Sandifer & Lynn, 1980, Joshi and Diwan, 1992). Later, the technique was extended and adopted to closed thelycum and unisex of shrimp species by Lumare (1981), Laubier and Ponticelli (1981), Muthu & Laxminarayan (1984), Lin & Ting (1986), Hillier (1989), Lin & Hanyu (1990), Pratoomchat *et al.*, (1993), Petersen *et al.*, (1996), Pawar and Iftekhar Mohiuddin (2000), Peixoto *et al.*, (2004), Bart *et al.*, (2006), Arce *et al.*, (1999) and Arnold *et al.*, (2012). The technique of artificial insemination is now routinely used in several leading laboratories to accomplish fertilization in adverse biotic and abiotic conditions

(AQUACOP 1983). Yet failures in reproduction are quite common due to factors such as local environmental conditions and species availability.

Worldwide shrimps are collected either from the wild or offspring from wild-caught brood stock for culture activity in captive condition. This practice is risky due to the reason that wild-caught shrimp may be carriers of pathogens, including viruses. Several of these viruses have devastated the global shrimp farming industry in recent years, resulting in the emergence of novel production systems that rely on pathogen free shrimps (McIntosh 1999; Moss 1999). A significant disadvantage in culturing wild-caught shrimp is the inability of the farmer to benefit from domestication and genetic improvement of stocks. Many penaeid shrimp possess characteristics that are amenable to selective breeding, including the ability to close the life cycle in captivity, a short generation time, and high fecundity. To overcome these difficulties in recent past many places in the world, particularly in Asian countries introduced the technique of artificial insemination. Most commercial shrimp hatcheries for most of times rely on natural matings of shrimps to produce larvae (Yano *et al.*,1988). Advantages of natural mates over artificial insemination are a greater number of nauplii produced per spawn and decreased labour costs. However, the male is unknown and this may be important information for selective breeding programs. The male may be identified by DNA fingerprinting if it is undesirable to use artificial insemination technique (Moore *et al.*, 1999; Hetzel *et al.*, 2000), but this approach requires sophisticated technologies that typically are unavailable to shrimp farmers. Arce *et al.*, (1999) reported that in some places in the breeding programs, specific males are mated with specific females by artificial insemination to produce half/full-sib families for estimation of genetic parameters, including heritability estimates, phenotypic and genetic variation, and phenotypic and genetic correlations.

Artificial insemination has been claimed as a way to overcome the lack of mating in closed thelycum species (Lumare, 1981; Lin and Ting, 1986; Peeters and Diter, 1994). Moreover, similar spawning performance using naturally mated and artificially inseminated females has been previously reported for *F. paulensis* (Petersen *et al.*, 1996). Several workers have done experiments with artificial insemination using unisex system rather than conventional systems (Lin and Ting, 1986; Pratoomchat *et al.*, 1993; Petersen *et al.*, 1996). Peeters and Diter (1994) argued that artificially inseminated females of *F. indicus* performed equally well either in the presence or absence of males. Nevertheless, most studies on artificial insemination have been conducted at an experimental level rather than at large-scale conditions. In this chapter efforts have made to summarise the works carried out by various workers on artificial insemination of the shrimps and other crustaceans and also improvement in the technology.

7.2 CURRENT STATUS OF ARTIFICIAL INSEMINATION IN SHRIMP AND OTHER CRUSTACEANS

Establishment of sperm bank and artificial insemination are widely practiced in animal husbandry and controlled breeding programmes (Leverage *et al.*, 1972). Such attempts among crustaceans aimed to improve gamete quality and propagation of species in culture, especially in commercially important groups of crustaceans. The attempts in this direction were made by Uno and Fujita (1972), Clarke *et al.*, (1973) and Sandifer and Smith (1979). The initial attempts involved mechanical extrusion of spermatophore, a process which often injured the males. The electro-ejaculation technique of extruding spermatophore from males simplified artificial insemination technique in crustaceans to a certain extent. This technique was adopted by different researchers in penaeid shrimps (Laubier-Bonichon and Ponticelli, 1981; Lumare, 1981 and Muthu and Laxminarayana, 1984) as well as palaemonid prawns (Sandifer and Lynn, 1980). Tave and Brown (1981) further refined the technique by use of gill irrigator and restraining device for the receptive females. The importance of stockpiling and exchanging the selected male genetic material in Crustacea, have led some researchers for evolving methods for preservation of spermatophore for different durations. Artificial insemination using cryopreserved spermatophore had been undertaken in the lobster, *Homarus americanus* by Kooda Cisco and Talbot (1983) and in the freshwater prawn *Macrobrachium rosenbergii* by Chow (1982) and Chow *et al.* (1985). Joshi and Diwan (1992) carried out artificial insemination studies successfully using electro-ejaculation technique in freshwater prawn *Macrobrachium idella.* The best possible permutations and combinations to get maximum viable yield of larvae have been explored. The possibility of using refrigerated spermatophore for artificial insemination has also been tapped by them. Pawar and Iftekhar Mohiuddin (2000) carried out the work on artificial insemination on two of the most commonly occurring penaeid shrimps species viz. *Metapenaeus affinis* and *Metapenaeus brevicornis* both of which possess a 'closed-thelycum' using the technique of electro-ejaculation without gill irrigator. They observed that after performing artificial insemination in these two shrimps the latency period ranged between10-16 days, while spawning occurred within 10-12 days of spermatophore transfer. They recorded three partial spawning viz., two in *M. affinis* and one in *M. brevicornis* with an average spawning and hatching rates of 30% and 72.3% respectively. Average percentage survival from first nauplius to one-day old post-larva was very less (3.43%). (Table 1 and 2)

Bart *et al.*, (2006) (Fig 1 and 2) while working on artificial insemination of *P. monodon* with cryopreserved spermatophores observed that manual stripping of male shrimps is convenient and efficient way to obtain good quality sperm for

Table 1. Details of Spermatophore implantation in the females of *Metapenaeus affinis* (Tag no1-7) and *M. brevicornis* (Tag no 8-10).

Tag No.	Body Weight (g)	Total Length (mm)	Date of Moulting	Date of Insemi-nation	Spermato-phores Implanted (no.)	Remarks
1.	15.0	122	29/1/94	31/1/94	One	Sperm mass held intact. Mortality recorded after 3 weeks. No spawning recorded.
2.	15.0	126	20/1/94	22/1/94	One	Female recorded dead on the second day of insemination.
3.	16.5	130	15/1/94	17/1/94	One	Appendages found damaged due to (accidental) excessive glue application. No spawning recorded up to 1 month.
4.	17.2	140	19/2/94	21/2/94	Two	Sperm mass held intact. No spawning recorded despite full ovarian development.
5.	20.1	130	6/3/94	8/3/94	One	Partial spawning with few viable eggs recorded.

Table 2. Spawning larval development details in the artificially inseminated females of *M. affinis* (Tag no 5-7) and *M. brevicornis* (tag no 6-8).

Tag No.	Body Weight (g)	Total Length (mm)	Date of Insemi-nation	Spawning			Days from N1 to PL1	PL 1 Obtained	
				Date	No of eggs	Hatching rate %		Number	Survival
5	20.1	130	8/3/94	21/3/94	3000	66	11	98	3.3
6	26.0	141	4/3/94	16/3/94	5400	70	12	223	4.1
7	28.0	145	10/3/94	21/3/94	a	Nil	Nil	Nil	Nil
8	12.5	110	5/4/94	18/4/94	3600	81	11	106	29

(**Source:** Pawar and Iftekhar mohiuddin., 2000).

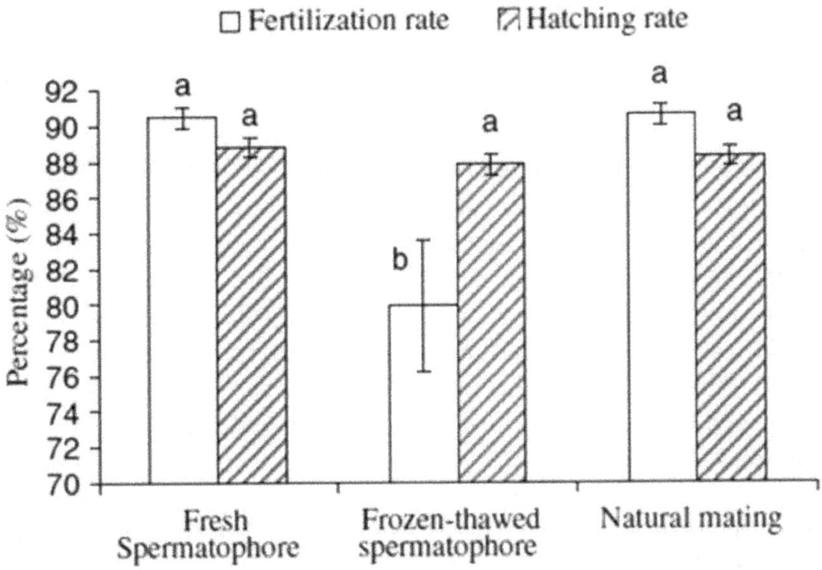

Fig. 1. Fertilization and hatch rates for artificial insemination using fresh spermatophore, frozen-thawed spermatophore and natural mating. The spermatophore was cryopreserved with 5% dimethyl sulphoxide. Different letters indicate significant difference among groups ($P < 0.05$). (**Source:** Bart *et al.*, 2006)

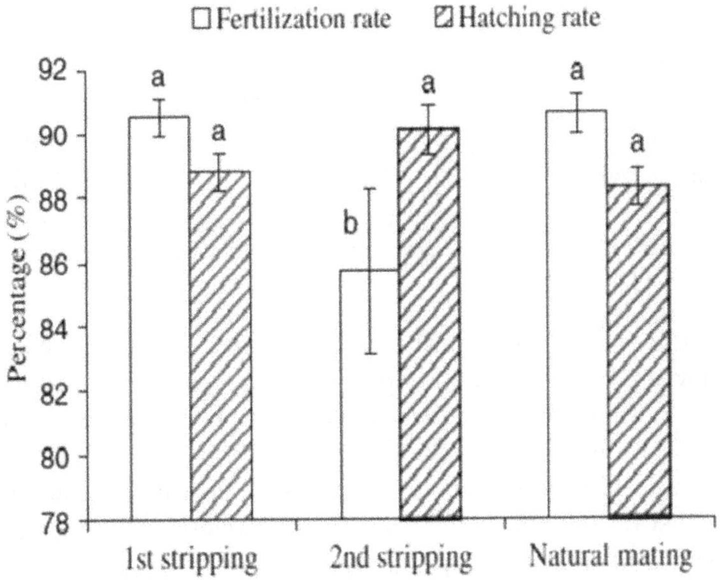

Fig. 2. Fertilization and hatch rates for artificial insemination using fresh spermatophore collected during first and second stripping and from natural mating. Different letters indicate significant difference among groups ($P < 0.05$). (**Source:** Bart *et al.*, 2006)

artificial insemination. High fertilization and hatch rates (80%) obtained using cryopreserved spermatophore indicated that the cryopreservation protocol used in this study for *P. monodon* spermatophore could be used for commercial shrimp hatcheries.

Peixoto *et al.*, (2004) worked on the influence of artificial insemination on the reproductive performance of the shrimp *Farfantepenaeus paulensis* in conventional and unisex maturation systems. They reported that in the group of unilaterally eyestalk ablated female shrimps when artificially inseminated with one compound spermatophore soon after moulting, the percentage of fertilized spawns increased from 26% before the use of artificial insemination to 57% afterwards (table 3). Further they mentioned that the reproductive performance of inseminated females held in conventional or unisex maturation system showed no significant differences.

Table 3. Reproductive performance parameters (mean FS.D.) of artificially inseminated *F. paulensis* females in conventional and unisex maturation systems.

	Conventional	Unisex
Number of inseminations	64	78
Number of spawns	57	68
Fertilized spawns (%)	61	54
Spawns/intermolt period	1.7 F 1.1	1.6 F 0.7
Eggs/spawning event	146,203F44,309	144,123F49,078
Eggs/female	211,636F156,960	186,305F98,317
Total egg production	4,867,630	4,098,700
Fertilization rate (%)	68.4F24.6	75.1F18.4
Hatching rate (%)	59.3F21.1	64.5F25.0

(**Source:** Silvio *et al.*, 2004)

Arce *et al.*, (1999) mentioned in their report that about breeding program of shrimp where Pacific white shrimp *Litopenaeus vannamei* are selected for rapid growth and resistance to Taura Syndrome Virus. They mentioned that shrimps are produced from maternal half-sib families by mating one female with two different males. The artificial insemination technique they used to produce these families relied on the removal of both spermatophores from a single male and the application of the spermatophores over the thelycum of a ripe female (Fig 3, 4 & 5). The female was then placed in a spawning tank where fertilized eggs were liberated. If previously inseminated female developed ripe ovaries before the 2-wk period elapsed, she was inseminated again with spermatophores from a different male. In an effort to maximize the number of half-sib families and reduce the time that families are produced, a different artificial insemination technique was used to produce paternal half-sib families. With this technique, each of the two spermatophores from a single male was manually extruded and placed on the thelycum of two different females. This later technique resulted in a significantly higher spawning success (84% vs. 58%) and females produced significantly more viable nauplii per spawn (24,400 nauplii vs 8,500 nauplii). Importantly, the time to produce selected families was reduced from 14 to 9 days, and the number of half-sib families increased. These improvements have significant implications for a selective breeding program (Arce *et al.*, 1999).'

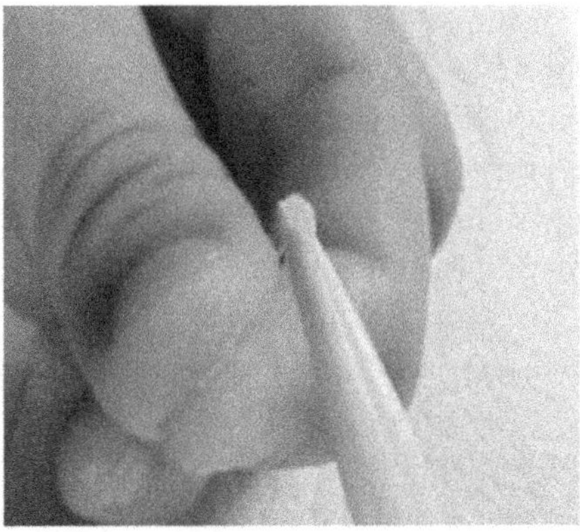

Fig 3. Liberating sperm mass from the spermatophore of a male *L. vannamei* broodstock. (**Source:** Arce *et al.*, 1999)

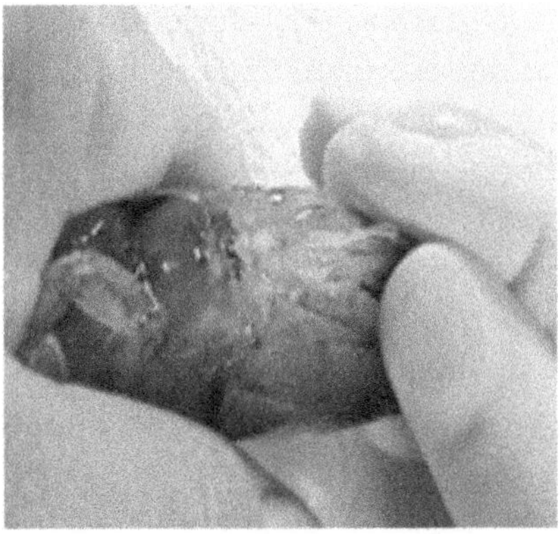

Fig 4. Completed artificial insemination via single. (**Source:** Arce *et al.*, 1999)

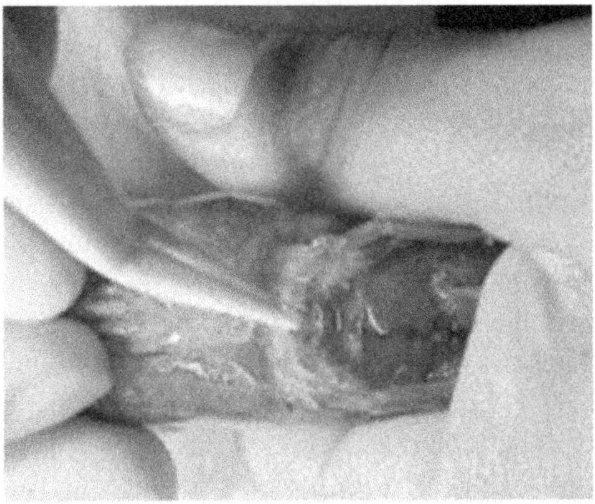

Fig 5. Placement of sperm mass into the thylecum of a female *L. vannamei* broodstock. (**Source:** Arce *et al.*, 1999)

Improving the reproductive performance of domesticated brood stock of shrimps remains a high priority particularly for black tiger shrimp. One of the main issues with many domesticated brood stocks of shrimps is low rates of egg fertilization which is believed in part to compromised male fertility (Pongtippatee *et al.*, 2007). Male fertility is the actual critical requirement to assess the reproductive performance in culture technologies. In addition, male fertility measures can be used to remove potential poor performers from commercial breeding programs and monitor the development of stocks over time. In context of this aspect Arnold *et al.*, (2012) carried out the studies to determine the relationship between male fertility and egg fertilization in *P. monodon*, by employing a novel approach of discriminating the fertilization influence of males. In this method the female brood stock of shrimps were artificially inseminated with one spermatophore each from a pair of males. It was observed that female shrimps undergone 22 successful spawning involving selected pairs from 33 male shrimps. Further they have estimated the proportion of embryos fertilized by each paired males by individual genotyping of embryos. The non-inseminated, twin spermatophore has been used to estimate measures of male fertility, sperm number and the number of normal sperm of each inseminated spermatophore. Using this protocol it has been shown that the egg fertilization potential of males could not be estimated simply by total sperm number or number of the normal sperm and there was extremely high variability in fertilization between paired males. Chotipuntu *et al.*, (2013) while assessing reproductive aspects of SPF *P. monodon* grown in closed culture captivity carried out studies on artificial insemination to find out reproductive performance of the shrimp grown in closed culture captivity (Table 4,5 & 6). In their study12 month -old female shrimp was manipulated to mature using eyestalk ablation and artificial insemination techniques. Spawning characteristics of the ablated shrimp found to vary among individuals. A number of ablated female exhibited multiple spawning up to 3 consecutive times with a single insemination. Numbers of egg and percentages of hatch apparently decreased in the later spawning in the group of shrimp spawned 3 consecutive times with single insemination. Egg counts in each spawning varied from 79,000 to 260,000 showed no particular pattern of relationship with body weight. Hatching feasibility varied from nil to 85%. This study demonstrated the promising potential in producing specific pathogen free broodstock in tropical climate environment using the technique of artificial insemination. Most of manipulated female exhibited multiple spawning with a single artificial insemination, and feasible offspring. However, it is suggested that further research studies are needed to enhance growth performance and survival rate. The quality of spawn in terms of viable spermatophore and ovarian development is yet to be improved (Chotipuntu *et al.*, 2013).

Table 4. Single spawning of individual *P. monodon* subjected to single insemination

Individual	Number of egg (×1,000)	Hatch (%)
Q13/1	98	21
Q14/2	200	85
Q14/2*	320	20
Q14/5	79	37
Q15/1	213	30
Q20/2	109	33
Q20/3	140	18
Q21/3	150	50
Q21/3*	129	38
Mean	**160**	**37**
SD	**75**	**21**

*These shrimp were inseminated twice, each insemination performed single spawning (**Source:** Chotipuntu *et al.*, 2013)

Table 5. Double spawning of individual *P. monodon* subjected to single insemination. Values in each row with same letter are not significantly different ($p > 0.05$).

Individual Code No.	1st spawn		Time interval (Day)	2nd spawn	
	Egg (×1,000)	Hatch (%)		Egg (×1,000)	Hatch (%)
Q13/2	220	0	4	280	0
Q13/2*	269	7	3	184	10
Q13/4	260	0	4	140	0
Q13/7	220	19	3	200	0
Q15/2	180	11	4	160	40
Q15/4	210	10	4	250	40
Q15/5	158	25	5	79	24
Mean	**206[a]**	**13**	**4**	**166[a]**	**21**
SD	**39**	**10**	**1**	**64**	**20**

* This shrimp was inseminated twice, each insemination performed double spawning. (**Source:** Chotipuntu *et al.*, 2013)

Table 6. Triple spawning of individual *P. monodon* subjected to single insemination. Values in each row with same letter are not significantly different ($p > 0.05$).

Individual Code No.	1st spawn		Time interval (Day)	2nd spawn		Time interval (Day)	3rd spawn	
	Egg (×1,000)	Hatch (%)		Egg (×1,000)	Hatch (%)		Egg (×1,000)	Hatch (%)
Q13/4*	250	0	3	200	0	3	3	180
Q13/7*	221	45	3	210	18	3	3	160
Q15/2*	187	20	3	230	35	3	3	158
Q15/2**	139	15	3	145	35	3	3	70
Q15/4*	220	10	4	249	40	3	4	239
Q13/4*	250	0	3	200	0	3	3	180
Q13/7*	221	45	3	210	18	3	3	160
Mean	**203^{a}**	**18**	**3**	**207^{a}**	**26**	**3**	**161^{a}**	**12**
SD	**42**	**17**	**0**	**39**	**17**	**0**	**61**	**22**

*These shrimp were inseminated more than once but performed triple spawning at second insemination.

** This shrimp was inseminated more than once but performed triple spawning at third insemination.

(Source: Chotipuntu *et al.*, 2013)

During mating and insemination in the wild there is no control over the mating pair and so over their progeny. On the contrary, by adopting artificial insemination technique, selective breeding could be achieved. The electro-ejaculation of spermatophore has been reported as the most effective and simple method of spermatophore retrieval in freshwater prawn *Macrobrachium* sps. (Sandifer and Lynn, 1980). Joshi and Diwan (1992) reported that electrical stimulus of 4.5 volts is found to be sufficient for extrusion of spermatophore in male *M. idella.* In fact a stimulus of 4.5 volts was without any ill effect, when the electro- ejaculation was attempted once in 24 hrs. When electrical stimulus of more than 5 volts was applied, terminal ampoules became blackish in colour probably due to tissue lysis. The electrical stimulus needed for ejaculation is reported to vary from species to species. For example 2 volts in *Palaemonetes*, 5-6 volts in *M.* rosenbergii (Sandifer and Lynn, 1980), and 5 volts in *P. japonicus* (Lumare, 1981). However, in the lobster, *H. americanus* (Kooda-cisco and Talbot, 1983) and the sand lobster *Thenus orientalis* (Silas and Subramoniam, 1987) electrical stimulus of 12 volts has been reported to be essential for electro-ejaculation of spermatophores.

A few reports describing the virility of male prawns and lobster are available. Sandifer and Lynn (1980) have reported in *M. rosenbergii* that six males were electro-ejaculated on 12 consecutive days without any ill effects and each time a male could extrude spermatophore after 24 hrs recovery period. In *T. orientalis* a 12 hrs recovery period has been reported (Silas and Subramoniam, 1987). In male *M. idella*, the capacity to extrude spermatophore on consecutive days has been reported to be directly related to size and robustness of the male. The largest males of size group 91-95 mm are found to be most virile, extruding spermatophore on an average 13 times in 15 days period. Such reports depicting the virility of male in relation to body size is not available (Joshi and Diwan, 1992).

Artificial insemination attempt has been found to be successful only when placement of spermatophore was conducted 2.5 to 4 hrs after the pre-spawning moult. Placement of spermatophore immediately after pre- spawning moult led to failure may be that the female was too soft to withstand the stress of handling, resulting in death. It has been also reported that delay (more than 4 hrs) for placement of spermatophore, after pre-spawning moult also resulted in failure of artificial insemination, probably due to hardening of moulted female and dislodging of the spermatophore (Joshi and Diwan, 1992). Of the 52 artificial insemination trials on *M. idella*, using complete spermatophore, 34 were successful. Failure in 18 trials may be due to the stress developed during handling. Another reason failure was small size of receptive females, which were delicate to handle (Joshi and Diwan, 1992) (Table 7 & 8). Sandifer and Lynn (1980) working on identical lines reported success in fertilizing females in 11 out of the total 18 trials. It has been reported that frequent handling of females during artificial insemination act resulted in delayed spawning and loss of manually placed spermatophore leading to failure in artificial insemination (Joshi and Diwan, 1992). Identical observations were recorded by Sandifier and Lynn (1980) for *M. rosenbergii.*

Table 7. Assessing proper time duration for spermatophore placement in Female *M. idella*

Trial No.	No. of replicate	Time interval between pre-spawning moult and placement of spermatophore (hrs)	Result	Remarks
1	1	0.5	Unsuccessful	Female died within 2 hrs after spermatophore placement
2	1	1.0	Unsuccessful	Female too soft. Sperm-atophore dislodged. Female died next day.
3	1	1.5	Unsuccessful	Female still soft. Sperm-atophore dislodged and female oviposited unfertilized eggs 5 hrs after pre-spawning moult.
4	1	2.0	Unsuccessful	Spermatophore dislodged. No oviposition. Ovary got resorbed.
5	1	2.5	Successful	Spermatophore retained. Oviposition 5.5 hrs after pre-spawning moult.
6	3	3.0	Successful	All females oviposited 6 hrs after pre-spawning moult.
7	1	3.5	Successful	Female oviposited fertilized eggs, 5 hrs after pre-spawning moult.
8	1	4.0	Successful	Female oviposited fertilized eggs 5.5 hrs after pre spawning moult.
9	1	4.5	Unsuccessful	Female too hard. Spermntophore dislodged.Female died next day.
10	1	5.0	Unsuccessful	Female oviposited unfertilized eggs before artificial insemination.
11	1	5.5	Unsuccessful	Female oviposited unfertilized eggs before artificial insemination.
12	1	6.0	Unsuccessful	Female oviposited unfertilized eggs before artificial insemination.

Experimental Conditions: Salinity: 6%0, Temperature: 28-29°C, pH: 8-8,2

Successful: Female spawned fertilized eggs, embryo development was normal and healthy normal larvae hatched out after an incubation period of 11-13 days.
(**Source:** Joshi and Diwan 1992)

The gill irrigator used ensured continuous supply of water to the branchial cavity of female and provided better chances of success in artificial insemination. Tave and Brown (1981) have reported that after using the gill irrigator and restraining device during spermatophore transfer, 88% of the females that received spermatophore have spawned and finally released healthy larvae. Lumare (1981) on the other hand performed the artificial insemination without any device and could achieve only limited success. From the earlier attempts on artificial insemination by different workers (Sandifer and Smith, 1979; Sandifer and Lynn 1980; Lumare, 1981; Bray *et al.*, 1982; Lin and Ting, 1984 and Silas and Subramoniam, 1987, Joshi and Diwan, 1992) it is evident that discovery of electroejaculation technique to extrude spermatophores in the males have simplified the artificial insemination protocol to a great extent. Attempts to inseminate 2 females with spermatophore from single male yielded a marginal success. The larval yield after such inseminations is found to be low, compared to larval yield from a female inseminated with complete spermatophore. Similar attempts to fertilize two females with the sperm mass from one male were undertaken by Sandifer and Lynn (1980) in *M. rosenbergii*, achieving success in 50% of the trials.

When 4 females of *M. idella* were inseminated simultaneously, using spermatophore from single male only a marginal success could be achieved. The larval yield was also very low. Difficulty experienced in these trials was non-adhering of the piece of sperm cord, lost its sickness within a short period and consequently the sperm mass got dislodged from the sperm receptacle before spawning occurred (Joshi and Diwan, 1992). Earlier, Sandifer and Smith (1979) examined the possibility of inseminating 4 females simultaneously with sperm mass from single male and encountered similar problems. Though the attempts to inseminate 2 and/ or 4 females simultaneously with the sperm mass of single male could yield only marginal success in *M. idella*, after some refinements, this technique could prove highly successful (Joshi and Diwan, 1992). Sandifer and Lynn (1980) also observed that the males of *M. rosenbergii* were aggressive and pugnacious and generally experienced higher mortality in laboratory holding tanks. Joshi and Diwan (1992) reported that in captive matured females of *M. idella* (matures after unilateral eyestalk ablation) artificial insemination attempts were successful and 60% of such successful attempts have also been reported for other species of *Macrobrachium*. In the shrimp *P. monodon*, Lin and Ting (1984) have reported successful artificial insemination of unilaterally ablated and matured females. Misamore and Browdy (1997) carried out studies on evaluating hybridization potential between *P. setiferus* and *P. vannamei* through natural mating, artificial insemination and in vitro fertilization. They have reported that in artificial insemination trials, 60% of the intraspecific inseminations were fertile with mean fertilization rates of 9.21% for *P. setiferus* and 11.6% for *P. vannamei*. No interspecific crosses were fertile.

Table 8. Artificial insemination trials with fresh spermatophores in *M idella.*

Experiment no	Female source	Details of Experiment	No. of trials^{+}	No. of success^{++}	No. of failure average		Larval yield
					A*	B**	
3A	Wild unablated	Spermatophore from one male used to inseminate one female (i.e. both sperm cord used)	52	34	6	12	3950
3B	Wild unablated	Spermatophore from one male used to inseminate two females simulatenously. (Each female receiving single sperm cord)	66	14	8	44	3180
3C	Wildunablated	Spermatophore from one male used to inseminate four females simulateneously. (Each femalereceiving half sperm cord)	12	2	2	8	2352
3D	Matured incaptivity by unilateral eyestalkablation	Spermatophore from one male used to inseminate one female	10	6	1	3	4365

Experimental conditions: Salinity: 6%0, Temperature: 28-31 °C, pH: 7.8-8.2

Trials+ : Attempts made on artificial insemination of a sexually receptive female.

Success++ : Releasing of healthy zoea larvae after successful completion of incubation period

Failure: A : Eggs were fertilized by the artificially placed spermatophore but not viable.

B : Eggs were not fertilized by the artificially placed spermatophore.

(**Source:** Joshi and Diwan, 1992)

Attempts on artificial insemination using refrigerated spermatophore in *M. idella* showed that the spermatophore could remain as active as freshly extruded ones when refrigerated for 24 hrs at 6°C but with further storage the fertilizability as well as viability was found to decrease. The larval yield also declined (Joshi and Diwan, 1992). Working on identical line in *M. rosenbergii* Sandifer and Lynn (1980) concluded that the spermatophore could be stored under refrigeration for 24 hrs without losing their activity. Chow (1982) preserved the spermatophore of *M. rosenbergii* in Ringer's solution in refrigerator at 2°C and concluded that spermatophores retained their viability upto 4 days. In *M. idella*, the spermatophores refrigerated at 6°C retained their viability for 72 hrs. However, the larval yield declined with increasing storage period (Joshi and Diwan, 1992) (Table 9).

Table 9. Artificial insemination trails with refrigerated spermatophores in *M idella*

Sr No.	Duration of spermatophore refrigeration (hrs)	No. of trials	No. of success++	No. of failures		Average larval yield
				A*	B**	
1	24	8	5	1	2	3948
2	48	8	2	2	4	3110
3	72	6	1	1	4	2860
4	96	5	1	0	4	2320
5	120	5	0	0	5	–

Experimental conditions: salinity: 6%0, Temperature: 28-31 °C, pH : 8 to 8.3 Spermatophores stored in refrigerator at 6°C.

Trails+ : S"Attempts made on artificial insemination of a sexually receptive female

Success++ : Releasing of healthy zoea larvae after successful completion of incubation period

Failure : A" Eggs fertilized by the artificially placed spermatophore were not viable.

B** : Eggs were not fertilized by the artificially placed spermatophores.

(**Source:** Joshi and Diwan, 1992)

Chow (1982) reported that the spermatophores, when preserved for longer time lost the protective and adhesive matrix and, subjected to damage and propagation of infection with bacilli resulting in fast degeneration of sperm mass. Bart *et al.*, (2006) to develop an appropriate cryopreservation protocol for spermatophores in *P. monodon*, carried out artificial insemination studies to assess the fertilizing ability of fresh and post-thaw spermatophores. They reported that the mean fertilization rate for artificial insemination using post thaw spermatophore was 79.9%, lower than the fertilization rates observed for artificial insemination using fresh spermatophore and natural mating. Mean hatch rates for fresh spermatophore, frozen-thawed spermatophore and natural mating observed were 88.8%, 87.8% and 88.3% respectively. This is the first study to report high fertilization and hatch rates from cryopreserved spermatophores using artificial insemination of spermatophores before spawning.

Artificial insemination results in high yield, in terms of nauplii per spawn in penaeid shrimp (Alfaro, 1993). Artificial insemination has been successively done in many decapod crustaceans by researchers like Sandifer and Lynn (1980) in *P. monodon*; Chow *et al.*, (1982) in *M. rosenbergii;* Bray *et al.*, (1983) in *P. setiferus*; Aiken *et al.*, (1984) in *P. monodon*; Lin and Ting (1985) in *P. monodon;* Talbot *et al.* (1986) in *H. americanus;* Joshi and Diwan (1992) in *M. idella*; Perez *et al.*, (2001) in *L. vannamei* and Coman *et al.*, (2006) in *P. monodon* in the laboratories. However, this technique is not practiced in most of the commercial hatcheries due to the unawareness of its high potentialities. The technique involves the extraction of spermatophores, and then its placement in the thelycum, which is a complicated process and needs utmost care. A small mistake in the technique may lead to the failure of the whole attempt. One of the biggest problems in spermatophores transfer is the amount of time needed to position the spermatophore properly. Shrimps are easily stressed by short periods out of water and stressed condition for longer period leads to the unsuccessful attempt. Strong healthy undamaged spermatophore enables a quick artificial insemination. Usually the spermatophore extrusion method determines the condition of the spermatophores. Sandifer *et al.*, (1984) suggested that, the use of electrical stimulation for obtaining spermatophores is an adequate method for penaeid shrimps. This technique has been widely used by many scientists for the ejaculations of spermatophores from various decapod crustaceans like lobsters (Kooda-Cisco and Talbot, 1983) and shrimps (Sandifer *et al.*, 1984; Rosas *et al.*, 1993; Alfaro and Lozano, 1993; Shoji 1997; Diwan & Shoji Joseph, 2001; Diwan & Shoji Joseph 2005).

7.3 ACQUISITION OF SPERMATOPHORES

Three methods of extraction for the acquisition of spermatophores are: (A) Micro-surgery techniques (B) Squeezing method (C) By electro-ejaculation techniques.

a. Micro-surgery Technique

In this method spermatophores are taken out by carefully dissecting the coxal part of the 5th pereiopods. Shrimps are kept ventral side up and held carefully and strongly, so that the jerk could be prevented during the time of dissection.

The gill irrigator in conjunction with a restraining device (Tave and Brown, 1981) is used to reduce the stress during the dissections. Coxal muscle is cut open, making the terminal ampoule exposed and by gentle squeezing, the spermatophores are extruded from it. Extreme care is taken to minimize the wound and loss of haemolymph. Dissections are made as quickly as possible to remove the spermatophores and the shrimps are then released in a separate tank containing filtered seawater provide with aeration. Wounded shrimps kept separately, fed with clam meat and observed for their generation of the spermatophores.

b. Squeezing Method

Here spermatophores are squeezed out of genital pore by gentle pressure around the coxae of the fifth pereiopods. No wounds are made during the process of extraction. However, this method required skilful experience for extracting whole spermatophore undamaged. The same gill irrigator described above is used here also to reduce the stress during the extrusion. Shrimps, which are used for the spermatophore extrusion, are kept separately in tanks filled with fresh seawater, and observations are made on their survival and regeneration of the spermatophores.

c. Electro-ejaculation Techniques

The electrical stimulation device is an improved model of that described by Sandifer and Lynn (1980). It consists of an electrical transformer to reduce voltage, a rheostat, a voltmeter, and two electrodes Fig.6. The shrimps are held ventral side up in a plastic tray and the delivery tubes of the gill irrigator is placed in each branchial cavity, and a continuous flow of water bathed each set of gills. The electrodes are placed near the gonopores at the base of the 5th pereiopods and a stimulus of 6 to 12V is applied, usually for 1 to 2sec. The stimulus is expected to cause the musculature surrounding the terminal ampule to contract, expelling a single spermatophore from each gonopore Fig.7. The expelled spermatophore Fig. 8 is removed with a wooden needle and used for further studies. In cases where the spermatophores are partially expelled, it is removed by applying gentle pressure at the coxal region for the complete ejection and the completely extruded spermatophores are taken out. In certain cases partially expelled spermatophores were taken out using forceps with blunt ends to avoid the damage during its transfer.

Among the three methods used for the extrusion of spermatophores, it is found that electro-ejaculation is the best method, as it does not harm the shrimps or the spermatophores. Therefore, this technique is followed for the acquisitions of spermatophores for the further investigations in *P. monodon*.

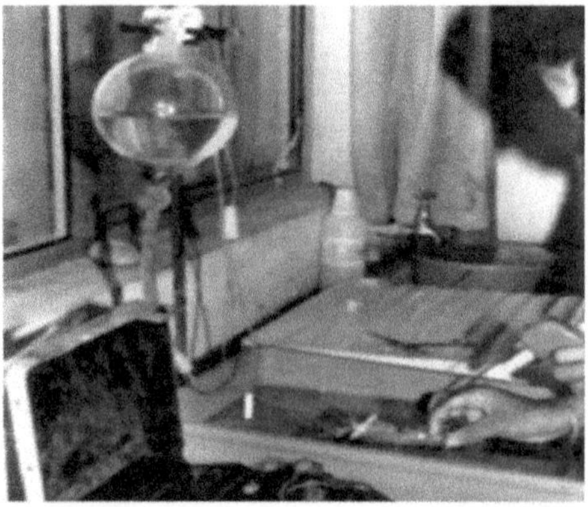

Fig. 6. Electro-ejaculation technique with Gill irrigator
(**Source:** Joshi and Diwan, 1992)

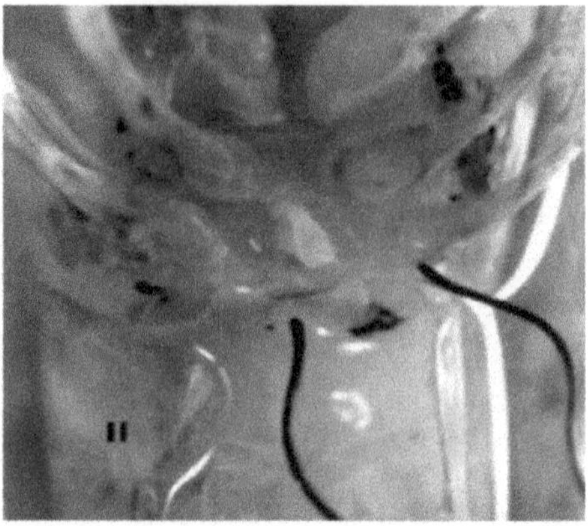

Fig.7. Extrusion of spermatophore through electrical stimulus.
(**Source:** Joshi and Diwan, 1992)

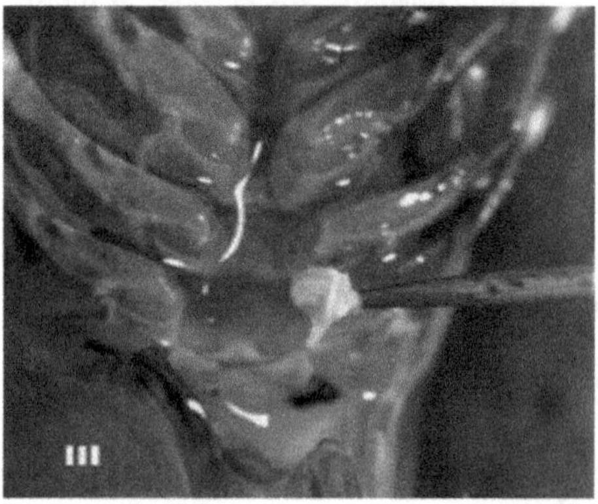

Fig. 8. Extruded spermatophore (Spm) from terminal ampoule
(**Source:** Joshi and Diwan, 1992)

7.4 CONCLUSION

Artificial insemination, basically is a tool employed for improvements in the phenotype and genotype characters, growth rates, disease resistance, and a number of other related aspects. Problems such as fragmentary nature of knowledge, and difficulty in synchronization of mating behaviour of some commercially important decapods can be successfully tackled through artificial insemination. Artificial insemination can thus be useful in increasing production from a maturation of shrimp in captive condition and thereby assist in closing the life cycle of such species. This is believed to be helpful in the domestication efforts for shrimp stocks and also in propagating them to new areas by controlled reproduction. Wild-caught shrimp may be carriers of pathogens, including viruses. Several of these viruses have devastated the global shrimp farming industry in recent years, resulting in the emergence of novel production systems that rely on pathogen free shrimps. To overcome these difficulties in recent past many places in the world, particularly in Asian countries introduced the technique of artificial insemination. Artificial insemination has been claimed as a way to overcome the lack of mating in closed thelycum species. In some places in the breeding programs, specific males are mated with specific females by artificial insemination to produce half/ full-sib families for estimation of genetic parameters, including heritability estimates, phenotypic and genetic variation, and phenotypic and genetic correlations. The possibility of using refrigerated spermatophore for artificial insemination has also been tapped. High fertilization and hatch rates (80%) obtained using cryopreserved spermatophore indicated that the cryopreservation protocol used in this study for *P. monodon* spermatophore could be used for commercial shrimp hatcheries. It has been reported that frequent handling of females during artificial insemination act resulted in delayed spawning and loss of manually placed spermatophore leading to failure inartificial insemination. Artificial insemination technique using refrigerated/ cryopreserved spermotophores in penaeid shrimps and other crustaceans has shown high fertilization and hatch rates which indicates the possibility of commercial applications of this technique for enhancing larval production similar to animal husbandry.

References

Aiken, D.E., S.L. Waddy, K. Moneland and S.M. Polar, 1984. Electrically induced ejaculation and artificial insemination of the american lobster *Homarus americanus*. *J. Crust. Biol.*, 4(4): 519-527.

Alfaro, J. 1993. Reproductive quality evaluation of male *Penaeus stylirostris* from a grow out pond, *J. World. Aquacult. Soc.*, 24(1): 6-11.

Arce, S M, Shaun M Moss and B J Argue, 1999. Spawning and Maturation of *Aquaculture* Species: Proceedings of the twenty-eighth UJNR *Aquaculture* Panel Symposium, Kihei, Hawai'i, November 10-12, 1999. Christine Tamaru. [*et al.*,], Editors. Kihei, Hawaii: University of Hawai'i Sea Grant College Program, 2000. UJNR Technical Report; no 2.

Arnold, S.J., Coman, G.J., Burridge, C., Rao, M., 2012. A novel approach to evaluate the relationship between measures of male fertility and egg fertilization in *Penaeus monodon*. *Aquaculture* 338–341, 181–189.

Bart Amrit N, Sudarhma Choosuk & Dhirendra P Thakur, 2006. Spermatophore cryopreservation and artificial insemination of black tiger shrimp, *Penaeus monodon* (*Fabricius*). *Aquaculture Research*, 37, 523^528.

Bray, W.A Chamberlain,G.W., ilnd Lawerance. AL, 1982. Increased larval production of *Penaeus setiferus* by artificial insemination during sourcing cruises. *World Maricult. Soc.*13: 123-133.

Bray, W.A., G.W. Chamberlain and A.L. Lawrence, 1983. Increased larval production of *Penaeus setiferus* by artificial insemination during sourcing cruises. *J. World Maricult. Soc.*, 13: 123-133.

Chow. S., 1982. Artificial insemination using preserved spermatophores in the palaemonid shrimp *Macrobrachium rosenbergii. Bull. Jap. Soc. Sci. Fish.*, 48(2): 1693-1695.

Chow. S., Taki, Y., and Ogaswara, Y., 1985. Cryopreservation of spermatophore of the fresh water shrimp. *Macrobrachium rosenbergii. Biol. Bull.,* 168:471-475.

Chotipuntu P, Wuthisuthimethavee S, Direkbusrakom S, Songtuay S, 2013. Reproductive aspects of SPF *Penaeus monodon* grown in closed culture captivity. *Walailak Journal of Science and Technology* (WJST); 10(3):227-36.

Clarke, W.H. Jr., Talbot, P., Neal, R.A., Mock, C.R., and Salser. B.R., 1973. In vitro fertilization with non-motile spermatozoa of the brown shrimp *Penaeus aztecus. Mar. Biol.*, 22: 353-354.

Coman G.J., S.J. Arnold, S. Peixoto, P.J. Crocos, F.E. Coman and N.P. Preston, 2006. Reproductive performance of reciprocally crossed wild-caught and tank-reared *Penaeus monodon* broodstock *Aquaculture*, 252: 372–384.

Diwan, A. D. and Shoji Joseph 2001. Cryopreservation of spermatophores of the marine shrimp *Penaeus indicus* H. Milne Edwards *Journal of Aquaculture in the Tropics.* 15(1): 35–43.

Diwan, A.D. and Shoji Joseph 2005. Cryopreservation of gametes embryos and larvae, Synopsis on penaeid shrimp *P. indicus* (eds. A. D. Diwan and M. J. Modayil) Narendra Publishers New Delhi pp 226–240.

Hetzel, D.J.S., P.J. Crocos, G.P. Davis, S.S. Moore and N.G. Preston, 2000. Response to selection and heritability for growth in the Kuruma prawn, *Penaeus japonicus. Aquaculture.* 181: 215-223.

Hillier AG, 1984. Artificial conditions influencing the maturation and spawning of sub-adult *Penaeus monodon* (*Fabricius*). *Aquaculture*; 36, 179-84.

Joshi, V.P. and A.D. Diwan 1992. Artificial insemination studies *Macrobrachium idella*. In: Silas E.G. (Ed). Freshwater prawns. Kerala Agricultural University. Trichur. Pp. 110-118.

Kooda-Cisco. M. and Talbot, P., 1983. A technique of electrically stimulating extrusion of spermatophores from the lobster, *Homarus americanus*, *Aquaculture*. 30:221-227.

Laubier-Bonichon. A. and Ponticelli. A., 1981. Artificial laying of spermatophores on females of the shrimp *Penaeus japonicus*. Bate. Poster Paper. World Conference on *Aquaculture*. Venice, Italy.

Leverage, W.E, Valerio, D.A, Schultz, A.P, Kingsubry, E. and Doray, C, 1972. Comparative study on the freeze preservation of spermatozoa, primate, bovine and human, Lab. *Anim. Sci.*, 22:882-889.

Lin. M.N. and Ting. Y.Y., 1984. Studies on the artificial insemination and fertilization of grass shrimp. *Peneaus monodon*. Oral presentation In Proceedings or the First International Conference on the culture of penaeid prawns/shrimps. Ilolo city, Philipines SEAFDEC *Aquaculture* dept.

Lin, M-N., & Ting, Y-Y. 1985. Studies on the artificial insemination and fertilization of grass shrimp, *Penaeus monodon* (Abstract only). In Taki Y., Primavera J.H. and L Lobrera J.A. (Eds.). Proceedings of the First International Conference on the Culture of Penaeid Prawns/Shrimps, 4-7 December 1984, Iloilo City, Philippines (p. 167). Iloilo City, Philippines: *Aquaculture* Department, Southeast Asian Fisheries Development Centre.

Lin, M-N. and Ting, Y-Y. 1986. Spermatophore transplantation and artificial fertilization in grass shrimp. *Bull. Jap. Soc. Fish.* 52:585-589.

Lin, M-N. and Hanyu, I. 1990. Improvements on artificial insemination in gravid females of closethelycum *Penaeus penicillatus*. In: R. Hirano and I. Hanyu (eds,) *The Second Asian Fisheries Forum*, Asian Fisheries Society, Manila, Philippines, 627pp.

Lumare. F, 1981. Artificial reproduction of *Penaeus Japonicus*. Bate as a basis for the production of eggs and larvae. *J. World Maricult. Soc.*, 12 (2): 335-344.

McIntosh, R.P. 1999. Changing paradigms in shrimp farming: 1. General description. Global *Aquaculture* Advocate. 2:40-47.

Misamore, M. and C.L. Browdy. 1997. Evaluating the hybridization potential between *P. setiferus* and *P. vannamei* through natural mating, artificial insemination and in vitro fertilization. *Aquaculture* 150:1-10, (6).

Moore, S.S., V. Whan, G.P. Davis, K. Byrne, D.J.S. Hetzel and N. Preston. 1999. The development and application of genetic markers for the Kuruma prawn, *Penaeus japonicus*. *Aquaculture*. 173: 19-32.

Moss, S.M., Argue, B.J. and Arce, S.M., 1999. Genetic improvement of the Pacific white shrimp, *Litopenaeus vannamei*, at the Oceanic Institute. *Glob. Aquac. Advocate* 2(6): 41–43.

Muthu. M.S., and Laxminarayan, A., 1984. Artificial insemination *Penaeus monodon*. Curr. Sci., 53(200): 1075-1077.

Pawar Ravindra and M. Iftekhar Mohiuddin, 2000. Artificial insemination in penaeid shrimps. *Journal of the Indian Fisheries Association*. 27, 19-25.

Peeters, L., Diter, A., 1994. Effects of impregnation on maturation, spawning, and ecdysis of female shrimp *Penaeus indicus*. *J. Exp. Zool.* 269, 522– 530.

Persyn, H.O. 1977. Artificial insemination of shrimp. U.S. Patent 4, 031,855, 4 pp.

Perez-Velazquez, M., Bray, W.A., Lawrence, A.L., Gatlin III, D.M. and Gonzalez-Felix, M.L., 2001. Effect of temperature on sperm quality of captive *Litopenaeus vannamei* broodstock. *Aquaculture*, 198: 209–218.

Petersen, R.L., Beltrame, E., Derner, R., 1996. Inseminacion artificial en Penaeus paulensis (Pe´rez-Farfante, 1967). *Rev. Investig.*, 215– 219.

Peixoto Silvio, Ronaldo O. Cavalli, Dariano Krummenauer, Wilson Wasielesky, Fernando D'Incao, 2004. Influence of artificial insemination on the reproductive performance of *Farfantepenaeus paulensis* in conventional and unisex maturation systems. *Aquaculture* 230, 197–204.

Pongtippatee P, Vanichviriyakit R, Chavadej J, Plodpai P, Pratoomchart B, Sobhon P, Withyachumnarnkul B, 2007. Acrosome reaction in the sperm of the black tiger shrimp *Penaeus monodon* (Decapoda, Penaeidae). *Aquaculture* Research 38: 1635–1644.

Pratoomchat, B.S., Piyatiratitivorakul and P. Menasveta, 1993. Sperm quality of pond reared and wild-caught *Penaeus monodon* in Thailand. *J. World Aquacult. Soc.*, 24(4): 530-540.

Rosas, C., Sanchez, A., Chimal, M.E., Saldana, G., Ramos, L. and Soto, L.A., 1993. The effect of electrical stimulation on spermatophore regeneration in white shrimp *Penaeus setiferus*. *Aquatic Living Resources*, 6(2), pp.139-144.

Sandifer, P.A. & T.I.J. Smith, 1979. Possible significance of variations in the larval development of Palaemonid shrimp. *J. Exp. Mar. Biol. Ecol.* 39:55-64, (20).

Sandifer. P.A. and Lynn, J.W, 1980. Artificial insemination of caridean shrimp. In "Advances in Invertebrate Reproduction". W.H. Clark Jr. and T.S. Adams (Eds.) *Elsevier*, North Holland. Inc., 271-278.

Sandifer, P.A., A.L. Lawrence, S.G. Harris, G.W. Chambeerlain, A.D. Stocks and Bray 1984. Electrical stimulation of spermatophore expulsion in marine shrimp *Penaeus* spp. *Aquaculture*, 41: 181-187.

Silas. M.R. and Subramoniarn. T, 1987. A new method of eleclroejaculation of spermatophore from the sand lobster. *Thennus orientalis.* (Abstract). The Firsl Indian Fisherie. Forum, Mangalore. Asian Fisheries Society In dilin branch (Pub1).

Talbot, P., C. Thaler and P. Wilson 1986. Artificial insemination of the American lobster. *Gam. Res.*, 14: 25-31.

Tave. D. and Brown, A. Jr., 1981. A new device to help facilitate manual spermatophore transfer in penaeid shrimp (A brief technical note). *Aquaculture.* 25: 299-301.

Uno. Y. and Fujita, M, 1972. Studies on the experimental hybridization of freshwater shrimps, *Macrobrachium nipponence* and *M. formosense.* 2nd International Ocean Development Conference. Tokyo, Japan (abstract).

Yano, L., Kanna, R.A., Oyama, R.N., Wyban, J.A., 1988. Mating behaviour in the penaeid shrimp *Penaeus vannamei. Mar. Biol.* 97, 171– 175.

SECTION-III

GROWTH PHYSIOLOGY AND BIOTECHNOLOGY

CHAPTER 8

PHYSIOLOGY OF GROWTH AND MOULTING

8.1 INTRODUCTION

In the life history of the shrimp, growth is achieved only through periodic shedding of the cuticular sheath covering the body. Thus, shedding of the cuticle is the most important physiological event in these animals. During moult cycle many cyclic changes occur in structural, biochemical and physiological processes. This dynamic event continues throughout the life span of the shrimp linking almost all biological activities with this process. In scientific shrimp farming aiming at high production, adequate knowledge on the physiology of growth is imperative. Pioneer workers like Drach in 1939 and Scheer in 1960 initiated the classic studies on crustacean moulting physiology. In subsequent years a great deal of information has been added to this particular field by number of workers (see review by Fingerman, 1987). In spite of these great contributions only very limited attention have been paid to the moulting physiology of natantians, especially penaeids. In this chapter much of the information reviewed is from the works of Vijayan (1988) on physiological aspects of moulting in *F. indicus* along with the work by others.

8.2 MOULT CYCLE

Classification and Characteristics

To understand the growth and moult linked processes in detail, it is necessary to understand the classification of moulting process. Drach in 1939 was the first to establish the concept of classification. Earlier workers developed moult staging schemes based on histological changes in the exoskeleton (Passano, 1960). But later, many workers have modified this scheme (Vijayan 1988). According to this new scheme the criterion is to identify the characteristics of new setae that are formed in the tissues of appendages to replace those lost with the old exoskeleton at the time of ecdysis. Procedure for identifying moult stages in pleocyematans is well documented but detailed staging of the complete moult cycle in dendrobranchiates especially among penaeids has not been widely reported. Other than the work of Longmuir (1983) in *Penaeus merguiensis* and Smith and Dall (1984) in *P. esculantus*, moult staging studies made by Dall (1965), Schafer (1967), Huner and Colvin (1979) and Pudadera *et al.*, (1984) furnish only limited information. Some preliminary works on moult staging of *F. indicus* have been carried out by Read (1977) and Diwan and Usha (1985).

Moult staging has also been carried out by studying the histological changes in the cuticle but such studies on penaeids are again limited. Similarly, Moulting and Growth investigation on moult cycle duration with respect of each moult stage and mechanism of exuviations are not much explored except for the work of Longmuir (1983) Smith and Dall (1984). Based on morphological changes associated with setal development and epidermal retraction in the uropod and histological characteristics of cuticle, the entire moulting cycle for *F. indicus* and other shrimp is classified broadly into three main stages i.e. postmoult, intermoult and premoult stage. The postmoult is further divided into two stages namely early postmoult A and late postmoult B, while the intermoult has only one stage C and the premoult stage D is subdivided into stages DO, D1', D1", D1'" and D23. The determinative characters of the various moult stages are described in Table 1 (Fig. 1).

Cuticle histology

Histologically the cuticle of early postmoult stage A has only preexuvial layers of epicuticle and exocuticle. Epidermal cells are distinct and elongated with an average cell height of about 40μm. The cuticle is generally soft and flexible. In the stage B in addition to epicuticle and exocuticle there is a development of one more layer *viz*, endocuticle which lies below the exocuticle. The height of epidermal cells in this stage reduces to around 17μm. In the intermoult stage C all the three layers of the cuticle are well differentiated. The epidermal cells become small and not differentiated properly. The height of cells at this stage measures around 7μm. In the late premoult stages D'" and D23 there is a formation of new cuticle under the old one. The new cuticle has again two preexuvial layers under the old cuticle, exocuticle and epicuticle. The endocuticle of the old cuticle disintegrates slowly. The epidermal cells become prominent and increase in size (Fig. 1).

Table 1. Criterion for the moult staging in *F. indicus* on the basis of setogenesis.

Moult stage	Characteristic features	Approximate duration in hrs	Average % of duration	
A	Body soft and slippery to touch. Rostrum flexible. Granular protoplasmic matrix continuous in new setae. Setal cones and cuticular nodes absent.	3-7	2.1	Postmoult
B	Parchment like integument well developed. Setal articulation and poorly developed cuticular nodes. Setal cones absent.	16-22	8.35	Inter-moult 18.35%
C	Rigid exoskeleton with firm rostrum.Presence of well defined setal cones and cuticular nodes.	36-48	18.35	
D0	Retracted epidermis with an amber coloured zone at the tip of the uropod.	24-36	13.05	
D1'	Setal invagination and scalloped epidermis.	48-72	26	
D1"	Protoplasm condenses in the region of the setae. Setal invagination deepened, marking the appearance of new setal walls.	24-48	15.2	Premoult 71 %
D1'''	New setae appears in the uropod matrix with double wall. Setal invagination completed. Shafts visible at the tip of the setae. Proximal part of the blunt. Well developed shafts and barbules.	20-36	11.9	
D2-3	Newly developed new setae appears in the matrix as 'tube in tube' structures and barbules.	8-14	4.75	
E	Rejection of the old cuticle			

(**Source:** Vijayan, 1988)

8.3 MOULT CYCLE DURATION

In one complete moult cycle of *F. indicus*, duration of premoult is the longest, comprising 71% of the moult cycle. While the intermoult period is the shortest, being around 10.5% and that of postmoult 18.0%. Duration of moult again varies depending upon biotic and abiotic factors influencing the process. Age

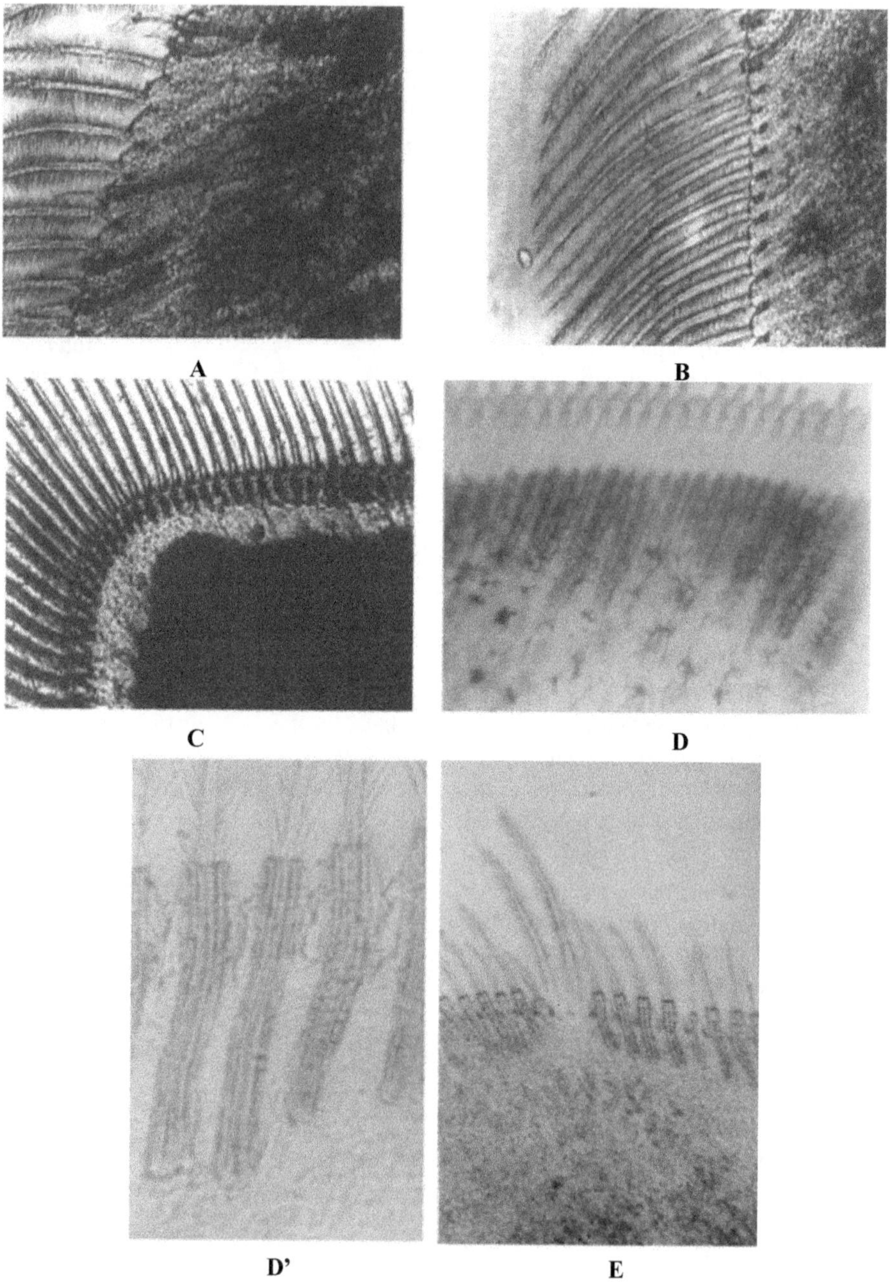

Fig. 1. Moult stage A – Setae with granular cytoplasm and setal articulation × 100.
Moult stage B – Cuticular node and setal lumen seen prominently × 100.
Moult stage C– Setal cones are seen distinctly × 100.
Moult stage DO– Retracted epidermis is seen clearly × 100.
Moult stage DO – Retracted epidermis is seen clearly × 100.
Moult stage E – Setal eversion × 100. (**Souce:** Vijayan, 1988)

and size of the animal also determines the duration. For example juvenile *F. indicus* of size 30-40 mm has average moult cycle duration of 96 hrs. Whereas adult of size 60-80 mm has duration of about 180 hrs and of size above 80 to120 mm has 240 hrs of moulting (Table 2).

Table 2. Moult cycle duration for different size groups of *F. indicus*

Size group (mm)	Number of animals	Average premoult duration (hrs)	Average moult cycle duration (hrs)
30-40	15	76±14	96±16
60-80	15	130±20	180±24
80-120	15	165±24	240±48

(**Source:** Vijayan, 1988)

8.4 MECHANISM OF EXUVIATION

Exuviation process has been carefully monitored in *F. indicus* by Vijayan (1988). The report describes that exuviation generally takes place between 00 to 04 hrs of the day. Shrimps in late premoult stage are very active in midnight hours and during this time they perform several activities like swimming, jumping, flicking, rolling and rotating movements. Rapid propulsion of the animal with the help of pereiopods along with flexing of the body convexly at the cephalo thorax abdomen joint is very common. During exuviations animals stretch their body vertically with the help of the arch centred on the third abdominal segment. This period of intense locomotion helps the animals in the removal of the old exoskeleton. During the actual shedding process, old carapce is thrown out separately of the cephalothorax. They then flick violently lifting the body out of the abdominal and ventral cephalothorax portion of the old exoskeleton. Following the moult they lay on side for about 530 minutes before regaining movement.

8.5 MOULTING METABOLISM

Passano (1960) and Yamaoka and Scheer (1970) while reviewing the principal physiological characteristics of the typical decapod moult cycle gave primary importance to the accumulation of organic reserves in different tissues of the animal body. Later several crustacean workers have confirmed profound changes in the organic contents of major body tissues in relation to the moult process. Hepatopancreas or the digestive gland is the primary organ for storage of organic reserves, while haemolymph plays a secondary role as a storage site. The organic reserves like protein, lipid and carbohydrates are important not only as a source of material for the construction of exoskeleton but also for the required energy during the process of moulting. Proper utilization of these reserves ensures the success of the moult as well as good growth at each moult cycle.

Considerable amount of work concerning the mobilization of organic reserves in relation to moult cycle is available in crustaceans but in penaeid shrimps the investigations are limited. For example Bursey and Lane (1971) have described haemolymph protein of *Penaeus duorarum* in relation to different stages of moult cycle. Later, Read and Caulton (1980) and Barclay *et al.*, (1983) have reported the mobilization of protein in relation to moult cycle in *F. indicus* and *P. esculentus* respectively. Utilization of lipid reserves in relation to moult cycle is known in *Metapenaeus monoceros* (Madhyastha and Rangnekar, 1974). Read and Caulton (1980) and Barclay and others (1983) have also followed the lipid reserves of *F. indicus* and *P. esculentus* as a function of moult stages. Moult linked variations of glucose and glucosamine content in the haemolymph of *Metapenaeus* sp. has been studied by Dall (1965), Diwan and Usha (1987) and Vijayan and Diwan (1993). Other than this, there is no information on behavioural pattern of several organic constituents in shrimps particularly *F. indicus.*

F. indicus also shows cyclic accumulation of organic reserves in selected major tissues at the time of each moult process. The general trend shows that the major metabolites reduce their levels during the post moult stages and an increase in levels during pre moult. Protein, RNA, lipid, glycogen and glucose in different tissues such as haemolymph, muscle and hepato pancreas follows this trend. The rise in protein levels particularly in haemolymph prior to moult is considered to be due to the resorption of organic material from chitin protein complex of the old exoskeleton and synthesis of protein from available amino acids particularly when the animal ceases to feed. Decreased protein concentration in the post moult stages is related to dilution of haemolymph by water which enters after the moult. Similarly, premoult increase in RNA levels also reflects the high turnover rate of protein synthesis, which helps in the building up of tissue after moulting. Even some workers have mentioned high RNA/protein ratio in early premoult stage but later the ratio reduces because of protein breakdown as the animal stops feeding in late premoult and early postmoult phases. The rise in the level of lipid particularly in hepatopancreatic tissue during premoult period is report to be due to accumulation process. From the hepato pancreatic tissue the lipid is released into the haemolymph for further transportation for its use in neo cuticle synthesis. This is the reason attributed for high lipid level in haemolymph at the time of premoult. Cholesterol also shows similar trend as lipid. It is suggested that in shrimps cholesterol may be circulated in the body as an unesterified sterol, perhaps as a lipo-protein complex, implicated in the transport of metabolites. The role of cholesterol in the synthesis of chitin is well known. Similarly the importance of tissue glycogen and haemolymph glucose in the synthesis of chitin and as an energy sources in intermediary metabolism is also well established. Increase in their levels in premoult stages and depletion in postmoult stages in *F. indicus* indicate that these organic constituents are mobilized for the growth of post- exuvial layers of the new integument (Bliss, 1983) (Table 3).

Other organic constituents like DNA in tissues, glucosamine in haemolymph, chitin in cuticle and water level in tissues have different trends. These constituents increase in their levels during postmoult and lower their values in premoult. It is reported that the higher DNA levels in the premoult and postmoult is due to cell division occurring at the time of rebuilding of the cuticle. In the biosynthesis of neocuticle, glucosamine is the immediate precursor for chitin formation. Therefore, levels of glucosamine and chitin are always higher in the late premoult and immediately after the moult, but subsequently the levels reduce as the hardening of cuticle advances. Water is the most important factor in the moulting process particularly in expanding the body volume. When there is rise in water level of tissues, particularly during postmoult stages, there is depletion in major organic reserves. Conversely at the time of accumulation of organic reserves with advancement of hardening of cuticle, there is a corresponding decrease in water content also. This clearly indicates the replacement of water by tissue growth during the periods between moults which in turn is considered as the actual growth in the animals, when there is an increase in the dry weight of the body (Bliss, 1983).

8.6 MINERALS AND MOULT CYCLE

The crustacean exoskeleton is extensively mineralized with calcium carbonate in the form of calcite, which is the principal inorganic component of the exoskeleton. Magnesium and phosphate salts are in relatively lower levels. The mineral load in the exoskeleton is in a constant state of flux, since most of the decapods have to mineralize the newly built exoskeleton in the preparation for the next moult (Passano, 1960). Extensive information is available on cuticular mineralization of heavily mineralized decapods such as the crayfishes, lobsters and crabs (Travis, 1965~ Mills and Lake 1976, Vigh and Dendinger 1982, Greenway, 1985). However, there is a paucity of information on natantian shrimps which are less mineralized. Some work on calcium, magnesium and phosphorus levels in relation to moult cycle is available on penaeid shrimps (Dall, 1965~ Bursey and Lane, 1971~ Welinder, 1974 and Huner *et al.* 1979 b~ Vijayan and Diwan, 1996).

Table 3. Variations in some of the organic constituents during different moult stages of shrimp, *F. Indicus*.

	Moult stages		A	B	C	D0	D1'	D1''	D2-3
Protein	Haemo-lymph (mg/ml)	N	7	7	7	7	7	7	7
		X	27.8	37.05	56.86	84.18	99.72	76.14	61.09
		SD	(2.09)	(3.51)	(6.05)	(7.89)	(5.02)	(4.99)	(3.28)
	Muscle (mg/100mg dry weight)	N	7	7	7	7	7	7	7
		X	38.90	47.89	51.45	53.88	63.30	54.21	40.95
		SD	(2.66)	(5.10)	(5.68)	(6.22)	(4.41)	(5.34)	(5.74)
	Hepato-pancreas (mg/00mg dry weight)	N	7	7	7	7	7	7	7
		X	8.39	12.10	14.05	15.37	18.35	16.33	14.42
		SD	(0.66)	(1.68)	(1.01)	(1.00)	(1.18)	(0.70)	(1.08)
RNA	Muscle (µg/mg) wet wt.	N	7	7	7	7	7	7	7
		X	3.31	5.35	7.70	12.33	16.98	13.83	11.05
		SD	(0.48)	(0.45)	(1.02)	(1.16)	(1.91)	(1.03)	(2.36)
RNA	Hepato-pancreas (µg/mg)	N	7	7	7	7	7	7	7
		X	11.43	25.37	31.77	45.33	55.8	41.88	34.33
		SD	(3.02)	(3.13)	(2.41)	(3.34)	(4.56)	(2.08)	(3.45)

[Table Contd.

Contd. Table]

	Moult stages		A	B	C	D0	D1'	D1'''	D2-3
DNA	Muscle (µg/mg) wet wt.	N	7	7	7	7	7	7	7
		X	5.35	4.38	2.6	2.43	2.4	3.13	4.45
		SD	(0.49)	(0.39)	()0.8	()0.57	(0.46)	0.71)	(0.48)
	Hepato-pancreas (µg/mg)	N	7	7	7	7	7	7	7
		X	4.77	4.85	3.67	3.6	2.8	3.47	4.35
		SD	(0.54)	(0.77)	(0.43)	(0.47)	(0.55)	(0.48)	(0.64)
Lipid	Haemo-lymph (mg/ml)	N	7	7	7	7	7	7	7
		X	7.43	8.51	8.82	9.35	9.70	9.78	10.34
		SD	(0.34)	(0.37)	(0.17)	(0.31)	(0.37)	(0.22)	(0.51)
	Muscle (mg/100mg dry weight)	N	7	7	7	7	7	7	7
		X	2.72	3.61	3.99	6.20	7.49	5.37	3.92
		SD	(0.38)	(0.49)	(0.28)	(0.42)	(0.35)	(0.31)	(0.23)
	Hepato-pancreas (mg/100mg dry weight)	N	7	7	7	7	7	7	7
		X	13.02	18.60	30.43	41.72	50.14	53.25	45.72
		SD	(2.07)	(1.37)	(3.71)	(1.61)	(1.7)	(1.92)	(1.27)

[Table Contd.

Contd. Table]

	Moult stages								
			A	B	C	D0	D1'	D1'''	D2-3
Glycogen	Haemo- lymph (µg/ml)	N	7	7	7	7	7	7	7
		X	370.83	2.78	319.83	344.5	396.57	525.83	650.59
		SD	(27.74)	(9.53)	(7.55)	(8.34)	(35.34)	(25.75)	(29.49)
	Muscle (mg/g)dry wt.	N	7	7	7	7	7	7	7
		X	7.24	6.28	8.47	8.88	10.87	15.14	17.37
		SD	(41)	(0.56)	(0.68)	(0.52)	(0.97)	(1.54)	(1.38)
	Hepato- pancreas (mg/g dry wt)	N	7	7	7	7	7	7	7
		X	27.43	13.00	15.53	19.61	21.52	30.59	35.41
		SD	(3.94)	(3.94)	(1.40)	(1.69)	(1.69)	(0.90)	(1.91)
Glucose →	Haemo-lymph (µg/ml)	N	7	7	7	7	7	7	7
		X	269.43	245.67	332.25	518.75	638.83	658.83	671.67
		SD	(10.15)	(9.20)	(13.08)	(10.96)	(25.57)	(25.57)	(35.71)
Gluco- samine →	Haemo-lymph (µg/ml)	N	7	7	7	7	7	7	7
		X	153.67	118.57	71.14	25.18	21.28	56.17	199.43
		SD	(18.06)	(5.44)	(18.21)	(7.10)	(3.82)	(20.10)	(16.78)

[Table Contd.

Contd. Table]

Moult stages			A	B	C	D0	D1’	D1’”	D2-3
Chitin	Exo-skeleton (µg/ml)	N	7	7	7	7	7	7	7
		X	27.83	22.02	18.92	18.44	18.73	16.43	13.48
		SD	(1.66)	(1.47)	(1.13)	(1.38)	(1.39)	(1.112)	1.190)
	Muscle (%)	N	7	7	7	7	7	7	7
		X	78.52	76.41	75.61	74.81	73.8	74.89	75.41
		SD	(1.11)	(0.81)	(0.49)	(0.67)	(1.36)	(0.54)	(1.26)
Water →	Hepato-pancreas (%)	N	7	7	7	7	7	7	7
		X	78.86	72.07	69.31	65.03	61.93	64.22	61.35
		SD	(2.92)	(3.03)	(3.17)	(5.93)	(3.77)	(3.4)	(3.33)

(**Source:** Vijayan, 1988)

The extent to which hardening of the cuticle takes place with calcification varies greatly not only between different crustacean, but also between different regions of the same animal. Investigations dealing with the distribution of important minerals in the exoskeleton of crustacean have not been explored properly particularly in penaeids. Similarly property of the exoskeleton during moult cycle always depends upon mobilization of minerals from other tissues of the body like haemolymph, hepato pancreas and muscle. Hence periodic replacement of exoskeleton brings many changes in body tissues including mineral contents. Considerable work dealing on these aspects is available on crustaceans other than penaeids (Greenway, 1976~ Wright 1980~ Sheets and Dendinger, 1983).

Among the various minerals calcium dominates in *F. indicus* in distribution and total amount present in whole of the exoskeletal parts. The swimmerets have very less amount of calcium (29 mg/g) whereas in the rostrum the amount is very large (394 mg/g) (table 4).

Table 4. Distribution of Calcium, Magnesium and Phosphorus in the major exoskeletal areas of *F. indicus*.

	Area	Calcium (mg/g)	Magnesium (mg/g)	Phosphorus (mg/g)
1.	Rostrum	393.75 ±35.82	28.91 ±4.8	10.27 ±2.7
2.	Upper region of Carapace	176.32 ±17.44	18.75 ±3.76	10.41 ±1.03
3.	Lower region of Carapace	82.81 ±12.50	9.44 ±2.57	12.87 ±2.01
4.	Upper abdomen	136.35 ±18.10	12.55 ±1.99	13.71 ±1.11
5.	Lower abdomen	165.09 ±19.27	14.43 ±2.33	14.42 ±2.05
6.	Telson	138.72 ±26.8	9.71 ±1.36	11.76 ±1.15
7.	Antenna	153.59 ±23.07	7.62 ±0.98	13.08 ±2.04
8.	Antennule	148.64 ±13.4	8.04 ±1.02	11.64 ±1.40
9.	Mouth parts &Maxillipeds	152.86 ±22.72	10.69 ±2.12	9.93 ±0.99
10.	Walking legs	278.14 ±26.68	14.23 ±2.08	10.37 ±1.04
11.	Swimmerets	29.32 ±8.92	3.99 ±0.49	14.48 ±1.22
12.	Uropod Mean Exoskeletal Concentration	58.55 ±11.31 159.52	4.87 ±1.01 11.94	11.95 ±1.07 12.07

(**Source:** Vijayan, 1988)

While in other body parts (Fig. 2) viz. carapace, abdomen and telson the calcium level is around 183 mg/g. Calcium level present in the appendages is comparatively less. Variations in the mineral content also occur between the two areas of same exoskeletal regions. For example, calcium levels are high in the upper

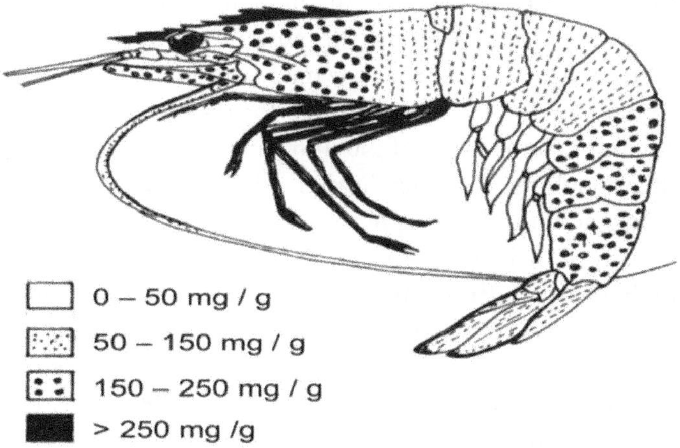

Fig. 2. The pattern of calcium distribution in the exoskeleton of *F. indicus.* (**Source:** Vijayan and Diwan, 1996).

region of carapace than the lower. Similar variation exits in upper and lower regions of abdominal cuticle. The reasons for such regional differences in the exoskeletal Ca content in crustaceans can be attributed possibly due to the conservation of Ca by restricting distribution to areas where it is most beneficial in terms of rigidity. Besides, environmental factors and nature of the habitat of animals also play important role in distribution pattern of minerals. The pattern of distribution of magnesium (Mg) on the exoskeletal areas is again similar to that of calcium but the amount present (in terms of quantity) is always less. In swimmerets Mg contents is around 4 mg/g and in rostrum the level is 29 mg/g. In other exoskeletal parts average Mg content is around 15 mg/g. The body appendages have comparatively less amount of Mg. Similar to calcium, variations in the Mg content also occur between the two areas of same exoskeletal regions. In upper region of carapace Mg content is high (19 mg/g) whereas in the lower regions of abdominal cuticle similar variations occur. Unlike Ca and Mg, phosphorus has no definite pattern in its distribution over the exoskeletal region which probably indicates that phosphorus has less role in hardening of the cuticle. However the total amount of phosphorus in exoskeletal region ranges between 10-14 mg/g (Vijayan and Diwan, 1996) (table 4).

There exists a good relationship between utilization of some important minerals from tissues and moult cycle in *F. indicus.* Calcium is mobilized to the maximum extent in body metabolism at the time of moulting process and the demand for calcium is particularly high in all crustaceans due to their periodic shedding of exoskeleton. In *F. indicus* there is initial rise in calcium levels of haemolymph, muscle and hepato pancreatic tissues in the premoult stages. Between the postmoult and intermoult stages, as the animal advances the process of hardening the cuticle the level of tissue calcium falls down and reaches its minimum in the intermoult stage. A rise in haemolymph calcium concentration is cited as an evidence for resorption of calcium from the old exoskeleton prior to the event of moulting and the postmoult fall is due to its utilization after moulting in mineralization of the newly formed exoskeleton. Haemolymph magnesium and phosphorous have similar role as calcium and hence there is a similarity in their trend. In muscle tissue calcium and magnesium levels build up slowly in ascending way during post moult to premoult period. Immediately after the moulting process there is a sudden reduction in the levels. In contrast to Ca and Mg, the fall in phosphorous content after moulting process is not very drastic (Bliss, 1983, Vijayan and Diwan, 1996). Passano (1960) established the importance of hepatopancreas as a temporary storage site for inorganic substances reabsorbed from the old exoskeleton prior to the event of ecdysis. The reabsorbed substances are then utilized effectively from hepatopancreas immediately after the moult till animal achieves normalcy of intermoult phase. Accumulation of Ca and Mg in hepatopancreas takes place in *F. indicus* from intermoult to premoult stages and this built up storage is maintained even after commencement of moulting. The real fall in the levels occur only when the minerals start mobilizing for hardening of the cuticle in the late postmoult stages. Even the hepatopancreas phosphorus has same trend as Ca and Mg. The Ca, Mg and P content in exoskeletal tissue have also cyclic pattern with the moulting process i.e. building up the levels from postmoult to intermoult and then decline in the levels slowly from premoult to moult. Mineralization and demineralization of exoskeletal tissues always depend upon many factors within the animal and surrounding environment (Table 5).

8.7 MOULTING AND ENVIRONMENTAL FACTORS

Many biotic and abiotic factors of the environment exert their significant effect on physiological system of the animal. Crustaceans are not an exception to these effects because of their distribution in a wide range of ecological habitat. Moulting process in case of crustaceans for most of the time is under the influence of some or other environmental factors.

Abiotic factors

Prominent abiotic factors which influence the moulting process directly or indirectly are temperature, salinity, pH and light. Influence of temperature on frequency of moult and growth rate of crabs and lobsters has been well studied (Diwan and Nagabhushanam, 1975~ Aiken and Waddy, 1975~ Haefner

Table 5. Variations in the concentration of Calcium, Magnesium and Phosphorous during different moult stages of shrimp, *F. indicus*.

	Moult stages								
	Tissue		**A**	**B**	**C**	**D0**	**D1'**	**D1"**	**D2-3**
Calcium	Haemo-lymph (mg/ml)	N	8	8	8	8	8	8	8
		X	1.738	1.232	0.834	0.826	0.781	1.346	1.884
		±SD	0.158	0.228	0.0062	0.099	0.078	0.134	0.234
	Muscle (mg/g)	N	8	8	8	8	8	8	8
		X	1.745	1.778	2.608	3.627	4.628	4.686	4.737
		±SD	0.366	0.331	0.426	0.450	0.501	0.553	0.631
	Hepato-pancreas (mg/g)	N	8	8	8	8	8	8	8
		X	14.202	11.674	7.506	8.052	8.395	10.623	12.7
		±SD	2.222	2.044	1.223	1.354	1.144	1.607	2.39
	Exo-skeleton (mg/ml)	N	8	8	8	8	8	8	8
		X	51.167	99.742	161.592	166.642	168.984	153.850	147.657
		±SD	9.362	13.804	9.583	7.850	6.820	9.986	6.605
Magnesium	Hemo-lymph (mg/ml)	N	8	8	8	8	8	8	8
		X	0.422	0.416	0.176	0.163	0.155	0.276	0.328
		±SD	0.013	0.021	0.011	0.002	0.009	0.009	0.017
	Muscle (mg/g)	N	8	8	8	8	8	8	8
		X	0.887	0.749	1.339	1.446	1.481	1.486	1.437
		±SD	0.029	0.015	0.046	0.035	0.019	0.027	0.031

[Table Contd.

Contd. Table]

	Moult stages								
	Tissue		A	B	C	D0	D1'	D1"	D2-3
Magnesium	Hepato-pancreas (mg/g)	N	8	8	8	8	8	8	8
		X	4.184	3.319	2.758	2.663	2.715	3.669	3.867
		±SD	0.029	0.248	0.176	0.152	0.184	0.126	0.580
	Exo-skeleton (mg/g)	N	8	8	8	8	8	8	8
		X	7.232	9.561	12.474	14.077	14.358	13.149	11.223
		±SD	0.147	0.495	0.409	0.534	0.536	0.499	0.726
Phosphorous	Haemo-lymph (mg/ml)	N	8	8	8	8	8	8	8
		X	0.098	0.072	0.050	0.044	0.042	0.081	0.108
		±SD	0.006	0.008	0.008	0.006	0.005	0.009	0.006
	Muscle (mg/g)	N	8	8	8	8	8	8	8
		X	5.462	5.231	4.722	4.437	5.096	5.628	-(r.907
		±SD	0.507	0.432	0.316	0.283	0.325	0.312	0.774
	Hepato-pancreas (mg/g)	N	8	8	8	8	8	8	8
		X	1.460	1.635	1.799	1.808	2.077	1.867	1.756
		±SD	0.090	0.055	0.105	0.094	0.107	0.084	0.123
	Exo-skeleton (mg/g)	N	8	8	8	8	8	8	8
		X	19.565	18.565	12.475	13.793	14.349	15.265	26.319
		±SD	0.544	3.054	2.389	1.262	1.467	1.764	4.009

(**Source:** Vijayan, 1988)

and Van Engel, 1975~ Winget *et al.*, 1976). Temperature influence on growth and moulting of other caridean shrimps has been also reported (Richard, 1978). *F. indicus* which is used extensively in the brackish water shrimp farming suffers significant stress from environmental factors, like temperature, salinity and pH throughout their life cycle. Among the various environmental factors, temperature effect is very much pronounced on moult cycle and growth. Moulting frequency and growth of *F. indicus* is faster at 31°C when compared to other temperatures like 26°,32. 5° and 35°C (Table 8). Though a positive relationship exists between moulting frequency and temperature, the growth of animals is not linear with rise in temperature. For example, moulting frequency increases or moult cycle duration shortens at 35°C but increase in weight in terms of growth is less at this temperature. Conan (1985) has mentioned that within a specific temperature range, interval between the ecdys is usually shortened with a corresponding increase in tissue growth until a threshold level of temperature is reached. Beyond this threshold temperature, moulting of the animal becomes rather erratic. This is true for *F. indicus*. The effect of higher temperature resulting enhanced growth can be due to accelerated metabolism. Similarly lower growth rate and moulting frequency can be anticipated due to slow metabolism. Ideal temperature condition for good growth of *F. indicus* is around 29-31°C and 31°C is a thermal threshold temperature. Like temperature, salinity also plays an important role in controlling physiological state of animal through its osmotic effects. (Table 6)

Table 6. Effect of temperature on moulting and growth in *F. indicus* juveniles

		Temperature (°C)			
		26	**31**	**32.5**	**35**
Increase in length (mm)	N	12	12	12	12
	X	6.9	11.9	8.33	4.4
	SD	(1.2)	(0.8)	(0.9)	(0.8)
Increase in wet weight (mg)	X	162.3	204.7	159.9	56.4
	SD	(7.0)	(2.6)	(2.4)	(3.0)
Moult cycle duration (hrs)	X	138.0	96.0	96.0	84.0
	SD	(10.8)	(0.0)	(0.0)	(12.5)

(**Source:** Vijayan, 1988)

Salinity effects on growth and survival of shrimps have been described by several authors (Gunter, 1961~ Venkataramiah *et al.*, 1972~ Raj and Raj, 1980). Brackish and intertidal environments are probably the most demanding and stressful aquatic biotope where shrimps like *F. indicus* complete the growth phase of their life. Salinity of the environment constantly influences the moulting and growth process in shrimps particularly at larval and juvenile phase of the life cycle. In juvenile *F. indicus* moulting and growth is reported to be faster at optimum salinity of 15%. *F. indicus* exposed to lower (5%) and higher (45%) salinities are reported to have developed signs of stress and muscle necrosis. In salinities other than optimum range, moulting process occurs but the growth increment per moult is less. It is described by Kinne, (1970) that most of the euryhaline animals though survive and moult in wide range of salinities, the actual growth takes place in restricted narrow range of salinity. This may be true for juvenile *F. indicus*. The effect of salinity on the growth of *F. indicus* post larvae and juveniles was also studied by Nair and Kutty (1975) and they reported considerable variations in the growth rate of these stages in different salinity conditions. The growth rate was high in 10 ppt saline condition for post larvae and for juveniles, 30 ppt salinity was favourable. In another investigation it has been found out that for adult *F. indicus* the highest mean weight gain of 37 mg/day occurred in 25% salinity followed by 15% and 5% and lowest weight gain was only 4 mg/day in salinity at 45%. The trend of survival rates also showed a similar pattern. The higher growth and survival rates recorded at lower salinities probably indicate the animal's preference for these salinities for better growth which in turn reflects on their efficiency of consumption and utilization of food (Raj and Raj, 1980) (Table 7 & Fig. 2).

Table 7. Effect of salinity on moulting and growth in *F. indicus* juveniles

		Salinity %				
		5	**15**	**25**	**35**	**45**
Increase in	N	17	17	17	15	11
length	X	7.5	11.8	9.6	4.6	2.9
(mm)	SD	(0.8)	(0.6)	(0.8)	(0.7)	(1.0)
Increase in	X	133.4	212.5	181.7	75.7	9.5
wet weight (mg)	SD	(8.4)	(5.8)	(23.7)	(6.0)	(4.8)
Moult cycle	X	104.5	96.0	98.1	160.8.	30.9
duration (hrs)		(11.8)	(0.0)	(4.7)	(41.8)	(16.5)

(**Source:** Vijayan, 1988)

Literature pertaining to the influence of pH on growth and moulting in crustacean animals is scanty. Wickins (1984) while studying the pH effects on *P. monodon*, described the possibility of pH fall in natural environment to nearly neutral values following the influx of rain water with corresponding reduction in salinity and alkalinity which sometimes adversely affect the normal physiology of the animal. Havas (1981) has also mentioned that many aquatic organisms are physiologically unable to tolerate the condition of extreme pH variations. Extremes of lower and higher pH of medium form a limiting factor for growth and moulting. In case of *F. indicus* optimum pH of 7.8 to 8.2 is necessary for successful moult. Outside the optimum range, both higher and lower pH adversely affects the moulting physiology of the shrimp. It has been described that since one third of the shrimp exoskeleton is calcium carbonate, under normal conditions, a major share of calcium is taken up from the seawater together with bicarbonate possibly in exchange for H ions to maintain electrical neutrality. Therefore, successful mineralization of the exoskeleton will depend upon the optimum ionic content of surrounding waters and the ability to take up ions. Without proper mineralization of exoskeleton moulting process is not possible (Table 8)

Table 8. Effect of pH on moulting and growth in *F. indicus* juveniles

		pH		
		7± 0.2	**8±0.2**	**9±0.2**
Increase in length (mm)	N	11	17	12
	X	4.9	10.9	3.1
	SD	(0.8)	(0.9)	(0.3)
Increase in wet length (mm) (mg)	X	155.5	194.9	48.7
	SD	(2.6)	(0.39)	(4.5)
Moult cycle duration (hrs)	X	91.6	96.0	109.3
	SD	(9.7)	(0.0)	(12.6)

(**Source:** Vijayan, 1988)

Studies on the effect of light over moulting and growth of crustaceans are confusing due to the contradictory results. Good amount of information is available dealing with the effects of photo period on moulting process in crabs, crayfishes, lobsters, isopods and caridean shrimps (Diwan and Nagabhushanam, 1975~ Emerson, 1980). But in penaeids there is a paucity of knowledge dealing with such effects. Though it has been established that the change in photoperiod definitely influences the moulting process in several crustaceans, in *F. indicus* its influence appears to be insignificant. However, further elaborate studies are needed to confirm this relationship (Table 9).

Table 9. Effect of light on moulting and growth in *F. indicus* juveniles

		Light (hrs)		
		24	12	0
Increase in length (mm)	N	12	12	12
	X	9.0	9.2	10.1
	±SD	(0.9)	(1.0)	(1.0)
Increase in wet weight (mg)	X	180.7	181.3	183.3
	±SD	(3.5)	(3.2)	(3.4)
Moult cycle duration (hrs)	X	102.7	100.0	96.0
	±SD	(10.9)	(9.34)	

(**Source:** Vijayan, 1988)

Considerable literature is available on the effect of autotomy on moulting and growth in reptantians (Chittleborough,1975~ Nakatani and Otsu, 1979~ Hopkins, 1982) but studies on similar lines in natantians particularly on penaeids are negligible. Autotomy of body appendages in case of *F. indicus* has in fact no influence in acceleration of moulting. The general hypothesis to explain this mechanism is that the moulting is stimulated by severing the nerves of the appendages and therefore related to the number of nerves cut during autotomy (Table10). Though much has been described about the feeding and starvation effects on growth and moulting (Rao et al 1977~ Dawirs, 1984~ Anger et al. 1985) information is lacking covering these aspects in penaeid shrimps. Importance of feeding on growth of *P. japonicus* and *P. serratus* has been described by Cuzon *et al.*, (1980) and Papathanassion and King (1984) respectively. In the case of *F. indicus* it is reported that starvation ceases the moulting process and the shrimps starved at different stages of moult cycle have initially prolonged their interval period between moult stages prior to death (Table 11).

Table 10. Effect of leg autotomy on the moult cycle of *F. indicus*

Group	Number of animals	Operation	Premoult Period(hrs)	't' test
I	17	2 legs removed	120 ± 19.6	
II	17	4 legs removed	123.±16.6	
III	17	6 legs removed	120 ± 13.9	P > 0.05
IV	17	8 legs removed	116 ± 16.6	
V	17	10legsremoved	126.9±181	
VI	17	Intact Controls	120 ± 19.6	

(**Source:** Vijayan, 1988)

Table 11. Effect of starvation on the moult cycle of *F. indicus**

Group	Number of animals	Treatment	Intermoult (hrs)	Premoult (hrs)
I	17	Starved from Postmoult stages A & B	54.9 ± 11.7	219.4 ± 16.6
II	17	Starved from Intermoult & early premoult (Stage C & DO)	34.3 ± 12.8	233.1 ± 11.7
III	17	Starved from late Premoult (Stages D1''' & D2-3)	34.3 ± 12.8	219.4 ± 16.6
IV	17	Properly fed controls	30.9 ± 11.7	130.3 ± 12.8

(**Source:** Vijayan, 1988)

8.8 NUCLEIC ACID AND PROTEIN RELATION TO BODY SIZE

Growth in animals has traditionally been expressed as an increase in either length or live weight or dry weight or dry weight body mass. Bulow in 1970 has described a new method to determine the growth rate especially during the exponential phase of the growth of animals by measuring the RNA/DNA ratio in the cell or tissue. Buckley (1979) further mentioned that this ratio can be used as an index to determine the nutritional status of the animal at the time of collection. This kind of study has been made for the first time in penaeid shrimps particularly *F. indicus* of different size groups (Thomas and Diwan, 1990). Mention has been made that high amount of RNA/DNA concentration declined as the size of the shrimps increased (upto147 mm) denoting an inverse relationship. DNA content also showed similar trend as RNA and DNA content appeared to be a special feature. Hence in *F. Indicus* RNA/DNA ratio has been found useful in correlating growth (in terms of protein increase) only in small sized shrimps. The ratio between these factors is erratic as the size of the animal increases. Thus, there is no information available at present about RNA/DNA ratio and its relation to protein content in penaeids.

8.9 CONCLUSION

The moulting accounts for growth in crustacean animals and the entire process is governed by hormonal factors directly or indirectly. The hormonal factors such as MIH and MAH are synthesized in neurosecretory cells of different neuroendocrine centers of the central nervous system. This has been shown by correlating the secretory activity of the neurosecretory cells with that of moulting process by routine histological techniques and also through the ultrastructural studies. But extensive studies are still warranted in this field particularly at biomolecular levels to understand the precise mechanism of the synthesis of the hormones (active biochemical factors) involved in growth and moulting process. In such investigations introduction of immune -cytochemical techniques are needed to identify the nature and functions of the specific cell and its product. Though the existence of MIH, MAH and MH has been also proved through experimental studies, information is scarce on the biochemical nature of these products, their titres in circulating body fluid and physiological specificity in relation to moulting process. Therefore, this area requires further elaborate investigations. The evidence for the Y-- organ as the source of ecdysone or MH in controlling moult process has been established from a number of observations and experiments. Recent advanced analytical techniques like radio -immono assay (RIA) and organ culture, made possible direct proof of the ecdysteroid secretory nature of the Y- organ. There is need for similar works in other species of penaeids to consolidate and establish the role of Y- organ in moulting and other physiological events. Suspected involvement of other tissues in moult controlling function with cytological evidences as in Y-organ has been reported by many workers. One of such structures is mandibular or cephatic gland. Unfortunately we do not have information on the mandibular organs, particularly in penaeids. Though much has been said on the organic reserves and changes in metabolites in relation to moult cycle, there are many gaps in our knowledge with regard to oxidative enzymes as well as enzymes involved in cuticle formation. To understand the whole mechanism of cuticle formation in such calcareous animals, ultra structural studies are essential.

The significance of minerals not only in the cuticle formation but their requirement for other physiological events, has been reviewed by many workers. Much attention has been devoted to the cuticular mineralization of the heavily mineralized decapods. However, little attention has been paid on natantian shrimps which are less mineralized. There is scope for further work on these aspects.

References

Aiken, D.E. and S.L. Waddy. 1975. Temperature increase can cause hyperecdysonism in American lobsters (*Homarus americanus*) injected with ecdydsterone. *J. Fish. Res.* Board Can., 32: 1843-1845.

Anger, K., M.Storch, and J.M. Capuzzo. 1985. Effects of starvation on moult cycle and hepatopancreas of stage I lobster. *Helgol. Wiss. Meeresunters.*, 39: 107-116.

Barclay, M.C., W. Dall, and D.M. Smith. 1983. Changes in lipid and protein during starvation and the moulting cycle in the tiger shrimp, *Penaeus esculentus. J. Exp. Mar. Biol. Ecol.*, 68: 221-244.

Bliss, D.E. (ed) 1983. The biology of Crustacea. Vol. 5. Academic Press, New York, 471pp.

Buckley, L.J. 1979. Changes in RNA, DNAand protein during ontogenesis in "Winter floyunder" (*Pseudopleuronectes americanus*) and the effect of starvation~ *Fish Bull.* U.S. 77.

Bursey, C.R. and C.E. Lane. 1971. Ionic and Protein concentration changes during the moult cycle of *Penaeus duorarum. Comp. Biochem. Physiol.* 40A: 155- 162.

Chittleborough, R.G. 1975. Environmental factors affecting growth and survival of juvenile western rock lobster *Panulirus longipes.* Aust. *J. Mar. Freshwat. Res.*, 26: 177-196.

Conan, G.Y. 1985. Periodicity and phasing of moulting. In: Crustacean Tssues (Ed. by A.M Wenner), 73-99.

Cuzon, G., C. Cachu, J.F. Aldrin, J.L. Messager, G. Stephen, and M. Mevel. 1980. Starvation and its effects on metabolism of *Penaeus japonicus. Proc. World Maric. Soc.*, II, 410-423.

Dall, W. 1965a. Studies on the physiology of a shrimp, Metapenaeus sp. III. Composition and structure of the integument. Aust. *J. Mar. Freshwat. Res.*, 16: 13-23

Dawirs, R.R. 1984. Influence of starvation on larval development of *Carcinus maenas.* J. *Exp. Mar. Biol. Ecol.*, 80: 47-66.

Diwan A.D. and R. Nagabhushanam 1975. Moulting behaviour and its control in the freshwater crab, *Barytelphusa cunicularis* (Westwood 1836). *Rivista de Biologia.* 68:79-99.

Diwan, A. D. and T. Usha. 1985. Characterization of moult stages of *Penaeus indicus* based on developing uropod setae and some closely allied structures. *Indian J. Fish.*, 32: 275-279.

Drach, P. 1939. Mut et cycle d' intermue chez les crustaces decapods. *Ann. Inst. Oceanogr. Monaco.*, 19: 103-391.

Emmerson, W.D. 1980. Induced maturation of prawn *Penaeus indicus. Mar. Ecol. Prog. Serv.*, 2, 121-131.

Fingerman, M. 1987. The endocrine mechanism of crustaceans. *J. Crust. Biol.*, 7: 1-24.

Greenaway, P. 1985. Calcium balance and moulting in the crustacean. *Biol. Rev.*, 60: 425-454.

Gunter, G. 1961. Some relation of estuarine organism to salinity. *Limnol. Oceanogr.*,6: 182-190.

Haefner, P.A., Jr. and W. Van Engel. 1975. Aspects of moulting growth and survival of male rock crabs Cancer irroratus, in Chesapeake Bay. *Chesapeake Sci.*, 16: 253-265.

Havas, M. 1981. Physiological response of aquatic animals to low pH. Proc.~ symp. Acidic precipitation on Benthos, 49-65, 1980, North American Benthological society, Hamilton, New York.

Hopkins, P.M. 1982. Growth and regeneration pattern in fiddler crab, *Uca pugilator. Biol. Bull.*, 163: 301-319.

Huner, J.V. and L.B. Colvin. 1979. Observations on the moult cycles of two species of juvenile shrimp, *Penaeus californiensis* and *Penaeus stylirostris* (Decapoda: Crustacea). *Proc. Nat. shellfish Assoc.*, 69: 77-84.

Huner, J.V., L.B. Colvin and B.L. Reid. 1979. Postmoult mineralization of the exoskeleton of juvenile Californian brown shrimp, *Penaeus californiensis. Comp. Biochem. Physiol.,* 62A: 889-893.

Kinne, O. 1970. Salinity in: "Marine Ecology" (Ed. by Kinne. O), 821-995.

Longmuir, E. 1983. Setal development, moult-staging and ecdysis in the banana prawn *Penaeus merguiensis. Mar. Biol.*, 77: 183-90.

Madhyastha, M.N. and P.V. Rangnekar. 1976. Neurosecretory cells in the central nervour system of the prawn *Metapenaeus monoceros. Revistia. Biol.*, IXXIX, 133-140.

Mills, B.J. and P.S. Lake. 1976 The amount and distribution of calcium in the exoskeleton of the intermoult crayfish *Parastacoides tasmanicus* (Erichsom) and *Astacopsis fluviatilis* (Gray). *Comp. Biochem.* Physiol., 53A: 355-360.

Nakatani, I. And T. Otsu. 1979. The effects of eyestalk, leg, and uropod removal on the moulting and growth of young crayfish, *Procambarus clarkii. Biol. Bull.* (Woods Hole, MASS.), 157: 182-188.

Nair, S.R. and M. Kutty 1975 Note on the varying effect of salinity on the growth of the juvenile of *Penaeus indicus* from the Cochin backwaters. *Bull. Dept. Mar. Sci. Univ.* Cochin 7,1: 181-184.

Papathanassion, E. and P.E. King. 1984. Effect of starvation on the fine structure of the hepatopancreas in the common prawn *P. serratus*. *Comp. Biochem. Physiol.* 77A: 243-249.

Passano, L.M. 1960. Moulting and its control. In: Physiology of Crustacea, Vol.I, 473-536. Ed. T.H. Waterman. Academic Press, New York.

Pudadera, R., J.Llobrera, R. Caballero, and N. Aquino. 1984. Moult staging in adult *Penaeus monodon*. Presented in first International Conference on the culture of penaeid prawns/shrimps. Lloili City, Philippines.

Raj, P.R. and P.J.S. Raj 1980 Effect of salinity on growth and survival of three species of penaeid prawns. *Proc. Symp. Costl. Aquaculture*, 1: 236-243.

Rao, K.R.,C.J. Mohreherr, D.Reinschmidt, and M.Fingerman. 1977. Control of growth during proecdysis in the crayfish *Faxonella clypeata* (Hay, 1899). *Crustaceana,* 32: 256-264

Read, G.H.L. 1977. Aspects of lipid metabolism in *Penaeus indicus* Milne Edwards. M.Sc. T hesis, Univ. of Natal.

Read, G.H.L. and M.S. Caulton. 1980. Changes in mass and chemical composition during the moult cycle and ovarian maturation in immature and mature *P. indicus*. *Comp. Biochem., Physiol.*, 66A: 431-437.

Richard, P. 1978. Effect of temperature on growth and moulting of *Palemon serratus* in relation to their size. *Aquaculture*, 14: 13-22.

Schafer, H.J. 1967. The determination of somestages of the moulting cycle of Penaeus duorarum, by microscopic examination of the setae of the endopodites of pleopods. FAO *Fish. Re.*, 57: 381-391.

Scheer, B.T. 1960. Aspects of the intermoult cycle in natantians. *Comp. Biochem. Physiol.*, 1: 3-18.

Scheets, W.C.P. and J.E. Dendinger. 1983. Calcium deposition in the cuticle of the blue crab, *Callinectes sapidus*, related to external salinity. *Comp. Biochem. Physiol.*, 74A: 903-907.

Smith, D.M. and W. Dall. 1984. Moult staging in the tiger prawn, *Penaeus esculentus*. In: Second Aust. *Nat. Prawn Sem.* Publ. By NPS2, Clevel and, QLD, 85-93.

Thomas, G. and A.D. Diwan, 1990. Changes in nucleic acids and protein content in relation to body size in the prawn *Penaeus indicus* H. Milne Edwards. *Proc. Indian Acad. Sci.* pp 125-130.

Travis, D.F. 1965. The deposition of skeletal structures in the Crustacea. V. The histomorphological and histochemical changes associated with the development and calcification of the branchial exoskeleton in the crayfish, Orconectes virilise Hagen. *Acta histochemica,* 20: 193-222.

Venketaramiah, A., G.J. Lakshmi, and G.Gunter. 1972. The effect of salinity, temperature and feeding levels on the food conversion, growth and survival rates of shrimp *Penaeus aztecus.* In: Proc. 3rd Workshop World Mariculture Soc., 267-283.

Vigh, D.A. and J.E. Dendinger 1982. Temporal relationship of post moult deposition of calcium, magnesium, Chitin and protein in the cuticle of Atlantic crab, *Callinectes sapidus, Comp. Biochem. Physiol.*, 72A: 365-369.

Vijayan K.K. 1988. Studies on the physiology of moulting in the penaeid prawn, *Penaeus indicus* H. Milne Edwards Ph.D. thesis submitted to Cochin Uni. of Sci. and Tech., Cochin. 241.

Vijayan, K.K. and A.D. Diwan, 1993. Studies on the physiology of moulting in the penaeid prawn *Penaeus indicus* H. Milne Edwards. *CMFRI spl. Publ.*:56: 6-72.

Vijayan K.K. and A.D. Diwan 1996. Fluctuations in Ca, Mg and P levels in the haemolymph, muscle, hepatopancreas and exoskeleton during the moult cycle of the Indian while prawn *Penaeus indicus* (Decapoda: Penaeidae) *Comp Biochem. Physiol.* 113A: 91-97.

Welinder, B.S. 1974. The crustacean cuticle. I. Studies on the composition of cuticle. Comp. *Biochem. Physiol.*, 47A: 779-787.

Wickins, J.F., 1984. The effect of reduced pH on carapace calcium, strontium and magnesium levels in rapidly growing prawns (*Penaeus monodon Fabricius*). *Aquaculture*, 41,1:49-60.

Winget, R.K., C.E. Epifanio, T.Runnels, and P.Austin. 1976. Effect of diet and temperature on growth and mortality of the blue crab, *Callinectes sapidus*, maintained in a circulating culture system. *Proc. Natl. Shellfish. Assoc.*, 66: 29-33.

Wright, D.A. 1980. Calcium balance in premoult and postmoult Gammarus pulex (Amphipoda). *Freshwat. Biol.*, 10: 571-579.

Yamaoka, L.H. and B.T. Scheer. 1970. Chemistry of growth and development in the crustaceans.321-341. In: M.Florkin and B.T. Scheer (Editors), Chemical Zoology. Academic Press, New York.

CHAPTER 9

NEUROENDOCRINE CONTROL OF MOULTING AND GROWTH

9.1 INTRODUCTION

The entire process of moulting and its control by hormones in crustacean are well known. The term 'moulting' is now used most often to refer to the whole process of preparation for shedding of the old exoskeleton. A great deal of information is now available describing the hormonal basis of moulting in crustaceans (Quackenbush, 1986). It had long been known that removal of the eyestalks can lead to precocious moulting. However, mistery behind this phenomenon particularly hormonal involvement was not clear until Brown and Cunningham (1939) demonstrated the involvement of moult inhibiting hormone (MIH) produced in the sinus gland of eyestalk neuroendocrine complex in crayfishes. Since then investigations carried out by several crustacean workers established the classical hypothesis of hormonal control of moulting (Fingerman 1987). As per this hypothesis the moulting process is regulated by two hormones, one is moult inhibiting hormone (MIH) produced in the eyestalk neurosecretory cells (X organ) and the other moulting hormone (MH) secreted by a discrete pair of endocrine gland situated in cephalic region of the animals termed as 'Y' organ. In addition to these two specific hormones, possible involvement of a third one the moult accelerating hormone (MAH) from the central nervous system (CNS) in moult control has also been reported (McWhinne and Mohrherr, 1970). It is seen that virtually the entire physiology of the animals gets disturbed by eyestalk removal. Hence eyestalk factors have been implicated in the regulation of a large variety of physiological processes. Approximately ten active factors controlling various physiological events including moulting, reproduction and water volume regulation have been isolated (Newcomb, 1983) but unfortunately chemical identity of such factors is not yet known thoroughly. The existence of MIH in the eyestalk of several crustaceans has been demonstrated by many workers (review of Fingerman 1987) by using the well known technique of eyestalk ablation which resulted in precocious moulting. The target organ of the moult inhibiting hormone has been identified as Y organ by Passano and Jyssum (1963). The presence of 'Y' organ has been reported in many species of crustaceans and its role in the control of moulting process has been proved beyond doubt (Vijayan *et al.* 1993). Recent advanced analytical techniques like RIA and organ culture made possible direct proof of the ecdysteroid secretory nature of Y organ (Mattson and Spaziani, 1985). Besides 'Y' organ, involvement of other tissues like 'mandibular organ" in controlling moult function has become contradictory though some workers have correlated its secretory activity with that of moulting process (Borst *et al.*, 1985). Further research in this area is warranted. Although much work has been done on in moulting physiology in reptantians, Natantians in general and Penaeids in particular are yet to be studied. Descriptions of neuroendocrine components in penaeid shrimps are available for *Metapenaeus* sp. (Dall, 1965) *Penaeus japonicus* (Nakamura, 1974) *Metapenaeus monoceros* (Madhystha and Rangnekar, 1976) *F. indicus* (Mohamed and Diwan 1993) and *Penaeus monodon* (Nanda and Ghosh, 1985). Influence of eyestalk neuroendocrine factors on moulting process and the structure of Y organ and its role in the control of moulting process in *F. indicus* have been described by Vijayan (1988), Vijayan and Diwan (1993), Vijayan *et al.*, (1993).

In the life history of the shrimp, growth is achieved only through periodic shedding of the cuticular sheath covering the body. Thus, shedding of the cuticle is the most important physiological event in these animals. During moult cycle many cyclic changes occur in structural, biochemical and physiological processes. This dynamic event continues throughout the life span of the shrimp linking almost all biological activities with this process. In scientific shrimp farming aiming at high production, adequate knowledge on the physiology of growth is imperative. Pioneer workers like Drach in1939 and Sheer in 1960 initiated the classic studies on crustacean moulting physiology. In subsequent years a great deal of information has been added to this particular field by number of workers (see review by Fingerman and Nagabhushanam, 1997). In spite of these great contributions, only limited attention has been paid to the moulting physiology of natantians especially penaeids.

In recent years in order to understand the mechanism underlying the neuroendocrine control of moulting and growth process, number of workers have carried out in d epth studies using genomic and molecular methods. Emphasis has been given on identifying moult inhibiting hormone (MIH) gene, its structure and mechanism of action. Some workers have mentioned cellular mechanism of action of MIH on Y-organ. In this chapter attempts have been made to cover all these aspects.

9.2 GROWTH PATTERN IN SHRIMP

All crustaceans have an exoskeleton and ecdys is necessary for growth. During early stages of the life cycle, shrimps undergo several ecdysis before they attain juvenile size. As size increases, frequency of ecdysis is reduced. Female shrimp moult as a requirement for mating (Vijayan *et al.,* 1997). Thus, the control of growth is dependent on the control of moulting. The moulting process in crustaceans has been demonstrated by number of workers (Passano, 1960; Diwan & Nagabhushanam, 1975; Diwan & Usha, 1985; Vijayan & Diwan, 1993. The process is regulated through ecdysteroid hormones from the Y-organs using proecdysis and moult inhibiting hormone (MIH) produced by a group of neurosecretory cells (NSC) in the X-organ sinus gland complex of the eyestalks of shrimp (Diwan & Nagabhushanam, 1975; Aiken, 1980; Skinner 1985; Skinner *et al.,* 1985; Lachaise *et al.,*1993; Chang 1989; Chang 1993; Vernet *et al.,*1994; Charmantier 1997).

9.3 ROLE OF Y-ORGANS

Gabe (1953) observed a Y-organ in 58 species of malacostra can crustaceans. Since then, Y-organs have been identified and described in numerous species of crustaceans including shrimps (Charmantier-D *et al.,* 1994; Dall, 1965; Bourguet *et al.,* 1977; Vijayan *et al.,*1993). The Y-organ is generally located in the musculature of the anterior branchial chamber and appears as a compact mass in crabs or a less compact mass in crayfish and lobsters (Lachaise F *et al.,* 1993). The size of the Y-organ is found to be 60-300μm in isopods and amphipods and 1-3 mm in decapods. Histological studies revealed that the cells of the Y-organ show secretory activity in relation to the moult cycle (Vijayan *et al.,* 1993). The endocrine function of the Y-organ in control of moulting was suggested by Gabe (1953) Charmantier-D *et al.,* (1994); Dall, (1965) Bourguet *et al.,*(1977); Vijayan *et al.,*(1993) Gabe, (1956) and subsequently this was demonstrated through surgical experiments by Echalier (1955, 1956) in C*arcinus maenas,* and by Passano and Jyssum (1963), Maissiat, (1970a 1970b) and Blanchet (1974) in several other species of decapods, isopods and amphipods respectively. The first evidence of Y-organs controlling the moulting process through production of ecdysteroid hormones was demonstrated by injecting exogenous ecdysteroid (20-hydroxyecdysone) (Skinner, 1985). *In vitro* culture of the Y-organ provided direct evidence that it produces ecdysteroids in *Procambarus clarkia* 36 and *Penaeus vannamei* (Blais *et al.,* 1994).

The major secretory product of the Y-organs is an ecdy steroid which is synthesized from cholesterol19 and converted to 20-hydroxyecdysone by several tissues in crustacea. Kaeuser *et al.* 1990 reported lack of evidence on identification of the target organs of ecdy steroids and the hormone receptor complex, which results in increased RNA synthesis through certain genes. However, recently Durica *et al.*, (2002) while studying the characterization of crab (*Ucapugilator*) homologs of ecdy steroid (ECR) and retenoid–X (RXR) receptors during limb regeneration and oocytes maturation found that ovarian tissue is a potential target for hormonal control in crustacea. High levels of ecdy steroid in blood and tissue during premoult and reduced levels of same just prior to ecdysis have been reported by Snyder and Chang (1991) while working on the American lobster, *Homarus americanus.* Blais *et al.*, 1994 reported that in *P. vannamei* at late premoult ecdysteroid concentrations increase in the haemolymph following activation of the Y-organ while the concentration remains low during postmoult and inter moult. The rate of ecdysteroid synthesis is generally controlled by MIH from the X-organ sinus gland complex. Experiments with eyestalk ablated shrimp and with Y-organs incubated *in vitro* in presence of sinus gland extract indicated that the X-organ regulates the synthesis of ecdysone in the Y-organ. It has been reported *Palaemon serratus* and *Sicyoniaingentis* have low levels of ecdy steroids in eggs at extrusion and the levels increase towards hatching (Spindler, 1987). *Alpheus heterochaelis* exposed to a micromolar concentration of 20-OH-ecdysone for 5days showed that the winter moult cycle was shortened by 18 days or by 65% (Mellon DeF Jr & Greer, 1987). Vijayan *et al.*, (1993) demonstrated that the size and tinctorial affinity of Y-organ cells in *F. indicus* change during the moult cycle (Fig 1 & 2). Further, they have shown that the Y-organ removal in this shrimp inhibits the onset of premoult. Chan (1998) cloned a cDNA from the shrimp *Metapenaeus ensis* which encodes a nuclear receptor, homologous to the insect ecdysone-inducible E 75 gene. The amino acid sequence has all five domains of a nuclear receptor and it is expressed in the epidermis, eyestalk and nervous tissue of premoult shrimp.

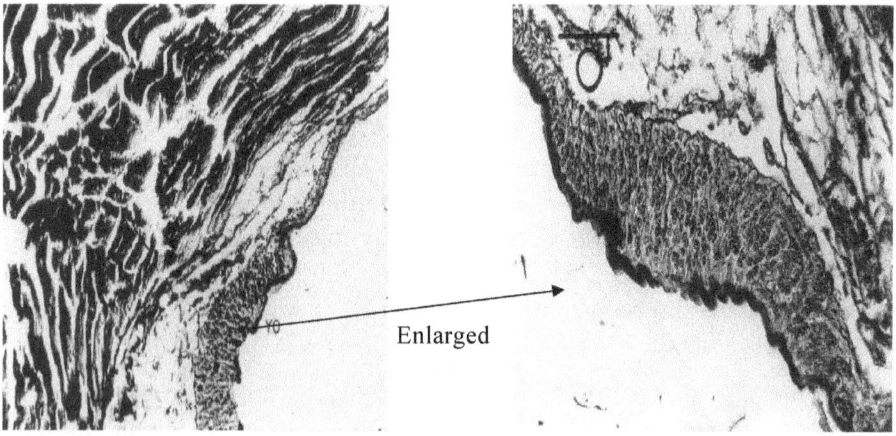

Fig. 1 & 2 Photomicrograph of the Y-organ of *F.indicus*. (a) X 100 (b) X 400 (**Source:** Vijayan *et al.*, 1993)

9.4 ROLE OF THE EYESTALK X-ORGAN SINUS GLAND COMPLEX

The eyestalks and their involvement in moult control were initially discovered by Zeleny in 1905. Hanström (1931) provided a detailed structural description of the X-organ. With the advancement of science in neuro endocrine aspects of crustaceans, many workers have described the X-organ-sinus gland complex that contains neurosecretory cells, in details (Fingerman & Nagabhushanam, 1997; Passano, 1960; Skinner, 1985; Nagabhushanam & Diwan, 1974; Mohamed & Diwan, 1991). Several types of neurosecretory cells in the X-organs containing neurosecretory granules in their cell bodies, axons and terminals have been mentioned by a number of workers using light and electron microscopy (Skinner, 1985; Mohamed & Diwan, 1991; Carlisle & Knowles, 1959; Gabe, 1967; Chang, 1985; Shoji, 1997). The distribution of neurosecretory cells in different parts of the optic ganglia has also been observed and depending on the location, the terminology adopted was the medulla terminalis ganglionic X-organ (MTGX), medulla externa ganglionic X-organ (MEGX) and medulla interna ganglionic X-organ (MIGX). The sinus gland is actually formed of axon endings of neurosecretory cells and glial cells lining haemolymph sinuses (Mohamed & Diwan, 1993). Exocytosis appears to be the dominant mechanism for the release of granules into the haemolymph (Cooke & Sullivan, 1982; Fingerman, 1992 Mohamed & Diwan, 1993; Skinner, 1985). In recent past, Huberman (2000) reviewed the chemical and functional aspects of various crustacean hormones and emphasized their importance in shrimp aquaculture.

9.5 STRUCTURE AND MECHANISMS OF ACTION OF MIH

In all penaeid shrimps like other crustaceans, moult inhibiting hormone (MIH) controls moulting by suppressing the synthesis/secretions of moulting hormone. In recent years, MIHs have been isolated and characterised for many crustacean species and it has been reported that MIH consist of 75-79 amino acids residues. These sequences are similar to crustacean hyperglycaemic hormone (CHH) and therefore MIH and CHH have been grouped together to form a peptide family. As on today, there have been no systematic studies on the structure-activity relationship of the CHH family peptides, although earlier some preliminary work has been carried out by few workers (Katayma *et al.*, 2002, Ohira *et al.*, 1999, 2003, Gu *et al.*, 2001, Aguilar *et al.*,1996, Chang *et al.*,1990). Katayma *et al.*, (2003) described the structure of MIH from penaeid shrimp, *M. japonicus* using NMR facility. They revealed that MIH is composed of five α-helices and does not contain β-structure and found out the functional sites of this hormone (Fig 3). To confirm this, various mutants of MIH were prepared and their moult inhibiting activities were assessed (Katayma *et al.*, 2004). All the peptides mutated at the putative functional sites exhibited circular dichroism spectra similar to the natural MIH, suggesting that the mutants retained their natural conformation regardless of the mutations. As expected, a majority of the mutants observed to be less active than the natural MIH (Katayama *et al.*, 2004). From these findings they (Katayama *et al.*,2004) have concluded that the functional site of MIH is located in the region containing the C- terminal ends of the N- and C- terminal α-helices and that Asn^{13}, Ser^{71} and Ile^{72} have been observed to be significant for conferring moult inhibiting activity. They claimed that this is the first report of studies on structure-activity relationship of the CHH family peptides. This kind of information will offer new insight on the molecular evolution of the CHH family peptides.

Very little known regarding molecular mechanism of MIH which inhibits the secretion of ecdysteroids from the Y-organ and controls the moulting process in crustaceans. Recently, the Halloween orthologues genes were discovered in the *Daphnia* data base, suggesting that Halloween gene candidates are generally used for the ecdysteroidogenesis in crustaceans (Rewitz and Gilbert, 2008). From the recent literature it has been observed that ecdysteroidogenic enzymes in the insects have been studied extensively (Orengo and Aguade, 2007), particularly the Halloween gene family including *phantom, disembodied, shadow* and *shade* which catalyse the final four steps of the ecdysteroid synthetic pathway (Gilbert, 2004., Truman, 2005). Number of workers have reported that these Halloween genes predominantly expressed in the peripheral organs such as fat body, midgut gland, Malpighian tubules and prothoracic glands (Warren *et al.*, 2002, 2004, Petryk *et al.*, 2003, Niwa *et al.*, 2004, 2005, Rewitz *et al.*, 2006). As most insects synthesize 3-dehydroecdysone in prothoracic glands, in some crustaceans including the shrimp *M. japonicus* also synthesize 3-dehydroecdysone as a main ecdysteroid from the Y-organ (Okumara *et al.*, 1989, Lachaise *et al.*, 1993). Asazuma *et al.*, (2009) carried out the investigation on presence of Halloween genes in this shrimp and whether their expression levels are regulated by MIH in the y-organ during moulting periods. From the findings they have reported that the transcription of *phantom* gene in the Y-organ may be regulated by the inhibitory mechanisms of MIH which indirectly indicates negative regulation of ecdysteroidogenesis at the transcriptional level.

Nakatsuji *et al.*, (2009) in their review paper on crustacean moult inhibiting hormone, discussed in detail regarding structure of the MIH peptide and gene. MIH transcript and peptide levels during the moult cycle, cellular mechanism of action of MIH and responsiveness of Y- organ to MIH. While discussing the structure of MIH it is reported that at least 26 MIH/MIH like sequences have been identified with primary structure which consists of 74 to 79 amino acids.

Further it is mentioned that the C-terminal end of MIH appears more viable than the N-terminal end. Recent investigation by number of workers have indicated the phenomenon of molecular polymorphism is present in type –ii peptides of penaeid shrimps. For example there are 4 MIH/MIH-like molecules identified so far in *P. monodon* (Krungkasem *et al.*, 2002, Yodmuang *et al.*, 2004), 3 for *P. japonicus* and *Fenneropenaeus chinensis* (Wang and Xiang, 2003, Ohira *et al.*, 2005, Yamano and Unuma, 2006), and 2 for *Litopenaeus vannamei* and *M. ensis* (Gu *et al.*,2002, Chen *et al.*,2007). Nakatsuji *et al.*,(2009) also reported that the penaeid peptides can be further divided in to two sub groups A and B. Peptides of sub group A have been reported to be more potent than the peptides of sub group B. Further it is mentioned that the peptides of different sub groups might diverge functionally, with peptides of subgroup A acting physiologically as MIH and those of sub group B playing as regulatory roles in addition to inhibition of moulting (Gu *et al.*,2002, Ohira *et al.*,2005, Nakatsuji *et al.*, 2009). It has been reported that further research is required to find out to determine whether there is functional divergence among different groups of peptides. This is due to recent studies of gene expression of MIH in different penaeid shrimps during the moult cycle which indicated that both peptides of group A and B act physiologically as MIH (Chen *et al.*, 2007). Webster *et al.*, (2012) in their review paper discussed in detail about CHH hormones controlling crustacean metabolism, osmoregulation, moulting and reproduction. They mentioned that from large number of crustacean species about 80 CHH peptide hormones have been fully identified by various methods but studies on their biological activities are limited. Recent work on MIH regulation of Y-organs has made it very clear that the whole process of regulation is not that simple and earlier concept that MIH solely regulates Y-organ is not correct due to the fact that MIH levels in the haemolymph have no correlation during different stages of moult cycle. Now the extensive studies on genomics of MIH and physiological evolution of CHH peptides has led to different situations in different species and so also the mechanism of MIH regulation in relation Y-organ differs. Further they have reported that to draw conclusive results it is necessary to have receptor studies, binding sites, tissue specific distribution and binding affinities of MIH and Y-organ to understand the whole process of ecdysteroidogenesis (Webster *et al.*, 2012) (Fig 5).

Das *et al.*, (2016) recently carried out the studies on transcriptome analysis of the moulting gland (Y-organ) from land crab, *Gecarcinus lateralis*. They reported that there is significant similarity in numerous signal transduction pathways in the YO transcriptomes of land crab and crayfish. Further it is mentioned that because of discovery of genes Y-organ regulation has become more complex. As the Y-organ gets activated and sustains moulting processes, it is assumed that the Y-organ is able to integrate a variety of extrinsic and intrinsic signals to effect an appropriate response. The decision to initiate moulting is controlled by MIH, but moulting can be interrupted or delayed by adverse conditions in early pre-moult. The decision to complete moulting is made when the animal transitions to mid-premoult, at which point moulting processes continue and the animal moults without delay (Das *et al.,* 2016).

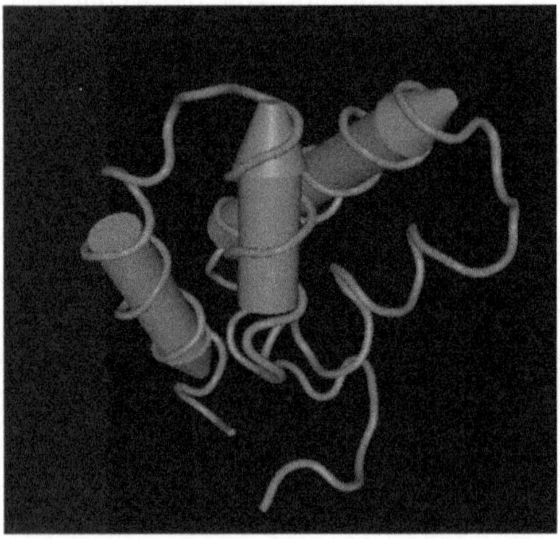

Fig. 3. The solution structure of molt-inhibiting hormone from the kuruma prawn *Marsupenaeus japonicus* (**Source:** Katayama *et al.,* 2003)

MIH Gene

Considerable amount of work has been carried out on several genes encoding MIH/MIH like peptides of number of penaeid shrimps and it is mentioned that the 3 exon/2 intron organization of these genes appears to be highly conserved (Chan *et al.,* 1998, Gu and Chan, 1998, Gu *et al.,* 2002, Yodmuang *et al.,* 2004, Chen *et al.,*2007). Nakatsuji *et al.,* (2009) carried out in-depth studies on MIH gene and described the intron and exon position of the signal peptide of penaeid genes. Further through the data analysis of the pattern of expression of MIH/MIH like peptides genes it has been reported that the transcript of subgroup of A-peptides of different penaeid shrimps are mainly expressed in the eyestalk ganglia, whereas those of subgroup B-peptides predominantly expressed in thoracic ganglia, abdominal ganglia, and brain. From these results it was hypothesized that peptides of different groups might play different physiological roles (Nakatsuji *et al.,* 2009).

Number of workers have investigated and reported MIH transcript abundance in the optic ganglia or MIH peptide levels in the sinus gland or haemolymph (Ohira *et al.,* 1997, Lee *et al.,*1998, Chung and Webster, 2003, Chen *et al.,*2007). Chen *et al.,*(2007) (Fig 4) determined two putative MIH transcripts in the optic ganglia of white shrimp *L. vannamei* and found that both these MIH levels were elevated during post moult, intermoult and early premoult period and then progressively declined, reaching low at the premoult stage. Further they have reported that during the same time (peak premoult period) ecdysteroid levels in haemolymph significantly elevated. Nakatsuji *et al.*, (2000, 2009) while working on MIH peptide levels in sinus gland and haemolymph of other crustaceans reported similar observations. On the contrary, Ohira *et al.*, (1997) while working on molecular cloning of moult inhibiting hormone in *P. japonicus* reported that MIH transcript levels in the eyestalk tissues did not change significantly during moult cycle. These discrepancies in the findings require further research by studying more species of crustaceans.

```
CCCGCCGTCTCAAGATACACTTCCACTTCAGACACCTTGCCTTTACTTCGTGTCTGGGTGTATTCTAATA
TGTACCGTCTAGCAATCGTAAGGGTTTAAGAGATTGTATATATATGGAATATCGTTCCTTAATAATTCGG
TGACCATCATTTTTTAAATTCTTCTCGGCTCTCCTCTAAAGACCTTTTTTTTCGTGGCAGAGGTCATGGT
TGCCGGTAATGACAGTACTATTTGCGACGAGCCTCTTCTTCGACACCGCCTCCGCCAGTCCCATCGACGG
GACGTGTCCAGGCAGAATGGGTAATCGTGAGATCTACAAGAAAGTTGATAGCGTTTGTAAGGACTGTGTT
AATATCTTCCGATTGCCAGAGCTGGAAGGCTTGTGTAGGTGAGTTTGCATATAATAAATAAAGGTTTATC
TACAAGCTGGTGTTATTAACGTTGAAATCTGTATGGAAACTGTCATATGCTTAGGATGTATTTGTGTATA
CTTTCTTGAGATATTTACTAAGTATAGCATGGAAAGCAGAATTATCCGCTAGAACATAAAAAAAAAAAAA
AAAAAATCAATGCATTCTTTTCCACTACAGAGATGAGTGCTTCATCAACGATTGGTTTCTATTTTGTGCG
AAGGCTGCCAAGCGGATGGACGAGATTGAAAATTTCAGAGTGTGGATAAGTATCCTGAACGCCTGAAAAT
GAGAGTGAGCCCAGAAGAACCCATTCCCTT
```

Fig. 4. *Litopenaeus vannamei* molt-inhibiting hormone MIH2 gene, complete cds (**Source:** Chen *et al.*, 2007)

9.6 CELLULAR MECHANISMS OF ACTION OF MIH

The available literature on cellular mechanisms of action of MIH indicate that cellular signalling pathways involving adenosine-3′, 5′-cyclic monophosphate (cAMP), guanosine-3′,5′-cyclic monophosphate (cGMP) or both play a significant role in MIH mediated regulation of ecdysteroidogenesis (Fig 5). Mattson and Spaziani (1985, 1986, 1987) while doing the studies on characterization of MIH action on Y-organ involving cyclic AMP found out that cAMP mediates MIH induced suppression of ecdysteroid production. However, studies conducted by Sedlmeier and Fenrich (1993) indicated that MIH induced suppression of ecdysteroid production is mediated by cGMP, and not by cAMP. Similar observations have been also reported by Nakatsuji *et al.*, (2006, a, b) while working on other crustacean species. These studies indicated that MIH suppresses ecdysteroidogenesis via a cGMP second messenger. Saidi *et al.*, (1994) while working on the mode of action MIH with the involvement of cAMP and cGMP mentioned that cGMP levels enhanced significantly during peak of premoult period and simultaneously there was also increase in cAMP

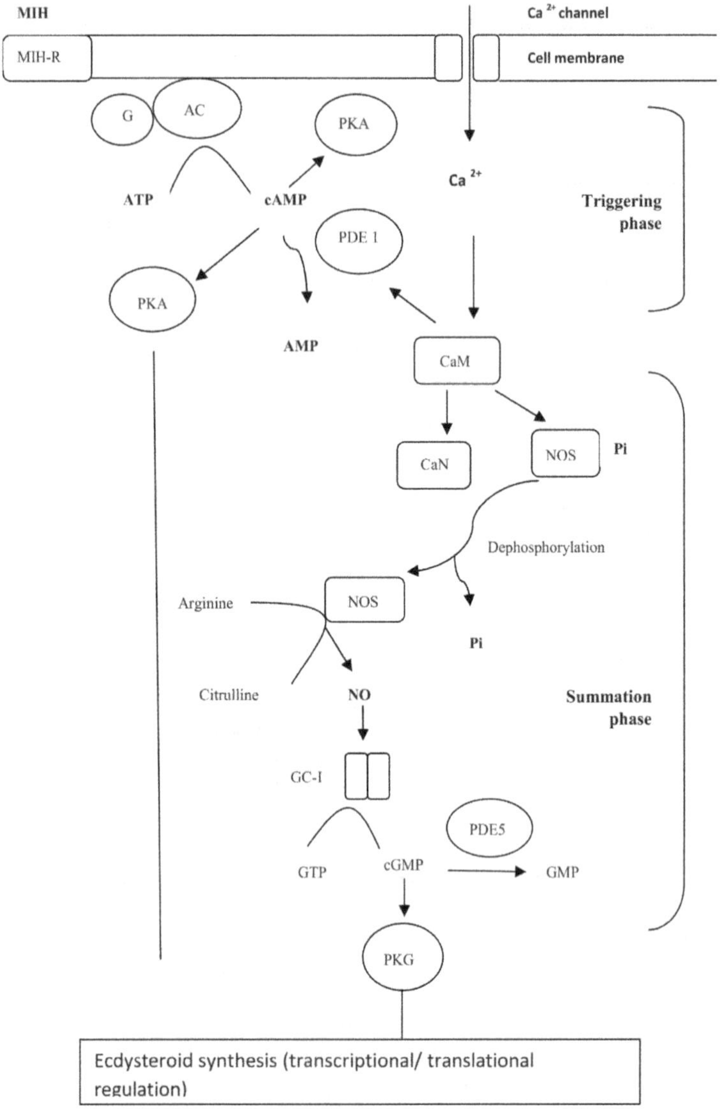

Fig. 5. Model of MIH signalling in crab Y-organs, proposed by Chang and Mykles. MIH binds to a G-protein coupled receptor (MIH-R), activating adenylyl cyclase (AC), resulting in cAMP production and protein kinase A (PKA) activation.Phosphorylation of enzymes via PKA may be important in constitutive control of ecdysteroid synthesis, but also by facultative regulation by phosphorylation of Ca^{2+} channels, leading to calmodulin (CaM) activation of nitric oxide synthase (NOS) either directly, or indirectly via calcineurin (CaN). Calmodulin can additionally abrogate increases in cAMP via activation of phosphodiesterase 1 (PDE1). A nitric oxide sensitive soluble guanylyl cyclase (GC-1) then increases cGMP synthesis, activating protein kinase G (PKG) which inhibits ecdysteroid synthesis. Phosphodiesterase 5 (PDE5) can abrogate increases in cGMP, potentially modulating this inhibitory pathway during for example, premoult, when the Y-organ becomesrefractory to MIH. (**Source:** Webster *et al.*, 2012)

which was not that significant, suggesting that both these cyclic nucleotides work cooperatively with each other. There is no much information as on today regarding MIH receptors. Some work which has been carried out in different crustaceans on cellular signalling in Y-organ can provide insight in to the nature of the MIH receptors. Nakatsuji *et al.*, (2009) mentioned that in their review paper that the MIH receptor has not been isolated and fully characterised for any crustaceans and the relevant work which has been carried out so far do not provide any conclusive evidence regarding the nature of the receptors and cell signalling pathways linked to the process of ecdysteroidogenesis.

9.7 FUNCTIONAL ASPECTS OF MOULTING HORMONES

The role of the optic ganglia of the eyestalk in the control of various physiological processes including growth and moulting in crustaceans has been described (Keller, 1992; Charmantier *et al.*, 1997; Mohamed & Diwan, 1991; Quackenbush, 1986; Beltz, 1988). It was demonstrated that eyestalk ablation in decapods leads to precocious moult, and moulting occurs several times following ablation, particularly in young animals. Reimplantation or injection of eyestalk extract in several crustaceans has been shown to reverse or inhibit the process of moulting (Diwan & Nagabhushanam, 1975). The impact of eyestalk removal on acceleration of the moulting/growth process is due to loss of MIH, which is predominantly present in the optic ganglia, particularly in younger stages of the lifecycles. The presence of MIH inhibits the moult promoting effect of Y-organs Passano, 1960; Charmantier, 1997; Cooke & Sullivan, 1982. Once the inhibitory effect of the eyestalk (MIH) on the moulting process was confirmed in several decapods, research was conducted to determine the chemical nature and other physiological functions of this hormone. MIH was then isolated and its peptidic nature established in several brachyuran and macruran species (Rao, 1965). In *Carcinus maenas* MIH was characterised (Webster, 1986) purified (Webster & Keller, 1986) and also sequenced (Webster & Dircksen, 1991). MIH has also been sequenced in *Homarus americanus* (Chang *et al.*,1990) and in *Callinectes sapidus* (Lee, *et al.*,1995). Aguilar *et al.*, 1996 isolated MIH from the eyestalk of the crayfish *Procambarus bouvieri* and compared its sequence with four other known peptides from *H. americanus, C. maenas, C. Sapidus* and *P. vannamei.* The lengths of these peptides vary between 72 and 78 residues and their molecular masses between 8 and 9 KDa. The first cloning of cDNA encoding a MIH like neuropeptide from *P. vannamei* was accomplished by Sun 1994. Further, it has been reported that MIH-like mRNA is found exclusively in the medulla terminalis ganglionic X-organ MTGXO of the eyestalk and in the brain the MIH-like gene transcript was detected in the neurosecretory cells (Sun, 1995). This could mean that a MIH-like neuropeptide could have specific functions in the nervous system in addition to its hormonal function. This neuropeptide has similarities in its amino acid sequence with other neuropeptides synthesized in the X-organ, i.e, the crustacean hyperglycemic hormone (CHH) and the vitellogenesis inhibiting hormone (VIH). A smaller neuropeptide (53-55 amino acids) with MIH activity was isolated from *P. bouvieri* (Huberman 2000 & Aguilar, 1989). This peptide was later sequenced and bioassayed (Aguilar *et al.*, 1996). Yang *et al.* (Yang W J, *et al.*, 1996) isolated and sequenced peptide with MIH activity from sinus glands of the shrimp *Marsupenaeus. japonicus. In vitro* experiments demonstrated that this peptide inhibited the synthesis of ecdysone in Y-organs of the crayfish *P. clarkii*. Four peptides with both MIH and CHH activity were isolated from the sinus gland of *P. vannamei* by Sefiani *et al.* (1996). It was hypothesized that different types of peptides that are detected in sinus glands display different activities, including the site responsible for MIH activity; but this hypothesis must be confirmed experimentally (Huberman, 2000). Mattson and Spaziani (1985) have shown that MIH is able to inhibit ecdy steroid secretion by Y-organs *in vitro* in a dose dependent manner. The injection of serotonin (5-OHtryptamine) lowers the haemolymph ecdy steroid levels and these effects are revoked by eyestalk ablation. In this experimental evidence Spaziani *et al.* (1994) proposed a cycle of MIH regulation sensory input via 5-HT that would release MIH and inhibit ecdy steroid synthesis and release by the Y-organs, and that MIH release would be subjected to a negative feedback by elevated ecdysteroid titers. Increase in the mitotic activity in the Y-organ cells after eyestalk ablation in *Pachygrapsus marmoratus* (Bressac, 1976) clearly indicated the possibility of enhanced steroidogenic activity. Mattson and Spaziani (1986) found that sinus gland extracts suppress Y-organ steroidogenesis. A conceptual model defining the mode of action of sinus gland extracts on steroid genesis by the Y-organ was proposed by Mattson and Spaziani (1987) and the same was later reviewed by many other workers (Webster & Keller, 1989; Lachaise, *et al.*, 1993). According to this model MIH increases intracellular cAMP, which activates a protein kinase that leads to inhibition of protein synthesis and steroidogenesis. Gu *et al.* (2000) while working on MIH in the shrimp *Metapenaeus ensis* showed that specific cells in three different clusters of the X-organ, the sinus gland and axonal tract of the eyestalk contain MIH. To test the moult inhibiting activity of MIH, shrimp of inter moult stage were injected with r-MIH (recombinant protein) and there was a significant increase in the moult cycle duration for the shrimp.

Hopkins (2012) in his review paper on brief history of crustacean neuroendocrinology has reported in details about crustacean neuropeptides that were analysed in the eyestalk neuroendocrine complex. He discussed and gave a historical progressive account in the development of crustacean neuropeptides, their origin, structure, properties and behaviour in relation to overall functions.

He mentioned two families of eyestalk neuropeptides i.e. one is crustacean hyperglycaemic hormone (CHH) and other is the pigmentary effector hormones (PDH). Besides CHH and PDH peptides, there are several other peptides which have been recently discovered and identified in crustacean central nervous system. Using sophisticated technologies like Fourier transformed mass spectrometry and with time of flight mass spectrometry and in silico transcriptome mining of available expressed sequence tags (EST) libraries, the presence of a large number of neuropeptides have been reported in the wide range of crustacean animals. Over 100 neuropeptides have been identified in decapod crustaceans and because of the presence of so many neuropeptides, there is a possibility of complexities involved in the crustacean hormonal pathway (Hopkins, 2012). In depth analysis of neuropeptides belonging CHH family, MIH, VIH and the PDH family has also been made and further it has been reported that still we lack the information about the number and roles of neuropeptides that control the ecdysteroid production.

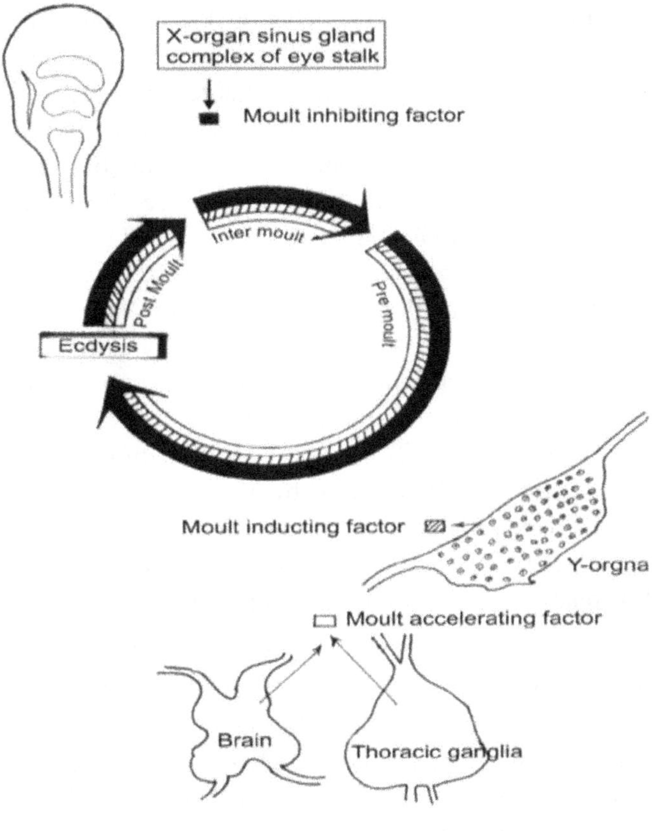

Fig. 6. Scheme of endocrine control of moulting process in *F. indicus*.
(**Source:** Vijayan, 1988)

9.8 MANDIBULAR ORGANS

The histology and ultrastructure of mandibular organs (MO) were described in a number of species of crustaceans (Charmantier and Charmantier, 1994; Waddy & Aiken, 2000). The main characteristics of cells of MO are the presence of a centrally located nucleus with abundant cytoplasm, a large number of mitochondria and smooth endoplasmic reticulum. Three types of cells have been identified in MO of *H. americanus* (Borst, 1994). The detailed structure of MO of *F. indicus* has been described by Vijayan and Diwan (Vijayan & Diwan 1994). There is sufficient evidence that terpenoids such as methyl farnesoate (MF) and farnesoic acid (FA), both secreted by the MO, have stimulatory effects on Y-organs and may influence reproductive activity. Methyl farnesoate has been detected in the MOs of *Libinia emarginata* (Borst, 1987; Laufer *et al.*,1987; Laufer, *et al.*,1987) and several other crustacean species (Borst & Laufer, 1990). The possible implications of MO and MF in the control of moult and growth have also been suggested (Vijayan & Diwan, 1993; Aiken, 1980; Skinner, 1985; Skinner *et al.*, 1985; Lachaise, 1993; Chang, 1985; Chang, 1993; Vernet & Charmantier, 1994; Charmantier *et al.*,1997). The nature of the MO inhibiting factor has been explored by many workers (Laufer *et al.*, 1986; Wainwright *et al.*,1996; Liu, *et al.*,1997). Eyestalk ablation causes hypertrophy of the gland and a higher density of smooth endoplasmic reticulum (Bazing, 1976; Hinsch, 1977). Tsukimura and Borst (1992) observed increased levels of MF following eyestalk ablation in *H. americanus*. Ablation of MO is difficult but transplantation of the MO reduces the length of the moult cycle in *Penaeus setiferus* (Yudin, 1980). While studying regulation of the crustacean MO Borst *et al.*, (2000) described the role of MO in controlling reproduction and development. Further, they mentioned (that methyl farnesoate (MF) production by MO is negatively regulated by neuropeptides from the sinus gland in the eyestalk. Two neuropeptides (MO-IH-1 and 2) have been isolated from the SG of the crab, *Cancer pagurus* that inhibit MF synthesis by MO of female crabs *in vitro*. It was revealed that the regulation of MO is complex and may involve several SG compounds (Borst, 2002). Further studies are needed to determine the exact role of MO in relation to growth and reproduction.

Quian-Ji *et al.*, (2007) while carrying out work on the application of juvenile hormone antagonist KK-42 on the performance of growth of shrimp *P. schmitti* found that there was increase in the body weight upto 167% in those shrimps treated with juvenile hormone antagonist KK-42 solution at a concentration of $1.95*10^{-4}$ than the control. Further they have observed that the number of acini of mandibular organ increased gradually as animals grew, and the diameter of acinus forming cells has been reported statistically bigger at the end of the experiment (120 days) than at other times. Methyl fernesoate (MF) the analog of JH secreted by the mandibular organ has also been demonstrated to play important roles in regulation of number of functions in crustaceans including growth, reproduction and behaviour (Wilder *et al.*,1995, Rodriguez *et al.*,2002 and Laufer *et al.*,2005)

9.9 FUTURE RESEARCH

In all crustaceans species, moulting and growth process are governed by ecdysteroids produced by paired endocrine glands, Y-organs that are located in the anterior cephalothoracic region. Crustacean hyperglycaemic hormones (CHH-neuropeptides) are potential modulators of various regulatory processes in the crustacean physiological system. Moult inhibiting hormone inhibits the synthesis and/or secretion of ecdysteroids from Y-organ and vitellogenin inhibitory peptide inhibits the reproduction. The most extensively accepted concept is that synthesis of ecdysteroids, derived from cholesterol is negatively regulated by moult inhibiting hormone synthesized in the medullar terminalis of X-organ and then transported to the sinus gland (Fig 6). Various studies have revealed that the quantity or the number of receptors on the Y-organ remains constant and during intermoult period, the synthesis of ecdysteroids favourably inactivated by MIH. Inhibition of synthesis of ecdysteroid by MIH is facilitated by activation of the specific transcription factor that impedes phantom expression. It has been reported that MIH predominantly binds to Guanylyl cyclase (GC-II) and/or G-protein coupled receptor on the Y-organ cells and suppresses the ecdysteroid biosynthesis by a cAMP-dependent activation of nitric oxide synthase (NOS) and NO-dependent guanylyl cyclase (GC-I). The existence of MIH was suggested by the fact that eye-stalk ablation leads to prompt upsurge in ecdysteroid titre in haemolymph and hence, stimulates precocious moulting. However, as crustacean hyperglycaemic hormone, gonad inhibiting hormone and mandibular organ inhibiting hormone are also synthesized in X-organ and secreted from sinus gland, and eye-stalk ablation leads to imbalances in all these hormone levels due to which affects other physiological processes in the body. For many years, crustacean endocrinologists developed a concept, in which moulting is considered to be regulated solely by the levels of MIH circulating in the haemolymph. If this so, then MIH levels should be high during intermoult and post-moult stage and low during premoult stage. However, recent findings show of measurements of MIH in the haemolymph during the moult cycle do not show this kind of pattern. During intermoult stage MIH level is not always high, and during mid-to late-premoult stages MIH levels can be higher than that during intermoult. This indicates that the hormonal control of moulting is very complex. As brief exposures of Y-organ to MIH have lasting effects particularly on production of ecdysteroidogenesis. It is thought that periodic releases of MIH are sufficient to keep the Y-organ inactive during intermoult. Y-organ from late premoult (stage D2–D3) animals are refractory to MIH which may result, at least in part, from increased degradation of cyclic nucleotides by phosphodiesterases. High levels of ecdysteroids in the haemolymph inhibit Y-organ ecdysteroid secretion, thus contributing to the rapid decrease in haemolymph ecdysteroids at the end of premoult. Future research should be directed towards understanding the signalling mechanisms underlying the transitions between physiological states. A biosystems approach is needed to fully understand the regulation of the Y-organ at transcriptional, translational, and post-translational levels, and how these interact with environmental signals. High throughput sequencing and mass spectrometry technologies now make it possible to obtain complete transcriptome and proteomic databases of the Y-organ and to quantify mRNA levels and protein levels and post-translational modifications at critical transitions during the moult cycle. A complete transcriptome will facilitate identification of proteins and phosphorylation sites using sophisticated technologies. This kind of research investigation will help in identifying genes in the MIH, and other signalling pathways, as well as identify other genes important in Y-organ regulation and function e.g. Phm and other Halloween genes. It may also identify candidate genes encoding the MIH receptor (Chang and Mykles, 2011).

From the available literature it is clear that, the cellular mode of action of MIH, is due to MIH-induced suppression of ecdysteroidogenesis which is mediated by a cGMP second messenger. However, the role for cAMP cannot be ruled out. Conclusive identification and thorough characterization of the MIH receptor will be critical steps in resolving these issues.

Changes in the responsiveness of Y-organs to MIH appear to play a significant role in the regulation of ecdysteroid production. Recent data indicate that changes in glandular responsiveness to MIH are due to changes in glandular phosphodiesterase activity. Because glandular phosphodiesterase activity, MIH is Ca+/calmodulin dependent, therefore, it seems very clear that an understanding of the mechanisms that underlie the regulation of free calcium levels in Y-organs is critical to a comprehensive understanding of the regulation of ecdysteroidogenesis.

As the gene structure of MIH is now known, there is a need to evaluate the application of gene editing techniques for accelerating the growth and moulting process in the shrimps.

References

Aiken D E, 1980. Molting and growth, in T*he biology and management of lobsters* Vol. 1 edited by J.S. Cobb and B.F. Phillips (Academic Press, New York), 91.

Aguilar M B., Falchetto R., Shabanowitz J, Hunt D F & Huberman, A, 1996. Complete primary structure of the moltinhibiting hormone (MIH) of the Mexican crayfish *Procambarus bouvieri* (Ortmann), *Peptides,* 17: 367

Asazuma H., S. Nagata, H. Nagasawa, 2009. Inhibitory effect of molt-inhibiting hormone on Phantom expression in the Y-organ of the kuruma prawn, *Marsupenaeus japonicus, Arch. Insect Biochem. Physiol.* 72 :220–233.

Bazing F, 1976. Mise en evidence des caracteres cytologiques des glandes steroidogenes dans les glanes mandibularies et les glandes Y du crabe *Carcinus maenas* (L) normal et epedoncule, *C.R. Acad. Sci. Paris,* 282: 739.

Beltz B S, 1988. Crustacean neurohormones in *Endocrinology of selected invertebrate Types,* Vol. 2 , edited by H. Laufer and R.G.H. Downer (Alan R. Liss, Inc., New York) 235.

Blanchet M F, 1974. Etude du controle hormonal du cycle d'intermue et de l'exuviation chez *Orchestia gammarella* par microcauterisation des organs Y suivie d'introduction d'ecdysterone, *C.R. Acad. Sci. Paris,* 278: 509.

Blais C, Sefiani M., Toullec J.Y. & Soyez D., 1994. *In vitro* production of ecdysteroids by Y-organs of *Penaeus vannamei* (Crustacea, Decapoda), Correlation with haemolymph titers, *Invert. Reprod. Dev,* 26: 3.

Borst D W, Tsukimura B, Laufer H. & Couch E.F, 1994. Regional differences in methyl farnesoate production by the lobster mandibular organ, *Biol. Bull,* 186 ,9.

Borst, D.W., M.Sinkus, and H.Laufer. 1985. Methyl farnesoate production by the crustacean mandibular organ. *Am. Zool.,* 25, 103A.

Borst D W, Laufer H, Landau M, Chang E S, Hertz W A, Baker F C & Schooley D A., 1987. Methyl farnesoate and its role in crustacean reproduction and development, *Insect Biochem,* 17: 1123.

Borst D W, Ogan J, Tsukimura B, Claerhout T & Holford K C, 2001. Regulation of the crustacean mandibular organ, *Am. Zool.* 41: 430.

Borst D W, Wainwright G & Rees H H, 2002. *In vivo* regulation of the mandibular organ in the edible crab, Cancer pagurus, *Proc R Soc Lond Ser B. Biol. Sci,* 269: 483.

Bourguet J.P, Exbrayat J.M, Trilles J P & Vernet, G, 1977. Mise en evidence et description de l'organe Y chez *Penaeus japonicus* (Bate, 1881) (Crustacea Decapoda, Natantia),. *C.R. Acad. Sci. Paris,* 285: 977.

Bressac, C, 1976. Effects de l'ablation des pedoncules oculaires sur les organs Y du crabe *Pachygrapsus marmoratus, C.R. Acad. Sci. Paris,* 282: 1873

Brown, F.A. and O. Cunningham. 1939. Influence of the sinus gland of crustaceans on viability and ecdysis. *Biol. Bull.,* 77: 104-114.

Carlisle D B & Knowles F, 1959. *Endocrine control in crustaceans,* (Cambridge University Press, New York).

Chan, T.Y. 1998 Shrimps and prawns. 851-971. In Carpenter, K.E. and V.H.Niem (eds). FAO species identification guide for fishery purposes. The living marine resources of the Western Central Pacific. Vol. 2. Cephalopods, crustaceans, holothurians and sharks. Rome, FAO. 687-1396.

Chang E S, 1985. Hormonal control of molting in decapod crustacea, *Am. Zool,* 25: 179.

Chang E S, 1989. Endocrine regulation of molting in Crustacea. *Rev. Aquat. Sciences* 1,131.

Chang E.S, Prestwich, G.D, & Bruce M.J, 1990. Amino acid sequence of a peptide with both molt-inhibiting and hyperglycemic activities in the lobster, *Homarus americanus, Biochem. Biophys. Res. Commun,* 171: 818.

Chang E S, 1993. Comparative endocrinology of molting and reproduction, Insects and crustaceans, *Annu. Rev. Entomol,* 38,161.

Chang ES, Mykles DL, 2011. Regulation of crustacean molting: a review and our perspectives. Gen Comp Endocrinol 172:323–330.

Charmantier-Daures M Charmantier G, 1994. Les organs endocrines. In: *Traite de Zoologie, Crustaces,*Vol. VII edited by I. J. Forest (Masson, Paris) 595.

Charmantier G, Charmantier-Daures M & Van Harp F, 1997. Hormonal regulation of growth and reproduction in crustaceans, in *Recent advances in marine biotechnology,* Vol.1 edited by M. Fingerman, R. Nagabhushanam, and Mary Frances Thompson (Oxford and IBH, New Delhi), 109

Chen, H.-Y., Watson, R.D., Chen, J.-C., Liu, H.-F., Lee, C.-Y., 2007. Molecular characterization and gene expression pattern of two putative molt-inhibiting hormones from *Litopenaeus vannamei. Gen. Comp. Endocrinol.* 151, 72–81.

Chung, J.S., Webster, S.G., 2003. Moult cycle-related changes in biological activity of moult-inhibiting hormone (MIH) and crustacean hyperglycemic hormone (CHH) in the crab, *Carcinus maenas. Eur. J. Biochem.* 270, 3280–3288.

Cooke I M & Sullivan R E, 1982. Hormones and neurosecretion, in *The Biology of crustacea* Vol. 3 edited by D.E. Bliss, H.L. Atwood and D.C. Sandeman (Academic Press, Orlando) 205.

Dall W, 1965. Studies on the physiology of a shrimp, *Metapenaeus sp.* (Crustacea, Decapoda, Penaeidae). II. Endocrines and control of molting. *Aust. J. Mar. Freshwater Res,* 16:1.

Das Sunetra, Natalie L. Pitts, Megan R. Mudron, David S. Durica, DonaldL.Mykles, 2016 Transcriptome analysis of the molting gland (Y-organ) from the black back land crab *Gecarcinus lateralis. Comparative Biochemistry and Physiology, Part D.* 1726–40.

Drach, P. 1939. Mut et cycle d' intermue chez les crustaces decapods. *Ann. Inst. Oceanogr. Monaco.,* 19: 103-391.

Diwan A D & Nagabhushanam R, 1975. The neurosecretory cells of the central nervous system of the freshwater crab *Barytelphusa cunicularis* (Westwood 1836). *Riv Biol* 68, 79.

Diwan A D & Usha T, 1985. Characterization of moult stages of *Penaeus indicus* based on developing uropod setae and some closely allied structures. *Indian J. Fish,* 32: 275.

Durica D S, Wu X., Anilkumar G., Hopkins P.M.,& Chung A C K, 2002. Characterization of crab EcR and RXR homologs and expression during limb regeneration and oocytes maturation, *Mol Cell Endocrinol,* 189: 59.

Echalier G, 1955. Role de l'organe Y dans le determinisme de la mue de *Carcinides (Carcinus) maenas L.,* (Crustace Decapode); experiences d'implantation, *C.R. Acad. Sci. Paris,* 240: 1581.

Echalier G, 1956. Effets de l'ablation et de la greffe de I'organe Ysur la mue de *Carcinus maenas L, Ann. Sci. Nat. Zool,* 11; 153.

Fingerman, M. 1987. The endocrine mechanisms of crustaceans. *J. Crust Biol.,* 7,1: 1-24.

Fingerman M, 1992. Glands and secretions, in *Microscopic anatomy of invertebrates,* Vol. 10 edited by F.W. Harrison (Wiley-Liss Inc. New York) 345.

Fingerman M & Nagabhushanam R, 1997. Role of neuroregulators in controlling the release of the growth and reproductive hormones in crustacea, in *Recent advances in marine biotechnology,* Vol. 1 edited by M, Fingerman, R. Nagabhushanam and Mary Frances Thompson (Oxford and IBH, New Delhi), 109.

Gabe M, 1953. Sur l'existence chez quelques Crustaces Malacostraces d'un organe comparable a la glande de mue des Insectes,. *C.R. Acad. Sci. Paris,* 237, 1111-13

Gabe M, 1956. Histologie comparee de la glande de mue (organe Y) des Crustaces Malacostraces, *Ann Sci. Nat. Zool,* 11:145.

Gabe M, 1967. *Neurosecretion.* (Gauthier-Villars, Paris).

Gilbert, L. I. 2004. Halloween genes encode P450 enzymes that mediate steroid hormone biosynthesis in Drosophila melanogaster. *Mol. Cell. Endocrinol.* 215, 1-10.

Gu P L & Chan S M, 1998. Cloning of a shrimp *(Metapenaeus ensis)* cDNA encoding a nuclear receptor superfamily member, an insect homologue of E75 gene, *FEBS Lett,* 436 395.

Gu P.L, Yu K.L & Chan S M, 2000. Molecular characterization of an additional shrimp hyperglycemic hormone : cDNA cloning, gene organization expression and biological assay of recombinant protein,. *FEBS Lett,* 472 :122.

Gu, P. L., Chu, K. H., and Chan, S. M, 2001. Bacterial expression of the shrimp molt inhibiting hormone (MIH): antibody production, immunocytochemical study and biological assay, *Cell Tissue Res.* 303, 129-136

Gu, P.-L., Tobe, S.S., Chow, B.-K., Chu, K.-H., He, J.-G., Chan, S.-M., 2002. Characterization of an additional molt inhibiting hormone-like neuropeptide from the shrimp *Metapenaeus ensis. Peptides* 23, 1875–1883.

Hanström B, 1931. Neue untersuchungen uber Sinnesorgane und Nervensystem der Crustacean, *I. Z. Morph. Okol. Tiere,* 23: 80.

Hinsch G W, 1977. Fine structural changes in the mandibular gland of the male spider crab, *Libinia emarginata* (L.) following eyestalk ablation, *J. Morph.* 154: 307

Hopkins, P.M., 2012. The eyes have it: a brief history of crustacean neuroendocrinology. Gen. Comp. Endocrinol. 175, 357–366.

Huberman A & Aguilar M B, 1989. A neuropeptide with moultinhibiting hormone activity from the sinus gland of the Mexican crayfish *Procambarus bouvieri* (Ortmann), *Comp. Biochem. Physiol,* 93: 299.

Huberman Alberto, 2000. Shrimp endocrinology. A review. Aquaculture 191. 191–208

Kaeuser G., Koolman J. & Karlson P, 1990. Mode of action of molting hormones in Insects, in *Morphogenetic Hormones of Arthropods*, Vol. 1 edited by A.P. Gupta (Rutgers University Press, New Brunswick), 361.

Katayama, H., Ohira, T., Nagata, K., Nagasawa, H., 2001. A recombinant molt-inhibiting hormone of the kurumaprawn has a similar secondary structure to a native hormone: determination ofdisulfide bond arrangement andmeasurements of circular dichroism spectra. *Biosci. Biotechnol. Biochem.* 65, 1832–1839.

Katayama, H., Ohira, T., Aida, K., Nagasawa, H., 2002. Significance of a carboxyl-terminal amide moiety in the folding and biological activity of crustacean hyperglycemic hormone. *Peptides* 23, 1537–1546.

Katayama, H., Nagata, K., Ohira, T., Yumoto, F., Tanokura, M., Nagasawa, H., 2003. The solution structure of molt-inhibiting hormone from kuruma prawn *Marsupenaeus japonicus. J. Biol. Chem.* 278, 9620–9623.

Katayama H,Tsuyoshi Ohira, Shinji Nagata and Hiromichi Nagasawa, 2004. Structure-Activity Relationship of Crustacean Molt-Inhibiting Hormone from the Kuruma Prawn *Marsupenaeus japonicus. Biochemistry 43,* 9629-9635

Krungkasem, C., Ohira, T., Yang, W.J., Abdullah, R., Nagasawa, H., Aida, K., 2002. Identification of two distinct molt-inhibiting hormone-related peptides from the giant tiger prawn *Penaeus monodon. Mar. Biotechnol.* 4, 132–140.

Keller R, 1992. Crustacean neuropeptides, structure, functions and comparative aspects, *Experientia,* 48, 439.

Lachaise F, Le Roux A, Hubert M & Lafont R, 1993. The molting gland of crustaceans, localization, activity and endocrine control (a review), *J. Crust. Biol,* 13, 198

Laufer H, Landau M, Borst D & Homola E, 1986. The synthesis and regulation of methyl farnesoate, a new juvenile hormone for crustacean reproduction in *Advances in invertebrate reproduction* 4, edited by M. Porchet J.C. Andries A. Dhianaut (*Elsevier Science*, Amsterdam), 135.

Laufer H, Landau M, Borst D.W. Homola E, 1987. Methyl farnesoate, its site of synthesis and regulation of secretion in a juvenile crustacean, *Insect Biochem,* 17 1123.

Laufer H, Homola E & Landau M, 1987. Control of methyl farnesoate synthesis in crustacean mandibular organs, *Am. Zool,* 27: 69.

Laufer, H., Demir, N., Pan, X., Stuart, J.D., Ahl, J.S., 2005. Methyl farnesoate controls adult male morphogenesis in the crayfish, *Procambarus clakii. J. Insect Physiol.* 51, 379–384.

Lee K J, Elton T S, Bei A K, Watts S A & Watson R D, 1995. Molecular cloning of a cDNA encoding putative moltinhibiting hormone from the blue crab, *Callinectes sapidus, Biochem. Biophys. Res. Commun,* 209: 1126.

Lee, K.J., Watson, R.D., Roer, R.D., 1998. Molt-inhibiting hormone mRNA levels and ecdysteroid titer during a molt cycle of the blue crab, *Callinectes sapidus. Biochem. Biophys. Res. Commun.* 249, 624–627.

Liu L, Laufer H, Wang Y & Hayes T, 1997. A neurohormone regulating both methyl farnesoate synthesis and glucose metabolism in a crustacean, *Biochem. Biophys. Res. Commun,* 237: 694.

Madhyastha, M.N. and P.V. Rangnekar. 1976. Neurosecretory cells in the central nervour system of the prawn *Metapenaeus monoceros. Revistia. Biol.*, IXXIX, 133-140.

Maissiat J, 1970a. Etude experimentale du role de "l'organeY" dans le determinisme endocrine de la mue chez l' Isopode Oniscoide *Porcellio dilatatus* Brandt, *C.R. Acad. Sci. Paris,* 270: 2573.

Maissiat J, 1970b. Anecdysis experimentale provoquee chez l'Oniscoide *Ligia oceanica* L. et retablissement de la mue par injection d'ecdysone ou reimplantation de glande maxillaire, *C.R. Soc. Biol,* 164:1607.

Mattson M P & Spaziani E, 1985. Characterization of moltinhibiting hormone (MIH) action on crustacean Y-organ segments and dispersed cells in culture and a bioassay for MIH activity, *J. Exp. Zool,* 236: 93.

Mattson M.P & Spaziani E, 1986. Regulation of crabs Y-organ steroidogenesis *in vitro,* evidence that ecdysteroid production increased through activation of cAMP-phophodiesterase by calcium-calmodulin, *Mol. Cell. Endocrinol,* 48: 135.

Mattson M P & Spaziani E, 1987. Demonstation of protein kinase C activity in crustacean Y-organs and partial definition of its role in regulation of steroidogenesis, *Molec. Cell. Endocr,* 49:159.

McWhinnie, M.A. and C.J. Mohrherr. 1970. Influence of eyestalk factors, intermoult cycle and season up on 14C- leucine incorporation into protein in the crayfish (*Orconectes virilis*). *Comp. Biochem. Physiol.,* 7: 1-14.

Mellon DeF Jr & Greer E, 1987. Induction of precocious molting and claw transformation in alpheid shrimps by exogenous 20-hydorxyecdysone, *Biol Bull,* 172: 350.

Mohamed K S & Diwan A D, 1991. Neuroendocrine regulation of ovarian maturation of the Indian White Prawn *Penaeus indicus* H. Milne Edwards *Aquaculture,* 98: 381.

Mohamed K S & Diwan A D, 1993. Neurosecretory cell types, their distribution and mapping in the central nervous system of the penaeid prawn *Penaeus indicus* H. Milane Edwards, *Bull Inst. Acad Sin,* 38.

Nagabhushanam R & Diwan A D, 1974. Neuroendocrine control of reproduction of female crab *Barytelphusa cunicularis, Marathwada University. J.Sci,* 6: 59.

Nakamura, K. 1974. Studies on the neurosecretion of the prawn *Penaeus japonicus* I. Positional relationship of the cells group located on the supra oesophageal and optic ganglion. *Mem. Fac. Fish. Kagoshina Univ.,* 23: 175-184.

Nakatsuji, T., Keino, H., Tamura, K., Yoshimura, S., Kawakami, T., Aimoto, S., Sonobe, H., 2000. Changes in the amounts of themolt-inhibiting hormone in sinus glands during the molt cycle of the American crayfish, *Procambarus clarkii. Zool. Sci.* 17, 1129–1136.

Nakatsuji, T., Sonobe, H., Watson, R.D., 2006a. Molt-inhibiting hormone-mediated regulation of ecdysteroid synthesis in Y-organs of the crayfish (Procambarus clarkii): involvement of cyclic GMP and cyclic nucleotide phosphodiesterase. *Mol. Cell. Endocrinol.* 253, 76–83.

Nakatsuji, T., Han, D.-W., Jablonsky, M.J., Harville, S.R., Muccio, D.D., Watson, R.D, 2006b. Expression of crustacean (Callinectes sapidus) molt-inhibiting hormone in *Escherichia coli*: characterization of the recombinant peptide and assessment of its effects on cellular signaling pathways in Y-organs. *Mol. Cell. Endocrinol.* 253, 96–104.

Nakatsuji Teruaki Chi-Ying Lee, R. Douglas Watson, 2009. Crustacean molt-inhibiting hormone: Structure, function, and cellular mode of action. *Comparative Biochemistry and Physiology, Part A* 152, 139–148.

Nanda, D.K. and P.K. Ghosh. 1985.The eyestalk neurosecretory system in the brackish water prawn, *Penaeus monodon* (*Fabricius*).Alight microscopical study. *J.Zool. Soc. India,* 37: 25-38.

Newcomb, R.W. 1983. Peptides in the sinus gland of *Cardisoma carnifex* isolation and aminoacid analysis. *J. Comp. Physiol.,* 153: 207-221.

Niwa R, Matsuda T, Yoshiyama T, Namiki T, Mita K, Fujimoto Y, Kataoka H, 2004. CYP306A1, a cytochrome P450 enzyme is essential for ecdysteroid biosynthesis in the prothoracic glands of Bombyx and Drosophila . *J Biol Chem*, 279: 35942–35949.

Niwa R, Sakudoh T, Namiki T, Saida K, Fujimoto Y, Kataoka H, 2005.The ecdysteroidogenic CYP CYP302A1 / disembodied from the silkworm, *Bombyx mori* , is transcriptionally regulated by prothoracicotropic hormone. *Insect Mol Biol*, 14: 563–571.

Ohira, T.,Watanabe, T., Nagasawa, H., Aida, K., 1997. Molecular cloning of amolt-inhibiting hormone cDNA from the kuruma prawn *Penaeus japonicus. Zool. Sci.* 14, 785–789.

Ohira, T., Nishimura, T., Sonobe, H., Okuno, A., Watanabe, T., Nagasawa, H., Kawazoe, I., and Aida, K, 1999. Expression of a recombinant molt-inhibiting hormone of the Kuruma prawn *Penaeus japonicus* in *Escherichia coli*, *Biosci. Biotechnol. Biochem.* 63, 1576-1581.

Ohira, T. Katayama, H., Aida, K., and Nagasawa, H.2003. Expression of a recombinant crustacean hyperglycemic hormone of the kuruma prawn *Penaeus japonicus* in methylotrophic yeast *Pichia pastoris*, *Fish. Sci.* 69, 95-100.

Ohira, T., Katayama, H., Tominaga, S., Takasuka, T., Nakatsuji, T., Sonobe, H., Aida, K., Nagasawa, H., 2005. Cloning and characterization of a molt-inhibiting hormonelike peptide from the prawn *Marsupenaeus japonicus. Peptides* 26, 259–268.

Okumura T, Nakamura K, Aida K, Hanyu I. 1989. Hemolymph ecdysteroid levels during moltcycle in the Kuruma prawn *Penaeus japonicus.* Nippon Suisan Gakk 55:2091–2098.

Orengo, D.J. and Aguadé, M., 2007. Genome scans of variation and adaptive change: extended analysis of a candidate locus close to the phantom gene region in Drosophila melanogaster. *Molecular biology and evolution*, 24, 5:1122-1129.

Passano L M, 1960. Molting and its control, in *The physiology of crustacea,* Vol. 1 edited by T.H. Waterman (Academic Press, New York), 473.

Passano L M & Jyssum S, 1963. The role of the Y-organ in crab proecdysis and limb regeneration, *Comp Biochem Physiol,* 9: 195

Petryk A, Warren JT, Marques G, Jarcho MP, Gilbert LI, Parvy J-P, Dauphin-Villemant C, O'Connor MB: Shade is the Drosophila P450 enzyme that mediates the hydroxylation of ecdysone to the steroid insect molting hormone 20-hydroxyecdysone. *Proc Natl Acad Sci USA 2003,* 100*: 13773–13778.*

Quackenbush L S, 1986. Crustacean endocrinology, a review, *Can. J. Fish. Aquat. Sci*, 43; 2271.

Quian J N, Fu, S.G., Xu, X.J. and He, J.T., 2007. A new and practical application of JH antagonist KK-42 to promoting growth of shrimp *Penaeus schmitti*. *Aquaculture*, 270,1:422-426.

Rao K R, 1965. Isolation and partial characterization of the moultinhibiting hormone of the crustacean eyestalk, *Experientia*. 21: 593.

Rodriguez, E.M., Lopez Greco, L.S., Medesani, D.A., Laufer, H., Fingerman, M., 2002. Effect of methyl farnesoate, alone and in combination with other hormones, on ovarian growth of the red swamp crayfish, *Procambarus clarkia*, during vitellogenesis. *Gen. Comp. Endocrinol.* 125, 34–40.

Rewitz KF, Rybczynski R, Warren JT, Gilbert LI, 2006. Identification, characterization and developmental expression of Halloween genes encoding P450 enzymes mediating ecdysone biosynthesis in the tobacco hornworm, *Manduca sexta . Insect Biochem Mol Biol,* 36: *188–199.*

Rewitz KF, Gilbert LI. 2008. Daphnia Halloween genes encode cytochrome P450s mediating the synthesis of the arthropod molting hormones: evolutionary implications. *BMC Evol Biol* 8:60.

Saïdi, B., de Besse, N.,Webster, S.G., Sedlmeier, D., Lachaise, F., 1994. Involvement of cAMPand cGMP in the mode of action of molt-inhibiting hormone (MIH) a neuropeptide which inhibits steroidogenesis in a crab. *Mol. Cell. Endocrinol.* 102, 53–61.

Sedlmeier, D., Fenrich, R., 1993. Regulation of ecdysone biosynthesis in crayfish Yorgans: I. Role of cyclic nucleotides. *J. Exp. Zool.* 265, 448–453.

Sefiani M., Le Caer J.P, & Soyez D, 1996. Characterization of hyperglycemic and molt-inhibiting activity from sinus glands of the penaeid shrimp *Penaeus vannamei, Gen. Comp. Endocrinol,* 103: 41

Skinner D M, 1985. Molting and regeneration, in *The Biology of crustacea,* Vol. 9 edited by D.E. Bliss and L.H. Mantel (Academic Press, New York), 43.

Skinner D M Graham D E Holland C A Mykles D LSoumoff C & Yamaoka L H, 1985. Control of molting in Crustacea, in *Crustacean issues, Factors in adult growth* Vol3, edited by A W Wenner (A A Balkema, Rotterdam), 3.

Scheer, B.T. 1960. Aspects of the intermoult cycle in natantians. *Comp. Biochem. Physiol.,* 1: 3-18.

Shoji J, 1997. Studies on reproductive endocrinology of penaeid prawn *Penaeus monodon* (*Fabricius*), Ph.D. Thesis, Cochin University of Science and Technology, Cochin.

Synder M J & Chang E S 199. Ecdysteroids in relation to the molt cycle of the American lobster *Homarus americanus.* I Hemolymph titers and metabolites, *Gen. Comp. Endocrinol,* 81: 133.

Spaziani E, Mattson M.P & Rudolph P H, 1994. Regulation of crustacean molt-inhibiting hormone, *Perspect Comp Endocrinol,* 243.

Spindler K D, Van Wormhoudt A, Sellos D & Spindler- Barth, M, 1987. Ecdysteroid levels during embryogenesis in the shrimp. *Palaemon serratus* (Crustacea, Decapoda), quantitative and qualitative changes, *Gen Comp Endocrinol,* 66: 116.

Sun P S, 1994. Molecular cloning and sequence analysis of a cDNA encoding a molt-inhibiting hormone-like shrimp *Penaeus vannamei, Mol. Mar. Biol. Biotechnol,* 3: 1.

Sun P S, 1995. Expression of the molt-inhibiting hormone-like gene in the eyestalk and brain of the white shrimp *Penaeus vannamei, Mol. Mar. Biol. Biotechnol,* 4: 262.

Truman, J.W., 2005. Hormonal control of insect ecdysis: endocrine cascades for coordinating behavior with physiology. *Vitamins and hormones*, 73:1-30.

Tsukimura B & Borst D W, 1992. Regulation of methyl farnesoate in the hemolymph and mandibular organ of the lobster, *Homarus americanus, Gen. Comp. Endocrinol,* 86: 597.

Vernet G & Charmantier-Daures M,1994. Mue, Autotomie et Regeneration, in *TraITE DE Zoologie, crustaces,* Vol. VII edited by J. Forest (Masson, Paris), 107.

Vijayan, K K, 1988. *Studies on the physiology of moulting in the penaeid prawn Penaeus indicus H Milne Edwards.* Thesis, Central Marine Fisheries Research Institute. Submitted to Cochin University of Science and Technology.

Vijayan K K & Diwan A D, 1993. Studies on the physiology of moulting in the penaeid prawn *Penaeus indicus* H. Milne Edwards. *CMFRI spl. Publ.* 56: 6.

Vijayan K K Mohamed K S & Diwan A D, 1993. On the structure and moult controlling function of the Y-organ in the prawn *Penaeus indicus* H. Milne Edwards, *J. World Aquaculture. Society,* 24: 516.

Vijayan K K & Diwan A D,1994. The mandibular organ of the prawn *Penaeus indicus* H. Milne Edwards and its in consequential role in moulting process, *J. Aqua. Biol,* 9:45.

Vijayan K K, Mohamed K S, & Diwan A D, 1997. Studies on moult staging, moulting duration and moulting behaviour in Indian white shrimp *Penaeus indicus.* H. Milne Edwards. (Decapoda, Penaeidae), *J Aqua. Trop,* 12 :53.

Waddy S L & Aiken D E., 2000. Endocrionology and the culture of homarid lobsters, Vol. 4 in *Recent advances in marine biotechnology* edited by M Fingerman, R. Nagabhushanam (Science Publishers, USA), 195.

Wainwright G, Webster S G, Wilkinson M C & Chung J S, 1996. Structure and significance of mandibular organ-inhibiting hormone in the crab, *Cancer pagurus.* Involvement in multihormonal regulation of growth and reproduction, *J. Biol. Chem,* 271: 12749.

Wang, Z.-Z., Xiang, J.-H, 2003. Cloning and analysis of three genes encoding type II CHH family neuropeptides from *Fenneropenaeus chinensis. Yi Chuan Xue Bao.* 30, 961–966.

Warren JT, Petryk A, Marques G, Jarcho M, Parvy J-P, Dauphin-Villemaont C, O'Connor MB, Gilbert LI, 2002: Molecular and biochemical characterization of two P450 enzymes in the ecdysteroidogenic pathway of Drosophila melanogaster . *Proc Natl Acad Sci* USA, 99: 11043–11048.

Warren JT, Petryk A, Marques G, Jarcho M, Parvy JP, Shinoda T, Itoyama K, Kobayashi J, Jarcho M, Li Y, O'Connor MB, Dauphin-Villemant C, Gilbert LI, 2004. Phantom encodes the 25-hydroxylase of Drosophila melanogaster and Bombyx mori: a P450 enzyme critical in ecdysone biosynthesis. *Insect Biochem Mol Biol,* 34: 991–1010.

Webster S G, 1986. Neurohormonal control of ecdysteroid biosynthesis by *Carcinus maenas* Y-organ *in vitro* and preliminary characterization of the putative molt-inhibiting hormone (MIH), *Gen. Comp. Endocrinol,* 61:237.

Webster S G & Keller R, 1986. Purification, characterization and amino acid composition of the putative molt-inhibiting hormone (MIH) of *Carcinus maenas* (Crustacea, Decapoda), *J. Comp. Physiol.* 156: 617.

Webster S.G & Keller R, 1989. Molt inhibiting hormone, in *Ecdysone, from chemistry to mode of action,* edited by J. Koolman (Theime Med. Publ., Stuttgart), 211.

Webster S G & Dircksen H, 1991. Putative molt inhibiting hormone in larvae of the shore crab *Carcinus maenas* L., an immunocytochemical approach, *Biol. Bull,* 180L: 65.

Webster Simon George, Rainer Keller b, Heinrich Dircksen, 2012. The CHH-superfamily of multifunctional peptide hormones controlling crustacean metabolism, osmoregulation, moulting, and reproduction. *General and Comparative Endocrinology* 175: 217–233.

Wilder, M.N., Okada, S., Fusetani, N., Aida, K., 1995. Hemolymph profiles of juvenoid substances in the giant freshwater prawn, *Macrobrachium rosenbergii* in relation to reproduction and molting. *Fish. Sci.* 61, 175–176.

Yang W J, Aida K., Terauchi A., Sonobe H. & Nagasawa H,1996. Amino acid sequence of a peptide with molt-inhibiting activity from the kuruma prawn. *Penaeus japonicus, Peptide,* 17: 197

Yamano, K., Unuma, T., 2006. Expressed sequence tags from eyestalk of kuruma prawn, *Marsupenaeus japonicus. Comp. Biochem. Physiol.* A 143, 155–161.

Yodmuang, S., Udomkit, A., Treerattrkool, S., Panyim, S., 2004. Molecular and biological characterization of molt-inhibiting hormone of *Penaeus monodon. J. Exp. Mar. Biol. Ecol.* 312, 101–114.

Zeleny C, 1905. Compensatory regulation, *J Exp Zool,* 2:1.

SECTION-IV

DISEASE CONTROL AND THERAPEUTICS

CHAPTER 10

IMMUNE SYSTEM AND MECHANISM OF DEFENCE

10.1 INTRODUCTION

The shrimp Aquaculture industry has grown remarkably over the past several years; it is because of growing international demand for shrimp and stagnating catches of wild shrimp. Initial hopes that farmed shrimp could provide an environmentally benign alternative to over-exploited wild stocks have, however, proven disappointing. Although global growth has been steady, the industry has experienced dramatic production crashes at national and sub-national scales associated with severe environmental damage. Shrimp continues to be the largest single commodity in value terms, accounting for about 15 percent of the total value of internationally traded fishery products in 2015. It is mainly produced in developing countries, and much of this production finds its way into international trade. However, as economic conditions improve in these countries, growing demand is leading to increased domestic consumption and hence lower exports. World farmed shrimp production volumes decreased in 2012 and particularly in 2013, mainly as a result of disease-related problems, such as early mortality syndrome in some countries in Asia and Latin America. The shrimp Aquaculture industry is growing continuously since 1980's but shrimp production is regularly and dramatically affected by series of problems linked not only to environmental degradation but also to infectious and non-infectious diseases. These problems are always causing hindrance in the growth of the shrimp industry. The causative agents of infectious diseases in shrimps are mainly viruses and bacteria belong to *Vibrionacea* (Bachere, 2000) and non-infectious diseases are often suspected to be the consequence of environmental degradation. Therefore, prevention and control of diseases has become the priority of all shrimp producing countries and this requires an integrated approach. In order to control the diseases, it is necessary to have the knowledge of shrimp immunity and understand the mechanism of defence involved. Innate immunity is the first line of defence against microbes and it is indispensable in preventing infections as well as in the development and regulation of the adaptive immune system. Innate immunity is based on the ability of genome-encoded proteins to recognize and bind microbial surface structures, which is followed by the activation of the innate immune response via various cell signalling pathways.

The entire gambit of shrimp immune system and mechanism of defence can be discussed under three sub headings i.e. Physical barriers, cellular and humoral defence system and shrimp immunostimulation. In shrimp and other crustaceans' physical barriers are the first line of defence and consists of a rigid exoskeleton which protects the animal from injury and microbial attacks (Guzman *et al.*, 2009). The exoskeleton is composed of calcium carbonate, carbohydrates and proteins, and contributes to different physiological processes associated with the immune response (Mylonakis and Aballay, 2005). There are reports of diffuse distribution of haemocyanin and catalytic phenol oxidation over the exocuticle and endocuticle of crustaceans; both are important immune response against microbes (Adachi *et al.*, 2005). However, the mechanisms involved in crustaceans cuticle hardening and the role of phenol oxidase are poorly understood.

In cell mediated immune defence system the shrimps have open circulatory system with haemolymph as blood which circulates through the body cavity and supply to all the tissues. In the haemolymph there are haemocytes and humoral components which are transported through haemolymph favouring their encounter with foreign particles (Rendon and Balcazar, 2003). The hematopoietic tissue (HPT) in the shrimp and other crustaceans is an extensive network of packed lobules located at the dorsal and dorsolateral sides of the stomach, close to the antennal artery and at the base of the maxillipedes. Hemocytes are produced within the walls of these tubules and released into the vessel lumens (Soderhäll *et al.*, 2003). The HPT of *Penaeus monodon* and other penaeid shrimp is located in different areas of organs like stomach, maxillipeds and antennal gland (Van de Braak *et al.*, 2002). The haemocytes in the penaeid shrimp have similar biological properties and functions as in vertebrate blood cells (Van de Braak, 2002). These cells participate in phagocytosis, encapsulation, and nodule formation, wound repair, clotting, and prophenol-oxidase activation. They also help the production of adhesion molecules, agglutinins and antimicrobial peptides (Destoumieux *et al.*, 1997; Bachere *et al.*, 2000). These cells also have inhibitory enzymes needed for regulating the proteolytic activity, preventing its over stimulation and the resultant tissue damage (Braak, 2002; Johansson *et al.*, 2000). There are three types of haemocytes i.e., hyalinocytes, granulocytes and semi-granulocytes. Hyalinocytes are small non-refractive cells, with a small nucleus relative to their cytoplasm, which have few or no cytoplasmic granules. Hyalinocytes have no phagocytic activity and the primary role of these cells is related to clotting and phagocytosis (Zhang *et al.*, 2006). Granulocytes have the smallest nucleus and a high number of cytoplasmic granules and they display phagocytic activity and store the enzyme pro-phenoloxidase. These cells may be stimulated by glucans, peptidoglycans (PG) and lipopolysaccharides to provoke exocytosis (Zhang *et al.*, 2006). Semi-granulocytes have a large numbers of small granules. These cells possess glucans receptors and their principal function involves phagocytosis, encapsulation and clotting (Martin and Graves, 2005; Zhang *et al.*, 2006).

The phagocytosis is a process by which certain cells engulf and destroy microorganisms and cellular debris. The process includes five steps (Fig 1): (1) invagination, (2) engulfment, (3) internalization and formation of the phagocyte vacuole, (4) fusing of lysosomes to digest the phagocytosed material, and (5) release of digested microbial products (Kondo *et al.*, 1998; Albores and Plascencia 1998). In fact the phagocytic cells destroy the internalized organisms by two mechanisms, one is by an aerobic process which uses NADPH, and reduces an oxygen electron to form the superoxide ion.

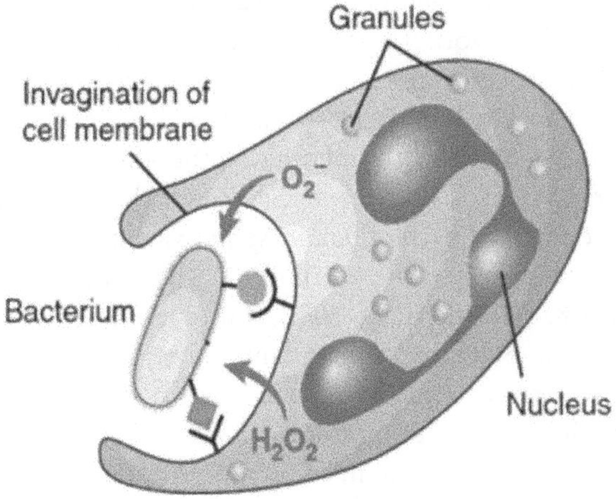

Fig. 1. Phagocytosis. (**Source:** Miller-Keane, 2000)

In shrimp haemocytes, the activation of the aerobic process has been demonstrated by the use of bacteria (*V. parahaemolyticus and V. vulnificus*) and surface microbial antigens, as both increase the phagocytic capacity of haemocytes to destroy pathogens (Itami *et al.*, 1998; Song and Huang, 2000; Cordova *et al.*, 2002). The second, anaerobic process is attributed to the action of diverse microbicidal enzymes, such as lysozyme and low molecular weight AMP (Nappi and Ottaviani, 2000). Encapsulation and nodule formation is also one of the defence mechanisms in the shrimp reported by several workers (Albores and Plascencia, 1998, Wang *et al.*, 2001, Braak, 2002). In this defence mechanism, semi granulocyte cells are responsible for the recognition of invading agents and their encapsulation with proteins that work as an opsonins associated to the pro phenol-oxidase activation system. Formation of hemolytical nodules in different tissue like gills and hepatopancreas has been reported to trap microorganisms that cannot be removed by phagocytosis (Wang *et al.*, 2001, Braak, 2002).

10.2 ANTIOXIDANT SYSTEM

In shrimp defence mechanism antioxidants factors play very important role and in recent past research is getting intensified in this particular area, since aquatic animals are susceptible to oxidative stress as a result of pathogen pressure and environmental perturbation (Castex *et al.*, 2010). Generally reactive oxygen intermediates and reactive nitrogen intermediates are generated in phagocytic vacuoles in the shrimp which are under stress. These molecules are capable of crossing cell barrier and damaging neighbouring cells (Nathan and Shiloh, 2000). In order to prevent this damage, antioxidant defence strategies have been developed including enzymatic substances which may neutralize the reactive intermediates molecules and repair the damage done to the cells (Nathan and Shiloh, 2000). Xu and Pan (2013) while working on bioflock impact on immune response and antioxidant status in the shrimp *L. vannamei* found that there was increase in the total antioxidant capacity in both plasma and the hepatopancreas and increased superoxide dismutase activity in the plasma of the shrimp which are grown in the bioflock culture system. These results indicated that the antioxidant system of the shrimp was enhanced by bioflock in some way. Similar observations have been made by other workers while working on the shrimps (Monostori *et al.*, 2009, Ju *et al.*, 2008, Xu and Pan, 2012, Yang *et al.*, 2010). Recent studies strongly suggest that an enhanced antioxidant activity facilitates shrimp immune defence functions (Castex *et al.*, 2009). Chiu *et al.*, (2007) mentioned that when *L. vannamei* were fed diet with probiotic *Lactobacillus plantarum*, there was reduction in respiratory burst in the haemolymph through increase in superoxide dismutase gene transcription. The high survival and improved growth of the shrimp culture in bioflock treatment systems has been reported by Xu and Pan (2012). However, further research is needed to verify the beneficial effects of bioflock serving as potential sources of immunostimulants and antioxidants.

Bugge *et al.*, (2007) reported the role of oxidative compounds in improving the immune system in shrimp and these compounds are produced in the haemocytes in response to invasion of pathogenic bacteria. This cellular response occurs during microbe phagocytosis. These compounds are superoxide anions (O_2), hydroxyl radicals (OH), H_20_2, reactive oxygen intermediates and reactive nitrogen intermediates. Their production is mediated through various enzymes.

The role of prophenoloxidase system (proPO) in strengthening immune system in the shrimp is well known in the literature. Generally granulocytes in the haemolymph are responsible for the synthesis storage and secretion of the proPO system. The process is activated by the molecules like β-glucan, peptidoglycans and lipopolysaccharides and once the process is activated, this leads to conversion of phenols to quinones which may help to kill pathogens and are used for melanin production (Lee *et al.*, 2004, Hellio *et al.*, 2007).

Melanisation and cytokines production also plays important role in the shrimp immune system (Mury, 2007, Nappi and Ottaviani, 2000). Melanin, a product of proPO system, have antibacterial properties that inhibit antigens (Holmblad and Soderhall, 1999) and protect the animals from pathogens. The cytokines which produced by the haemocytes also have antimicrobial properties and helps in protection of animals from invading pathogens. Yu *et al.*, (2003) worked on cytokines in the shrimp *P. monodon* and reported for the first time the correlation between heat shock protein (cytokines) and immune responses of the shrimp.

Clotting of haemolymph or coagulation process in the shrimp is to prevent the loss of haemolymph through cuts and wounds in the exoskeleton and to immobilise the invading pathogen (Yu *et al.*, 2005). Three types of haemolymph clotting processes have been reported in crustaceans (Guzman *et al.*, 2009). In first type there is a rapid haemocyte agglutination without plasma coagulation, in the second type there is a cellular aggregation with limited plasma coagulation and in the third type there is a limited cellular aggregation and lysis followed by plasma coagulation. It is reported that in the shrimp the third type process is involved in clotting protein cascade (Braak *et al.*, 2002).

There are many aspects of the crustacean immune system which are now comparatively well understood and documented by many workers (Smith and Chisholm, 1992; Soerhall and Cerenius, 1998; Holmblad and Soderhall, 1999; Sritunyalucksana and Soderhall, 2000). Lot of research work has been carried out on immunology of crab, crayfish and lobster but research on shrimps is still scanty and more emphasis is now given to understand the basic mechanisms involved (Bachere *et al.*, 1995; Sritunyalucksana *et al.*, 1999; Sritunyalucksana *et al.*, 2001). Smith and Chisholm (1992) and Smith and Chisholm (2001) while working on crustacean immune system mentioned that the circulating haemocytes play extremely important roles not only by direct sequestration and killing of infectious agents but also by synthesis of a battery of bioactive protein molecules. The haemocytes actually execute inflammatory-type reactions like phagocytosis, haemocyte clumping, production of reactive oxygen intermediates and the release of microbicidal proteins. Full immune reactivity is always achieved through interaction between haemocytes or their products. A key protein is peroxinectin and this is present in granulocytes of both shrimps and prawns (Sritunyalucksana *et al.*, 2001). Similarly, the fundamental role of phenoloxidase seems to be identical in the immune system of *Penaeus monodon* (Bachere *et al.*, 1995; Sritunyalucksana *et al.*, 1999) and other decapods. There are other antibacterial peptides known as penaeidins which have reported so far only in species of penaeid shrimp (Destoumieux *et al.*, 2000; Sritunyalucksana *et al.*, 2001). Under normal conditions, most immuno-reactive factors are stored within the haemocytes, usually in an inactive state. Their release and 'activation' occurs through regulated exocytosis following stimulation by the presence of non-self-molecules in the haemolymph. It has been reported that the 'clumping' behaviour of the haemocytes is influenced by some of the proteins from the cells. The biochemical and molecular signals involved in activation process are gradually being elucidated and a number of key proteins have been purified, sequenced and cloned. Many of these are associated with the prophenoloxidase activating system. Activation process which results in the generation of number of potent bioactive products helps in the process of phagocytosis, cell to cell adhesion and the formation of melanin deposits (Soderhall and Smith, 1986; Soderhall and Cerenius, 1998). The Fig. 2 illustrates immune responses of different crustacean including Penaeid shrimp after entry of pathogen in the hemocoel of the host.

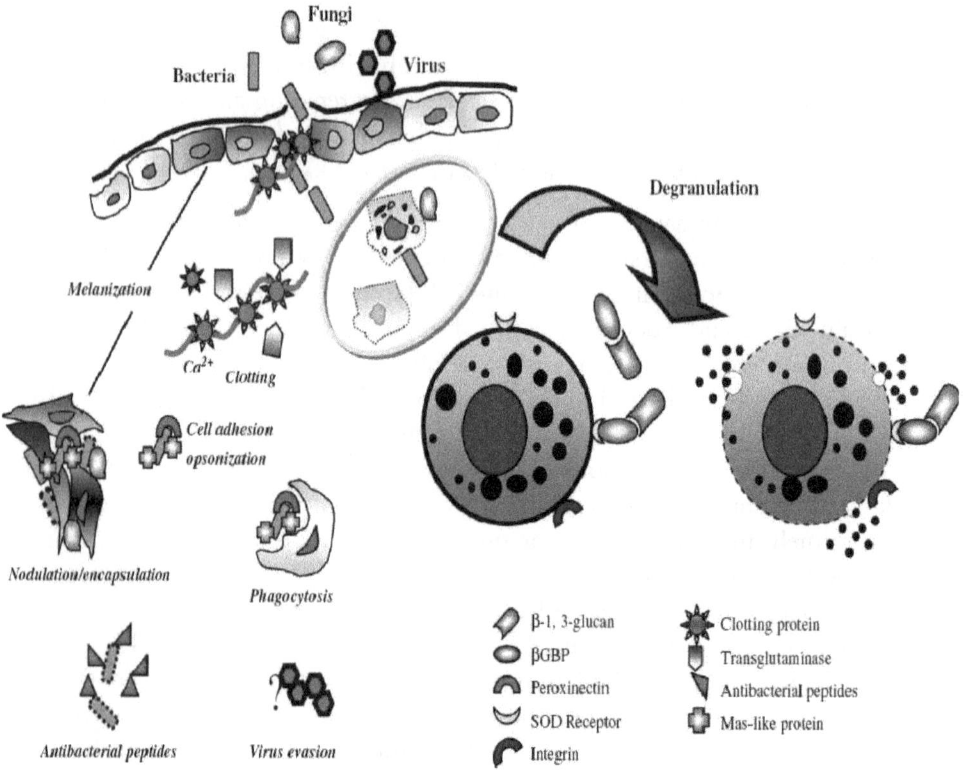

Fig. 2. Innate immune responses of crustaceans after entry of pathogen in hemocoel of host (**Source:** Jiravanichhpaisal *et al.*, 2006).

10.3 COMPONENTS OF CRUSTACEANS IMMUNE SYSTEM

Circulating haemocytes and haematopoietic cells

In crustacean animals when external molecules enter the body then their recognition take place by activating the haemocytes, which play an important role in host defence. According to numerous works dealing with the identification of haemocyte cell types, three types of cell have been identified viz., hyaline, semi granular and granular cells (Martin and graves, 1985). Omori *et al.*, (1989) based on their studies on morphological and cytological characteristics haemocytes in relation to different defence reactions, they attributed the involvement of hyaline cells with coagulation process. Likewise the granular and semi granular cells with the function of phagocytosis has been identified by Gargioni and Barracco, (1998) and the exocytosis of the prophenoloxidase activating system by Johansson and Soderhall, (1985). It is reported that the crustaceans contains many soluble defence molecules, like hemocyanins, lectins, and C-reactive proteins, in the haemolymph (Iwanaga *et al.*, 1998). The innate immune system in some of the crustaceans recognizes invading pathogens by using lectin-agglutinins reactions with different specificities components exposed on the surfaces of pathogens (Crouch *et al.*, 2000; Chiou *et al.*, 2000; Kawabata *et al.*, 2001). Five types of lectins have been identified in circulating haemocytes and haemolymph plasma (Kawabata and Iwanaga, 1999; Kawabata and Tsuda, 2002). These components function synergistically to form an effective host defence against invading microbes and foreign substances. There is another class of bacterial agglutinins called C-reactive protein present in the haemolymph of crustaceans involved in defence mechanism (Iwaki *et al.*, 1999). Three types of C-reactive proteins have been reported so far and they have been isolated and purified and also characterised (Iwaki *et al.*, 1999). There is another component called tachyplesin which is peptide antibiotic and present in plasma of haemolymph of crustaceans was found to be effective in inhibiting the growth of gram-negative and gram-positive bacteria and fungi (Iwanaga *et al.*, 1998; Morvan *et al.*, 1997). Likewise there are other two components called tachystatins and tachycitin (Kawabata *et al.*, 1996) which have been also reported in the haemolymph of crustaceans having the property of antibacterial and antifungal. In immediate defence system the non- self-recognition factors in the shrimp and other crustaceans activate several other components in the haemolymph. The pro-phenol oxidase (pro-PO) activating system actually consists of several proteins that participate in melanin formation, cytotoxic reactions, cell adhesion, encapsulation, and phagocytosis processes (Söderhäll *et al.*, 1994; Cannon *et al.* 2004). It is an efficient non-self-recognition system and is initiated by the recognition of peptidoglycans of bacteria and β-D-glucans of fungi. It has been reported that the pro-PO system of shrimp *P. leniusculus*, is composed of a protease cascade of recognition proteins, several serine protease and their zymogens. The active form of pro-PO, phenol oxidase, also known as tyrosinase, catalyses two successive reactions; the first is the hydroxylation of a monophenol to an O-diphenol, and the second is the oxidation of O-diphenol to O-quinone (Sugumaran, 2002).

The production of antimicrobial peptides is in fact widespread in all organisms from bacteria to plants and from invertebrates to vertebrates (Bachere *et al.*, 2000). However, until now in crustaceans the innate defence reaction involving the synthesis of antimicrobial peptides is less studied. In penaeid shrimp *P. vannamei* three antimicrobial peptides have been purified from the haemocytes and the plasma. It has been further reported that these proteins have been fully characterized and their cDNA cloned (Destoumieux *et al.*, 1997) and they were named as penaeidins, after the genus *Penaeus*. Further research being carried out to investigate the role of penaeidins in the immune response against pathogensas well as the characterization of other peptides in the shrimp.

Takahashi *et al.*, (1998) while working on efficacy of oral administration of fucoidan, a sulphated polysaccharide, in controlling white spot syndrome (WSS) in kuruma shrimp, *P. japonicus* found that fucoidan inhibits the attachment of WSSV to shrimp cells, resulting in prevention of WSSV infection. The inhibitory effect of sulphated polysaccharides (fucoidan and dextran sulphate) on the replication of enveloped viruses have proven earlier also (Baba *et al.*, 1988) and it was mentioned that they are potent prophylactic agents against infections of these viruses. The mechanism of action of these compounds has been attributed to the inhibition of virus adsorption to host cells. Chotigeat *et al.*, (2004) also studied the effect of fucoidan on the disease resistance in the shrimp *P. monodon.* They reported that crude fucoidan extracted from *sargassum polycystum* when administered in to the shrimp can reduce the impact of WSSV infection. Further they mentioned that fucoidan can also inhibit the growth of *Vibrio harveyi*, *Staphylococcus aureus* and *Escherichia coli* at minimal inhibition concentration of 12.0, 12.0 and 6.0 mg/ml, respectively.

Lio-po *et al.*, (2005) studied the anti-luminous vibrio factors associated with the 'green water' grow out culture of the tiger shrimp, *P. monodon.* In their study in order to prevent outbreaks of *luminous vibriosis* infection in culture operations of the shrimps, they used isolates of bacteria, fungi, phytoplankton and fish skin mucous. It was reported that among 85 bacterial isolates tested, 63 isolates inhibited the infection of the *Vibrio harveyi* pathogen after 24-48 h co-cultivation. Among fungi isolates also yielded similar results i.e. complete inhibition of *luminous vibrio* and the skin mucous from jewel tilapia also prevented the growth of *luminous Vibrio* in shrimp. This study indicates that the effectiveness of the 'green water' culture of tiger shrimp in preventing outbreaks of luminous Vibrios is in grow out ponds.

10.4 PROTEINS INVOLVED IN SHRIMP IMMUNO-DEFENCE MECHANISMS

Agglutinins

Albores, (1995) while working on proteins involved in shrimp immune defence mechanism mentioned that penaeid shrimps have protein molecules, cells and systems for defensive mechanisms to prevent invasions by microorganisms. Further it is reported that two kinds of proteins are involved in the recognition of microbial products, and their activation of cellular functions. The first group of proteins is named hemagglutinins or lectins which induces β-glucan though they are able to bind sugar residues. Several workers have reported agglutinating activities in plasma of *Penaeus monodon* (Ratanapo & Chulavatnatol 1990), *Penaeus stylirostris* (Albores *et al.*, 1992), *Penaeus californiensis* (Albores *et al.*, 1993a), *P. japonicus* (Bacheré *et al.*, 1995) and *Penaeus indicus* (Maheswari *et al.*, 1997). However, only the agglutinins from *P. monodon* and *P. californiensis* have been purified and their main properties studied. It is reported that this lectin, named monodin, induces the agglutination of *Vibrio vulnificus*, a major infective bacterium for shrimp (Ratanapo & Chulavatnatol 1992).

The second recognizing protein detected in shrimp plasma is named as β-glucan binding protein because it has the capacity to react with β-glucan. This protein has been purified from *P. californiensis*, *P. vannamei* and *P. stylirostris* plasma as a 100-kDamonomeric protein. The amino acid content of these shrimp β-glucan is nearly identical to the homologues from the freshwater crayfish, *P. leniusculus*. (Albores *et al.*, 1996). Unfortunately, studies on β-glucan binding proteins that stimulate the proPO system are not yet available.

In shrimp, phenoloxidase (PO) system actually promotes hydroxylation of phenols and oxidation of o-phenols to quinones, in response to foreign matter invading the hemocoele and during wound healing (Ashida & Yamazaki, 1990; Johansson & Söderhäll, 1989; Söderhäll, 1992; Söderhäll *et al.*, 1994). These quinones are subsequently transformed to melanin by non-enzymatic reactions. Although a direct antimicrobial activity has been described for melanin, the production of reactive oxygen such as superoxide anions and hydroxyl radicals during the generation of quinoids (Nappi *et al.*, 1995; Song & Hsieh, 1994) has also an important antimicrobial role.

The non-self-recognition factors activate several immediate defence systems mediated through the haemocytes. Bachere (2000) while discussing on shrimp immune system mentioned that the pro-PO activating system is one of the best studied immune systems in crustaceans. Numerous components associated with this immediate defence process, which leads to the reaction of melanisation, have been characterized and some functions in the pro-PO activating system have been determined and one of these functions is haemolymph coagulation. Clotting process and wound healing are also an important defence reaction by participating in the engulfing of foreign invading organisms (Kopacek *et al.*, 1993). It has been reported that in shrimps some components of these defence reactions have been identified and among these some of them like pro-PO activating system, a plasmatic clotting factor and α 2-microglobulin have been characterized by means of specific monoclonal antibodies (Bachere *et al.*, 1995, Rodriguez *et al.*, 1995). The defence reaction also include the haemolytic process of encapsulation, phagocytosis and the microbicidal mechanism based on the production of cytotoxic reactive oxygen intermediates (Song and Hsieh, 1994., Bachere *et al.*, 1995). In crustaceans, transglutaminase is an active enzyme secreted by haemocytes circulating in the haemolymph. It rapidly initiates polymerization of clotting proteins entrapping foreign particles and stopping bleeding when hosts are traumatized (Cerenius and Soderhall, 2011, Loof *et al.*, 2011). In shrimps where genes responsible for production of transglutaminase were silenced are found to be highly susceptible to microbial infection (Fagutao *et al.*, 2012, Maningas *et al.*, 2008).

10.5 USE OF IMMUNO-STIMULANTS FOR SHRIMP DISEASE CONTROL

In sustainable Aquaculture production, diseases are major issues especially for the farming of shrimps in coastal areas (Bachere, 2003). In intensive culture operations, shrimp species are often exposed to stressful conditions, eventually becoming more susceptible to microbial infections, especially in their larval stages (Smith *et al.*, 2003). In order to prevent occurrence of diseases, it is common practice to disinfect the water before use and to apply chemotherapy (e.g. antibiotics) (Vadstein, 1997). Yet such practices are undesirable since they promote the selection and dissemination of antibiotic-resistant bacteria in both the target organisms, as well as in the environment (Smith *et al.*, 2003). Nowadays, the use of preventive approaches, are becoming increasingly important by using various methods like vaccines, immunostimulants and probiotics. At many shrimp farms and hatcheries several antibiotics, vaccines, and chemotherapeutic agents as well as some immunostimulants have been used to prevent viral, bacterial, parasitic, and fungal diseases. These immunostimulants mainly facilitate the function of phagocytic cells and increase their bactericidal activities. Several immunostimulants also stimulate the natural killer cells, complement, lysozyme and antibody responses. Contrary to the vertebrates, which have both an innate and a specific immunesystem, the defence mechanism in shrimps depends mainly on a non-specific orinnate immune response to fight infectious diseases (Kurtz and Franz, 2003). In shrimp innate immune system presumably recognizes a wide diversity of pathogens, represented by common molecular patterns (e.g. lipopolysaccharides or peptidoglycans from bacteria, or β-glucans from bacteria and fungi) rather than structures specific for particular microbe (Soderhall *et al.*, 1996). In cases where disease outbreaks are cyclic and can be predicted, immunostimulants may be used to elevate the non-specific defence mechanisms, to reduce stress and mortalities, and to maintain the health of cultured organism (Raa, 2000). Although the exact mechanism is still not yet completely understood, several immunostimulants are being applied in shrimpcultures, to induce and build up protection against a wide range of diseases. The immunostimulants are generally obtained in large amounts, e.g. glucans from yeast (Anderson, 1997; Bagni *et al.*, 2000) or chitosan (Anderson, 1997; Bullock *et al.*, 2000) from shrimps shell meal. These ingredients can be given through dietary supplements due to the relatively low cost of their source. The use of immunostimulants, can improve the innate defence of the animal providing resistance to pathogens during periods of high stress (Conceicao *et al.*, 2004; Bagni *et al.*, 2000). The use of immunostimulants in vaccine formulations has given very good antibody responses. (Anderson, 1997; Figueras and Santarem, 1998; Kawakami, 1998; Romalde, 1999).

Heat shock proteins (HSP) are abundantly found in several animals performing cytoprotective functions (Srivastava, 2002). They are categorised into several families based on their molecular weight, such as HSP70, HSP90 etc. Among those HSPs, HSP70 is the best studied and there are increasing evidences available which suggest that HSP70 plays an important role in the regulation of the early innate immune responses. In this direction Hu *et al.*, (2014) carried out work on bacterial HSP70 as an efficient immune stimulator in *L. vannamei*. It is reported that bacterial HSP70 is an efficient immune stimulator in *L. vannamei*. Further they have found out that the expression of two types of immune related genes were quantitatively more when *L. vannamei* was injected with recombinant full length bacterial HSP70.

While working on the effects of lipopolysaccharides extracted from *Escherichia coli* and administration of dissolved ammonia on immune response in southern white shrimp, *Litopenaeus schmitti* Rodríguez-Ramos *et al.*, (2008) found that the lethal dose of lipopolysaccharides caused changes in phenol oxidase activity in plasma and nitric oxide and total haemocyte counts within first 24 hrs. High concentration of dissolved ammonia also caused a decrease in haemocytes by 66% within first 72 hrs when compared to the control. These results indicated that the shrimp is able to recognise the lipopolysaccharides and responded to his microbial elicitor, and that increased dissolved ammonia affected the number of circulating haemocytes.

Lectins are one of recognition receptors present in crustaceans that can bind to specific carbohydrates on the cell surface of microorganisms (Sharon and Lis, 2004). They play many roles in crustacean immunity such as cell adhesion (Lis and Sharon, 1998), glycoprotein clearance (Dodd and Drickamer, 2001), phagocytosis, nodule formation (Yu *et al.*, 2005), encapsulation (Ling and Yu, 2006), antibacterial or antiviral activity (Sun *et al.*, 2008; Zhao *et al.*, 2009), recognition of microbe associated molecular pattern (MAMPs) and activation of prophenoloxidase system (Yu *et al.*, 1999) (Fig 3). In this direction recently Thepnarong *et al.*, (2015) worked on molecular cloning of a C-type lectin with one carbohydrate recognition domain from *F. merguiensis* and its expression upon challenging by pathogenic bacterium virus. From the results obtained it is reported that the lactin may participate in recognition of invading pathogens such as bacteria and viruses, and play roles in the immune response of the shrimp.

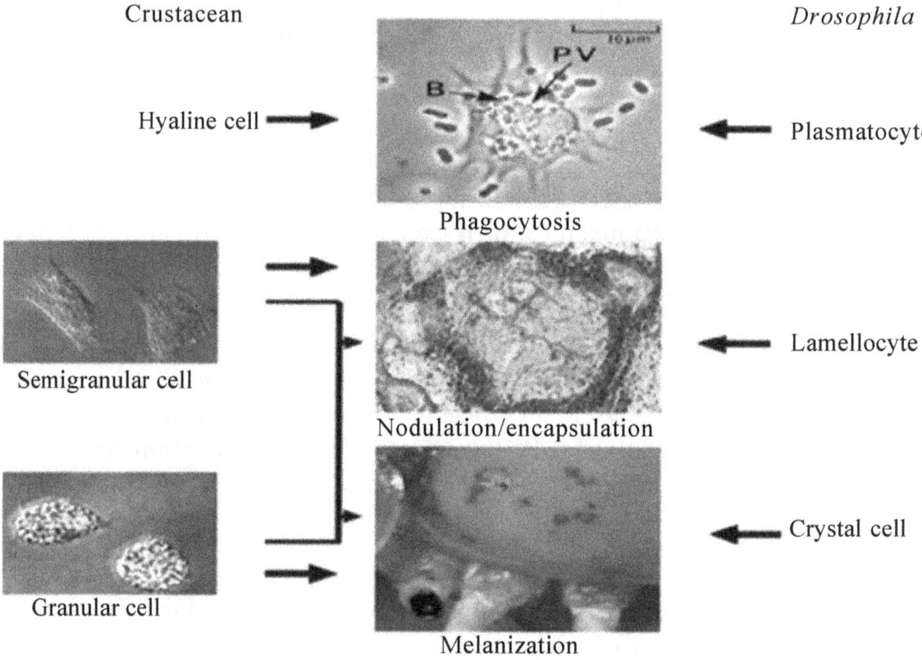

Fig. 3. Comparative characteristics of crustaceans and Drosophila haemocytes. Hyaline cells and plasmatocytes are the major phagocytic cells. Semigranular and lamellocytes take part in nodule formation and encapsulation. Granular and crystal cells store components of proPO system (**Source:** Jiravanichhpaisal *et al.*, 2006).

Menasaveta *et al.*, (2000) worked on immunity enhancement in the shrimp *P. monodon* using a probiont bacterium *Bacillus S11*. They reported that survival and growth of the shrimps increased after feeding them with *Bacillus S11* compared with non-treated shrimps. Bacillus S11 efficiently activated and increased engulfment of foreign particles (phagocytic activity) as measured by % of phagocytosis and phagocytic index in the haemolymph. Increase in phenol oxidase and antibacterial activities have also been noticed with increase in age of the shrimps. Further they have mentioned that such shrimps when challenged with pathogenic bacteria immune responses were found to be substantial and more pronounced in treated shrimps with Bacillus S11. These findings indicated that Bacillus S11 provided disease protection by activating both cellular and humoral immune defences.

Work on immunomodulation by carrageenan, (seaweed polysaccharide) in the white shrimp *L. vannamei* and its resistance against *Vibrio alginolyticus* has been carried out by Yeh and Chen, (2008). It is mentioned that total haemocyte count, phenol-oxidase activity, respiratory burst and phagocytic activity were significantly higher in shrimp that received carrageenan after 24 hrs. Further, it was mentioned that among different types of carrageenan used, the type IV was more effective as it exhibited higher immune ability of *L. vannamei* as well as resistance against *V. alginolyticus* infection. Immunostimulation of shrimp *P. monodon* through oral administration of *Vibrio bacterin* and yeast glucan has been done by Devaraja *et al.*, (1998). It was reported that Vibriocidal activity was found in the haemolymph and haemocyte lysate. Enhanced phenol-oxidase activity in the haemocyte has also been reported. When a combination of bacterin and glucan was administered the response was much higher compared to individual treatments. Harikrishnan *et al.*, (2011) worked on the impact of S*olanum nigrum* extract, a medicinal herb on enhancement of the immune response and disease resistance in the shrimp *P. monodon* against *Vibrio harveyi*. They reported that there was significant enhancement in the activity of phenol-oxidase, superoxide dismutase and glutathione peroxidase of *P. monodon* fed with different doses of *S. nigrum* extract enriched diet against *V. harveyi*. They also found enhancement in respiratory burst activity and phagocytic activity against the pathogen. Immune stimulant effects of a nucleotide-rich baker's yeast extract in the shrimp *M. japonicus* was carried out by Biswasa *et al.*, (2012) by examining expression of antimicrobial peptides such as penaeidin, crustin, and lysozyme genes. Further to confirm that the baker's yeast extract induced anti-microbial peptides, studies were carried out to find out the effect of oral administration on the resistance to *Vibrio nigripulchritudo* infection in the shrimp. The results indicated that after injection of baker's yeast extract and in fed shrimp the gene expression with regard to up regulation of all the above mentioned peptides were very significant in the lymphoid organ. It was also observed that resistance to bacterial pathogen in the shrimp was better due to higher percentage of survival when compared to the control ones. This indicates that the immunostimulatory effects of the nucleotide-rich baker's yeast extract on shrimp immune system supports its potential use in Aquaculture. Chen *et al.*, (2012) carried out the studies on dietary administration of a *Gracilaria tenuistipitata* extract on immune system of shrimp *L. vannamei* in response to ammonia stress. The immune parameters analysed were hyaline cells, granular cells including semi granular cells, total haemocyte count, phenol-oxidase activity, respiratory burst, superoxide dismutase activity and haemolymph protein levels. The findings indicated that shrimps fed with Gracilaria extract exhibited a protective effect against ammonia stress as evidenced by the earlier recovery of immune parameters. Li *et al.*, (2009) worked on the effect of dietary probiotic *Bacillus* and isomalto oligosaccharides on immune responses and resistance to WSSV in the shrimp *L. vannamei*. They observed that with the increasing doses of probiotic bacillus and feeding with isomalto oligosaccharides in shrimp's diet, has improved shrimp's survival rate and immune parameters generally increased whereas the counts of total viable bacteria and vibrio decreased. Recombinant anti-lipopolysaccharide factor isoform 3 and its impact on the prevention of vibriosis in the shrimp *P. monodon* has been worked out by Ponprateep *et al.*, (2009). The results showed that the survival rate in the shrimp was 100% after incubation with recombinant anti-lipopolysaccharide factor isoform 3. Further the effects of injection of anti-lipopolysaccharide factor isoform 3 were monitored on the expression of five genes related to the immune system i.e. phagocytosis activating protein, prophenoloxidase activating enzymes, cytosolic manganese superoxide dismutase, survivin and recombinant anti-lipopolysaccharide factor isoform 3 activity. The results indicated that the recombinant anti-lipopolysaccharide factor isoform 3 is likely to affect various defence pathways in protecting the shrimp bacterial pathogens. Maggioni *et al.*, (2004) made an evaluation of some haemato-immunological parameters in female shrimp *L. vannamei* submitted to unilateral eyestalk ablation and whose diet was supplemented with super doses of vitamin C, as a form of immunostimulation. The analysed parameters were total haemocyte count, protein concentration, agglutinating and phenol-oxidase activities in the serum and levels of glucose and lactate in the plasma. It was reported that the majority of haemato-immunological parameters did not exhibit any statistically significant change in relation to eyestalk ablation in normal shrimp group. There was only a significant reduction in the agglutinating activity on the third day following ablation. Whereas in the unilateral eyestalk ablated group fed with the high doses of vitamin C did not exhibit a decrease in the agglutinating activity on the third day after ablation. This could possibly suggest a potential immunostimulation in this parameter by vitamin C. Shanti and Vaseeharan (2014) worked on Alpha 2 macroglobulin (α2M) gene and their expression in response to *Vibrio parahaemolyticus* and WSSV pathogen when challenged in the shrimp *F. indicus*. It is mentioned that α2M is a non-specific protease inhibitor involved in host defence mechanism in invertebrates. By c-DNA cloning and sequencing, α2M gene has been identified from the haemocytes of *F. indicus*. Sequence comparison and phylogenetic analysis revealed that in the shrimp α2M displays the highest similarity with *L. vannamei* (96%). When challenged with *Vibrio harveyi* and WSSV infection, the shrimp significantly increased the α2M mRNA transcript up to 24h in the haemocytes. These results suggest that α2M is an important molecule in shrimp immune system.

Salinity is one of the most important environmental factors that directly affect the survival, growth, and physiological function of shrimps (Kumlu and Eroldogan 2000; de la Vega *et al.*, 2007) and the large-scale outbreaks of diseases in shrimp *Aquaculture* are also related to environment changes and unscientific farming activities. Decline in the salinity leads to osmotic imbalance directly and metabolic disorders in vivo, inducing outbreaks of diseases (Liu *et al.*, 2008; Moullac and Haffner, 2000; Cheng and Chen, 2000; Perazzolo *et al.*, 2002). Therefore, immuneresponse in shrimps under low salinity stress becomes an important factor in shrimp Aquaculture. The cellular and humoral immunity was found decreased when shrimps are transferred to low salinity environment (Li 2008). Number of workers have described the immune signalling pathway of crustaceans exposed to salinity stress and discovered that the neuroregulators concentration in haemolymph changed rapidly to trigger diverse immune activities via signal transduction pathways (Zatta, 1987; Cheng *et al.*, 2005; 2006). Immune function genes have also been identified to date, including immune recognition molecules, oxidative enzymes, non-digestive proteases, protease inhibitors, antimicrobial peptides, heat shock proteins and other immune effectors (Li *et al.*, 2014; Xu, 2010; Sookruksawong *et al.*, 2013). Digital gene expression studies in haemocytes of the shrimp *L. vannamei* in response to low salinity stress have been carried out by Zhao *et al.*, (2015). They reported that during low salinity stress, functional categorization and pathways of differentially expressed genes in the haemocytes revealed immune signalling pathways, cellular immunity, humoral immunity, apoptosis, cellular protein synthesis, lipid transport and energy metabolism were differentially regulated and concluded that this is a valuable information for better understanding of immune mechanisms in the shrimp under low sanity stress.

10.6 RNAI MACHINERY AND SHRIMP'S ANTIVIRAL FUNCTIONS & IMMUNITY

Though several methods are now being adopted to control the diseases in shrimp Aquaculture, in recent years with the advent of molecular biology, efforts are being made to develop powerful techniques in diseases management in the shrimps. Over the last decade RNA interference techniques have emerged in various plants and animals as critical regulators of many diverse biological functions including antiviral defences (Labreuche and Warr, 2013). Although this technique of RNAi machinery/or gene silencing process has been reported to be relatively well conserved in species of different phyla, in penaeid shrimp work has already initiated by several workers and there are supporting evidences now available regarding the existence of an intact and functional RNAI machinery. In shrimp, gene silencing RNAi was reported for the first time by Robalino *et al.*, (2004, 2005) who mentioned that the administration dsRNA in the shrimp *L. vannamei* was very effective in protection against viral infection. Subsequently, based on these observations RNAi technology has been increasingly used to unravel gene function in controlling viral infection in shrimps. The delivery of ds RNAs or siRNAs specific for viral genes in to shrimp has proved feasible and highly effective to suppress viral replications particularly those of WSSV, the yellow head virus (YHV), and the taura syndrome virus (TSV) (Labreuche and Warr, 2013). Recently, Labreuche and Warr (2013) reviewed in detail regarding antiviral functions of the RNAi machinery in penaeid shrimp. While discussing functional RNAi pathway they reported that only one Dicer gene has been identified in *P. monodon* and *M japonicus* and two Dicer like genes have been identified and cloned in *L. vannamei* (Yao *et al.*, 2010; Chen *et al.*, 2011; Su *et al.*, 2008)

Several key proteins in implicated in RNAi pathways have been characterized in *F. chinensis*, *M. japonicus* and *L. vannamei* and it has been reported that there is lot of similarity of these proteins among the species (Wang *et al.*, 2009 and 2012). These proteins belong to Argonaute family members and form the core of the RNA- induced silencing complex. As indicated by the different studies, considerable advances have been made in the characterization of RNAi machinery in penaeid shrimps. However, the discovery of different sets of Dicer and Argonaute proteins in *L. vannamei* suggest the potential existence of several RNAi pathways in shrimp (Labreuche and Warr, 2013). The mechanism of RNA interference technology to combat viral attack/antiviral immunity in the shrimp has not been worked out properly and our knowledge is scarce in comparison with other animals. However, several other studies have provided evidences clearly supporting a naturally antiviral role for RNA silencing in shrimp. Further research is needed in this area of field to discover the exact mechanism of controlling viral infection and transmission (Labreuche and Warr, 2013). Labreuche and Warr (2013) in their review paper mentioned about the role of miRNA in protecting the shrimp from viral infection and also about dsRNA induced antiviral immunity in the shrimps. It is reported that though the role miRNA in the antiviral defence network has gradually been proposed by some workers dwin *M. japonicus* after examining the expression levels of 35 identified miRNA upon WSSV infection, thirteen were reported to be upregulated and

15 down regulated after viral challenge, leading the authors to propose their putative implications in the shrimp immune response (Ruan *et al.*, 2011). Several reports are available in other organisms indicating similar modifications in the host miRNA expression profile following viral infection (Tian *et al.*, 2012; Li *et al.*, 2011; Du *et al.*, 2011).

Labreuche and Warr (2013) also mentioned in their paper about dsRNA induced immunity mechanisms in penaeid shrimps. It is reported that dsRNA sensing gets initiated in host immune system via comlex intracellular signalling pathways which results in the induction of inflammatory responses mediated by various transcription factors (Saito and Gale 2007; Loo and Gale, 2011; O'Neill LA and Bowie, 2010). In the shrimp's immune system such transcription factors earlier it was assumed that they were absent because they lack the respective receptors. But while using dsRNA-mediated gene silencing to find out the role of specific host genes in shrimp immune response, Robalino *et al.*, (2004) investigated that the administration of dsRNA in *L. vannamei* prevented the animal from viral infection. Later several workers have confirmed these results while working with several invertebrates (Hirono *et al.*, 2011; Robalino *et al.*, 2007; Shekhar and LU, 2009). In shrimps *P. japonicus* and *P. monodon* similar type of observations have been made (He *et al.*, Kongton K, 2011) which support the concept of the existence of innate immune system in the shrimps in general. To further understand the details of the innate responses to dsRNA in penaeid shrimps, several approaches have been made and among them major focus was the identification of the receptors involved in dsRNA recognition (Labreuche and Warr, 2013).

Argonaute is a family of protein that play central roles in small RNA-mediated gene silencing pathways. Leebonoi *et al.*, (2015) worked on a novel gonad specific argonaute protein which serves as a defence against transposons in the shrimp *P. monodon*. They have reported that this protein predominantly expressed in shrimp gonad and further demonstrated its role in protecting the shrimp genome against an invasion of transposons. This confirmation was due to no significant change in the expression of argonaute protein in response to either dsRNA or yellow head virus injection.

Injection of shrimp with non-specific double-stranded dsRNA of diverse length, is known to induce non-specific immunity and protect against lethal disease, although the mechanisms are unclear. To understand the effects of non-specific dsRNA in shrimp, Maralit *et al.*, (2015) carried out studies on kuruma shrimp *M. japonicus* with a dsRNA and a small interfering RNA (siRNA) that is not specific to any gene in the shrimp genome and then examined gene expression at 24 and 48 h with a microarray. The microarray results showed that many genes were up-regulated and some were down-regulated by dsRNA. In addition, injection also increased survival of the shrimp following WSSV challenge. They found that the up-regulated genes included genes for eight immune-related proteins: c-type lectin, hemocyte homeostasis-associated protein, viral responsive protein, fibrinogen-related protein, argonaute, Dicer, and heat shock protein. These results show that injection of shrimp with non-specific dsRNA hinders viral accumulation and prevents significant mortalities.

The work on protection of shrimp *L. vannamei* against the WSSV using recombinant a highly expressed protein called fortilin has been carried out by Sinthujaroen *et al.,* (2015). This fortilin protein was derived from the shrimp *P. monodon* infected with WSSV. It is mentioned that when *L. vannamei* were fed with fortilin in their diet there was reduction in the levels of WSSV infection. In shrimps, fortilin has been identified in the cDNA of various species (Wang *et al.,* 2009). Kulkarni *et al.,* (2014) worked on the evaluation of immune and apoptosis related gene using an RNAi approach in vaccinated shrimp *P. monodon* during oral WSSV infection. In their study they used RNA interference to elucidate the connection between two endogenous genes inhibitor of apoptosis and selected immune apoptosis related genes in orally vaccinated shrimp after WSSV infection. It is reported that the results provided certain important clues on the relationship with selected immune/apoptosis genes in orally vaccinated *P. monodon* during WSSV infection.

The shrimps are deficient in an adaptive immune system, so mostly they rely on innate immunity to defend themselves from invading microorganisms (Beutler *et al.,* 2004). As the first step of the innate immune response, pathogen recognition is mediated by a set of pattern recognition receptors, which can recognize highly conserved molecules on the cell surface of microorganisms while they are absent on the host cells, called as pathogen-associated molecular patterns (Janeway and Medzhitov, 2002) or microbe-associated molecular patterns (Huang *et al.,* 2013). Binding of these receptors to specific pathogen associated molecules activates specific signalling pathway that leads to rapid stimulation of humoral and cellular immune responses (Christophides *et al.,* 2004).

Recent advances in researches on shrimp immune pathway involved in WSSV genes regulation have been reviewed by Chen *et al.,* (2014). In their paper it is mentioned that so far researchers have paid more attention to the shrimp immune system, and mechanism of environmental stress. Among shrimp immune systems, humoral immunity is the most well studied and lot of literature is currently available on the subject. Studies so far made demonstrate that living organisms have developed strategies to cope with many kinds of environmental changes. However, when confront acute changes, they suffer stresses and this is true with shrimp also. The main damage of stresses is protein denaturation. In other words, these proteins lose their native, functional configuration. Therefore, stress responses may play a more direct role in the immune function of haemocytes besides insuring survival. Currently, the main pathways of the shrimp immune system identified are, toll like receptors (TLRs) pathway, immune deficiency (IMD) pathway, Janus kinase (JAK) - signal transducer and activator of transcription (STAT) pathway, and RNAi pathway. Meanwhile, shrimp unfolded protein response (UPR) is considered to be the core of its environmental stress resistance system. Besides these pathways, number of other pathways have also been discussed in detail and reviewed by Chen *et al.*, (2014). Their studies revealed a secret of WSSV: it activated and used a number of immune pathways as well as the UPR of shrimp, to increase its genes replication. Further they mentioned that these studies will also help us to understand the relationship between shrimp humoral immune response/UPR and WSSV infection. Deepika *et al.*, (2014) while working on toll pathways in the shrimp *P. monodon* investigated that myeloid differentiation factor 88 and tumour necrosis factor receptors are the key molecules in signalling pathways in the innate immune system of this shrimp in relation to WSSV infection. The responses of expression profiles of these key genes in normal and infected shrimps observed in their studies suggested that toll pathways as a whole play a crucial role in the immune response against the viruses in shrimp.

10.7 FUTURE RESEARCH AND PERSPECTIVES

Though considerable work has been carried out on shrimp immunity and defence mechanisms lot work is needed to understand thoroughly the role of immune system against pathogen. Knowledge regarding the molecular and cellular mechanisms of the immune system associated recognition and elimination of pathogens particularly in the shrimps is scarce. Innovative research in the new areas like genomics and their products, proteomics and its role in immune system and activation of genetic mechanisms associated with shrimp immune system is very much needed to evolve viable and powerful technologies to protect the animals from various harmful pathogens. Further studies are also required on proPO system, cell adhesion proteins, peroxinectins, glucan binding proteins and haemocyte formation/activation and their protein synthesis.

Earlier workers have done some work on shrimp immunology particularly on proteins and other immune products, their isolation and purification etc. however, it was not very clear how these products interact with immune system in the shrimps. Therefore, in future we need to focus our studies on isolation and purification of immune products, ability of quantitative detection of individual mRNAs or proteins during immunostimulation or infection of disease, the characterization of microbial or viral structures and their products associated with infectious diseases, and target organs during course of infection. Research is also required on the mechanism of the formation of haemocytes and activation process in to pathogens. To prevent the diseases and management of disease control in the shrimp Aquaculture are extremely difficult and to mitigate this kind of situation there is need to develop viable and powerful techniques using the modern scientific knowledge based on molecular biology, genomics and proteomics. There is necessity to make an integrated approach in the research to evolve new immune system models in the shrimps and other crustaceans which will possibly have a significant impact on the understanding on the immune system in general and shrimp in particular.

References

Adachi, K., Endo, H., Watanabe, T., Nishioka, T. and Hirata, T. 2005. Hemocyanin in the exoskeleton of crustaceans: enzymatic properties and immunolocalization. *Pigment Cell Res.* 18: 136-143.

Anderson D P,1997. Adjuvants and immunostimulants for enhancing vaccine potency in fish. *Dev Biol Stand*; 90:257-65.

Ashida M, Yamazaki H, 1990. Biochemistry of the phenoloxidase system in insects: with special reference to its activation. In Ohnishi E, Ishizaki H (eds) Molting and metamorphosis., Springer-Verlag: Berlin; 239-265.

Baba M, R Snoeck, R Pauwels, and E de Clercq, 1998. Sulfated polysaccharides are potent and selective inhibitors of various enveloped viruses, including herpes simplex virus, cytomegalovirus, vesicular stomatitis virus, and human immunodeficiency virus. Antimicrob Agents Chemother. Nov; 32,11)\: 1742–1745.

Bachère E, E. Mialhe, J. Rodriguez, 1995. Identification of defence effectors in the haemolymph of crustaceans with particular reference to the shrimp *Penaeus japonicus* (Bate): prospects and applicationsn. *Fish & Shellfish Immunol.*, 5. 597–612.

Bachère E, 2000. Introduction: Shrimp immunity and disease control. *Aquaculture* 191: 1–3, pg no 3–11.

Bachere, E., Destoumieux, D. and Bulet, P. 2000. Penaeidins, antimicrobial peptides of shrimp, a comparison with other effectors of innate immune. *Aquaculture*. 191: 71-88.

Bachere E, 2003. Anti-infectious immune effectors in marine invertebrates: potential tools for disease control in larviculture. *Aquaculture*. 227: 427-38.

Bagni M, Archetti L, Amadori M, Marino G, 2000. Effect of long-term oral administration of an immunostimulant diet on innate immunity in sea bass (Dicentrarchus labrax). *J Vet Med Ser*; 47: 745-5.

Beutler, B., 2004. Innate immunity: an overview. *Mol. Immunol.* 40, 845–859.

Biswasa G, 1, H. Korenagaa, 1, R. Nagamineb, T. Konob, H. Shimokawac, T. Itamia, M. Sakaia, 2012. Immune stimulant effects of a nucleotide-rich baker's yeast extract in the kuruma shrimp, *Marsupenaeus japonicus. Aquaculture.*366–367, 40–45.

Buggé D.M., Hégaret, H., Wikfors, G.H. and Allam, B. 2007. Oxidative burst in hard clam *(Mercenaria mercenaria)* haemocytes. *Fish & Shellfish Immunol.* 23:188-196.

Bullock G, Blazer V, Tsukuda S, Summerfelt S, 2000. Toxicity of acidified chitosan for cultured rainbow trout (Oncorhynchus mykiss). *Aquaculture*; 185; 273-80.

Cannon, J. P., Haire, R. N., Rast, J. P. and Litman, G. W, 2004. The phylogenetic origins of the antigen-binding receptors and somatic diversification mechanisms. *Immunol. Rev.*; 200:12-22.

Castex M., Lemaire, P., Wabete, N., Chim, L, 2009. Effect of dietary probiotic *Pediococcus acidilactici* on antioxidant defences and oxidative stress status of shrimp *Litopenaeus stylirostris. Aquaculture* 294, 306–313.

Castex M., Lemaire, P., Wabete, N., Chim, L, 2010. Effect of probiotic *Pediococcus acidilactici* on antioxidant defences and oxidant stress of *Litopenaeus stylirostris* under Vibrio nigripulchritudo challenge. *Fish & Shell fish Immunology*. 28, 622–631.

Cerenius L, K. Söderhäll, 2011. Coagulation in invertebrates. *J. Innate Immun.*, 3, pp. 3–8

Chen Y-H, Jia X-T, Zhao L, Li C-Z, Zhang S, Chen Y-G, Weng SP, He JG, 2011. Identification and functional characterization of Dicer2 and five single VWC domain proteins of *Litopenaeus vannamei. Dev Comp Immunol* ; 35:661-71

Chen Yu-Yuan, Su Sing Sim, Siau Li Chiew, Su-Tuen Yeh, Chyng- Hwa Liou, Jiann-Chu Chen, 2012. Dietary administration of a *Gracilaria tenuistipitata* extract produces protective immunity of white shrimp *Litopenaeus vannamei* in response to ammonia stress. *Aquaculture* 370–371, 26–31.

Chen Y, Li X, He J, 2014. Recent Advances in Researches on Shrimp Immune Pathway Involved in White Spot Syndrome Virus Genes Regulation. *J Aquac Res Development* 5: 228.

Cheng W, Chen JC, 2000. Effects of pH, temperature and salinity on immune parameters of the freshwater prawn *Macrobrachium rosenbergii*. *Fish & Shellfish Immunol*;10:387e91.

Cheng W, Chieu HT, Tsai CH, Chen JC, 2005. Effects of dopamine on the immunity of white shrimp *Litopenaeus vannamei*. *Fish & Shellfish Immunol.* 19,4: 375-85.

Cheng W, Chieu HT, Hob MC, Chen JC, 2006. Noradrenaline modulates the immunity of white shrimp Litopenaeus vannamei. *Fish & Shellfish Immunol.* 21,1: 11.

Chiu C.H., Guu, Y.K., Liu, C.H., Pan, T.M., Cheng, W, 2007. Immune responses and gene expression in white shrimp, *Litopenaeus vannamei*, induced by *Lactobacillus plantarum*. *Fish & Shellfish Immunology* 23, 364–377.

Chiou,S. T., Chen, Y. W., Chen, S. C., Chao, C. F. and Liu, T. Y. Isolation and characterization of proteins that bind to galactose, lipopolysaccharide of *Escherichia coli*, and protein A of *Staphylococcus aureus* from the hemolymph of Tachypleus tridentatus. *J. Biol. Chem.*2000; 275: 1630-1634.

Chotigeat Wilaiwan, Suprapa Tongsupa, Kidchakan Supamataya, Amornrat Phongdara, 2004. Effect of Fucoidan on Disease Resistance of Black Tiger Shrimp. *Aquaculture*. 233, 1–4, 23–30.

Christophides, G.K., Vlachou, D., Kafatos, F.C., 2004. Comparative and functional genomics of the innate immune system in the malaria vector Anopheles gambiae. *Immunol. Rev.* 198, 127–148.

Conceicao LEC, Skjermo J, Skjak-Bræk G, Verreth JAJ. Effect of an immunostimulating alginate on protein turnover of turbot (*Scophthalmus maximus L.*). *Fish Physiol Biochem* 2004; 24: 207-12.

Cordova Campa, A.I., Hernandez-Saavedra, N.Y., Philippis, R. De and Ascencio, F. 2002. Generation of superoxide anion and SOD activity in haemocytes and muscle of American white shrimp (*Litopenaeus vannamei*) as a response to β-glucan and sulphated polysaccharide. *Fish & Shellfish Immunol.* 12: 353-366.

Crouch, E., Hatshorn, K. and Ofek, I, 2000. Collectons and pulmonary innate immunity. *Imm. Rev.*; 173: 52-56.

De la Vega E, Degnan BM, Hall MR, Wilson KJ, 2007. Differential expression of immune-related genes and transposable elements in black tiger shrimp (*Penaeus monodon*) exposed to a range of environmental stressors. *Fish & Shellfish Immunol.* 23,5:1072-88.

Deepika A, K. Sreedharan, Anutosh Paria, M. Makesh, K.V. Rajendran, 2014. Toll-pathway in tiger shrimp (*Penaeus monodon*) responds to white spot syndrome virus infection: Evidence through molecular characterisation and expression profiles of MyD88, TRAF6 and TLR genes. *Fish & Shellfish Immunology* 41, 441e454.

Destoumieux D, P. Bulet, D. Loew, A. Van Dorsselaer, J. Rodriguez, E. Bachère Penaeidins: 1997. A new family of antimicrobial peptides in the shrimp *Penaeus vannamei* (Decapoda) J. Biol. Chem., 272, pp. 28398–28406.

Destoumieux D, M. Munoz, P. Bulet, E. Bachère, 2000. Penaeidins, a family of antimicrobial peptides from penaeid shrimp (Crustacea, Decapoda) *Cellular and Molecular Life Sciences*, 57, pp. 1260–1271.

Devaraja, TN, Otta SK, Shubha G, Karunasagar I, Tauro P, Karunasagar I, 1998. Immunostimulation of shrimp through oral administration of *Vibrio* bacterin and yeast glucan. *In* Flegel TW (ed) Advances in shrimp biotechnology. National Center for Genetic Engineering and Biotechnology, Bangkok.

Dodd, R.B., Drickamer, K., 2001. Lectin-like proteins in model organisms: implications for evolution of carbohydrate binding activity. *Glycobiology* 11, 71–79.

Du, P., Wu, J., Zhang, J., Zhao, S., Zheng, H., Gao, G., Wei, L. and Li, Y., 2011. Viral infection induces expression of novel phased microRNAs from conserved cellular microRNA precursors. *PLoS pathog*, 7,8: 1002176.

Fagutao F. F, M.B.B. Maningas, H. Kondo, T. Aoki, I. Hirono, 2012. Transglutaminase regulates immune-related genes in shrimp *Fish & Shellfish Immunol.*, 32. 711–715.

Figueras A, Santarem MM, Novoa B, 1998. Influence of the sequence of administration of beta-glucans and a Vibrio damsela vaccine on the immune response of turbot (Scophthalmus maximus L.). *Vet Immunol Immunopathol*; 64:59-68.

Gargioni R, M.A. Barracco, 1998. Haemocytes of the palaemonids *Macrobrachium rosenbergii* and *M. acanthurus*, and of the penaeid *Penaeus paulensis J. Morphol*, 236. 209–221.

Guzman Gabriel Aguirre, Jesus Genaro Sanchez-Martinez, Angel Isidro Campa-Cordova, Antonio Luna-Gonzalez, Felipe Ascencio, 2009. Penaeid Shrimp Immune System. *Thai J. Vet. Med*. 39,3: 205-215.

Harikrishnana Ramasamy, Chellam Balasundaramb, Sundaram Jawaharc, Moon-Soo Heoc,, 2011. *Solanum nigrum* enhancement of the immune response and disease resistance of tiger shrimp, *Penaeus monodon* against *Vibrio harveyi*. *Aquaculture*. 318, 1–2, 67–73.

He N, Qin Q, Xu X, 2005. Differential profile of genes expressed in hemocytes of white spot syndrome virus-resistant shrimp (*Penaeus japonicus*) by combining suppression subtractive hybridization and differential hybridization. *Antivir Res*; 66:39-45.

Hellio, C., Bado-Nilles, A., Gagnaire, B., Renault, T. and Thomas-Guyon, H. 2007. Demonstration of a true phenoloxidase activity and activation of a ProPO cascade in Pacific oyster, Crassostrea gigas (Thunberg) in vitro. *Fish & Shellfish Immunol.* 22:433-440.

Hirono, I., Fagutao, F.F., Kondo, H. and Aoki, T., 2011. Uncovering the mechanisms of shrimp innate immune response by RNA interference. *Marine biotechnology*, 13,4.622-628.

Holmblad, T. and Soderhäll, K. 1999. Cell adhesion molecules and antioxidative enzyme in a crustacean, possible role in immunology. *Aquaculture*. 172: 111-123.

Hu Bing, Le Hong Phuoc, Patrick Sorgeloos, Peter Bossier, 2014. Bacterial HSP70 (DnaK) is an efficient immune stimulator in *Litopenaeus vannamei*. *Aquaculture*. 418–419, 87–93.

Huang Hai-Hong, Xiao-Lin Liu, Jian-Hai Xiang, Ping Wang, 2013. Immune response of *Litopenaeus vannamei* after infection with *Vibrio harveyi*. *Aquaculture*. 406–407, 115–120.

Itami T, M. Asano, K. Tokushige, K. Kubono, A. Nakagawa, N. Takeno, H. Nishimura, M. Maeda, M. Kondo, Y. Takahashi, 1998. Enhancement of disease resistance of kuruma shrimp, *Penaeus japonicus*, after oral administration of peptidoglycan derived from *Bifidobacterium thermophilum*. *Aquaculture*, 164:277–288.

Iwaki, D., Osaki, T., Mizunoe, Y., Wai, S. N., Iwanaga, S. and Kawabata, S, 1999. Functional and structural diversities of Creactive proteins present in horseshoe crab hemolymph plasma. *Eur. J. Biochem*. 264: 314-326.

Iwanaga S, S.-I. Kawabata, T. Muta, 1998. New types of clotting factors and defence molecules found in horseshoe crabhaemolymph: their structures and functions *J. Biochem*, 123, 1–15.

Janeway, C.A. and Medzhitov, R. 2000. Viral interference with IL-1 and Toll signaling. *Proc. Natl. Acad. Sci. U.S.A.* 97: 10682-10683.

Jiravanichhpaisal P, B.L. Lee, K. Söderhäll, 2006. Cell-mediated immunity in arthropods: Hematopoiesis, coagulation, melanization and opsonization. *Immunobiology*, 211,4.213–236.

Johansson M.W, K. Söderhäll, 1985. Exocytosis of the prophenoloxidase activating system from crayfish haemocytes *J. Comp. Physiol.*, 156, pp. 175–181.

Johansson M.W, K. Söderhäll, 1989. Cellular immunity in crustaceans and the proPO system Parasitol. Today, 5, pp. 171–176.

Johansson, M., Keyser, P., Sritunyalucksana, K. and Soderhäll, K. 2000. Crustacean haemocytes and haematopoiesis. *Aquaculture*. 191: 45-52.

Ju, Z.Y., Forster, I., Conquest, L., Dominy, W, 2008. Enhanced growth effects on shrimp (*Litopenaeus vannamei*) from inclusion of whole shrimp floc or floc fractions to a formulated diet. *Aquaculture* Nutrition 14, 533–543.

Kawabata, S., Tokunaga, F., Kugi, Y., Motoyama, S., Miura, Y., Hirata, M. and Iwanaga, S. 1996. Limulus factor D, a 43-kDa protein isolated from horseshoe crab hemocytes, is a serine protease homologue with antimicrobial activity. *FEBS Lett.*; 398: 146-150.

Kawabata, S. and Iwanaga, S,1999. Role of lectins in the innate immunity of horseshoe crab. *Dev. Comp. Immunol.* 23: 391- 400.

Kawabata, S., Beisel, H. G., Huber, R., Bode, W., Gokudan, S., Muta, T., Tsuda, R., Koori, K., Kawahara, T., Seki, N., Mizunoe, Y., Wai, S. N. and Iwanaga, S. 2001. Role of tachylectins in host defence of the Japanese horseshoe crab *Tachypleus tridentatus*. *Adv. Exp. Med. Biol.*; 484: 195-202.

Kawabata, S. and Tsuda, R, 2002. Molecular basis of non-self recognition by the horseshoe crab tachylectins. *Biochem. Biophys. Acta,*; 1572: 414-421.

Kawakami, H., Shinohara, N., Sakai, M, 1998. The non-specific immunostimulation and adjuvant effects of *Vibrio anguillarum* bacterin, M-glucan, chitin or Freund's complete adjuvant in yellow tail *Seriola quinqueradiata* to *Pasteurella piscicida* infection. *Fish Pathol.*; 33: 287–292.

Kondo, M., Itami, T., Takahashi, Y., Fujii, R. and Tomonaga, S. 1998. Ultrastructural and cytochemical characteristics of phagocytes in kuruma prawn. *Fish Pathol.* 33: 421-427.

Kongton K, Phongdara A, Tonganunt-Srithaworn M, Wanna W. Molecular cloning and expression analysis of the interferon-g-inducible lysosomal thiol reductase gene from the shrimp *Penaeus monodon. Mol Biol Rep* 2011; 38: 3463-70.

Kopacek P, L. Grubhoffer, K. Söderhäll, 1993. Isolation and characterization of a hemagglutinin with affinity for lipopolysaccharides from plasma of the crayfish *Pacifastacus leniusculus. Dev. Comp. Immunol.*, 17, 407–418.

Kulkarni Amod D, Christopher M.A. Caipang, Viswanath Kiron, Jan H.W.M. Rombout, Jorge M.O. Fernandes, Monica F. Brinchmann, 2014. Evaluation of immune and apoptosis related gene responses using an RNAi approach in vaccinated *Penaeus monodon* during oral WSSV infection. *Marine Genomics* 18, 55–65.

Kumlu M., O.T. Eroldogan, M. Aktas, 2000. Effects of temperature and salinity on larval growth, survival and development of *Penaeus semisulcatus. Aquaculture*, 188, pp. 167–173.

Kurtz J, Franz K. Evidence for memory in invertebrate immunity. *Nature* 2003;425: 37-8.

Labreuche Yannick, Gregory W. Warr, 2013. Insights into the antiviral functions of the RNAi machinery in penaeid shrimp. *Fish & Shellfish Immunology*, 34 :1002-1010.

Le Moullac G, Haffner P, 2000. Environmental factors affecting immune responses in Crustacea. *Aquaculture*. 191:121-31.

Lee, M.H., Osaki, T., Lee, J.Y., Baek, M.J., Zhang, R., Park, J.W., Kawabata, S., Soderhäll, K. and Lee, B.L. 2004. Peptidoglycan recognition proteins involved in 1,3- -D-glucan-dependent prophenoloxidase activation system of insect. *J. Biol. Chem.* 279: 3218-3227.

Leebonoi W, Suchitraporn S, Sakol P, Apinunt U, 2015. A novel gonad-specific Argonaute 4 serves as a defence against transposons in the black tiger shrimp *Penaeus monodon. Fish & Shellfish Immunology* 42, 280-288.

Li E, Chen L, Zeng C, Yu N, Xiong Z, Chen X, Jian G. Qin, 2008. Comparison of digestive and antioxidant enzymes activities, haemolymph oxyhemocyanin contents and hepatopancreas histology of white shrimp, *Litopenaeus vannamei*, at various salinities. *Aquaculture*. 274 (1):80-86.

Li J, Beiping Tana, Kangsen Maia, 2009. Dietary probiotic *Bacillus* OJ and isomaltooligosaccharides influence the intestine microbial populations, immune responses and resistance to white spot syndrome virus in shrimp (*Litopenaeus vannamei*). *Aquaculture* 291: 1–2, 35–40.

Li Y, Li J, Belisle S, Baskin CR, Tumpey TM, Katze MG, 2011. Differential microRNA expression and virulence of avian, 1918 reassortant, and reconstructed 1918 influenza A viruses. Virology; 421:105-113

Li E, Wang S, Li C, Wang X, Chen K, Chen L, 2014. Transcriptome sequencing revealed the genes and pathways involved in salinity stress of Chinese mitten crab, *Eriocheir sinensis. Physiol Genomics.* 46:177-190.

Lio-Po Gilda D, Eduardo M. Leaño, Ma. Michelle D. Peñaranda, Annie U. Villa-Franco, ChristopherD. Sombito, Nicholas G. Guanzon Jr, 2005. Anti-luminous *Vibrio* factors associated with the 'green water'grow-out culture of the tiger shrimp *Penaeus monodon. Aquaculture.* 250: 1–2, 1–7.

Ling E and Yu XQ, 2006b. Hemocytes from the tobacco hornworm *Manduca sexta* have distinct functions in phagocytosis of foreign particles and self dead cells. *Dev Comp Immunol.*;30:301–309.

Lis, H., Sharon, N., 1998. Lectins: carbohydrate-specific proteins that mediate cellular recognition. *Chem. Rev.* 98, 637–674.

Liu HJ, Pan LQ, Hu FW, 2008. The appraisement of immune ability of *Litopenaeus vannamei* under the change of salinity. *Trans Oceanol Limnol;* 159,2: 159-166.

Loo YM, Gale MJ, 2011. Immune signaling by RIG-I-like receptors. *Immunity*; 34:680-192.

Loof. T G, O. Schmidt, H. Herwald, U. Theopold, 2011. Coagulation systems of invertebrates and vertebrates and their roles in innate immunity: the same side of two coins *J. Innate Immun.*, 3, 34–40.

Maggioni Daniela S, Edemar R. Andreatta, Elizabeth M. Hermes, Margherita A. Barracco, 2004. Evaluation of some hemato-immunological parameters in female shrimp *Litopenaeus vannamei* submitted to unilateral eyestalk ablation in association with a diet supplemented with superdoses of ascorbic acid as a form of immunostimulation. *Aquaculture*, 241:1–4, 501–51.

Maningas M.B.B, H. Kondo, I. Hirono, T. Saito-Taki, T. Aoki, 2008. Essential function of transglutaminase and clotting protein in shrimp immunity. *Mol. Immunol.*, 45,1269–1275.

Maheswari R, P. Mullainadham, M. Arumugan, 1997. Characterisation of a natural haemagglutinin with affinity for acetylated aminosugars in the serum of the marine prawn, *Penaeus indicus Fish & Shellfish Immunol*, 7,17–28.

Martin, G.G. and Graves, B. 2005. Fine structure and classification of shrimp haemocytes. *J. Morphol.* 185: 339-348.

Maralit BA[1], Komatsu M, Hipolito SG, Hirono I, Kondo H , 2015. Microarray Analysis of Immunity Against WSSV in Response to Injection of Non-specific Long dsRNA in Kuruma Shrimp, *Marsupenaeus japonicus. Marine Biotechnology*. Volume 17, 4: 493-501.

Monostori, P, Wittmann, G., Karg, E., Túri, S., 2009. Determination of glutathione and glutathione disulfide in biological samples: an in-depth review. *Journal of Chromatography. B, Analytical Technologies in the Biomedical and Life Sciences* 877, 3331–3346.

Miller-Keane Encyclopedia and Dictionary of Medicine, Nursing, and Allied Health, Seventh Edition. © 2003 by Saunders, an imprint of Elsevier.

Morvan, A., Iwanaga, S., Comps, M. and Bachere, E, 1997. *In vitro* activity of the Limulus antimicrobial peptide tachyplesin I on marine bivalve pathogens. *J. Invertebr. Pathol.*; 69: 177-182.

Mury Barillas- C. 2007. CLIP proteases and *Plasmodium* melanization in *Anopheles gambiae. Trends Parasitol.* 23: 297-299.

Mylonakis, E. and Aballay, A. 2005. Worms and flies as genetically tractable animal models to study host-pathogen interactions. *Infect. Immun.* 73: 3833-3841.

Nappi A J, Vass E, Frey F, Carton Y, 1995. Superoxide anion generation in *Drosophila* during melanotic encapsulation of parasites. *Eur J Cell Biol*; 68:450-456.

Nappi, A.J. and Ottaviani, E. 2000. Cytotoxicity and cytotoxic molecules in invertebrates. *BioEssays.* 22: 469-480.

Nathan, C. and Shiloh, M.U. 2000. Reactive oxygen and nitrogen intermediates in the relationship between mammalian hosts and microbial pathogens. *Proc. Natl. Acad. Sci.* U.S.A. 97: 8841-8848.

Omori S.A, G.G. Martin, J.E. Hose, 1989. Morphology of haemocyte lysis and clotting in the ridgeback prawn *Sicyonia ingentis. Cell Tissue Res.*, 255, 117–123.

O'Neill LA, Bowie AG. Sensing and signaling in antiviral innate immunity. *Curr Biol* 2010; 20:R328-333

Pan LQ, Zhang LJ, Liu HY, 2007. Effects of salinity and pH on ion-transport enzyme activities, survival and growth of *Litopenaeus vannamei* postlarvae. *Aquaculture*. 273:711-720.

Perazzolo LM, Gargioni R, Ogliari P, Barracco MAA, 2002. Evaluation of some hemato-immunological parameters in the shrimp *Farfantepenaeus paulensis* submitted to environmental and physiological stress. *Aquaculture*. 214:19-33.

Ponprateep S, Kunlaya Somboonwiwat, Anchalee Tassanakajon, 2009. Recombinant anti-lipopolysaccharide factor isoform 3 and the prevention of vibriosis in the black tiger shrimp, *Penaeus monodon. Aquaculture*. 289, 3–4: 219–224.

Raa J, 2000. The use of immune-stimulants in fish and shellfish feeds. In: Cruz-Sua´rez L, *et al.*, editors. Avances en Nutricion Acuýcola V. Memorias del V Simposium Internacional de Nutricio´n Acuý´cola November 19-22. Me´rida, Yucata´n, Mexico.

Ratanapo S, Chulavatnatol M Monodin, 1990. A new sialic acidspecific lectin from black tiger prawn (*Penaeus monodon*). *Comp Biochem Physiol*; 97 B: 515-520.

Ratanapo S, 1992. Chulavatnatol M. Monodin-induced agglutination of *Vibrio vulnificus*, a major infective bacterium in black tiger prawn (*Penaeus monodon*). *Comp Biochem Physiol*; 102B:855-859.

Rendon, L. and Balcazar, J.L. 2003. Inmunologia de camarones: Conceptos basicos y recientes avances. *Revista AquaTIC*. 19: 27-33.

Rengpipat Sirirat, Sombat Rukpratanporn, Somkiat Piyatiratitivorakul, Piamsak Menasaveta, 2000. Immunity enhancement in black tiger shrimp (*Penaeus monodon*) by a probiont bacterium (*Bacillus* S11) *Aquaculture*. 191, 4, 271–288.

Robalino, J., Browdy, C.L., Prior, S., Metz, A., Parnell, P., Gross, P. and Warr, G. 2004. Induction of Antiviral Immunity by Double-Stranded RNA in a Marine Invertebrate. *J. Virol.* 78: 10442-10448.

Robalino J, Bartlett T, Shepard E, Prior S, Jaramillo G, Scura E, Chapman RW, Gross PS, Browdy CL, Warr GW, 2005. Doublestranded RNA induces sequence-specific antiviral silencing in addition to nonspecific immunity in a marine shrimp: convergence of RNA interference and innate immunity in the invertebrate antiviral response? *J Virol.*79:13561-71.

Robalino J, Bartlett TC, Chapman RW, Gross PS, Browdy CL, Warr GW, 2007. Double-stranded RNA and antiviral immunity in marine shrimp: inducible host mechanisms and evidence for the evolution of viral counter-responses. *Dev Comp Immunol* ;31:539-47.

Rodriguez J, V. Boulo, E. Mialhe, E. Bachère, 1995. Characterisation of shrimp haemocytes and plasma components by monoclonal antibodies. *J. Cell Sci.*, 108: 1043–1050.

Rodríguez-Ramos Tania, Georgina Espinosa, Jorge Hernández-López, Teresa Gollas-Galván, Jeannette Marrero, Yaisel Borrell, Maria E. Alonso, Ubaldo Bécquer, Maray Alonso, 2008. Effects of *Echerichia coli* lipopolysaccharides and dissolved ammonia on immune response in southern white shrimp. *Aquaculture.* 274, 1, 118–125.

Romalde JL, Magarinos B, Toranzo AE, 1999. Prevention of streptococcosis in turbot by intraperitoneal vaccination: a review. *J Appl Ichthyol*; 15:153-8.

Ruan L, Bian X, Ji Y, Li M, Li F, Yan X. Isolation and identification of novel microRNAs from Marsupenaeus japonicus. *Fish & Shellfish Immunol* 2011; 31: 334e40.

Saito T, Gale Jr M, 2007. Principles of intracellular viral recognition. *Curr Opin Immunol*; 19:17-23

Shanthi Sathappan , Baskaralingam Vaseeharan, 2014. Alpha 2 macroglobulin gene and their expression in response to GFP tagged *Vibrio parahaemolyticus* and WSSV pathogens in Indian white shrimp *Fenneropenaeus indicus. Aquaculture.* 418–419,48–54.

Sharon, N., Lis, H., 2004. History of lectins: from hemagglutinins to biological recognition molecules. *Glycobiology* 14, 53–62.

Shekhar M, Lu Y, 2009. Application of nucleic-acid-based therapeutics for viral infections in shrimp *Aquaculture. Mar Biotechnol* (NY);11:1-9.

Sinthujaroen Patuma, Moltira Tonganunt-Srithaworn, Lily Eurwilaichitr, Amornrat Phongdara, 2015. Protection of Litopenaeus vannamei against the white spot syndrome virus using recombinant Pm-fortilin expressed in *Pichia pastoris. Aquaculture.* 435, 450–457.

Smith V, Brown J, Hauton C, 2003. Immunostimulation in crustaceans: does it really protect against infection? *Fish & Shellfish Immunol*; 15: 71-90.

Smith V.J, Chisholm J.R.S, 1992. Non-cellular immunity in crustaceans *Fish & Shellfish Immunol.*, 2, pp. 1–31.

Smith VJ, Chisholm JRS, 2001. Antimicrobial proteins in crustaceans. *Adv Exp Med Biol*; 484:95–112.

Soderhall, K and Smith, V.J., 1986. The prophenoloxidase activating cascade as a recognition and defence system in arthropods. In: Gupta, A.P. _Ed.., Humoral and Cellular Immunity in Arthropods. Wiley, New York, 251–285.

Soderhall K. Biochemical and molecular aspects of cellular communication in arthropods. *Boll Zool.* 1992; 59:141-151.

Soderhall K, Cerenius L, Johansson M W, 1994. The prophenoloxidase activating system and its role in invertebrate defence. Ann NY *Acad Sci.*; 712:155-161.

Söderhäll K., L. Cerenius, M.W. Johansson, 1996. The prophenoloxidase system in invertebrates. K. Söderhäll, I. Sadaaki, G. Vasta (Eds.), New Directions in Invertebrate. Immunology, SOS Publications, Fair Haven, 229–253.

Söderhäll K, L. Cerenius, 1998. Role of the prophenoloxidase- activating system in invertebrate immunity. *Curr. Opin. Immunol.*, 10: 23–28.

Soderhäll, I., Bangyeekhun, E., Mayo, S. and Soderhäll, K. 2003. Hemocyte production and maturation in an invertebrate animal; proliferation and gene expression in hematopoietic stem cells of *Pacifastacus*m *leniusculus. Dev. Comp. Immunol.* 27: 661-672.

Song Y.L, Y.T. Hsieh, 1994. Immunostimulation of tiger shrimp (*Penaeus monodon*) haemocytes for generation of microbicidal substances: analysis of reactive oxygen species *Dev. Comp. Immunol.*, 18, pp. 201–209.

Song, Y.L. and Huang, C.C. 2000. Aplications of immunostimulant to prevent shrimp diseases. In: Resent advances in marine biotechnology. 1st ed. M. Fingerman and R. Negabhusanam (eds). Playmouth: Science Publishers Inc.: 173-187.

Sookruksawong S, Sun F, Liu Z, Tassanakajon A, 2013. RNA-Seq analysis reveals genes associated with resistance to Taura syndrome virus (TSV) in the Pacific white shrimp *Litopenaeus vannamei. Dev Comp Immunol.*41:523-33.

Sritunyalucksana K, L. Cerenius, K. Söderhäll, 1999. Molecular cloning and characterization of prophenoloxidase in the black tiger shrimp *Penaeus monodon. Dev. Comp. Immunol.*, 23, pp. 179–186.

Sritunyalucksana, K. and Söderhäll, K, 2000. The proPo and clotting system in crustaceans. *Aquaculture*, 191:53-69.

Sritunyalucksana K., K. Wonhsuebantati, M.W. Johansson, K. Söderhäll, 2001. Peroxinectin, a cell adhesion protein associated with the proPO system from the black tiger shrimp, *Penaeus monodon. Dev. Comp. Immunol.*, 25: 353–363

Srivastava P., 2002. Roles of heat-shock proteins in innate and adaptive immunity. *Nat. Rev. Immunol.*, 2:185–194.

Sun, Y.-D., Fu, L.-D., Jia, Y.-P., Du, X.-J., Wang, Q., Wang, Y.-H., Zhao, X.-F., Yu, X.-Q.,Wang, J.-X., 2008. A hepatopancreas-specific C-type lectin from the Chinese shrimp *Fenneropenaeus chinensis* exhibits antimicrobial activity. *Mol. Immunol.* 45, 348–361.

Su J, Oanh DTH, Lyons RE, Leeton L, van Hulten MCW, Tan S-H, Song L, Rajendran KV, Walker PJ, 2008. A key gene of the RNA interference pathway in the black tiger shrimp, *Penaeus monodon*: identification and functional characterisation of Dicer-1. *Fish & Shellfish Immunol*; 24:223-33.

Sugumaran, M, 2002. Comparative biochemistry of eumelanogenesis and the protective roles of phenoloxidase and melanin in insects. *Pigment Cell Res.* 15: 2-9.

Takahashi, Y., Uehara, K., Watanabe, R., Okumura, T., Yamashita, T., Omura, H., Yomo, T., Kanemitsu, A., Kawano, T., Narasaka, H., Suzuki, N., Itami, T., 1998. Efficacy of oral administration shrimp in Japan. In: Flegel, T.W. (Ed.), Advances in Shrimp Biotechnology. National Center for Genetic Engineering and Bio- technology, Bangkok,171–173

Thepnarong Supattra, Phanthipha Runsaeng, Onnicha Rattanaporn, Prapaporn Utarabhand, 2015. Molecular cloning of a C-type lectin with one carbohydrate recognition domain from *Fenneropenaeus merguiensis* and its expression upon challenging by pathogenic bacterium or virus. *Journal of Invertebrate Pathology* 125 1–8.

Tian F, Luo J, Zhang H, Chang S, Song J, 2012. MiRNA expression signatures induced by Marek's disease virus infection in chickens. *Genomics*; 99:152-59

Van de K Braak, 2002. Haemocytic defence in balck tiger shrimp (*Penaeus monodon*). PhD Degree. Wageningen University, Netherland, 159

Vadstein O, 1997. The use of immunostimulation in marine larviculture: possibilities and challenges. *Aquaculture*; 155:401-17.

Van de Braak C.B.T, M.H.A. Botterblom, N. Taverne, W.B. Van Muiswinkel, J.H.W.M. Rombout, W.P.W. Van der Knaap, 2002. The roles of haemocytes and the lymphoid organ in the clearance of injected *Vibrio* bacteria in *Penaeus monodon* shrimp. *Fish & Shellfish Immunology*, 293–309.

Vargas-Albores F, Guzman-Murillo MA,Ochoa J.L, 1992. Size dependent haemagglutinating activity in the haemolymph from sub-adult blue shrimp (*Penaeus stylirostris* Stimpson). *Comp Biochem Physiol.*; 103A:487-491.

Vargas-Albores F. M.A. Guzman, J.L. Ochoa, 1993. An anticoagulant solution for haemolymph collection and prophenoloxidase studies of penaeid shrimp (*Penaeus californiensis*) *Comp. Biochem. Physiol.*, Part A: *Mol. Integr. Physiol.*, 106:299–303.

Vargas-Albores F, 1995. The defence system of brown shrimp (*Penaeus californiensis*): humoral recognition and cellular responses *J. Mar. Biotechnol.*, 3, :153–156.

Vargas-Albores F, F. Jimenez-Vega, K. Söderhäll, 1996. A plasma protein isolated from brown shrimp (*Penaeus californiensis*) which enhances the activation of prophenoloxidase system by β-1, 3-glucan. *Dev. Comp. Immunol*, 20: 299–306.

Vargas-Albores, F. and Yepiz-Plascencia, G. 1998. Shrimp immunity: A review. Trends *Comp. Biochem. Physiol.* 5: 195-210.

Wang R, S.Y. Lee, 1. Cerenius, K. Söderhäll, 2001. Properties of the prophenoloxidase activating enzyme of the freshwater crayfish *Pacifastacus leniusculus Eur. J. Biochem.*, 268 ,4. 895–902.

Wang S., Zhao, X.-F., Wang, J.-X., 2009. Molecular cloning and characterization of the translationally controlled tumor protein from *Fenneropenaeus chinensis. Mol. Biol. Rep.* 36, 1683–1693.

Wang S, Liu N, Chen A-J, Zhao X-F, Wang J-X, 2009. TRBP homolog interacts with eukaryotic initiation factor 6 (eIF6) in *Fenneropenaeus chinensis. J Immunol*;182:5250-8.

Wang S, Chen AJ, Shi LJ, Zhao XF, Wang JX, 2012. TRBP and eIF6 homologue in *Marsupenaeus japonicus* play crucial roles in antiviral response. *PLoS One*;7:30057.

Xu Q, 2010. Expressed sequence tags from cDNA library prepared from gills of the swimming crab, Portunus trituberculatus. *J Exp Mar Biol Ecol.* 394: 105-15.

Xu, W.J., Pan, L.Q., 2012. Effects of bioflocs on growth performance, digestive enzyme activity and body composition of juvenile *Litopenaeus vannamei* in zero-water exchange tanks manipulating C/N ratio in feed. *Aquaculture* 356–357: 147–152.

Xu Wu-Jie and Pan Lu-Qing, 2013. Enhancement of immune response and antioxidant status of *Litopenaeus vannamei* juvenile in biofloc-based culture tanks manipulating high C/N ratio of feed input. *Aquaculture*. 412–413:117–124.

Yao X, Wang L, Song L, Zhang H, Dong C, Zhang Y, Qiu L, Shi Y, Zhao J, Bi Y, 2010. A Dicer-1 gene from white shrimp *Litopenaeus vannamei*: expression pattern in the processes of immune response and larval development. *Fish & Shellfish Immunol*;29: 565-70.

Yang, S.P., Wu, Z.H., Jian, J.C., Zhang, X.Z., 2010. Effect of marine red yeast *Rhodosporidium paludigenum* on growth and antioxidant competence of *Litopenaeus vannamei*. *Aquaculture* 309, 62–65.

Yeh Su-Tuen, Chen Jiann-Chu , 2008. Immunomodulation by carrageenans in the white shrimp *Litopenaeus vannamei* and its resistance against *Vibrio alginolyticus*. *Aquaculture*. 30, 22–28.

Yu, X.Q., Gan, H., Kanost, M.R., 1999. Immulectin, an inducible C-type lectin from an insect, Manduca sexta, stimulates activation of plasma prophenol oxidase. *Insect Biochem. Mol. Biol.* 29, 585–597.

Yu CJ, Lin YF, Chiang BL and Chow LP, 2003 Proteomics and Immunological Analysis of a Novel Shrimp Allergen, Pen m 2. *J. Immunol.*170445–453.

Yu, X.Q., Tracy, M.E., Ling, E., Scholz, F.R., Trenczek, T., 2005. A novel C-type immulectin-3 from Manduca sexta is translocated from hemolymph into the cytoplasm of hemocytes. *Insect Biochem. Mol. Biol.* 35, 285–295.

Zatta P, 1987. Dopamine, noradrenaline and serotonin during hypoosmotic stress of Carcinus maenas. *Mar Biol* ;96:479e81.

Zhang, Z.F., Shao, M. and Ho Kang, K. 2006. Classification of haematopoietic cells and haemocytes in Chinese prawn *Fenneropenaeus chinensis*. *Fish & Shellfish Immunol.* 21: 159-169.

Zhao, Z.Y., Yin, Z.X., Xu, X.P., Weng, S.P., Rao, X.Y., Dai, Z.X., Luo, Y.W., Yang, G., Li,

Z.S., Guan, H.J., Li, S.D., Chan, S.M., Yu, X.Q., He, J.G., 2009. A novel C-type lectin from the shrimp *Litopenaeus vannamei* possesses anti-white spot syndrome virus activity. *J. Virol.* 83, 347–356.

Zhao Qun, Luqing Pan, Qin Ren, Dongxu Hu, 2015. Digital gene expression analysis in hemocytes of the white shrimp *Litopenaeus vannamei* in response to low salinity stress. *Fish & Shellfish Immunol* 42, 400-407.

CHAPTER 11

DISEASES IN SHRIMP AND CURRENT DIAGNOSTIC METHODS

11.1 INTRODUCTION

Aquaculture is one of the fastest growing food production sectors in the world (Subasinghe *et al.,* 1998). Global Aquaculture production attained another all-time high of 90.4 million tonnes in 2012 (US$144.4 billion), including 66.6 million tonnes of food fish and 23.8 million tonnes of aquatic algae, with estimates for 2013 of 70.5 million and 26.1 million tonnes, respectively (FAO, 2014). According to the *FAO (2015)*, global production of farmed shrimp increased from 3.4 million tonnes in 2013 to 3.6 million tonnes in 2014. Asian producers had the lion share at 3 million tonnes, whereas production in the Americas was estimated at 671 000 tonnes. With production increasingly shifting from black tiger to vannamei shrimp in VietNam, Indonesia, and India, the farming of vannamei shrimp in Asia increased from 2.12 million tonnes in 2013 to 2.37 million tonnes in 2014. Subsequently, black tiger production in the region suffered, decreasing from 744000 tonnes in 2013 to 635000 tonnes in 2014. Considering these factors, the year-on-year rise in Asian farmed shrimp production was 145 000 to 150 000 tonnes in 2014. However, diseases out breaks have caused serious economic losses in several countries. According to Lightner (2009) Aquaculture losses in Asia alone due to shrinp diseases are around $3, 500 milliom. Therefore, health management is of major importance in Aquaculture. Lightner (2011) while discussing the status of shrimp diseases and advances in shrimp health management mentioned that world's shrimp production was directly or indirectly dependant on wild shrimp stocks for the seed stock used to populate its farms. In USA the practice of collecting shrimp seed is from wild resource. While in Asia the shrimp seed is obtained from wild sourced brood stock. Further it is reported that the seed collected from wild source provided the shrimp industry with little protection from significant losses due to infectious diseases.

Shrimp farming has become one of the most important food production industries of the world and is one of the most lucrative and widely traded Aquaculture products, generating huge amount of revenue and foreign exchange, employing millions of people. The explosive growth of shrimp farming has been supported by national government, international agencies and private investors to generate valuable foreign exchange and employment. Shrimp farming is highly beneficial to local communities as well as national economies of developing countries. Though the growth of this industry and its importance in the coastal economy is impressive, the Aquaculture of shrimp has not certainly been without problems. Catastrophic infectious disease hit shrimp farming since mid-1990s causing devastating losses in shrimp production countries. The enormous concentration of animals and their coprophagous behavior imposed by an intensive culture have triggered the development of disease outbreaks, which are often explosive and sometimes leads to the loss of the complete stock.

The development of the commercial culture of shrimps has been accompanied by the occurrence of diseases of infectious and noninfectious etiologies. Many of the important diseases are caused by organisms that are part of the normal micro-flora and fauna. These organisms are opportunistic pathogens that cause disease only under conditions that favor them over the host. Many organisms in this category are ubiquitous, and most have been recognized and/or reported from each of the major penaeid culture areas of the world. Included among this category of pathogens are the filamentous bacteria, the peritrich protozoans, the invasive bacteria and the fungi. Among the most important disease-causing agents are the penaeid viruses. These penaeid viruses may once have been limited in their geographic distribution in wild stocks, but they have become widespread in penaeid culture facilities. With the advent of commercial penaeid hatcheries, the shipment of brood stock and post larvae from these culture facilities to others in different geographic regions has often resulted in the spread of these agents outside their normal range in wild populations. With respect to disease agents, the Global Aquaculture Alliance (GAA) survey revealed that 60% of losses are due to viruses and about 20% due to bacteria.

Other important diseases of cultured shrimps are related to the nutritional, physical, and toxic disease syndromes. The ascorbic acid deficiency syndrome called "black death" is the best understood nutritional disease. Among the physical diseases occurring in penaeid culture, gas bubble disease and tail cramp are probably the most common. Important toxic disease syndromes include aflatoxicosis and red disease (which may be due to mycotoxins); hemocytic enteritis (due to certain species of filamentous blue-green algae, especially Schizothrix calcicola) and toxic syndromes due to toxic algal blooms.

It has been reported that the first serious outbreak of disease in the shrimps occurred during mid 1980's in Taiwan and the epidemic was *monodon baculovirus* (MBV) (Flegel *et al.*, 2008). This was followed by infectious hypodermal and hematopoietic necrosis virus (IHHNV) in USA (Lightner, 1996), yellow head virus (YHV) in Thailand (Flegel, 1997), and taura syndrome virus (TSV) again in the USA (Brock *et al.*, 1997). During the year 1993-2003 when shrimp industry was still struggling with MBV, IHHNV, YHV and TSV outbreaks, the industry faced with another disaster with the arrival of white spot syndrome virus (WSSV). After its first appearance in China in 1992, it spread rapidly around Asia creating devastating losses at many places in the world (Flegel *et al.*, 2008). Gunalan *et al.*, (2014) while doing the studies on disease occurrence in *L. vannamei* of pond culture system of different regions of India reported the occurrence of six diseases in the shrimp including Black gill disease, TSV, IHHNV, White muscle disease, White gut disease, and Muscle cramp disease. The symptoms of each disease and their possible cure were also described in their report in details. Therefore, disease control and health management are of major importance in Aquaculture and for growth of the shrimp industry. In this chapter efforts have been made to cover major diseases caused by microorganisms and also other diseases related to fungi, parasites. Attempts have also been made to report the latest development in disease diagnostic tools and methods.

11.2 BACTERIAL DISEASES

Bacterial diseases of shrimp have been observed for many years and number of worker have noticed that bacterial infection usually occurs when shrimp are weak or shrimps cultured in unhygienic condition. Even sometimes normal shrimps also get infected if conditions favour presence and abundance of a particularly harmful bacterium. Shrimp body fluids are most often infected by the bacterial group named Vibrio. Infected shrimps show discoloration of the body tissues in some instances, but not in others. The clotting function of the blood and wound repair process, slows down or lost during some infections. Bacteria also invade the digestive tract. A typical infection in larval animals is seen throughout the digestive system. In larger animals, infection becomes obvious in the digestive gland after harmful bacteria gain entry to it, presumably via connections to the gut. Pond reared shrimps occasionally die in large numbers because of diseased digestive glands. The specialized cells that line the inside of the tubules arc particularly fragile and arc easily infected. Tubules progressively die and darken. This kind of disease manifestation is seen in recent reports of rickettsia infection. Cells of the digestive gland tubules are severely damaged as rickettsia invade and develop therein. If infected by bacteria capable of using shell for nutrition, the exoskeleton will demonstrate erosive and blackened areas. These bacteria typically attack edges or tips of exoskeleton parts, but if break occurs in the exoskeleton the bacteria are quick to enter and cause damage. Filamentous bacteria are commonly found attached to the cuticle, particularly fringe areas beset with setae. When infestation is heavy, filamentous bacteria may also be present in large quantity on the gill filaments (Johnson, 1995). Several workers have reviewed the bacterial diseases in shrimp and the range of problems associated with it leading to mass mortalities. Over 20 species of bacteria associated with shrimps have been recognized, some of these are human pathogens (eg. *Vibrio cholera, V. parahaemolyticus* and *V. vulnificus*) while some species are pathogens of aquatic animals (eg. *V. harveyi, V. spendidus, V. penaecida, V. anguillarium, V. parahaemolyticus, V. vulnificus*) (Otta *et al.*, 1999, 2001). Considerable work has been carried out on luminous bacteria like *V. harveyi* mostly found in coastal and marine waters, in association with surface and gut of marine and estuarine organisms and also in shrimp pond water and sediment (Orndorff and Colwell, 1980, Otta *et al.*, 1999, 2001). There are reports of causing mass mortalities in shrimps due to the presence of *V. harveyi* (Fig. 1) (Lavilla-Pitogo *et al.*, 1990, Karunasagar *et al.*, 1994). It has been reported that certain strains of *V. harveyi* had high LD_{50} for *P. monodon* larvae (Karunasagar *et al.*, 1994). Others have reported that strains of this bacterial species virulent to *P. monodon* formed a separate cluster in protein profile but no virulence factors have been established in this species (Liu *et al.*, 1996, Harris and Owens, 1999). Brown gill syndrome in *P. monodon* has been reported due to the presence of *V. harveyi* (Pasharawipas *et al.*, 1998) but they further mentioned that presence of this bacteria was not that critical for shrimp pathogenicity.

Filamentous bacteria such as *Leucothrix mucor, Thiothrix sp, Flexibacter sp., lavobacterium*, and *Cytophaga sp* have been reported in shrimps causing infection particularly at the larval stages. Discoloration of gills, low growth and feeding and enhanced mortality are common signs of the disease. High degree of infection may lead to necrosis in gill tissue (Karunasagar *et al.*, 2005).

11.3 VIRAL DISEASES

More than 20 viruses have been recognised among the farmed shrimps as causative agents of various diseases in penaeid shrimps (Table 1). These viruses can be classifies as members of Parvoviruses, Baculoviruses, Picornaviruses, Toga like viruses and some of the newly identified virus families. Of these 7 viral pathogens have been listed by Office International des Epizootics (OIE) (Table 2) considering the extent of damage caused by these viral pathogens (OIE, 2004). Some of the important viral pathogens affecting the shrimp culture globally are listed and discussed in this chapter.

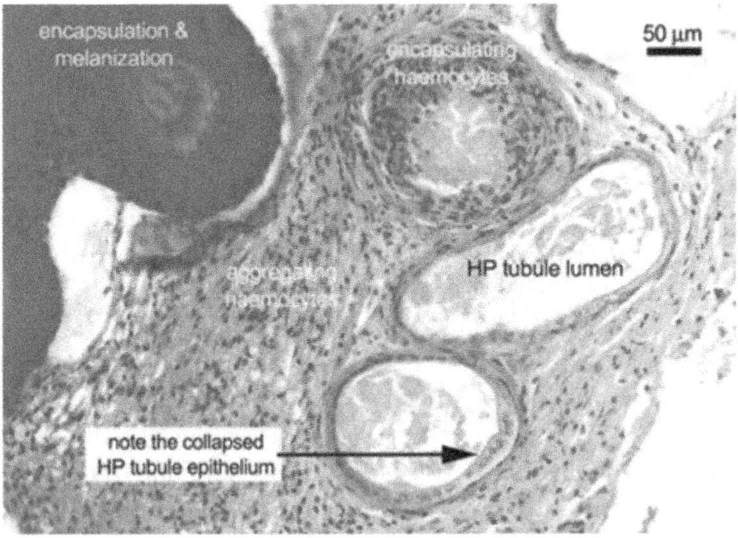

Fig. 1. Hematoxylin and eosin stained tissue section of the hepatopancreas of the black tiger shrimp *Penaeus monodon* shrimp specimen heavily infected with Vibrio. Note the massive aggregation of hemocytes and the melanized granule formed by encapsulating hemocytes. (**Source:** Alday-Sanz, Victoria, ed. *The shrimp book*. Nottingham University Press, 2010)

White spot Syndrome Virus (WSSV)

Number of workers have identified WSSV in the several shrimps as one of the most serious disease problems which have affected the shrimp production all over the world (Takahashi *et al.*, 1994, Chou *et al.*, 1995, Wongteerasupaya *et al.*, 1995, Flegel, 1997, Karunasagar *et al.*, 1997, 2005, Hsu *et al.*, 1999). Chou *et al.*, (1995) reported the occurrence of this virus for the first time in 1982 in the shrimp *P. japonicus* in Taiwan. Since then there are numbers of reports about the occurrence of WSSV causing mortalities of shrimps and consequent serious damage to the shrimp industry (Chou *et al.*, 1995, Wongteerasupaya *et al.*, 1995, Karunasagar *et al.*, 1997, 2005, Hossain *et al.*, 2001). This virus was given many different names like white spot baculo virus (Wang *et al.* 1995), systemic ectodermal and mesodermal baculovirus (Wonteerasupaya *et al.* 1995), *Penaeus monodon* non-occluded baculovirus II (Wongteerasupaya *et al.*, 1996), *Penaeus monodon* non- occluded baculovirus III, Chinese baculovirus, hypodermal and haematopoietic necrosis baculovirus, penaeid rod shaped DNA virus, epithelium envelope baculovirus of *F. chinensis*, lymphoid cell nuclear baculo virus and non-occluded shrimp virus. The name white spot syndrome virus (WSSV) was subsequently adopted. Lightner (1996) gave a detailed account of this virus which is known to affect most commercial important species of penaeid shrimps including *P. monodon, P. japonicus, P. indicus, P. chinensis, P. merguiensis, P. aztecus,*

P. stylirostris, P. vannamei, P. duorarum, and *P. setiferus*. Even in wild shrimps such as *P. semisulcatus, Metapenaeus dobsoni, M. monoceros, M. elegans*, and *P. stylifera* presence of this virus has been reported by number of workers (Lo *et al.*, 1996, Hossain *et al.*, 2001, Chakraborthy *et al.*, 2002).

Durand *et al.*, (1997) worked on the ultrastructure and morphogenesis of WSSV and mentioned that the WSSV is a rod-shaped, double stranded DNA virus with very large genome in the order of 300kb. Sequence analysis of WSSV genomic DNA and the comparison of sequence data has indicated that the WSSV is unique not showing any homology with any known virus (Lo *et al.*, 1997). WSSV replication takes place in the nucleus and is first indicated by chromatin margination and nuclear hypertrophy. Viral morphogenesis begins by the formation of membranes *de novo* in the nucleoplasm and by the elaboration of segmented, empty, long tubules. These tubules break into fragments to form naked empty nucleocapsids. After that, membranes envelop the capsids leaving an open extremity. The nucleo proteins, which have a filamentous appearance, enter the capsid through this open end. When the core is completely formed, the envelope narrows at the open end and form the apical tail of the mature virion (Durand *et al.*, 1997).

The principal clinical signs of WSSV syndrome include lethargy, anorexia, and presence of white spots on the exoskeleton and epidermis (Fig 2). The target tissues of WSSV are generally of mesodermal and ectodermal origin including connective and epithelia1 tissues, lymphoid organs, antennal gland, ovary, hematopoietic nodules, haemocytes, the gills, epidermis, foregut (stomach), striated muscles, and nerves (Wongteerasupaya *et al.*, 1995, Durand *et al.* 1996). The other signs of the disease include rapid reduction in food consumption, loose cuticle and, generalized red dish to pink discoloration (Nakano *et al.*, 1994, Durand *et al.*, 1997, Karunasagar *et al.*, 1997, Otta *et al.*, 1999). WSSV infected shrimp show characteristic cytopathological changes such as, hypertrophic cells, lysed and necrotic cells, cells with hyper trophy and intra nuclear basophilic inclusions, exhibiting a very high level of WSSV infection. Infected tissues appear degenerated with necrotic, atrophied and hypertrophied cells. The cellular integrity of the organs are found to be destroyed completely by WSSV infection. Because of the rapid multiplication of the virus in the target organs the above pathogenesis sets in rapidly. Once the cells are damaged inevitably the animals succumb to mortality. This is further aggravated by the cannibalistic nature of the animal itself. Shrimp feed on the infected dead animals and thus ingest the viral particles. Once the viral particles enter the digestive system, the disease process sets in resulting the spread of the virus in the culture environment in geometric proportion. This is one of the reasons why this virus spreads so rapidly in the culture ponds resulting in the complete destruction of the crop within a fortnight (Alwandi

Table 1. Viral pathogens causing diseases in farmed shrimps of Eastern and Western hemisphere.

Name of the virus	Reference
DNA viruses	
• **Parvo viruses**	
Infectious hypodermal and hematopoieticnecrosis virus (IHHNV)	Lightner *et al.*, 1983
Hepatopancreatic parvo virus (HPV)	Lightner, 1988
Lymphoidalparvo-likevirus(LPV)	Owens *et al.*, 1991
• **Baculo viruses**	
Baculo virus penaei-type (PvSNPV-type sp.) (BP-type)	
– BP from the Gulf of Mexico	Couch, 1974
– BP from Hawaii	Brock *et al.*, 1986
– BP from the Eastern Pacific	Lightner *et al.*, 1985
– Hypodermal and hematopoietic necrosis baculovirus (HHNBV)	Huang *et al.*, 1995
Penaeus monodon -type (PmSNPV-type sp) (MBV-type)	
– MBV from Southeast Asia	Lightner, 1988
– MBV from Italy	Lightner *et al.*, 1985
– *Penaeusplebejus* baculo virus (PBV)	Lester *et al.*, 1987
Baculo viral midgut gland necrosis–types (BMN-type)	
– BMNfrom *Penaeus japonicus* in Japan	Sano *et al.*, 1981
– Type C baculovirus of *P. monodon* (TCBV)	Brock and Lightner, 1990
– White spot syndrome baculo viruses (WSSV)	Wongteerasupaya *et al.*, 1995
• **Large Bacuol like viruses**	
– Haemocyte-infecting non occluded baculo virus (PHRV)	Owens, 1993
• **Irido viruses**	
Shrimp irido virus (IRDO)	Lightner and Redman, 1991
RNA viruses	
• **Picorna virus**	
Taura syndrome virus (TSV)	Lightner *et al.*, 1995
• **Reo viruses**	
– Type III reo-like virus (REO-III)	Tsing and Bonami, 1987
– Type IV reo-likevirus(REO-IV)	Adams and Bonami, 1991
• **Toga virus**	
Lymphoid organ vacuolization virus (LOVV)	Lightner *et al.*, 1994
• **Rhabdo virus and Rhabdo like viruses**	
– Yellow head virus of *P. monodon* (YHV/YBV)	Bonami *et al.*, 1992
– Rhabdo virus of penaeid shrimp (RPS)	Nadala *et al.*, 1992

(**Source:** Alwandi and Santiago, 2008)

Table 2. Viral pathogens listed by the Office Internationaldes Epizootics (OIE)

Diseases of crustaceans listed by the OIE
• Taura syndrome TSV (viral/ penaeid shrimp)
• White spot disease WSD (viral/ penaeid shrimp and other decapod crustaceans)
• Yellow head disease YHV (viral/ penaeid shrimp)
• Tetrahedral baculo virosis (Baculo virus penaei)
• BP (viral/ penaeid shrimp)
• Spherical baculo virosis (*Penaeus monodon*- type baculo virus)
• MBV (viral/ penaeid shrimp)
• Infectious hypodermal and haematopoietic necrosis IHHNV (viral/ penaeid shrimp)
• Spawner- isolated mortality virus disease (viral/penaeid shrimp)

(**Source:** Alwandi and Santiago, 2008)

and Santiago, 2008). Lo *et al.*, (1997) reported that since WSSV can infect ovary, vertical transmission of the virus from the brood stock to eggs is possible. Such WSSV infected shrimp brood stock may produce infected /uninfected larvae (Karunasagar *et al.*, 1997). Alwandi and Santiago, (2008) while working on *P. monodon* reported that WSSV infected Y-organ tissue appeared degenerated with necrotic, atrophied and hypertrophied cells with basophilic inclusions of WSSV. Further they mentioned that the cellular integrity of the Y-organ was found to be destroyed completely by WSSV infection. Similar observations were also made by Wang *et al.* (1998) in the WSSV infected cuticular epidermis of *M. japonicus*.

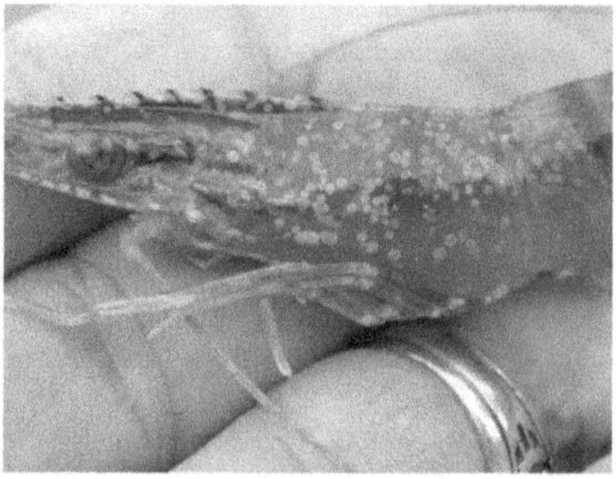

Fig. 2. White shrimp infected with WSSV, visible white spots on epidermis.

Transmission of WSSV

1. **Horizontal transmission:** Chang *et al.* (1998) mentioned that WSSV can be transmitted horizontally by feeding on infected tissue or via water. Postlarvae, juvenile and sub adult penaeids are susceptible to WSSV infection and infections have been found in all life stages of penaeids. Alwandi and Santiago (2008) reported that horizontal transmission of WSSV from the affected shrimp farms to the neighbouring ecosystem has created a realistic scenario in which the receiving ecosystem carries the WSSV load in the form

of live or dead tissues, dead and decomposed tissues and free virions. WSSV virions can remain infective in the decaying tissues or in detritus up to 4 d, contrary to the common belief that free virus cannot survive in natural waters more than 24h. This virus could be transmitted to benthic crustaceans and other fauna through different feeding pathways such as filter feeding, detritus feeding, and predation (Alwandi and Santiago, 2008).

2. **Passive Transmission:** Viruses can also pass into the digestive tracts of other invertebrates, other than crustaceans and can persist in the alimentary canal, potentially making the animal a passive carrier of the virus. When these passive carriers are consumed by the shrimp, they can potentially infect the shrimp with WSSV. Vijayan *et al.,* (2005) indicated that WSSV could be transmitted through polychaete worms to infect *Penaeus monodon* broodstock. Polychaete worms accumulated the viral pathogen in their digestive tract through feeding. Though the virus itself is not infecting the worms, worms remain potent in their digestive tracts and acts as a passive vector of WSSV in aquatic systems. When these worms are fed to broodstock shrimp in the hatchery, the shrimp ingest the virus in the worms causing patent WSSV infection.
3. **Vertical Transmission:** Various investigator shown that the virus is transmitted from the broodstock to the offspring. Though the exact mechanism of this transmission is not known, field experience reveals that this vertical transmission takes place. Lo *et al.,* (1997) reported that since WSSV can infect ovary, vertical transmission of the virus from the brood stock to eggs is possible. Such WSSV infected shrimp brood stock may produce infected /uninfected larvae (Karunasagar *et al.,* 1997).

Monodon Baculovirus (MBV)

Lightner and Redman (1981) while working on diseases of *P. monodon* reported for the first time the presence of baculovirus in this particular shrimp. They further mentioned that it is a nuclear polyhedrosis virus of the family Baculoviridae. Rohrmann (1986) reported that this virus has double stranded circular DNA genome of 80-100 × 10^6Da within a rod shaped, enveloped particle often found occluded within protenaceous bodies. Later several workers have reported the presence of this virus in *P. merguiensis, P. semisulcatus, P. kerathurus, P. vannamei, P. esculentus, P. penicillatus, P. indicus,* and *Metapenaeus ensis* (Johnson and Lightner, 1988, Lightner, 1988, Chen *et al.,* 1989, Ramasamy *et al.,* 1995, Vijayan *et al.,* 1995, Karunasagar *et al.,* 1998). It has been observed that MVB is common wide spread pathogen and after infection severe mortalities have been recorded particularly in post larvae and juvenile stages (Nash *et al.,* 1988). Growth retardation and hepatopancreatic damage have been recorded in MBV infected shrimp *P. monodon* by Lightner *et al.,* (1983), and Chang and Chen, 1994).

There are reports that MBV is generally found in mixed infection with other pathogens including viruses, bacteria and parasites (Karunasagar *et al.*, 1998, Umesha, 2003). Transmission of MBV occurs only horizontally through faecal route and because of this, eggs and larvae in the hatcheries get infected with MBV (Chen *et al.*, 1992, Natividad and Lightner, 1992). Therefore to eliminate MBV from the hatchery is to identify carrier brood stock individually and discard contaminated batches of larvae (Fegan *et al.*, 1991).

Hepatopancreatic Parvovirus (HPV)/ Densovirus

Lightner and Redman in 1985 reported for the first time presence of HPV in post larvae of *P. chinensis*. Later the same virus was reported in *P. monodon* by Flegel and Sriurairatana in 1993. It was later reported by many workers in several penaeid species in many parts of the world including Asia, Africa, Australia and North and South America (Paynter *et al.*, 1985, Colorni *et al.*, 1987, Brock and Lightner, 1990, Lightner and Redman, 1992, Lightner, 1996). Clinical diagnostic characteristics of infected shrimps with HPV generally show atrophy of hepatopancreas anorexia, poor growth rate, and reduced preening activities (Lightner and Redman, 1985, Chen, 1992, Lightnter *et al.*, 1992, Sukhumsirichart *et al.*, 1999) (Fig 3). Bonami *et al.*, (1995) and Sukhumsirichart *et al.*, (1999) have isolated and characterised HPV form *P. chinensis* and *P. monodon* respectively and it was observed that both these viruses comprise unenveloped, icosahedral particles of approximately 22 to 24 nm diameter. The nucleic acid chain of both viruses is single stranded DNA although genome size both these viruses was different.

Infectious Hypodermal Hematopoietic Necrosis Virus (IHHNV)

Lightner *et al.*, (1983) detected IHHN virus for the first time in juvenile *P. stylirostris* from Hawaii in 1981. Subsequently this virus has been detected in a number of other penaeid species in all over the world (Flegel 1997). Bonami *et al.*, (1990) while working on the characterization of the IHHNV and hematopoietic necrosis virus of penaeid shrimp reported the structure of IHHNV. They mentioned that IHHNV is small, icosahedral non enveloped virus containing single stranded linear DNA genome approximately 4.1 kb in length. Nunan *et al.*, (2000) sequenced the complete genome of IHHNV and reported that this genome contains 3 large open reading frames. One of the frames which encompass approximately 50% of the genome encode a polypeptide of 666 amino acids and other two frames encode 343 amino acids and 329 amino acids polypeptide respectively (Shike *et al.*, 2000) (Fig 4). Infection of this virus has been noticed in shrimps *P. vannamei* and *P. monodon* without any mortality symptoms. Some workers have noticed runt – deformity syndrome due this virus and substantial economic losses in shrimp Aquaculture (Bell and Lightner, 1984, Kalagayan *et al.*, 1991).

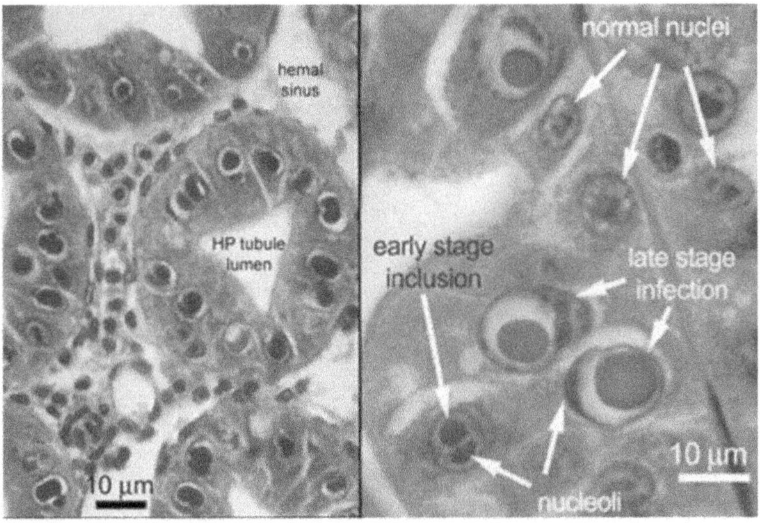

Fig. 3. Hematoxylin and eosin stained tissue section of the hepatopancreas of the black tiger shrimp *Penaeus monodon* shrimp specimen heavily infected with a parvovirus formerly called hepatopancreatic parvovirus (HPV) but now called a densovirus in the proposed genus Hepanvirus (Tijssen and 2008). Note the lack of massive hemocyte aggregation and melanization despite the large number of viral infected cells characterized by large basophilic intranuclear inclusions. (**Source:** Alday-Sanz, 2010.)

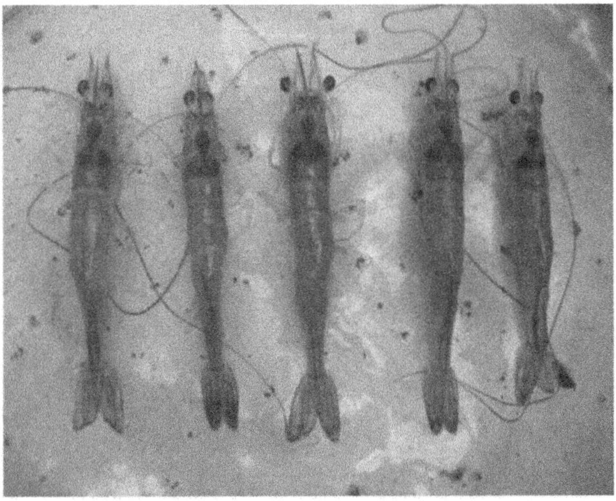

Fig. 4. IHHNV affected shrimps. (**Source:** Gunalan *et al.*, 2014)

Yellow Head Virus (YHV)

Yellow head virus was first detected in pond reared *P. monodon* in central Thailand in 1990 (Cowley *et al.*, 2000, Mayo, 2002). From central Thailand, the disease spread to southern areas on the eastern and western coasts of the Gulf of Thailand (Chantanachookin *et al.* 1993). Yellow head virus (YHV) has since been reported from many shrimp-farming countries in Asia, including India, Indonesia, Malaysia, the Philippines, Sri Lanka, Vietnam and Taiwan (Mohan *et al.* 1998; Wang & Chang 2000). Yellow-head-related viruses have also been reported in *P. monodon* from Thailand and in *P. japonicus* from Taiwan (Wang *et al.* 1996; Soowannayan *et al.* 2003). A survey of shrimp samples from Mozambique, India, Thailand, Malaysia, Indonesia, the Philippines, Vietnam, Taiwan and Australia has revealed that YHV is one of at least six genetic lineages (genotypes) of related viruses that occur commonly in farmed *P. monodon* throughout the Indo-Pacific region (Wijegoonawardane *et al.* 2008a). Shrimp infected with YHV is characterised by light yellow colouration of the dorsal cephalothorax area and pale appearance (Fig 5). Generally shrimps die within a few hours of developing colour and mortality takes place within 3-5 days (Flegel *et al.*, 1995). YHV has also been shown to infect and cause diseases in *P. vannamei* and *P. stylirostris* (Lu *et al.*, 1994). Yellow head virus infects tissues of ectodermal and mesodermal origin and causes severe necrosis, particularly in the lymphoid organ and gills, (Chantanachookin *et al.* 1993; Lu *et al.* 1995). The YHV is a rod-shaped, single-stranded RNA (ssRNA) virus (40–60 nm ×<“150–200 nm) with a helical nucleocapsid and prominent knob-like surface projections (Chantanachookin *et al.* 1993; Wongteerasupaya *et al.* 1995a,b; Nadala *et al.* 1997; Tang & Lightner 1999). The YHV virions contain three major structural proteins. The nucleocapsid contains the ssRNA genome and the nucleoprotein which encapsidates and protects

Fig. 5. Gross signs of yellow head disease (YHD) are displayed by the three *Penaeus monodon* on the left. (**Source:** T.W. Flegel, 1997)

the RNA. The envelope comprises a tri-laminar lipid membrane and two glycoproteins, gp116 and gp64, which penetrate the membrane to form the spike-like projections on the virion surface. Although yellow head disease outbreaks have been reported only in *P. monodon*, YHV infection has been reported in several other commonly farmed species of penaeid shrimp, including *P. vannamei*, *P. stylirostris*, *P. merguiensis*, *P. setiferus*, *P. aztecus* and *P. duorarum* (Chantanachookin *et al.* 1993; Lu *et al.* 1994; Lightner *et al.* 1998; de la Rosa-Velez *et al.* 2006).

A yellow-head-like virus has also been reported in *P. japonicus* in Taiwan (Wang *et al.* 1996). There is evidence of natural and/or experimental YHV infection in wild penaeid and palemonid shrimp found within ponds or nearby canals (Flegel *et al.* 1997b; Longyant *et al.* 2006). However, the severity of disease varies, with some susceptible shrimp displaying mild signs and surviving up to 30 days after experimental infection (Longyant *et al.* 2005). Yellow head virus can be transmitted horizontally by injection, immersion or ingestion of infected shrimp tissue and by co-habitation with infected shrimp (Lu *et al.* 1994; Flegel *et al.* 1995; Lightner *et al.* 1998; Walker *et al.* 2001; Longyant *et al.* 2006). There is no direct evidence that YHV is transmitted vertically.

Gill Associated Virus (GAV)

Spann and Lester, (1997) reported the presence of GAV in juvenile, adult cultured and wild spawners of *P. monodon* causing large mortalities. Cowley and Walker, (2002) worked out the complete genome sequence of gill associated virus of *P. monodon* and mentioned that GAV is a rod shaped, enveloped RNA nodavirus closely related to the YHV. Spann *et al.*, (1997) mentioned that the shrimp *P. monodon* infected with GAV displays pink to red color of the body and appendages and pink to yellow coloration of the gills. Other signs of disease include lethargy, lack of appetite, secondary fouling and tail rot. Spann *et al.*, (2000) shown that the shrimps viz *P. monodon, P. japonicus, P. esculentus,* and *P. merguiensis* are susceptible to GAV infection and develop disease.

Baculoviral Midgut Gland Necrosis Virus (BMNV)

Sano *et al.*, (1981) for the first time reported the presence of BNMV in the hatchery reared larvae of *P. japonicus*. High mortality rate was observed in infected larval stages of the shrimp with BMN disease. Infected shrimp shows the signs of white turbid midgut gland with remarkable cell necrosis (Takahashi *et al.*, 1998) hypertrophied nuclei, diminished nuclear chromatin, nuclear dissociation and the absence of occlusion bodies (Sano *et al.*, 1981, 1984 1985, Momoyama,

1983). Sano and Momoyama (1992) have developed a technique of rinsing eggs to prevent the shrimp larvae form becoming infected with BMNV.

Baculovirus Penaei (BP)

Baculovirus penaei was first reported by Couch in 1974 in pink shrimp *P. duorarum*. There are reports of causative effects this virus leading to mortalities in shrimp particularly at the post-larval and early juvenile stages (Couch, 1991, Lightner and Redman, 1991, 1992). Lightner and Redman (1998) mentioned that BP infects only the hepatopancreas and midgut epithelial cells and the route of infection observed was through faecal contamination of spawned eggs from BP infected adult spawners (Lightner, 1996).

Taura Syndrome Virus (TSV)

Fig. 6. Healthy *Litopenaeus vannamei* (above); *L. vannamei* infected with Taura syndrome virus (below) **Source:** https://en.wikipedia.org/wiki/Taura_syndrome (article by Herman Gunawan)

Taura syndrome virus was reported as new disease in shrimp *P. vannamei* near the mouth of the Taura River in Ecuador in June 1992 (Jiminez 1992). Since its discovery by 1996, this virus (TSV) had spread in to the major shrimp growing regions of Peru and north-eastern Brazil, throughout the Pacific and Caribbean coasts of Central America, and to Florida and Texas in the USA (Lightner 1996). It had also spread to *P. vannamei* breeding programs in Hawaii by 1994, (Brock *et al.* 1995; Brock *et al.*, 1997). In 1998, TSV spread to Taiwan through the brood stock of *P. vannamei* imported from Central and South America (Tu *et al.* 1999; Yu & Song 2000). By 2004, because of large scale farming of *P. vannamei* TSV had become endemic in most major shrimp-farming countries in Asia (Sunarto *et al.* 2004; Van 2004; Nielsen *et al.* 2005; Do *et al.* 2006). The impact of TSV on the shrimp-farming industry in the America was estimated to be \$US1–2 billion up to 2001 (Lightner 2003). TSV disease represents a serious problem in culture of *P. vannamei* due to high mortality rate and consequence economic losses (Lightner *et al.*, 1997). Three distinct diagnostic features are found in the infected shrimp i.e. moribund shrimp displaying an overall pale reddish coloration, multifocal melanized cuticular lesions and remain persistently infected throughout its remaining life span (Lightner,

1996, Hasson *et al.*, 1997) (Fig 6). In recent years Walker and Mohan (2009) while reviewing viral disease emergence in shrimp Aquaculture discussed in detail about various viral diseases in the shrimp, its origin, impact and effectiveness of health management strategies.

Taura syndrome virus is a small, non-enveloped, positive-sense ssRNA virus with icosahedral architecture (31–32 nm diameter). The TSV virions comprise the RNA genome, three major capsid proteins, and a minor protein (Bonami *et al.* 1997; Mari *et al.* 2002). The functions of the individual capsid proteins are not yet known, (Senapin & Phongdara 2006).

Although *P. vannamei* is the major host of TSV, experimental infections have demonstrated susceptibility to infection in several other penaeid shrimp species, including *P. stylirostris*, *P. setiferus*, *P. aztecus*, *P. duorarum*, *P. chinensis* and *P. monodon* (Overstreet *et al.* 1997; Srisuvan *et al.* 2005). However, susceptibility to disease varies. Clinical signs and mortalities have been reported following natural or experimental infection of *P. stylirostris*, *P. schmitti*, *P. setiferus*, *P. monodon* and *M. ensis*. Taura syndrome virus can be transmitted horizontally by injection, ingestion of infected tissue or exposure to infected shrimp or shrimp carcasses (Brock *et al.* 1995; Hasson *et al.* 1995; Lotz 1997; Lotz *et al.* 2003; Srisuvan *et al.* 2006a). Exposure to chronically infected shrimp can result in TSV transmission in the absence of disease (Lotz *et al.* 2003). Vertical transmission of TSV has not been demonstrated experimentally (Lightner & Redman 1998a; Dhar *et al.* 2004).

Lymphoid Organ Parvolike Virus (LOPV)

This particular virus was first detected in Australia by Owens *et al.*, (1991) in the cultured shrimps *P. monodon, P. merguiensis,* and *P. esculentus.* Infected shrimps showed multinucleated giant cell formation in their hypertrophied lymphoid organs (Ownes *et al.*, 1991). Further it was reported that these giant cells showed mild nuclear hypertrophy and marginated chromatin. Lightner *et al.*, (1987) also made identical observation earlier in *P. monodon* collected from Taiwan. Munday and Owens (1998) reported mild mortality in the shrimps infected with LOPV. LOPV is still poorly understood viral disease of penaeid shrimps (Karunasagar *et al.*, 2005).

Lymphoid Organ Vacuolization Virus (LOVV)

Spann *et al.*, (1995) reported this virus in the shrimp *P. monodon* and further mentioned that this is a rod shaped enveloped RNA virus. Bower *et al* (1994) also found this virus in the shrimps of Asian and Australian regions. There is no much information about the prevalence and pathogenicity of LOVV in penaeid shrimp (Karunasagar *et al.*, 2005), however, Flegel *et al.*, (1995) and Bonami *et al.*, (1992) carried out transmission electron microscopy and histological studies of the lymphoid organs of the infected shrimp and stated that the histological changes are similar to that found in YHV infections.

Rhabdovirus of Penaeid Shrimp (RPS)

Lu *et al.*, (1991) reported the presence of rhabdovirus from an IHHNV infected shrimp *P. stylirostris* and further mentioned that this is a bullet shaped ssRNA virus. Lightner *et al.*, (1996) also reported this virus from the American penaeid shrimps *P. vannamei* and *P. stylirostris.* It was observed that when shrimps were simulated with this virus there were no signs of any clinical abnormality and also no mortality (Karunasagar *et al.*, 2005).

Infectious Myonecrosis Virus (IMNV)

Infectious myonecrosis (IMN) is the most newly identified major viral diseases of shrimp. It was first recognized in 2002 in farmed *P. vannamei* in north-eastern Brazil (Lightner *et al.* 2004). Lightner *et al.*, (2004) further mentioned that due to IMN infection in shrimp, the shrimp industry suffered losses of millions of dollars in Brazil. By 2004, infectious myonecrosis virus (IMNV) had spread to other regions of north-eastern Brazil and was subsequently detected in *P. vannamei* collected from East Java in Indonesia in May 2006 (Senapin *et al.* 2007). It is likely that the virus was introduced to Indonesia in broodstock imported from Brazil. By April 2007, IMNV had reached *P. vannamei* farming regions in south and north-east Sumatra (Walker and Mohan, 2009). Infectious myonecrosis virus is a small (40 nm) non-enveloped, non-segmented double-stranded RNA (dsRNA) virus with icolahedral symmetry (Poulos *et al.* 2006).

Simulation infection studies have demonstrated that *P. vannamei*, *P. stylirostris* and *P. monodon* are all susceptible to IMNV infection, but only *P. vannamei* was shown to be susceptible to disease with the isolate used (Tang *et al.* 2005). Infectious myonecrosis virus primarily infects skeletal muscle, but there is also evidence of infection in the lymphoid organs, hindgut and gills, and the virus has been detected in phagocytic cells within the hepatopancreas and heart (Tang *et al.* 2005). Lightner *et al.*, (2004) reported that shrimps affected with IMN infection show extensive white necrotic areas in the striated muscles, especially in the distal abdominal segments and tailfan. Further this progression of myonecrosis is accompanied with hemocytic infiltration and fibrosis. Senapin *et al.*, (2013) while working on *P. vannamei* from Indonesia observed that whitening of muscle tissue in the shrimp was due to stress induced muscle cramps and from viral infections caused by IMNV or nodavirus. This they studied using immune histochemical reactions for both IMNV and nodavirus. However, further they mentioned that presence of both these viruses at a time is more lethal to the shrimp than single IMNV infection (Fig 7).

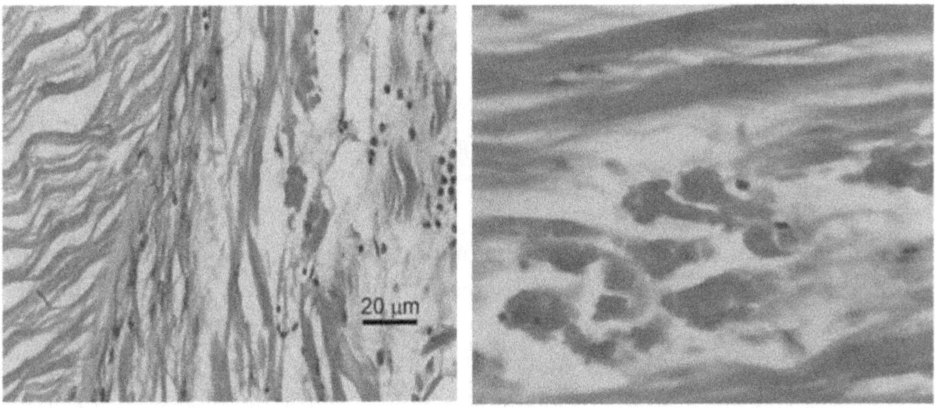

Fig. 7. Comparison of histopathology for IMNV (left) and muscle cramp (right) showing the absence of hemocyte aggregation and viral inclusions in muscle cramp. (**Source:** Alday-Sanz, 2010.)

11.4 FUNGAL DISEASES

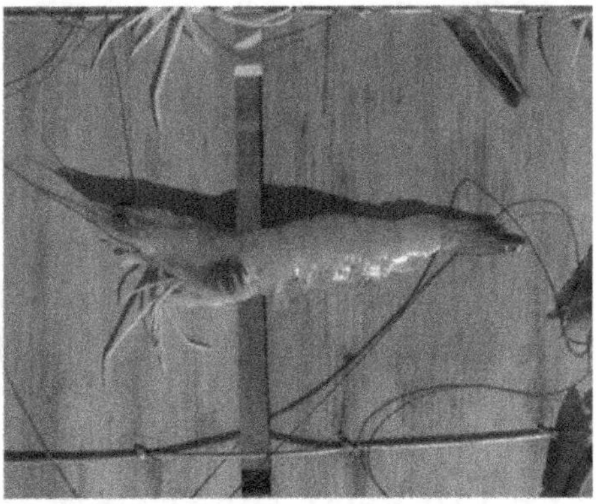

Fig. 8. Black gill disease *L.vannamei* (**Source:** Reuters November 1, 2013. REUTERS/Barry Gooch/SCDNR/Handout via Reuters)

Several fungi are known as shrimp pathogens. Two groups commonly infect larval shrimp, whereas another attacks the juvenile or larger shrimp. The most common genera affecting larval shrimp are *Lagenidium callinectes* and *Sirolpidium.* The protozoea and mysis stages are generally affected with clinical signs such as lethargy and mortality. Fungal spores and mycelia are observed in affected tissues, particularly gill and appendages. *Lagenidium marina* and *Sirolpidium parasitica* infections have been reported by Gopalan *et al.*, (1980) in *P. monodon*. Ramasamy *et al.*, (1996) reported larval stages mortalities of *P. monodon* particularly at nauplii, zoea and mysis stages due to fungal infection. The method of infection requires a thin cuticle such as that characteristic of larval shrimp. The most common genus of fungi affecting larger shrimp is *Fusarium*. It is thought that entry into the shrimp is gained via cracks or eroded areas of the cuticle. *Fusarium* may be identified by the presence of canoe-shaped macroconidia that the fungus produces. Fusariosis and black gill disease (Fig. 8) caused by *Fusarium sp* may affect all developmental stages of penaeid shrimp. Fusarium species like *F. solani*, and *F. moniliformae* have been reported to be opportunistic pathogens that may lead to high mortalities (Karunasagar *et al.*, 2005).

11.5 PARASITIC DISEASES

Karunasagar *et al.*, (2005) while reviewing microbial diseases in shrimp mentioned about the parasitic diseases, particularly about prozoans, ciliates, gregarians and microsporidia and their impact on the shrimp. With regard to protozoan, at high levels of infection, protozoans may induce gill obstruction leading to anorexia, reduced growth, locomotion and increased susceptibility to other infections. Protozoa such as *Zoothamnium, Epistylis, Vorticella, Anophrys, Acineta sp, Lagenophrys* and *Ephelota* may encounter as external parasites in the shrimps (Karunasagar *et al.*, 2005) particularly at the early developmental stages. Further it has been reported that the ciliates protozoans such as *Paranophrys spp* and *Parauronema spp* may cause mortalities in larvae and juveniles of the shrimps. Gregarian protozoans have reported as endoparasitic infecting the shrimp. These parasites generally have two hosts viz, a mollusc or an annelid worm and crustaceans. Gregarians observed in the shrimp include *Nematopsis spp, N. litopenaeus, Paraphioidina scolecoide, Cephalobolus liotopenaeus, C. petiti* and *Cephaloidophoridae stenai.* These parasites may cause reduced absorption of food from the gut and occasionally intestinal blockage. It has been reported that microsporidia protozoans like *Agmasoma sp, Microsporidium sp* may invade the muscle, heart, gonad, gills or hepatopancreas of the shrimps and lead to opacity of the tissues which affect external appearance of the shrimp and marketability (Karunasagar *et al.*, 2005).

11.6 CURRENT DIAGNOSTIC METHODS

Concomitant with the growth of the shrimp culture industry, recognition of disease in the shrimp has gained lot of importance, especially those caused by infectious agents. The most important diseases of shrimp have had viral or bacterial etiologies, but a few important diseases have fungal and protozoan agents as their cause. Diagnostic methods for these pathogens include the traditional methods of morphological pathology, bioassay methods, traditional microbiology, and the application of serological methods. In order to reduce the heavy losses in the revenue in the shrimp Aquaculture industry due to outbreak of diseases rapid progress has been made in disease diagnostic methods during last 15-20 years. Therefore, the need for rapid, sensitive diagnostic methods has led to the application of modern biotechnology to penaeid shrimp disease. The shrimp industry now has modern diagnostic genomic probes with nonradioactive labels for viral pathogens like IHHNV, HPV, TSV, WSSV, MBV, and BP. Additional genomic probes for viruses, for bacterial pathogens and microsporidia have also been developed. Highly sensitive detection methods for some pathogens that employ DNA amplification methods based on the polymerase chain reaction (PCR) now exist, and more PCR methods are being developed for additional agents. These advanced molecular methods promise to provide quick and reliable diagnostic and research tools to an industry reeling from catastrophic epizootics. For the detection of shrimp viruses, World Animal Health Organization (the OIE-Office-International des Epizootics) recommends polymerase chain reaction (PCR) based methods (Chaivisuthangkura *et al.*, 2014).

The protocol and methods that are used to identify the presence of WSSV in the shrimps are both conventional and molecular. Conventional methods are to do the microscopy and histology of the infected tissues and identify the virus, whereas rapid molecular methods involve gene probes and polymerase chain reaction (Durand *et al.*, 1996, Wongteerasupaya *et al.*, 1996, Nunan and Lightner, 1997, Takahashi *et al.*, 1996, Lo *et al.*, 1996). Use of Nested PCR method has been reported by Lo *et al.*, (1996) which has added advantage of increased sensitivity of detection of this virus (Hossain *et al.*, 2004). Now the PCR, Nested PCR and RT-PCR techniques are being commonly used to screen shrimp larvae before stocking in to the ponds. There are other molecular and immunological methods such as *in-situ* hybridization (Lightner, 1997, Wang *et al.*, 1998, Chen *et al.*, 2000), dot blot nitrocellulose enzyme immunoassay (Nadala and Loh, 2000), ELISA (Nadala *et al.*, 1997, Sahul Hameed *et al.*, 1998) and western blotting (Nadala *et al.*, 1997, Magbanua *et al.*, 2000) used to detect WSSV in shrimp and carrier species. (Karuansagar *et al.*, 2004). Dante *et al.*, (2006) developed a duplex polymerase chain reaction (PCR) protocol for the simultaneous detection of two shrimp viruses, namely, WSSV and MBV infecting *P. monodon* in Philippines. The method was designed for screening postlarval samples with dual infections. The developed protocol was able to generate a 211 bp amplicon which is highly specific for WSSV and a 361 bp amplicon specific for MBV. In addition to its high specificity and sensitivity, it is reported that the developed duplex PCR offers an efficient and rapid tool for screening shrimp viruses since both WSSV and MBV can be diagnosed in a single reaction.

Recently Flegel *et al.*, (2008) while reviewing shrimp disease control, mentioned that though PCR and real time (RT)-PCR methods have been very important in helping to control the spread of major shrimp diseases but they have the disadvantage of requiring sophisticated equipment and highly trained personnel. Further they reported that they have developed a new technique of lateral flow chromatographic immunodiagnostic strips similar to common drug-store pregnancy test, for detecting shrimp virus. Further, they mentioned that using these strips, unskilled farm personnel can easily diagnose shrimp disease outbreaks at the pond site. The strips are relatively cheap and detect the virus within 10 minutes (Flegel *et al.,* 2008).

For MBV detection in the shrimps besides conventional histopathological method, rapid and sensitive molecular techniques like PCR (Vickers *et al.,* 1992, Chang *et al.,* 1993, Lu *et al.,* 1993) and genomic probes have been developed (Vickers *et al.,* 1993, Poulos *et al.,* 1994). ELISA test and Nested PCR are also well known techniques for detection of MBV in the shrimps (Hsu *et al.,* 2000, Otta *et al.,* 2003). For HPV detection also gene probes and sensitive PCR techniques are being used (Bonami *et al,* 1995, Mari *et al.,* 1995, Pantoja and Lightner, 2000, Phromjai *et al.,* 2002). Detection of IHHNV is usually done by histopathology (Lightner, 1996), however, in-situ hybridization and PCR methods provide the highest detectable sensitivity of this virus in the shrimps (Lightner *et al.,* 1994). Tang and Lightner (2001) developed RT-PCR technique using a flurogenic 5' nuclease assay detector for detecting this virus in the shrimps. Kim *et al.,* (2011) carried out the studies on the detection of IHHNV in *Litopenaeus vannamei* shrimp cultured in South Korea by using PCR assays. YHV infections in the shrimp can be diagnosed histologically by appearance of massive systemic necrosis in the tissues of ectodermal mesodermal origin (Boonyaratpalin *et al.,* 1993, Chantahachookin *et al.,* 1993). For detection of YHV number of methods like dot blot nitrocellulose enzyme immunoassay (Lu *et al.,* 1996, Nadala and Loh, 2000) western blot technique (Nadala *et al.,* 1997), reverse transcriptase PCR (Wongteerasupaya *et al.,* 1997) and gene probe (Tang and Lightner, 1999) have been developed. Cowley *et al.,* (2000) developed RT-nested PCR technique for detection of GAV in *P. monodon* in Queensland. Similarly a fluorescent antibody technique has been developed in Japan for rapid detection of BNMV (Sano *et al.,* 1981, 1985, Momoyama, 1988). For detection of BP virus besides conventional method of histopathology, advanced techniques like gene probes and immunodiagnostic tool like ELISA have been developed (Lightner *et al.,* 1992, Bruce *et al.,* 1994). A PCR based detection method has also been in use (Wang *et al.,* 1996).

Detection of TSV in the shrimps is generally carried out by gene probes (Lightner 1996, Hasson *et al.,* 1997, Mari *et al.,* 1998) and RT-PCR technique (Nunan *et al.,* 1998). Sukanya *et al.,* (2006) while working on molecular detection of TSV along with WSSV in the shrimp *P. vannamei* in Thailand reported that TSV infection in shrimps was diagnosed by RT-PCR using the OIE recommended

9195/9992 primers. Positive shrimp gave the expected amplicon of 231 bp. For detection of RPS in the shrimp a streptovidin biotin-enhanced nitrocellulose enzyme immunoassay method was developed by Nadala *et al.*, (1992).

Chaivisuthangkura *et al.*, (2014) developed immunological–based assays for specific detection of shrimp viruses. They developed several Monoclonal antibodies (MAbs) specific to the shrimp viruses particularly for WSSV, YHV, IMNV, TSV, HPV, and BP and used as an alternative tool in various immunoassays such as enzyme linked immunosorbent assay, dot blotting, Western blotting and immunohisto-chemistry. Further they mentioned that some of these MAbs were developed into immune-chromatographic strips tests (Fig. 9) for the detection of these viruses and into a dual strip test for the simultaneous detection of WSSV/YHV. The strip test has the advantage of speed, as the disease can be detected within 15 minutes.

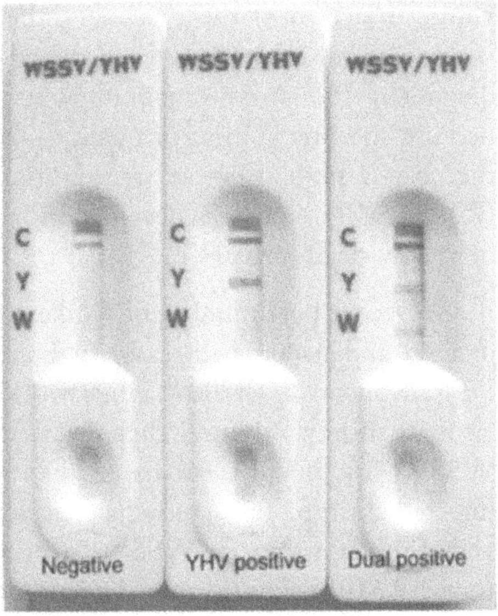

Fig. 9. Example of a lateral-flow, immunochromatographic strip test for the presence of white spot syndrome virus and yellow head virus at pre-patent or patent levels of infection.
(**Source:** Chaivisuthangkura *et al.*, 2014)

11.7 LOOP MEDIATED ISOTHERMAL AMPLIFICATION (LAMP) TECHNIQUE

Loop Mediated Isothermal Amplification technique is a novel sensitive and rapid method which can be applied for disease diagnosis in aquaculture. In comparison to the Polemerase Chain Reaction (PCR) technology, in which the reaction is carried out with a series of alternate temperature steps or cycles, isothermal amplification is carried out at a constant temperature and does not require PCR machine. In LAMP technique the target sequence is amplified at a constant temperature of 60 -65°C using either two or three sets of primers and a polemerase with high strand displacement activity in addition to a replication activity. In this technique four different primers are typically used to identify 6 different regions on the target gene, which adds highly to the specificity. An additional pair of "loop primers" can further accelerate the reaction. Due to the specific nature of action of these primers, the amount of DNA produced in LAMP is considerably higher than PCR based amplification. Despite its wide use as a diagnostic tool in the laboratory, its application at the farm level is limited, because the preparation of quality DNA templates is still highly technical, and the cost is high. Reverse transcription loop-mediated isothermal amplification (RT-LAMP) assay was developed for detecting the structural glycoprotein gene of yellow head virus (YHV).

In recent past number of workers have used LAMP technique for detection of viral and bacterial diseases in Penaeid shrimps as an alternative method to PCR. Kono *et al.*, (2004) while working on detection of white spot syndrome virus in shrimp reported that the LAMP method was successful in identifying WSSV virus in different tissues like heart, stomach and lymphoid organ from infected shrimp. This study has developed a diagnostic procedure which is rapid and highly sensitive for WSSV detection in shrimp. Kiatpathomchai *et al.*, (2008) used LAMP technology for detection of Taura syndrome virus (TSV) in shrimp successfully with high specificity and sensitivity. In this particular technique they followed 60 minutes reverse transcription LAMP (RT- LAMP) for amplification of TSV cDNA using biotin labeled primer combined with chromatographic lateral flow dipstick (LFD) for rapid and simple visual detection of TSV specific amplicons. Puthawiboon *et al.*, (2009) while working on myonecrosis virus in shrimp used RT-LAMP technology for detection of this virus successfully. However they mentioned that this RT-LAMP-LFD gave negative results with nucleic acid extracts from normal shrimp and from shrimp infected with other viruses including infectious hypodermal hematopoietic necrosis virus (IHHNV), *monodon baculovirus* (MBV), a *hepatopancreatic parvo virus* from *P. monodon* (PmDNV), WSSV, yellow head virus (YHV), TSV, *Machrobrachium rosenbergii* nodavirus (MrNV) and Gill associated virus (GAV). Jaroenram *et al.*, (2009) also reported successful use of RT-LAMP LFD technique for identification of WSSV in the shrimp *Penaeus monodon*. Nimitphak *et al.*, (2010) reported that successful use of LAMP combined with LFD technique for rapid and sensitive detection of *Penaeus monodon nucleopolyhedrovirus* (PemoNPV). However they further mentioned that the LAMP-LFD method gave negative test result for shrimp infected with other common shrimp DNA viruses including *Penaeus monodon* densovirus (PmDNV), WSSV and *Penaeus stylirostris* densovirus (PstDNV). Sappat *et al.*, (2011) also reported detection shrimp TSV successfully by using LAMP method. In this study they used turbidimetric end-point detection method with spectroscopic measurement of LAMP by-product: magnesium pyrophosphate. The device incorporated a heating block that maintained an optimal temperature of 63°C for the duration of the RT-LAMP reaction. Further they mentioned that this method showed negative results with other shrimp viruses except for TSV. He and Xu (2011) reported the use of multiplex LAMP for detection of two Penaeid shrimp viruses namely WSSV and IHHNV which are the major viral pathogens of penaeid shrimp worldwide. They further reported that the multiplex LAMP method offers an efficient, convenient and rapid tool for screening penaeid shrimp viruses and will be useful for the control of these viruses in shrimp. Suebsing *et al.,* (2013) reported an emerging micropsordian parasite namely *Enterocytozoon hepatopenaei* which is linked to recent losses caused by white faeces syndrome (WFS) in cultured black tiger shrimp *P. monodon* and whiteleg shrimp *L. vannamei*. For detection of these micropsordian paracytes they developed LAMP assay method combined with colorimetric nanogold for rapid, sensitive and inexpensive detection of this parasite. Arunrut *et al.,* (2013) developed a novel strategy for the detection of reverse transcription loop-mediated isothermal amplification (RT-LAMP) products derived from infectious myonecrosis virus (IMNV), causes a serious myonecrosis in *Penaeus* (*Litopenaeus*) *vannamei*, by using a ssDNA-labeled with gold nanoparticle (AuNP) probe. This technique relies on a self-aggregation method, when the AuNP aggregation is induced by an increasing of salt concentrations with visual detection. The presence of IMNV-LAMP target prevented an AuNP aggregation and a solution remained as pink color of AuNP, while non-complementary targets cannot prevent AuNP aggregation, resulting in a visible colour change to purple color after addition of salt. Further they mentioned that this assay can be adapted easily for rapid detection of other shrimp infectious diseases agents at low-cost with robust reagents and using a simple colorimetric detection method. Amalea Dulcene *et al.,* (2014) used LAMP protocols for the detection of the two most common shrimp pathogens, white spot syndrome virus (WSSV) and Vibrio spp., in *L. vannamei* in the Philippines. By using this protocol they showed that the LAMP assay was faster and 10 times more sensitive than polymerase chain reaction in detecting WSSV and was more efficient than the traditional microbiological method in diagnosing vibriosis. Overall, the results indicated that a LAMP protocol, which is more convenient, highly sensitive, faster, and more practical, has been effectively utilized to detect WSSV and vibriosis in selected Philippine shrimp farm.

11.8 LATER FLOW IMMUNO-ASSAY (LFIA) FOR WSSV DETECTION

White spot syndrome virus is serious viral pathogen and responsible for severe mortality in shrimp. Various diagnostic methods are available to detect WSSV but farmer's friendly and reliable diagnostic tool is still not available. Later Flow Immuno Assay (LFIA) was developed based on antibody raised against recombinant protein of WSSV. The sample pad, conjugate release matrix and absorbent pad (MDI, India) were assembled according to manufacturer's instructions on a plastic supported with nitrocellulose membrane (0.8 μm) with 2 mm overlap. The sample pad was placed anterior to the glass fiber pad at the sample application well, and the absorbent pad was placed at the posterior end, next to the control line for collecting the excess reagent. The conjugate release matrix was soaked with pAb –AuNPs conjugate and dried for 1 h at 37°C. The conjugate release matrix was placed between sample pad and nitrocellulose membrane with a overlap of 2 mm. Test (T) and control (C) lines were printed using anti-WSSV pAb (1 mg/mL) and protein A (1 mg/mL) respectively on nitrocellulose membrane (NC) using automated antibody printing machine (MDI, India). After fabrication, assay strips were cut in to a size of 4 mm by using a programmable strip cutter (MDI, India) and enclosed in a plastic cassette (Fig. 10). (Yoganandhan, *et al.*, 2004, Syed Musthaq, *et al.*, 2006)

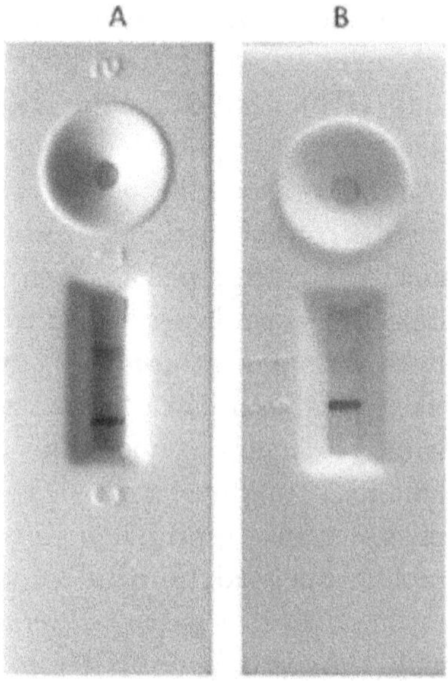

Fig. 10. LFIA strips. Gill sample homogenate from WSSV-infected (A) and uninfected (B) *L. vannamei* applied to the test strip. T: test line; C: control line. (**Source:** Yoganandhan, *et al.*, 2004, Syed Musthaq, *et al.*, 2006)

In a test solution containing WSSV, the virus binds to the pAb-AuNPs conjugate in the LFIA and migrates in the nitrocellulose membrane. This complex interacts with the antibody coated at the 'test' line on the membrane and forms a sandwich. The unbound pAb conjugated with AuNPs moves across the test line and is captured at the 'control' line due to high affinity of protein A to antibody (Fig. 10). LFIA showed that serial dilution of WSSV-infected sample at a protein concentration of 50, 25, 12.5, 6.25 and 3.125μg/mL yielded positive results up to 6.25 μg/mL. So the detection limit of LFIA was 6.25 μg/mL of total protein. The same serial dilutions of samples after DNA extraction were examined by one step PCR. The PCR was able to detect WSSV in the sample containing protein at the level of 3.125 μg/mL. Although the LFIA was sensitive little less than one step PCR, it is much more convenient as results obtained very quickly (<15min) without need of any expensive equipment. It is farmer's friendly and reliable. To determine the specificity of the assay, the LFIA was further tested with different shrimp viruses such as IHHNV, MBV and HPV. The developed LFIA do not show any cross reactivity with other shrimp viruses as evidenced from the appearance of single line in control zone of the LFIA only.

Time course infectivity experiment was conducted to collect WSSV-infected samples at different time intervals to study the use of LFIA for early detection of WSSV. The LFIA detected WSSV in haemolymph sample after 3 h after infection. At 6 h after infection, the WSSV was detected in haemolymph and gill samples. The WSSV was detected in all samples except eyestalk after 12, 24 and 36 h after infection. It was detected in all the samples tested at moribund stage. Copy number of WSSV was quantified in the sample prepared from WSSV-infected shrimp by quantitative PCR and was diluted 10-fold with PBS. The diluted samples containing 10^7 to 10^3 copies of WSSV were screened for WSSV by LFA. The results revealed that the LFA detected more than 2400 copies of WSSV per µL. This product would be commercialized soon for the benefit of shrimp farmers.

11.9 BIOINFORMATICS AND PREDICTION OF SHRIMP DISEASES

WSSV is one of the most catastrophic pathogens that are destructive to shrimp *Aquaculture* activity in commercial shrimp farms Liu *et al.,* (2009). In order to control this virus, more and more studies have been carried out in shrimps in last decade. Lan *et al.,* (2006) reported the discovery of transcriptional profile of WSSV genes in shrimp with DNA microarray and some early genes. Numbers of workers have identified many host genes and proteins responding to WSSV infection through large scale approaches (Liu *et al.,* 2011, Wang *et al.,* 2006, Zhao *et al.,* 2007, Li *et al.,* 2013). From these studies, it is observed that a lot of host genes and proteins were found upregulated or downregulated after WSSV infection. However, there is a paucity of information on the direct interaction between WSSV and the host proteins for proper understanding the pathogenesis of WSSV in shrimp. Under such situation, bioinformatic analysis will provide a highly effective approach for identifying genes and proteins involved in WSSV/shrimp interaction based on the public protein-protein interaction (PPI) databases (Sun *et al.,* 2014).

Sun *et al.,* (2014) while working on bioinformatic prediction of WSSV-Host protein-protein interaction in shrimp *F. chinensis* mentioned that with the genome data of WSSV (van Hulten *et al.,* 2001, Yang *et al.,* 2001) and abundant transcriptome data from shrimp, it becomes possible to identify WSSV/shrimp interacting proteins with developed bioinformatic techniques. Exploitation of protein-protein interaction information with bioinformatic approaches provides an effective way to analyse high-throughput experimental data in distinct organisms (Pattin and Moore, 2009, Remmerie *et al.,* 2011). The predicted interaction between shrimp protein and WSSV protein which might be either independent interaction or synergistic interaction, provide information on possible invasion approaches during WSSV infection to host cells. Moreover, these interactions could also lead to intracellular signalling pathways initiated by the host proteins.

References

Adams, J.R. and J.R. Bonami. (Eds). 1991. *Atlas of Invertebrate viruses*. CRC Press, F.L. Bacca Raton, 1-53.

Alday-Sanz, Victoria, ed. *The shrimp book*. Nottingham University Press, 2010.

Alwandi S V and T C. Santiago, 2008. Microbial diseases: In physiology of reproduction breeding and culture of tiger shrimp *P monodon*. (authors: A D Diwan, S Joseph, S Ayyappan), Narendra publishing house, New Delhi. 217-243.

Amalea Dulcene D. Nicolasora, Benedict A. Maralit, Christopher Marlowe A. Caipang, Mudjekeewis D. Santos, Adelaida Calpe and Mary Beth B. Maningas, 2014. Utilization of loop-mediated isothermal amplification (LAMP) technology for detecting White Spot Syndrome Virus (WSSV) and Vibriospp. in *Litopenaeus vannamei* in selected sites in the Philippines. *Philippine Science Letters*. 7:2.

Arunrut N, Suebsing R, Withyachumnarnkul B, Kiatpathomchai W, 2014. Demonstration of a Very Inexpensive, Turbidimetric, Real-Time, RT-LAMP Detection Platform Using Shrimp Laem-Singh Virus (LSNV) as a Model. *PLoS ONE* 9,9:108047.

Bell TA, Lightner DV, 1984. IHHN virus: infectivity and pathogenicity studies in *Penaeus stylirostris* and *Penaeus vannamei*. *Aquaculture*; 38: 185–194.

Bonami, J.R., M. Brehelin, J. Mari, B. Trumper and D.V. Lightner. 1990. Purification and characterization of IHHN virus of penaeid shrimps. *J. Gen. Virol.* 71: 2657.

Bonami, J.R., D.V. Lightner, R.M. Redmanand B.T. Poulos, 1992. Partial characterization of the togavirus (LOVV) associated with histopathological changes of the lymphoid organ of penaeid shrimps. *Dis. Aquat. Org.* 14:145-149.

Bonami JR, Mari J, Poulos BT, Lightner DV, 1995. Characterization of hepatopancreatic parvo-like virus, a second unusual parvovirus pathogenic for penaeid shrimps. *J Gen Virol*; 76 (Pt 4): 813-817

Bonami JR, Hasson KW, Mari J, Poulos BT, Lightner DV, 1997. Taura syndrome of marine penaeid shrimp: characterization of the viral agent. *J Gen Virol*; 78,2: 313-319.

Boonyaratpalin, S., K. Supamattaya, J. Kasornchandra, S. Direkbusracom, U. Eckpanithanpong and C. Chantanachookin, 1993. Non occluded baculo like virus, the causative agent of yellow-head disease in black tiger shrimp (*Penaeus monodon*). *Fish Pathol.* 28:103-119.

Bower SM, Mcgladdery SE, Price IM, 1994. Synopsis of infectious diseases and parasites of commercially exploited shellfish. In: Faisal M, Hetrick FM (eds) *Annual Review of Fish Diseases* Vol. 4, Elsevier Science Ltd, USA, p 1-199.

Brock J.A.,Nakagawa L.K.,Van Campen H.,Hayashi T.& Teruya S, 1986. A record of Baculovirus penaeifrom *Penaeus marginatus* Randall in Hawaii. *J. Fish Dis.* 9, 353–355.

Brock, J.A. and D.V. Lightner, 1990. Diseases of crustacea. Diseases caused by microorganisms. *In* O. Kinne (ed.) *Diseases of Marine Animals.* 3. John Wiley and Sons, New York, 245.

Brock, J.A., Gose, R., Lightner, D.V, Hasson, K.W, 1995. An overview on Taura Syndrome, an important disease of farmed *Penaeus vannamei.* In: Browdy, C.L., Hopkins, J.S. (Eds.), Swimming through troubled water, Proceedings of the special session on shrimp farming, *Aquaculture* '95. San Diego. World *Aquaculture* Society, Baton Rouge, LA, USA,. 84-94.

Brock, J.A., Gose, R.B., Lightner, D.V., Hasson, K.W, 1997. Recent developments and an overview of Taura Syndrome of farmed shrimp in the Americas. In: Flegel, T.W., MacRae, I.H. (Eds.), *Diseases in Asian Aquaculture III*, Proceedings from the Third Symposium on Diseases in Asian *Aquaculture*, Bangkok. Fish Health Section, Asian Fisheries Society, Manila, 275-283.

Cowley JA, Dimmock CM, Spann KM, Walker PJ, 2000. Gill associated virus of *Penaeus monodon* prawns: an invertebrate virus with ORF1a and ORF1b genes related to arteri and corona viruses. *J Gen Virol*; 81: 1473-1484.

Chakraborty A, Otta SK, Joseph B, Sanath Kumar, Hossain Md S, RFF I, Venugopal MN, Karunasagar I, 2002. Prevalence of white spot syndrome virus in wild crustaceans along the coast of India. *Curr Sci* 82,11:1392-1397.

Chaivisuthangkura P, Longyant S, Sithigorngul P, 2014. Immunological- based assays for specific detection of shrimp viruses. *World J Virol.* 3,1: 1-10.

Chantanachookin C, Boonyaratanapalin S, Kasornchandra J, Direkbusarakom S, Ekpanithanpong U, Supamataya K, Siurairatana S, Flegel TW, 1993. Histology and ultrastructure reveal a new granulosis-like virus in *Penaeus monodon* affected by "yellow-head" disease. *Diseases of Aquatic Organisms*; 17: 145–157.

Chang PS, Lo CF, Kou GH, Lu CC, Chen SN, 1993. Purification and amplification of DNA from *Penaeus monodon*-type baculovirus (MBV). *J Invert Pathol* 62,2: 116-120.

Chang PS, Chen SN, 1994. Effect of *P. monodon* type baculovirus (MBV) on survival and growth of *Penaeus monodon* Fabricus. *Aquat Fish Manage* 25,3:311-317.

Chang, P. -S., H. -C. Chen, and Y. -C. Wang. 1998a. Detection of whitespot syndrome virus associated baculovirusin experiment lobsters by in situ hybridization. *Aquaculture* 164: 233-242.

Chaivisuthangkura P, Longyant S, Sithigorngul P. Immunological-based assays for specific detection of shrimp viruses. *World J Virol.* 3,1: 1-10.

Chen SN, Chang PS, Kou GH, 1989. Observation on pathogenicity and epizootiology of *Penaeus monodon* Baculovirus (MBV) in cultured shrimp in Taiwan. *Fish Pathol* 24: 189-195

Chen SN, Chang PS, Kou GH, 1992. Infection route and eradication of *Penaeus monodon* baculovirus (MBV) in larval gaint tiger prawns, *Penaeus monodon* In: Fulks W, Main KL (eds) Diseases of cultured penaeid shrimp in Asia and the United States. The Oceanic Institute, Makapuu Point, Honolulu, Hawaii, 177-184.

Chen LL, Lo CF, Chiu YL, Chang CF, Kou GH, 2000. Natural and experimental infection of white spot syndrome virus (WSSV) in benthic larvae of mud crab *Scylla serrata. Diseases of Aquatic Organisms* 40: 157-161.

Chou, H.Y., C.Y. Huang, C.H. Wang, H.C. Chiang and C.F. Lo, 1995. Pathogenicity of a baculovirus infection causing White Spot Syndrome in cultured penaeid shrimp in Taiwan. *Diseases of Aquatic Organisms* 23:165-173.

Couch, J.A, 1974. An epizootic nuclear polyhedrosis virus of pink shrimp: ultrastructure, prevalence and enhancement. *J. Invertebr. Pathol.* 24:311-331.

Couch J. A, 1991. Baculoviridae, Nuclear polyhedrosis viruses. Part 2. Nuclear polyhedrosis viruses of invertebrates other than insects. In: Adams JR, Bonami JR (eds) *Atlas of invertebrate viruses*, CRC Press, Boca Raton, FL, 205-226.

Colorni A, Samocha T, Colorni B, 1987. Pathogenic viruses introduced into Israeil mariculture systems by imported penaeid shrimp. *Bamidgeh* 39: 21-28.

Cowley J A, Dimmock C M, Spann K M, Walker P J, 2000. Detection of Australian gill associated virus (GAV) and lymphoid organ virus (LOV) of *Penaeus monodon* by RT- nested PCR. *Diseases of Aquatic Organisms* 39: 159-167.

Cowley J A, Walker P J, 2002. The complete genome sequence of gill associated virus of *Penaeus monodon* prawns indicates a gene organization unique among nodo viruses. *Arch Virol* 147:1977-1987.

Dante Karlo T. Natividad, Maria Veron P. Migo, Juan D. Albaladejo, Jose Paolo V. Magbanua, Nakao Nomura, Masatoshi Matsumura, 2006. Simultaneous PCR detection of two shrimp viruses (WSSV and MBV) in postlarvae of *Penaeus monodon* in the Philippines. *Aquaculture* 257, 1–4, 142–149.

De la Rosa-Velez, Y. Cedano-Thomas, J. Cid-Becerra, J.C. Mendez-Payan, C. Vega-Perez, J. Zambrano-Garcia, J.R. Bonami, 2006. Presumptive detection of yellow head virus by reverse transcriptase-polymerase chain reaction and dot-blot hybridization in *Litopenaeus vannamei* and *L. stylirostris* culture on the northwest coast of Mexico *J. Fish Dis.*, 29: 717–726.

Dhar A K, Cowley J A, Hasson K W, Walker P J, 2004. Genomic organization, biology, and diagnosis of Taura syndrome virus and yellow head virus of penaeid shrimp. *Advances in Virus Research* 63: 347–415.

Do JW, Cha SJ, Lee NS, Kim YC, Kim JW, Kfm JD, Park JW, 2006. Taura syndrome virus from *Penaeus vannamei* shrimp cultured in Korea. *Diseases of Aquatic Organisms* 12: 171–17.

Durand, S., D.V. Lightner, L.M. Nunan, R.M. Redman, J. Mari and J.R. Bonami, 1996. Application of gene probes as diagnostic tools for white spot baculovirus (WSBV) of penaeid shrimp. *Dis. Aquat. Org.* 27:59-66.

Durand, S., D.V. Lightner, R.M. Redman and J.R. Bonami, 1997. Ultrastructure and morphogenesis of White spot syndrome Baculovirus (WSSV). *Dis. Aquat. Org.*29:205–211.

Fegan DF, Flegal TW, Sriurairatana S, Waiakrutra M, 1991. The occurrence, development and histopathology of monodon baculovirus in *Penaeus monodon* in Southern Thailand. *Aquaculture.* 96:205–217

Flegel TW, Sriurairatana S, 1993. Black tiger prawn disease in Thailand. *Tech. Bull., Am.* Soyabean Assoc, Vol. AQ39/1993/ 3, Sect. B, 1-30.

Flegel TW, Fegan DF, Sriurairatana S, 1995. Environmental control of infectious shrimp diseases in Thailand. In: Shariff M, Arthur JR, Subasinghe RP (eds) *Diseases in Asian Aquaculture* II. Fish Health Section, Asian Fish. Soc., Manila, 65-79.

Flegel, T.W, 1997. Special topic review: Major viral diseases of the black tiger prawn (*Penaeus monodon*) in Thailand. *World J. Microbiology and Biotechnology* 13:433-442.

Flegel T W, Sriurairatana S, Wongteerasupaya C, Boonsaeng V, Panyim S, Withyachumnarnkul B, 1997b. Progress in characterization and control of yellow-head virus of *Penaeus monodon. In:* Flegel TW, Menasveta P, Paisarnrat S *(eds)* Shrimp *Biotechnology in Thailand,*. 71–78. *National Center for Genetic Engineering and Biotechnology, Bangkok.*

Flegel, T.W., Lightner, D.V., Lo, C.F. and Owens, L. 2008. Shrimp disease control: past, present and future, pp. 355-378. In Bondad-Reantaso, M.G., Mohan, C.V., Crumlish, M. and Subasinghe, R.P. (eds.). Diseases in Asian Aquaculture VI. Fish Health Section, Asian Fisheries Society, Manila, Philippines 505.

Gopalan U K, Meenakshikunjamma P P, Purushan K S, 1980. Fungal infection in the tiger prawn (*Penaeus monodon*) and in other crustaceans from the Cochin backwaters. *Mahasagar Bull Natl Inst Oceanogr* 13 : 359-365.

Gunalan B, P. Soundarapandian, T. Anand, Anil S. Kotiya, Nina Tabitha Simon, 2014. Disease Occurrence in *Litopenaeus vannamei* Shrimp Culture Systems in Different Geographical Regions of India. *International Journal of Aquaculture.* 4,4: 24-28.

Harris LJ, Owens L, 1999. Production of exotoxins by two luminous *Vibrio harveyi* strains known to be primary pathogens of *Penaeus monodon* larvae. *Diseases of Aquatic Organisms* 38 : 11-22

Hasson K.W., D.V. Lightner, B.T. Poulos, R.M. Redman, B.L. White, J.A. Brock, and J.R. Bonami, 1995. Taura Syndrome in *Penaeus vannamei*: Demonstration of a viral etiology. *Diseases of Aquatic Organisms* 23:115-126.

Hasson KW, Redman RM, Mari J, Lightner DV, 1997. Lesion development in *Penaeus vannamei* juveniles infected with Taura syndrome virus: determination by *in situ* hybridization with TSV-specific genomic probes. *Book of abstracts,* Linking science to sustainable industry development, Ann. Int. Con. and Exp. of the World *Aquaculture.* Soc.,19-23, 1997, Seattle, WA, 204-205.

He Lin, Hai-sheng Xu, 2011. Development of a multiplex loop-mediated isothermal amplification (mLAMP) method for the simultaneous detection of white spot syndrome virus and infectious hypodermal and hematopoietic necrosis virus in penaeid shrimp. *Aquaculture*. 311, (1–4):94–99.

Hossain Md.Shahadat, Anirban Chakraborty, Biju Joseph, S.K. Otta, Indrani Karunasagar, Iddya Karunasagar, 2001. Detection of new hosts for white spot syndrome virus of shrimp using nested polymerase chain reaction *Aquaculture*. 198, 1–2, 1–11.

Hossain , Md. Sh., Otta, S.K., Chakraborty, A.,Kumar, H. S., Karunasagar, I.andKarunasagar, I.,2004.Detection of WSSV in cultured shrimps, captured brooders, shrimp postlarvae and water samples in Bangladesh by PCR using different primers. *Aquaculture*, 237, 59–71.

Hsu H C, Liu K F, Su M S, Kou G H, 1999. Studies on effective PCR screening strategies for white spot syndrome virus (WSSV) detection in *Penaeus Smonodon* brooders. *Diseases of Aquatic Organisms* 39: 13-19.

Hsu YL, Wang KH, Yang YH, Tung MC, Hu CH, Lo CF, Wang CH, Hsu T, 2000. Diagnosis of *Penaeus monodon*-type baculovirus by PCR and by ELISA of occlusion bodies. *Diseases of Aquatic Organisms* 40: 93-99.

Huang J, X. L. Song, J. Yu, and C.H. Yang, 1995. Baculoviral Hypodermal and Hematopoietic Necrosis - study on the pathogen and pathology of the explosive epidemic disease of shrimp. *Mar. Fish Res.* 16: 1-10.

Jaroenram W, Wansika Kiatpathomchai, Timothy W. Flegel, 2009. Rapid and sensitive detection of white spot syndrome virus by loop-mediated isothermal amplification combined with a lateral flow dipstick. *Molecular and Cellular Probes*. 23,2: 65–70.

Jimenez R, 1992. Syndrome de Taura (Resuman). *Aquaculture del Equador* 1: 1-16

Johnson PT, Lightner DV, 1988. The rod-shaped nuclear viruses of crustaceans: gut-infecting species. *Diseases of Aquatic Organisms* 4: 123- 141.

Johnson S.K., 1995, *Hand book of shrimp disease. Aquaculture*, Department of Wildlife and fisheries service, Texas A and M University.

Kalagayan G., D. Godin, R. Kanna, G. Hagino, J. Sweeney, J. Wyban, and J. Brock, 1991. IHHN virus as an etiological factor in Runt-Deformity Syndrome of juvenile *Penaeus vannamei* cultured in Hawaii. *J. World Aquaculture Soc.* 22: 235-243.

Karunasagar I, Pai R, Malathi G R, Karunasagar I, 1994. Mass mortality of *Penaeus monodon* larvae due to antibiotic resistant *Vibrio harveyi* infection. *Aquaculture* 128: 203-209.

Karunasagar I., S. K. Otta and I. Karunasagar, 1997. Histopathological and bacteriological study of white spot syndrome of *Penaeus monodon* along the west coast of India. *Aquaculture* 153:9–13.

Karunasagar I., S. K. Otta, and I. Karunasagar, 1998. Disease problems affecting cultured penaeid shrimp in India. *Fish Pathol.* 33: 413-419.

Karunasagar I, Indrani Karunasagar, and R. K. Umesha 2005. "13. Microbial Diseases in Shrimp Aquaculture". Department of Fishery Microbiology, University of Agricultural Sciences, College of Fisheries, Mangalore, India.

Kim Ji Hyung, Casiano H. Choresca Jr., Sang Phil Shin, Jee Eun Han, Jin Woo Jun, Sang Yoon Han, Se, 2011. Chang Park Detection of infectious hypodermal and hematopoietic necrosis virus (IHHNV) in *Litopenaeus vannamei* shrimp cultured in South Korea. *Aquaculture*.313: 1–4, 161–164.

Kiatpathomchai W, Wansadaj Jaroenram, Narong Arunrut, Sarawut Jitrapakdee, T.W. Flegel, 2008. Shrimp Taura syndrome virus detection by reverse transcription loop-mediated isothermal amplification combined with a lateral flow dipstick. *Journal of Virological Methods*. 153,2: 214–217.

Kono T, Savan R, Sakai M, Itami T, 2004. Detection of white spot syndrome virus in shrimp by loop-mediated isothermal amplification. *J Virol Methods*.115,1:59-65.

Lan Y, Xu X, Yang F, Zhang X, 2006. Transcriptional profile of shrimp white spot syndrome virus (WSSV) genes with DNA microarray. *Arch Virol.* 151:1723–1733.

Lavilla-Pitogo, C.R., Baticados, M.C.L., Cruz-Lacierda, E.R., de la Pena, R. 1990. Occurence of luminous bacterial disease of *Penaeus monodon* larvae in the Philippines. *Aquaculture* 91:1-1

Lester, R. J G, A. Doubrovsky, J. L. Paynter, S.K. Sambhi and J.G. Atherton. 1987. Light and electron microscopic evidence of baculovirus infection in the prawn *Penaeus plebejus*. *Dis. Aquat. Org*. 3:217-219.

Li S. H, X. J. Zhang, Z. Sun, F. H. Li, and J. H. Xiang, 2013. "Transcriptome analysis onChinese shrimp *Fenneropenaeus chinensis* during WSSV acute infection," *Plos ONE*,. 8,3 :58627.

Lightner, D.V. and R.M. Redman, 1981. A baculovirus caused disease of the penaeid shrimp *Penaeus monodon*. *J. Invertebr. Pathol.* 38:299-302.

Lightner, D.V., R.M. Redman, and T.A. Bell, 1983. Infectious Hypodermal and Hematopoietic Necrosis a newly recognized virus disease of penaeid shrimp. *J. Invertebr. Pathol.* 42: 62 70

Lightner, D.V. and R.M. Redman, 1985. A parvo-like virus disease of penaeid shrimp. *J. Invertebr. Pathol.* 45:47-49.

Lightner DV, Hedrick RP, Fryer JL, Chen SN, Liao IC, Kou GH, 1987. A survey of cultured penaeid shrimp in Taiwan for viral and other important diseases. *Fish Pathology* 22*:* 127–140.

Lightner D.V, 1988. Diseases of cultured penaeid shrimp and prawns. pp. 8-127, in: Disease Diagnosis and Control in North American Marine *Aquaculture*, C.J. Sindermann & D.V. Lightner (eds.) *Elsevier*, Amsterdam, The Netherlands.

Lightner D V, Redman R M, 1991. Hosts, geographic range and diagnostic procedures for the penaeid virus diseases of concern to the shrimp culturists in the Americas. In: DeLoach PF, Dougherty WJ, Davidson MA (eds) Frontiers in shrimp Research. Elsevier, Amsterdam, 173-196.

Lightner, D.V., T.A. Bell, R.M. Redman, and L.A. Perez, 1992. A collection of case histories documenting the introduction and spread of the virus disease IHHN in penaeid shrimp culture facilities in Northwestern Mexico. *ICES Marine Science Symposia* 194: 97-105.

Lightner DV, Redman RM, 1992. Penaeid virus diseases of the shrimp culture industry of the Americas. In: Fast AW, Lester LJ (eds) Marine shrimp culture: Principles and Practices. *Elsiever,* Amsterdam, 569-588.

Lightner D V, Jones L S, Ware G W, 1994. Proceedings of the Taura syndrome workshop: executive summary, submitted reports, and transcribed notes. Univ Arizona, Tucson.

Lightner, D.V., R.M. Redman, K.W. Hasson, and C.R. Pantoja, 1995. Taura Syndrome in *Penaeus vannamei* (Crustacea: Decapoda): gross signs, histopathology and ultrastructure. *Diseases of Aquatic Organisms* 21: 53-59.

Lightner, D.V. (ed.) 1996. A handbook of shrimp pathology and diagnostic procedures for diseases of cultured penaeid shrimp. World *Aquaculture* Society, Baton Rouge, LA., USA, 305 p.

Lightner DV, Redman RM, Poulos BT, Nunan LM, Mari JL, Hasson KW, 1997. Risk of spread of penaeid shrimp viruses in the Americas by international movement of live and frozen shrimp. Revue Scientifique et Technique, Office International des Epizooties 16*:* 146–160.

Lightner, D.V. and Redman, R.M, 1998. Strategies for the control of viral diseases of shrimp in the Americas. *Fish Pathology* 33: 165-180

Lightner DV, Hasson KW, White BL, Redman RM, 1998. Experimental infection of western hermisphere Penaeid shrip with Asian white spot syndrome virus and Asian Yellow Head Virus. *J Aquat Anim Health* 10: 271- 281.

Lightner, D.V., S.V. Durand, R.M. Redman, L.L. Mohney, and K. Tang-Nelsonm, 2001. Qualitative and quantitative studies on the relative virus load of tails and heads of shrimp acutely infected with WSSV: implications for risk assessment. pp. 285-291, in C.L. Browdy and D.E. Jory (eds.), The New Wave, Proceedings of the Special Session on Sustainable Shrimp Culture, *Aquaculture* 2001. The World *Aquaculture* Society, Baton Rouge, LA, USA.

Lightner, D.V, 2003. The penaeid shrimp viral pandemics due to IHHNV, WSSV, TSV and YHV: history in the Americas and current status. *In:* Y. Sakai, J.P. McVey, D. Jang, E. McVey and M. Caesar (editors), Proceedings of the Thirty-second US Japan Symposium on *Aquaculture*. US-Japan Cooperative Program in Natural Resources (UJNR). U. S. Department of Commerce, N.O.A.A., Silver Spring, MD, USA. pp. 6-24.

Lightner D.V., Pantoja C.R., Poulos B.T., Tang K.F.J., Redman R.M., Pasos De Andrade T. & Bonami J.R. 2004. Infectious Myonecrosis: New Disease In Pacific Whiteshrimp. *Global Aquaculture Advocate*, 7, 85.

Lightner D.V., Redman R.M., Arce S.& Moss S.M. 2009. Specific Pathogen-Free Shrimp Stocks in Shrimp Farming Facilities as a Novel Method for Disease Control in Crustaceans. In: Shellfish Safety and Quality, Shumway S. & Rodrick G., eds. Woodhead Publishers, London, UK. 384–424.

Lightner, D.V, 2011. Status of shrimp diseases and advances in shrimp health management, pp. 121-134. *In* Bondad-Reantaso, M.G., Jones, J.B., Corsin, F. and Aoki, T. (eds.). *Diseases in Asian Aquaculture VII*. Fish Health Section, Asian Fisheries Society, Selangor, Malaysia. 385 pp.

Liu P C, Lu K K, Chen S N, 1996. Pathogenicity of different isolates of *Vibrio harveyi* in tiger prawn, *Penaeus monodon* . *Lett Appl Microbiol* 22: 413-416.

Liu H, K. Soderhall, and P. Jiravanichpaisal, 2009. "Antiviral immunity in crustaceans," *Fish and Shellfish Immunology*. 27, 2:79–88.

Liu H, R. Chen, Q. Zhang, H. Peng, and K.Wang, 2011. "Differential gene expression profile from haematopoietic tissue stem cells of red claw crayfish, *Cherax quadricarinatus*, in response to WSSV infection," *Developmental and Comparative Immunology*. 35. 7:716–724.

Lo, C.F., J.H. Leu, C.H. Ho, C.H. Chen, S.E. Peng, Y.T. Chen, C.M. Chou, P.Y. Yeh, C.J. Huang, H.Y. Chou, C.H. Wang and G.H. Kou, 1996. Detection of baculovirus associated with White Spot Syndrome (WSBV) in penaeid shrimps using polymerase chain reaction. *Diseases of Aquatic Organisms* 25:133-141.

Lo, C.F., C.H. Ho, C.H. Chen, K.F. Liu, Y.L. Chiu, P.Y. Yeh, S.E. Peng, H.C. Hsu, H.C. Liu, C.F. Chang, M.S. Su, C.H. Wang and G.H. Kou, 1997. Detection and tissue tropism of white spot syndrome baculovirus (WSBV) in captured brooders of *Penaeus monodon* with a special emphasis on reproductive organs. *Diseases of Aquatic Organisms*. 30:53-72.

Longyant, S., Sithigorngul, P., Chaivisuthangkura, P., Rukpratanporn, S., Sithigorngul, W. and Menasveta, 2005. Differences in susceptibility of palaemonid shrimp species to yellow head virus (YHV) infection. *Diseases of Aquatic Organisms*. 64:5-12.

Longyant S, Sattaman S, Chaivisuthangkura P, Rukpratanporn S, Sithigorngul W, Sithigorngul P, 2006. Experimental infection of some penaeid shrimps and crabs by yellow head virus (YHV). *Aquaculture* 257*:* 83–91.

Lotz, J.M, 1997. Effect of host size on virulence of Taura virus to the marine shrimp *Penaeus vannamei* (Crustacea: Penaeidae). *Diseases of Aquatic Organisms* 30: 45-51.

Lotz JM, Flowers AM, Breland V, 2003. A model of Taura syndrome virus (TSV) epidemics in *Litopenaeus vannamei*. *Journal of Invertebrate Pathology* 83*:* 168–176.

Lu, Y., L.M. Tapay, J.A. Brock, and P.C. Loh, 1994. Infection of the Yellow Head Baculolike virus (YBV) in two species of penaeid shrimp *Penaeus stylirostris* (Stimpson) and *Penaeus vannamei* (Boone) *Journal Fish Diseases* 17: 649-656.

Lu Y, Tapay LM, Loh PC, Brock JA, Gose R, 1995. Development of a quantal assay in primary shrimp cell culture for yellow head baculovirus (YBV) of penaeid shrimp. *Journal of Virological Methods* 52*:* 231–236.

Lu CF, Nadala ECB Jr, Brock JA, Loh PC, 1991. A new virus isolate from infectious hypodermal and haematopoietic necrosis virus (IHHNV) infected penaeid shrimp. *J Virol Methods* 31 : 189- 196

Lu CC, Tang KFG, Kou GH, Chen SN, 1993. Development of a *Penaeus monodon*-type (MBV) DNA probe by polymerase chain reaction and sequence analysis. *J Fish Dis* 16: 551-559

Lu Y, Tapay L M, Loh P C, 1996. Development of nitrocellulose enzyme immunoassay (NC–EIA) for the detection of yellowhead virus from penaeid shrimp. *J Fish Dis* 19: 9-13.

Mari J, Lightner D V, Poulos B T, Bonami J R, 1995. Partial cloning of the genome of an unusual parvovirus (HPV): use of gene probes in disease diangosis. *Diseases of Aquatic Organisms* 22: 129-134.

Mari, J., J.R. Bonami, and D.V. Lightner. 1998. Taura Syndrome of penaeid shrimp: cloning of viral genome fragments and development of specific gene probes. *Diseases of Aquatic Organisms* 33: 11-17.

Mari, J., B.T. Poulos, D.V. Lightner, and J.R. Bonami. 2002. Shrimp Taura Syndrome Virus: genomic characterization and similarity with members of the genus Cricket paralysis-like viruses. *J. Gen. Virol.* 83: 915-926.

Magbanua FO, Natividad KT, Migo VP, Alfafara CG, De La Pena FO, Miranda RO, Albaladejo JD, Nadala ECB Jr, Loh PC, Tapay LM, 2000. White spot syndrome virus (WSSV) in cultured *Penaeus monodon* in the Philippines. *Diseases of Aquatic Organisms* 42: 77-82.

Mayo MA, 2002. A summary of taxonomic changes recently approved by ICTV. *Arch Virol* 147:1655-1656.

Mohan CV, Shankar KM, Kulkani S, Sudha PM, 1998. Histopathology of cultured shrimp showing gross signs of yellow head syndrome and white spot syndrome during 1994 Indian epizootics. *Diseases of Aquatic Organisms* 34: 9–12.

Momoyama K, 1983. Studies on baculoviral mid-gut gland necrosis virus of kuruma shrimp (*Penaeus japonicus)*, presumptive diagnostic technique. *Fish Pathol* 17: 263-268.

Momoyama K, 1988. Infection source of baculoviral mid-gut gland necrosis (BMN) in mass production of kuruma shrimp larvae, *Penaeus japonicus. Fish Pathol* 23: 105-110.

Munday BL, Owens L, 1998. Viral diseases of fish and shellfish in Australian mariculture. *Fish Pathol.* 33,4:193-200.

Nash G, Anderson IG, Shariff M, 1988. Pathological changes in the tiger prawn *Penaeus monodon* Fabricius, associated with culture in brackish water ponds developed from potentially acid sulphate soils. *J Fish Dis.* 11:113-123.

Nadala ECB Jr, Lu Y, Loh PC, Brock JA, 1992. Infection of *Penaeus stylirostris* (Boone) with a rhabdovirus isolated from *Penaeus spp. Fish Pathol* 27: 143-147.

Nadala ECB Jr, Tapay LM, Cao S, Loh PC, 1997. Detection of yellow head virus and Chinese baculovirus in penaeid shrimp by the western blot technique. *J Virol Meth* 69: 39-44.

Nadala ECB Jr, Loh P C, 2000. Dot-blot nitrocellulose immunoassay for the detection of white spot virus and yellow head virus of penaeid shrimp. *J Virol Meth* 84: 175-179.

Nakano, H., H. Koube, S. Umezawa, K. Momoyama, M. Hiraoka, K. Inouye, and N. Oseko. 1994. Mass mortalities of cultured Kuruma shrimp, *Penaeus japonicus*, in Japan in 1993: Epizootiological survey and infection trails. *Fish Pathol.* 29: 135-139.

Natividad JM, Lightner DV, 1992. Prevalence and geographical distribution of MBV and other diseases in cultured gaint tiger prawn (*Penaeus monodon*) in The Philippines. In: Fulks W, Main KL (eds) Diseases of cultured penaeid shrimp in Asia and the United States. The Oceanic Institute, Makapuu Point, Honolulu, Hawaii, 139-160.

Nielsen, L., Sang-oum, W., Cheevadhanarak, S. and Flegel, T.W. 2005. Taura syndrome virus (TSV) in Thailand and its relationship to TSV in China and the Americas. *Diseases of Aquatic Organisms.* 63:101-106.

Nimitphak T, Watcharachai Meemetta, Narong Arunrut, Saengchan Senapin, Wansika Kiatpathomchai, 2010. Rapid and sensitive detection of *Penaeus monodon* nucleopolyhedro virus (PemoNPV) by loop-mediated isothermal amplification combined with a lateral-flow dipstick. *Molecular and Cellular Probes.* 24, 1:1–5.

Nunan LM & Lightner DV, 1997. Development of a non- radioactive gene probe by PCR for detection of white spot syndrome virus (WSSV). *J Virol Methods* 63:193 –201.

Nunan, L.M., B.T. Poulos, and D.V. Lightner, 1998. The detection of White Spot Syndrome Virus (WSSV) and Yellow Head Virus (YHV) in imported commodity shrimp. *Aquaculture* 160: 19-30.

Nunan, L.M., B.T. Poulos, and D.V. Lightner. 2000. Use of polymerase chain reaction (PCR) for the detection of Infectious Hypodermal and Hematopoietic Necrosis Virus (IHHNV) in penaeid shrimp. *Mar. Biotechnol.* 2: 319-328.

Otta SK, Karunasagar I, Karunasagar I, 1999. Bacterial flora associated with shrimp culture ponds growing *Penaeus monodon* in India. *J Aqua Trop* 14,4: 309-318.

Otta SK, Karunasagar I, Karunasagar I, 2001. Bacteriological study of shrimp, *Penaeus monodon* Fabricius hatcheries in India. *J Appl Ichthyol* 17: 59-63

Otta SK, Karunasagar I, Karunasagar I, 2003. Detection of Monodon Baculo Virus (MBV) and White Spot Syndrome Virus (WSSV) in apparently healthy *Penaeus monodon* from India by polymerase chain reaction. *Aquaculture* 220: 59-69.

Orndorff, S.A. and Colwell, R.R. 1980. Distribution and identification of luminous bacteria from the Sargasso Sea. *Appl Environ Microbiol* 39, 983–987.

Overstreet, R.M., D.V. Lightner, K.W. Hasson, S. McIlwain, and J. Lotz. 1997. Susceptibility to TSV of some penaeid shrimp native to the Gulf of Mexico and southeast Atlantic Ocean. *J. Invertebr. Pathol.* 69: 165-176.

Owens L, De Beer S, Smith J R, 1991. Lymphoidal parvo-like virus from Australian penaeid prawns. *Diseases of Aquatic Organisms.* 11: 129-228.

Owens L, 1993. Description of the first haemocytic rod-shaped virus from a penaeid prawn. *Diseases of Aquatic Organisms.* 16:217-221.

Pantoja CR, Lightner DV, 2000. A non-destructive method based on the polymerase chain reaction for detection of hepatopancreatic parvovirus (HPV) of penaeid shrimp. *Diseases of Aquatic Organisms* 39: 177-182.

Pasharawipas T, Sriurairatana S, Direkbasarakom S, Donayadol Y, Thaikua S, Ruangpan L, Flegel TW, 1998. Luminous *Vibrio harveyi* associated with tea brown gill syndrome in black tiger shrimp. In: Flegel TW (ed) *Advances in Shrimp Biotechnology*, National Centre for Genetic Engineering and Biotechnology, Bangkok, 213-216.

Pattin K. A. and J. H. Moore, 2009. “Role for protein-protein interaction databases in human genetics,” *Expert Review of Proteomics*, 6, 6: 647–659.

Paynter J L, Lightner D V, Lester R J G, 1985. Prawn virus from juvenile *Penaeus esculentus*. In: Rothlisberg P C, Hill B J, Staples D J (eds) Second Australian National Prawn Seminar, NPS2, Cleveland. Australia, 61-64.

Poulos B T, Mari J, Bonami J R, Redman R M, Lightner D V, 1994. Use of non-radioactively labeled DNA probes for the detection of a baculovirus from *Penaeus monodon* by *in situ* hybridization on fixed tissue. *J Virol Meth* 49: 187-194.

Poulos, Kathy F. J. Tang, Carlos R. Pantoja, Jean Robert Bonami and Donald V. Lightner Bonnie T, 2006. Purification and characterization of infectious myonecrosis virus of penaeid shrimp. *Journal of General Virology*. 87: 987–996.

Phromjai, J., Boonsaeng, V., Withyachumnarnkul, V. and Flegel, T.W. 2002. Detection of hepatopancreatic parvovirus in Thai shrimp *Penaeus monodon* by in situ hybridization, dot blot hybridization and PCR amplification. *Diseases of Aquatic Organisms*. 51:227-232.

Puthawibool T, Saengchan Senapin, Wansika Kiatpathomchai, Timothy W. Flegel, 2009. Detection of shrimp infectious myonecrosis virus by reverse transcription loop-mediated isothermal amplification combined with a lateral flow dipstick. *Journal of Virological Methods*. 156, 1–2: 27–31.

Ramasamy P, Brennan GP, Jaykumar R, 1995. A record and prevalence of monodon baculovirus from postlarval *Penaeus monodon* in Madras, India. *Aquaculture* 130: 129-135.

Ramasamy P, Rajan PR, Jayakumar R, Rani S, Brenner GP, 1996. *Lagenidium callinectes* (Couch, 1942) infection and its control in cultured larval Indian tiger prawn, *Penaeus monodon* Fabricus. *J Fish Dis* 19: 75-82.

Remmerie N. T. de Vijlder, K. Laukens, Dang TH, Lemière F, Mertens I, Valkenborg D, Blust R, Witters E, 2011 "Next generation functional proteomics in non-model plants: a survey on techniques and applications for the analysis of protein complexes and post-translational modifications," *Phytochemistry*, 72, 10:1192–1218.

Rohrmann GF, 1986. Polyhedrin structure. *J Gen Virol.* 67: 1499-1513.

Sahul Hameed AS, Anilkumar M, Stephen Raj ML, Jayaraman K, 1998. Studies on the pathogenicity of systemic ectodermal and mesodermal baculovirus and its detection in shrimp by immunological methods. *Aquaculture* 160: 31-45

Sano T, Nishimura T, Oguma K, Momoyama K, Takeno N, 1981. Baculovirus infection of cultured kuruma shrimp *Penaeus japonicus* in Japan. *Fish Pathol* 15: 185-191.

Sano T, Nishimura T, Fukuda H, Hayashida T, Momoyama K, 1984. Baculoviral mid-gut gland necrosis (BMN) of kuruma shrimp (*Penaeus japonicus*) larvae in Japanese intensive culture systems. *Helgolander Meeresunters* 37: 255-264.

Sano T, Nishimura T, Fukuda H, Hayashida T, Momoyama K, 1985. Baculoviral infectivity trials on kuruma shrimp larvae, *Penaeus japonicus*, of different ages. In: Ellis AE (ed) *Fish and shellfish pathology*, Academic Press, London, 397-403.

Sano T, Momoyama K, 1992. Baculovirus infection of penaeid shrimp in Japan. In: Fulks W,

Main KL (eds) Diseases of cultured penaeid shrimp in Asia and the United States. The Oceanic Institute Makapuu Point Honolulu Hawaii. 169-174.

Senapin S *and* Phongdara A, 2006. Binding of shrimp cellular proteins to Taura syndrome viral capsid proteins VP1, VP2 and VP3. *Virus Research* 122*:* 69–77.

Senapin S, Phewsaiya K, Briggs M, Flegel T W, 2007. Outbreaks of infectious myonecrosis virus (IMNV) in Indonesia confirmed by genome sequencing and use of an alternative RT-PCR detection method. *Aquaculture* 266*:* 32–38.

Sappat A, Jaroenram W, Puthawibool T, Lomas T, Tuantranont A, Kiatpathomchai W. 2011. Detection of shrimp Taura syndrome virus by loop-mediated isothermal amplification using a designed portable multi-channel turbidimeter. *J Virol Methods*; 175, 2:141-8.

Shike H, Dhar AK, Burns JC, Shimizu C, Jousset FX, Klimpel KR, Bergoin M, 2000. Infectious hypodermal and hematopoietic necrosis virus of shrimp is related to mosquito brevidenso viruses. *Virology*; 277: 167-177.

Soowannayan, C., Flegel, T.W., Sithigorngul, P., Slater, J., Hyatt, A., Cramerri, S., Wise, T., Crane, M.S.J., Cowley, J.A., McCulloch, R.J. and Walker, P.J., 2003. Detection and differentiation of yellow head complex viruses using monoclonal antibodies. *Diseases of aquatic organisms*, 57, 3:193-200.

Spann KM, Vickers JE, Lester RJG, 1995. Lymphoid organ virus of *Penaeus monodon* from Australia. *Diseases of Aquatic Organisms* 23: 127- 134.

Spann KM, Lester RJG, 1997. Special topic review: viral deseases of penaeid shrimp with particular reference to four viruses recently found in shrimps from Queensland. *World J Microbiol Biotech* 13: 419-426.

Spann KM, Cowly JA, Walker PJ, Lester RJG, 1997. A yellow head virus from *Penaeus monodon* cultured in Australia. *Diseases of Aquatic Organisms* 31: 169-179.

Spann K M, Donaldson R A, Cowley J A, Walker P J, 2000. Differences in susceptibility of some penaeid prawn species to gill-associated virus (GAV) infection. *Diseases of Aquatic Organisms* 42: 221- 225.

Srisuvan T, Tang K F, Lightner D V, 2005. Experimental infection of *Penaeus monodon* with Taura syndrome virus (TSV). *Diseases of Aquatic Organisms* 67*:* 1–8.

Srisuvan T, Pantoja C R, Redman R M, Lightner D V, 2006a. Ultrastructure of the replication site in Taura syndrome virus (TSV)-infected cells. *Diseases of Aquatic Organisms* 73: 89–101.

Subasinghe R, Bartly D M, Mcgladdery S, Barg U, 1998. Sustainable shrimp culture development : biotechnological issues and challenges. In : Flegel TW (ed) *Advances in shrimp*

biotechnology : National Centre for Genetic Engineering and Biotechnology, Bangkok, p 13-18.

Suebsing R, Prombun P, Srisala J, Kiatpathomchai W, 2013. Loop-mediated isothermal amplification combined with colorimetric nanogold for detection of the microsporidian Enterocytozoon hepatopenaei in penaeid shrimp. *J Appl Microbiol.*114,5:1254-63.

Sukanya P, Janenuj W, Meena S, Nareerat V, 2006. The molecular detection of Taura syndrome virus emerging with White spot syndrome virus in penaeid shrimps of Thailand. *Aquaculture.* 260, 1–4, 77–85.

Sukhumsirichart W, Wongteerasupaya C, Boonsaeng V, Panyim S, Sriurairatana S, Withyachumnarnkul B, Flegel T W, 1999. Characterization and PCR detection of hepatopancreatic parvovirus (HPV) from *Penaeus monodon* in Thailand. *Diseases of Aquatic Organisms.* 38: 1-10.

Sun Zheng, Shihao Li, Fuhua Li and Jianhai Xiang, 2014. Bioinformatic Prediction of WSSV-Host Protein-Protein Interaction. *BioMed Research International*, ID 416543, 9.

Sunarto A, Widodo , Taukhid , Koesharyani I, Supriyadi H, Gardenia L, Rukmono, D, 2004. Current status of transboundary fish diseases in Indonesia: occurrence, surveillance, research and training. *In:* Lavilla-Pitogo CR, Nagasawa K *(eds)* Transboundary Fish Diseases in Southeast Asia: Occurrence, Surveillance, Research and Training, *pp.* 91–121. *SEAFDEC Aquaculture Department, Tigbauan, Iloilo.*

Syed Musthaq, S., Yoganandhan, K., Sudhakaran, R. Rajesh Kumar. S and Sahul Hameed, A.S. 2006. Neutralization of white spot syndrome virus of shrimp by antiserum raised against recombinant VP28. *Aquaculture*, 253, 98-104.

Takahashi Y, Itami T, Kondo M, Maeda M, Fujii R, Tomonaga S, Supamattaya K, Boonyaratpalin S, 1994. Electron microscopic evidence of bacilliform virus infection in kuruma shrimp (*Penaeus japonicus*). *Fish Pathol*; 29: 121-125.

Takahashi Y, Itami T, Maeda M, Suzuki N, Kasornchandra J, Supamattaya K, Khongpradit R, Boonyaratpalin S, Kondo M, Kawai K, Kusuda R, Hirono I, Aoki T, 1996. Polymerase chain reaction (PCR) amplification of bacilliform virus (RV-PJ) DNA in *Penaeus japonicus* Bate and systemic ectodermal and mesodermal baculovirus (SEMBV) DNA in *Penaeus monodon* Fabricus. *J Fish Dis* 19: 399-403.

Takahashi Y, Itami T, Maeda M, Kondo M, 1998. Bacterial and viral diseases of kuruma shrimp (*Penaeus japonicus*) in Japan. *Fish Pathol* 33,4: 357-364.

Tang, K.F.J., and D.V. Lightner. 1999. A Yellow-Head virus gene probe: application to in situ hybridization and determination of its nucleotide sequence. *Diseases of Aquatic Organisms.* 35: 165173.

Tang K F J, Pantoja C R, Poulos B T, Redman R M, Lightner D V, 2005. *In situ* hybridization demonstrates that *Litopenaeus vannamei, L. stylirostris* and *Penaeus monodon* are susceptible to experimental infection with infectious myonecrosis virus (IMNV). *Diseases of Aquatic Organisms* 63: 261–265.

Tsing, A. and J.R Bonami, 1987. A new viral disease of the shrimp, *Penaeus japonicus* Bate. *J Fish Dis*, 10: 139-141.

Tu C, Huang HT, Chuang SH, Hsu JP, Kuo ST, Li NJ, Hsu TL, Li MC, Lin SY, 1999. Taura syndrome in Pacific white shrimp *Penaeus vannamei* cultured in Taiwan. *Diseases of Aquatic Organisms*; 38: 159–161.

Umesha KR, Uma A, Otta SK, Karunasagar I, Karunasagar I, 2003. Detection by PCR of hepatopancreatic parvovirus (HPV) and other viruses in hatchery-reared *Penaeus monodon* postlarvae. *Diseases of Aquatic Organisms* 57: 141-146.

Van V K, 2004. Current status of transboundary fish diseases in Vietnam: occurrence, surveillance, research and training. *In:* Lavilla-PitogoCR, NagasawaK *(eds)* Transboundary Fish Diseases in Southeast Asia: Occurrence, Surveillance, Research and Training. 221–227. *SEAFDEC Aquaculture Department, Tigbauan, Iloilo.*

van Hulten, M.C., Witteveldt, J., Peters, S., Kloosterboer, N., Tarchini, R., Fiers, M., Sandbrink, H., Lankhorst, R.K. and Vlak, J.M. 2001. The white spot syndrome virus DNA genome sequence. *Virology.* 286:7-22.

Vickers JE, Lester RJG, Spradbrow PB, Pemberton JM, 1992. Detection of *Penaeus monodon*-type baculovirus (MBV) in digestive glands of postlarval prawn using polymerase chain reaction. In: Shariff, M., Subhasinghe, R. P. and Arthur, J. R. (Eds.) *Diseases in Asian Aquaculture I. Fish Health Section Asian Fish Soc Manila.* 127-133.

Vickers JE, Paynter JL, Spradbrow PB, Lester RJG, 1993. An impression smear method for rapid detection of *Penaeus monodon*-type baculovirus (MBV) in Australian prawns. *J Fish Dis.* 16: 507-511.

Vijayan KK, Alavandi SV, Rajendran KV, Alagarswami K, 1995. Prevalence and histopathology of monodon baculovirus (MBV) infection in *Penaeus monodon* and *P. indicus* in shrimp farms in the south-east coast of India. *Asian Fish Sci.* 8,3: 267-272.

Vijayan, K.K., V. Stalin Raj, C. P. Balasubramanian, S. V. Alavandi, V. Thillai Sekhar,, and T.C. Santiago. 2005. Polychaete worms——a vector for white spot syndrome virus (WSSV). *Diseases of Aquatic Organisms* 63: 107-111.

Walker PJ, Cowley JA, Spann KM, Hodgson RAJ, Hall MR, Withychumnarnkul B, 2001. Yellow head complex viruses: transmission cycles and topographical distribution in the Asia- Pacific region. In: Browdy CL, Jory DE, editors. The new wave: Proceedings of the Special Session on Sustainable Shrimp Culture, *Aquaculture*. The World *Aquaculture* Society: Baton Rouge. 227–237.

Walker, P. J. and Mohan, C. V. 2009. Viral disease emergence in shrimp *Aquaculture*: origins, impact and the effectiveness of health management strategies. Reviews in *Aquaculture*, 1: 125–154.

Wang, C., Lo, C., Leu, J., Chou, C., Yeh, P., Chou, H., Tung, M., Chang, C., Su, M. and Kou, G.,1995. Purification and genomic analysis of baculovirus associated with white spot syndrome (WSBV) of *Penaeus monodon. Diseases of Aquatic Organisms.* 23: 239-242.

Wang SY, Hong C, Lotz TM, 1996. Development of PCR procedure for the detection of Baculovirus penaei in shrimp. *Diseases of Aquatic Organisms.* 25: 123-131.

Wang YC, Lo CF, Chang PS, Kou GH, 1998. Experimental infection of white spot baculovirus in some cultured and wild decapods in Taiwan. *Aquaculture*, 164: 221-231.

Wang B, F. Li, B.Dong, X. Zhang, C. Zhang, and J. Xiang, 2006. "Discovery of the genes in response to white spot syndrome virus (WSSV) infection in *Fenneropenaeus chinensis* through cDNA microarray," *Marine Biotechnolog*. 8, 5491–500.

Wang YC, Chang PS, 2000. Yellow head virus infection in the giant tiger prawn *Penaeus monodon* cultured in Taiwan. *Fish Pathology.* 35: 1–10.

Wijegoonawardane P.K.M., J.A. Cowley, T. Phan, R.A.J Hodgson, L. Nielsen, W. Kiatpathomchai, P.J. Walker, 2008a. Genetic diversity in the yellow head nidovirus complex. *Virology,* 380: 213–225.

Wongteerasupaya, C, J.E. Vickers, S. Sriurairatana, G.L. Nash, A. Akarajamorn, V. Boonsaeng, S. Panyim, A. Tassanakajon, B. Withyachumnarnkul, and T.W. Flegel. 1995. A non-occluded, systemic baculovirus that occurs in cells of ectodermal and mesodermal origin and causes high mortality in the black tiger prawn, *Penaeus monodon. Diseases of Aquatic Organisms* 21: 69-77.

Wongteerasupaya C, Wongwisansri S, Boonsaeng V, Panyim S, Pratanpipat P, Nash G L, Withyachumnarnkul B, Flegel T W, 1996. DNA fragment of *Penaeus monodon* baculovirus PmNOBII gives positive in situ hybridization with white spot viral infections in six penaeid shrimp species. *Aquaculture* 143: 23-32.

Wongteerasupaya C, Tongcheua W, Boonsaeng V, Panyim S, Tassanakajon A, Withyachumnarnkul B, Flegel TW, 1997. Detection of yellow head virus (YHV) of *Penaeus monodon* by RT-PCR amplification. *Diseases of Aquatic Organisms* 31: 181-186.

Yang F, He J, Lin X, Li Q, Pan D, Zhang X, Xu X, 2001. "Complete genome sequence of the shrimp white spot bacilliform virus," *Journal of Virology.* 75,23: 11811–11820.

Yoganandhan, K., Syed Musthaq, S., and Sahul Hameed, A.S. 2004. Production of polyclonal antiserum against recombinant VP28 protein and its application for the white spot syndrome virus in crustaceans. *Journal of Fish Diseases*, 27: 517-522.

Yu C I, and Song Y L, 2000. Outbreaks of Taura syndrome in Pacific white shrimp *Penaeus vannamei* cultured in Taiwan. *Fish Pathol* 35: 21–24.

Zhao Z, Z. Yin, S. Weng, Hao-Ji Guan, Se-Dong Li, Ke Xing, Siu-Ming Chan, Jian-Guo He, 2007. "Profiling of differentially expressed genes in hepatopancreas of white spot syndrome virus-resistant shrimp (*Litopenaeus vannamei*) by suppression subtractive hybridisation," *Fish and Shellfish Immunology*, 22, 5: 520–534.

CHAPTER 12

DISEASE CONTROL IN SHRIMPS AND THERAPEUTICS

12.1 INTRODUCTION

As per the FAO estimates half of the world's seafood demand will be met by aquaculture in 2020, as wild capture fisheries are overexploited and marine fish production is either declining or remained static. Shrimp aquaculture is widespread throughout the tropical world. It is in an industry set for a period of strongly growing demand, and is currently worth around US$10 billion. Shrimp aquaculture is a multi-billion dollar industry contributing a major income to several countries. The rapid growth of shrimp industry has led to an economic boom but, unfortunately, the outbreak of diseases has increased the economic risks and slowed the industry development (Flegel, 2006) (Fig. 1). The most important diseases of cultured penaeid shrimp, in terms of economic impact, have infectious aetiologies. Among the infectious diseases of cultured shrimp, certain virus-caused diseases are the most significant. The epidemics due to the penaeid viruses like WSSV and TSV, and to a lesser extent to IHHNV (Infectious Hypodermal and Hematopoietic Necrosis virus) and YHV (Yellow Head Virus), have cost the penaeid shrimp industry billions of dollars in lost crops, jobs and export revenue. The social and economic impacts of the epidemics caused by these pathogens have been profound in countries in which shrimp farming constitutes a significant industry.

The risk of disease in shrimp farming often increases with culture intensity and high stocking densities, and when polyculture is replaced by monoculture. High pond densities will facilitate the spread of pathogens between ponds. Shortage of clean water supply and insufficient waste removal lead to overloading of metabolites, environmental degradation, and to the shrimp becoming stressed by bad water quality, and thus more prone to becoming affected by disease. Excessive fluctuations in abiotic factors like oxygen, salinity, and temperature may also increase stress and susceptibility to disease. The location of farms in mangrove environments can lead to acidification that may directly, or indirectly, through release of heavy metals from the sediments, lower disease resistance. The use of hatchery-reared larvae will increase genetic uniformity and thus disease risk is reduced in comparison to the collection of wild larvae where selection has already favoured the most viable individuals. Transportation of seed larvae and brood stock from different environments will facilitate the spread of pathogens. Apart from the above factors, which are all dependant on the farming itself, contamination by pesticides and pollutants from agricultural fields and industrial activities may lower disease resistance of the shrimp, especially if combined with other environmental factors.

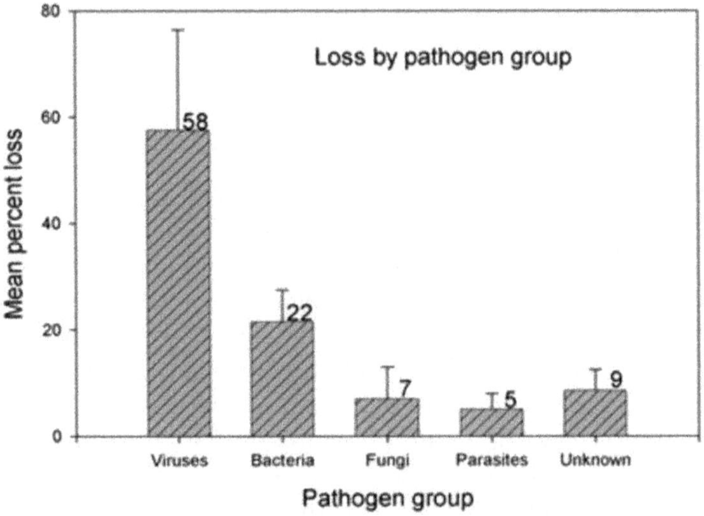

Fig. 1. Estimated losses to various pathogens in shrimp aquaculture according to the Global Aquaculture Alliance survey: (**Source:** Flegel, 2006a).

The sustainability and development of shrimp aquaculture are largely at stake as significant ecological and pathological problems are increasing in the vast majority of the shrimp producing countries. Prevention and control of diseases are now the priority for the durability of this industry. Within the past decade, intensification of the shrimp production, based on progress in culture technology, has increased but corresponding increase in understanding the scientific knowledge of shrimp physiology has not progressed much. Within this field, shrimp immunology is a key element in establishing strategies for the control of diseases in shrimp aquaculture.

12.2 DISEASE CONTROL AND THERAPEUTICS

Flegel *et al.*, (2008) while discussing shrimp disease control methods gave a very detailed account on new diagnostic methods and promising new directions based on new scientific discoveries. They have suggested use of various probiotics, immunostimulants, quorum sensing control of bacterial virulence, phage therapy, shrimp –virus interactions, shrimp vaccines, RNA interference, antibacterial and antiviral substances in shrimp, molecular epidemiology, and shrimp breeding and selection, as promising new possible methods for controlling various diseases in shrimps. While mentioning new diagnostic methods, they mentioned about the use of lateral flow chromatographic immunodiagnostic strips to easily diagnose shrimp disease outbreaks at the pond site. These strips are relatively cheap and can give results within 10 minutes. Alday *et al.*, (2009) while discussing the designing a biosecurity plan for shrimp farming, gave a detailed account on various aspects of management and precautions to be taken to prevent occurrence and spread of viral and bacterial diseases both in hatchery and culture systems. In their paper they dealt with important shrimp viruses and importance of persistent viral infection in various species of cultivated shrimps. They have mentioned that shrimps generally carry one or more viral pathogens as persistent infection for a long period without gross signs of disease. These viruses can be passed on to other shrimps resulting in disease formation or sometimes these viruses spread on to the brood stock shrimps and to their larvae and post-larvae. To avoid cross transfer of viruses, they suggested to rear individual species of shrimps separately and to prevent the entry of potential carrier species in to shrimp cultivation ponds by adopting appropriate biosecurity plan. Number of worker have given different plans and protocols for control of WSSV, YHV, TSV, Infectious **Myonecrosis Virus** (INMV), *Penaeus monodon* **densovirus** (PmDNV), and *Penaeus monodon* **polyhedron virus** (PmNPV).

Flegel *et al.*, (2006) reported that **WSSV** can cause high mortality among all cultivated penaeid shrimp species and in order to prevent the occurrence of WSSV they have suggested to use domesticated specific pathogen free stocks (SPF) in the hatchery and culture system for the development of sustainable aquaculture. Another suggestion is, for spawning and rearing of shrimps, use of individual specimen which is SPF is recommended. However, even with these precautions PCR testing of the post larvae has been recommended (Thakur *et al.*, 2002). Before stocking ponds with SPF post larvae, ponds should be prepared in a manner that eliminates all possible natural crustacean carriers of WSSV (Flegel *et al.*, 2006). With this kind of precautions, it is indicated that the probability of WSSV outbreaks can be significantly reduced (Alday *et al.*, 2009). Alday *et al.*, (2009) also reported that for detection of WSSV in shrimps, instead of using PCR kit it is suggested to use immunochromatographic test strips which is most convenient and inexpensive test and one step closer to nested PCR test. During adverse environmental conditions or when seasonal weather is unstable and things like rapid changes in temperature and salinity cannot be controlled, it is reported that during such times cultivation of shrimps should be delayed until favourable climate resumes (Liu *et al.*, 2006, Guan *et al.*, 2003). WSSV has been described as double-stranded DNA virus with a genome approximately 300 kb and first found in Japan (Inouye *et al.*, 1994, 1996) and later from other Asian countries.

For controlling **yellow head virus** in shrimps, similar protocols as mentioned for controlling WSSV are need to be adopted. Alday *et al.*, (2009) mentioned about one more factor concerning the spread of YHV in ponds i.e. YHV out breaks is possible through insects and through cannibalism. Therefore, it has been suggested that YHV spread can be prevented by covering the ponds with fine-mesh netting. This covering will protect the ponds from both insects and dropped shrimps. The dropped shrimps are generally dead and infected shrimps from the neighbouring ponds and the birds carry them in their efforts to eat and drop them partially or fully in the ponds leading to spread of infection (Alday *et al.*, 2009). Hence it is clear that dropped and moribund shrimp pose a serious risk for disease spread, not only for YHV but also for WSSV and other shrimp pathogens.

For the control of **Taura syndrome virus** (TSV) in shrimps similar protocols have been recommended as described for WSSV and YHV. However, severe disease losses from TSV have been reported only for *P. vannamei* in USA and later from Asian countries also (Wyban 2007). Number of workers have mentioned genetic composition of YHV indicating that it is a rod shaped, single stranded, positive sense RNA virus with spikes envelope and size of approximately 27 kb (Cowley and Walker, 2002, Sittidilokratna *et al.*, 2002, 2008). There are reports that TSV infection can be avoided by selecting the shrimp from genetically tolerant domesticated stocks (Erickson *et al.*, 2005, Alday *et al.*, 2009). It is also mentioned that in Asian aquaculture SPF tolerant stocks of shrimps often test positive for TSV indicating that they become infected without signs of disease (Alday *et al.*, 2009).

Poulos *et al.*, (2006) while working on **myonecrosis virus** in penaeid shrimps, described the structure of this virus as ds RNA of 40nm diameter with a genome of 7650bp. Infection of this virus has been reported in *P. vannamei* in Brazil in 2002 and the symptoms were whiting of the abdominal muscle of the shrimp accompanied by slow mortality. For controlling the outbreak of this disease some workers have suggested careful control over the import and development of SPF stocks against this pathogen. Whitening of the muscles in penaeid shrimp particularly in *P. vannamei* has been reported due to the presence of noda virus (Tang *et al.*, 2007). Ravi *et al.*, (2009) mentioned that different types of noda viruses not only cause muscle whitening but also high mortality in post-larvae of *P. monodon* and *P. indicus*.

There are also reports of controlling other less important viruses like **Laem singh virus (LSNV)**, *P. monodon* **densovirus (PmDNV)** and *P. monodon* **polyhedrovirus (PmNVP)** during shrimp hatchery operations and also at the time of shrimp cultivation (Alday *et al.*, 2009). Controlling methods of these viruses are more or less similar as described above. LSNV is an RNA virus with a genome size of approximately 5 kb. Alday *et al.*, (2009) suggested that LSNV should also be added to the list of pathogens for inclusion in the SPF list while developing the SPF free brood stocks of the shrimps. With regards to shrimp densoviruses, it has been reported that the genome size of these viruses differs depending upon the virus associated with particular shrimp species (Alday *et al.*, 2009). Flegel *et al.*, (2006) mentioned that this densovirus in *P. monodon* does not cause severe mortality, however, they found severe retardation of the growth in the shrimp. Like all other viruses, densovirus also is transmitted to post larvae through the brood stocks. Phromjai *et al.*, (2002) mentioned that densovirus infection is transmitted to the shrimps through aquatic insects also. With regard to polyhedrovirus in shrimps Alday *et al.*, (2009) mentioned that this is a bacilliform virus with double stranded DNA genome and it produces granules of polyhedral protein in the nuclei of haepatopancreatic and midgut cells of *P. monodon*. Transmission of this virus takes place in uninfected shrimp when they ingest polyhedron granules containing this virus through shrimp feces. Flegel *et al.*, (1997) reported high mortality in the shrimp hatchery due to this virus infection and also retardation of growth. Alday *et al.*, (2010) suggested that this virus can be eliminated from the hatcheries by proper washing of eggs and nauplii. This procedure will also help to develop SPF brood stock in captive condition against this viral pathogen.

Following are some of the promising new methods to be developed on priority basis for effective control of diseases among the shrimps particularly when they are produce and cultivated in captive condition.

12.3 PROBIOTICS

The probiotics contains live microbial feed supplements which helps in improving the production of inhibitory compounds, competition for nutrients, iron, adhesion sites, and enhancement of host immune response and degradation of harmful wastes like ammonia (Gatesoupe, 1999, Verschuere *et al.*, 2000, Karunasagar *et al.*, 2010). However, Flegel *et al.*, (2008) reported that adding probiotics to the cultivation ponds has neither helped in improving the quality of pond water nor it has benefitted in preventing the diseases in the shrimps. These observations were based on earlier works carried out by several workers (Boyd and Gross, 1998, Rengpipat *et al.*, 1998, 2000). Probiotic bacteria have been shown to be effective in enhancing survival, moulting rate and growth of tiger shrimp, *P. monodon* (Rengpipat *et al.*, 1998) and *L. vanamei* (Garriques and Arevalo, 1995), to reduce population of pathogenic Vibrio spp (Moriarty, 1998, Chythanya *et al.*, 2002, Karunasagar *et al.*, 2005), improve digestibility of food (Liu *et al.*, 2009) and stimulate immune system of *P. monodon* (Rengpipat *et al.*, 2000). In various studies, it has been reported that application of probiotics will be more effective either to larval rearing tanks, pond water or added to the feed (karunasagar *et al.*, 2010). Feed supplementation has been observed as more effective than direct addition to rearing water. Moriarty (1998) reported that in the field studies that have been carried out by using probiotics showed improved shrimp health, better production and economic returns to the farmers. In some hatchery studies, Decamp *et al.*, (2008) found that daily application of Bacillus probiotics resulted in better larval survival rate similar to that obtained with antibiotic application. Though controversy exists on beneficial effects of probiotics in control shrimp diseases, more research is needed to explore this aspect. One of the major difficulties in making use of probiotics on disease control is the wide range of environmental condition where shrimp is grown. This means commercial products of probiotics need to be adapted to a range of conditions suited to the situation to situations depending on the environment. Karunasagar *et al.*, (2010) mentioned that best probiotics are the ones that are produced with local isolates and one should also know the method of their selection and production. Further it is also mentioned that the amount of probiotics that need to be added to positively alter the microbial balance needs to be evaluated for cost efficiency before large scale applications.

12.4 PHAGE THERAPY

Phage therapy is also called as bacteriophages which means bacteria eater and lysis of the pathogenic bacterial host. Such phages are also called as "lytic phages". This means the actual use of natural bacterial viruses to control bacterial populations (Karunasagar *et al.*, 2010, Flegel *et al.*, 2008). The emergence of multidrug resistance in several bacteria pathogens has led to a renewed interest in phage therapy. Bacteriophages are widely distributed in the environment, particularly in aquatic environment (Skurnik and Strauch, 2006).The unique character of these bacteriophages is that they increase or decrease in numbers following the target population, therefore, they are self-regulated.

Several workers have successfully demonstrated the application of bacteriophages in therapy against fish pathogens and further it is mentioned that the bacteriophages are not only host specific but they can be genus, species and strain specific also and hence they kill only the target bacteria. This property of the bacteriophages therapy will help only to kill targeted bacteria and do suppress other useful commensal flora that are required to maintain the normal health of the animals (Nakai *et al.*, 1999, Park *et al.*, 2000, Nakai and Park 2002, Karunasagar *et al.*, 2010). Considerable work has been done on bacteriophages against the shrimp pathogen *V. harveyi*. Shivu *et al.*, (2007) tested *V. harveyi* phages against 180 isolates of bacteriophages from different geographical regions and reported that these strains were able to lyse 65-70% of the strains. Vinod *et al.*, (2006) while doing the work on isolation of *V. harveyi* bacteriophage and study its effect on controlling the luminous vibriosis in hatchery environment during rearing shrimp larvae and post-larvae of *P. monodon* found that larval survival was only 25% in control tanks whereas tanks treated with bacteriophages showed 85% survival. Further it is mentioned that bacteriophages treatment has brought down the counts of luminous bacteria in the tanks. Karunasagar *et al.*, (2007) also reported similar results 86-88% survival with bacteriophages treatment when compared to 65-68% with antibiotics. These studies indicated the potential for bacteriophages to be effective alternatives to antibiotics in shrimp larval health management. Karunasagar *et al.*, (2010) while explaining the mechanisms of host resistance to bacteriophages mentioned that bacteriophages produce group of enzymes called lysins and these enzymes lyse the host cells and control their growth. Loessner, (2005) has advocated the use of purified lysins rather than live phages for shrimp disease therapy. Further it has been mentioned that these enzymes kill only the species/ strains from which they are produced, that means lysins would kill only the pathogens with little or no effect on the normal bacterial flora (Loessner, 2005). From these observations it has been concluded that the use of lytic enzymes could be an effective mechanism for control of *V. harveyi* population and the treatment of disease outbreak caused by this agent (Karunasagar *et al.*, 2010).

12.5 IMMUNOSTIMULANTS

Anderson (1992) has defined immunostimulant as a chemical, drug or the action that enhances the non-specific defence mechanisms or the specific immune response. Further he stated that penaeid shrimps or other aquatic crustaceans depend to a large extent on non-specific mechanisms and therefore the immunostimulants can play a vital role in preventing or controlling the diseases. Number of workers have mentioned that immunostimulants are molecules derived from microbial cell wall or outer membranes with characteristic patterns such as repeating units e.g. β-glucans, lipopolysaccharides, peptidoglycans. These substances induce enhanced activities in non-specific defence mechanisms such as increased oxidation activity of neutrophils and macrophages, augmented phagocytosis and potentiating cytotoxic cells (Kajita *et al.*, 1992, Chen and Ainsworth, 1992, Anderson, 1992, Sakai *et al.*, 1993, Yoshida *et al.*, 1993, Dalmo and Bogwald, 2008, Karunasagar *et al.*, 2010).

Despite their relatively short life and lesser complexity, crustaceans in general and penaeid shrimps in particular have mechanisms to detect foreign matter in the form of pathogen or contaminant. They appear to recognize common characteristics present in bacteria and fungi, such as β-glucans, mannan-oligosaccharide (MOS) and lipopolysaccharides (LPS). Although these microbial components can directly activate defensive cellular functions such as phagocytosis, melanisation, encapsulation and coagulation, plasma recognition proteins amplify these stimuli. Beta glucan binding protein (BGBP) reacts with β-glucans and the glucan-BGBP complex induces degranulation and the activation of prophenoloxidase (proPO). This protein is present in all crustaceans studied so far and is highly conserved. Together with LPS-binding agglutinin, BGBP stimulates cellular function only after its reaction with β-glucans or LPS, resembling the secondary activities of vertebrate antibodies.

In shrimps haemocytes play important roles in the host immune response including recognition, phagocytosis, melanisation, cytotoxicity and cell-cell communication. Classification of the haemocyte types is based mainly on the presence of cytoplasmic granules into hyaline cells, semi granular cells, and granular cells. Each cell type is active in defence reactions, for example, the hyaline cells are chiefly involved in phagocytosis, and the semi granular cells are the cells active in encapsulation, while the granular cells participate in storage and release of the prophenoloxidase (proPO) system and cytotoxicity. The haematopoietic tissue has been described in several crustacean decapod species and shown to be the haemocyte-producing organ. Tentative stem cells have been shown to be present in this tissue. Using *in situ* hybridization, it was demonstrated that proPO is not present in the haematopoietic tissue of crustaceans which suggests that protein expression is different between circulating haemocytes and the cells in the haematopoietic tissue. Several quantitative, fast and easy procedures are being adapted to evaluate the expression of the immune response of shrimp. With regard to cellular parameters, the hemogram and two cellular mechanisms, the radical oxygen intermediates (ROIs) generated during post phagocytic events and phenol oxidase (PO) activity has been considered as potential markers. Concerning humoral parameters, the antibacterial activity of plasma and the concentration of plasma proteins can be considered as criteria of health status.

Emphasis has been placed on natural environment variations, chemical contaminants and physico-chemical changes, especially with regard to reared shrimp. Studies on the effects of environmental factors on immune function in marine Crustacean have concentrated, predominantly, on total and differential haemocyte counts, antibacterial activity, phagocytic activity, the release of reactive oxygen intermediates and the modulation of the prophenoloxidase (proPO) system.

Invertebrate immune system must rely on non-self-recognition molecules to ensure efficient defence responses against infectious pathogens that continuously threaten their survival. Lectins from the haemolymph of invertebrates, including crustaceans, have been regarded as potential molecules involved in immune recognition and microorganism phagocytosis. The precise mechanisms underlying non-self-recognition represent the basis to prevent and control infections as well as to stimulate animal resistance. This is particularly relevant for cultivated aquatic species, especially penaeid shrimps, which are frequently constrained by recurrent diseases that often provoke great economic losses.

The production of antimicrobial peptides is a first-line host defence mechanism of innate immunity. However, in spite of the importance of infectious diseases in crustaceans, few molecules displaying antimicrobial activities have been fully characterized in these invertebrates. The recent findings on the identification of a family of antimicrobial peptides, named penaeidins, in the shrimp *Penaeus vannamei*. The penaeidins are original, 5.5 to 6.6 kDa peptides which combine a proline-rich amino-terminal domain and a carboxyl-domain containing six cysteine engaged in three disulphide bridges.

Vibrio species have become a major source of concern for shrimp culture because of their close association with low survival rates in hatcheries or grow out ponds. New shrimp pathogens belonging to the Vibrio genus have been described although their virulence is not yet fully understood. Indeed, they may act as opportunistic agents in secondary infections or be true pathogens.

Direct nutrient-specific effects are documented for highly unsaturated fatty acid (HUFA), phosphatidyl choline (PC) and ascorbic acid (AA) (AscoSol-C, Neospark). Specific immunostimulants may also contribute to an improved stress resistance of the shrimp fry. Some emphasis is also placed on the role of the microflora during the hatchery rearing, the possible reduction of bacteria in the live food administered, and the use of probionts (spectraLac and UltraZyme) through the diet.

Due to their hydrophobic nature, lipids are transported in the haemolymph of shrimp by protein-lipid-complexes named lipoproteins. Since cholesterol (Ch) and polyunsaturated lipids must be provided by the diet, and they are stored mainly in the hepatopancreas; a special vehicle is necessary for their mobilization to other tissues. Two types of haemolymph lipoproteins have been isolated from penaeid shrimp. Non sex-specific lipoproteins present in males and females (LPI), and female-specific lipoproteins (LPII or Vg) that occur mainly in mature females undergoing ovarian maturation. These lipoproteins are of the high density and very high density types. Their lipids are predominantly phospholipids (PL), but sterols, diacylglycerols (DG), triacylglycerol, and hydrocarbons (HC) have also been reported.

The possible existence of a peculiar form of adaptive immunity in invertebrates is important for a better understanding of immunological evolution and for the development of vaccination strategies. These may be relevant in the control of infectious diseases, common under intensive farming of economically important crustaceans. Adaptive immunity has been assumed to be absent from invertebrates because they lack the immunoglobulin (Ig), T cell receptor (TCR) and Major histocompatibility complex (Mhc) high diversity molecules. Since adhesion Ig super family (SF) molecules, which in mammals are known to be involved in adaptive immune response, are present in invertebrates, it can be postulated that they may also be responsible for invertebrate adaptive immunity. However, because invertebrate IgSF molecules are not phylogenetically homologous to those of vertebrates, the existence of an anticipatory immunity has not been accepted in invertebrates.

12.6 IMMUNOSTIMULANTS AND THEIR INDUCTION OF HAEMOLYMPH FACTORS

The haemolymph factors associated with defence in crustaceans include lectins, agglutinins, precipitins, bactericidins, lysins and bacteriostatic substances (Smith and Chisholm, 1992) and induction of such haemolymph factors in shrimps has been reported in literature. Adams (1991) reported that bactericidins and other humoral factors are induced in black tiger shrimp, *P. monodon* within day 1 of injection of heat killed *Vibrio alginolyticus*. Devaraja *et al.*, (1998) also found that administration of *V. bacterin* orally through diet in *P. monodon* induced bactericidin. Enhancement of bactericidal activity in the haemolymph of *P.monodon* can be brought about through treatment with immunostimulants such as β-glucans, zymosan or *V. bacterin* (Sung *et al.*, 1996). Much higher titres of vibriocidal activity of hemolymph has been observed in *P. monodon* fed with a diet containing β- glucans and *V. bacterin* compared to diet with either component alone suggesting synergistic effect between these imunostimulants (Devaraja *et al.*, 1998). Besides substances such as bactericidins which directly contribute to the killing of the pathogens, crustaceans may have humoral factors which function as recognition molecules to recognize non-self-molecules and interact with cellular factors. Such molecules generally recognize cell wall components of microorganisms and are referred to as Pattern Recognition Proteins (PRPs) (Sritunyalucksana *et al.*, 1999; 2002; Roux *et al.*, 2002). Charoensapsri *et al.*, (2009) observed that RNA interference mediated suppression gene coding for proPO in *P. monodon* increases susceptibility to *V. harveyi* infection. Enhanced phenol-oxidaseactivities have been demonstrated in *P. monodon* treated with immunostimulants such as 1, 3 glucan by immersion (Sung *et al.*, 1996) or through feed (Devaraja *et al.*, 1998) and in the shrimp *P. japonicus* fed with diet containing lipopolysaccharide (Takahashi *et al.*, 2000).

Number of workers have reported cellular activities particularly the haemocytes and their induction by various immunostimulants. Enhanced phagocytic activity in the granulocytes of *P. monodon* fed with glucan and *P. japonicus* fed with peptidoglycan has been observed by Itami *et al.*, 1994, 1998). These two substances have been derived from the bacterium, *Bifidobacterium thermophilum.* Sung *et al.*, (1996) also found that *Vibrio* spp disappeared from shrimp (*P. monodon*) haemolymph after feeding with glucan and peptidoglycan. Further they have reported that in the same shrimp production of reactive oxygen has increased significantly when compared to control shrimps. Tseng *et al.*, (2009) found that in *L. vannamei* fed with diet containing 108 *Bacillus subtilis* E20 there is enhanced phenol oxidase activity, phagocytic activity and clearance against *Vibrio alginolyticus.* These observations indicate that immunostimulants using microbial cell wall components enhance the cellular component of non-specific defence in shrimps.

Considerable work has also been carried out on the effect of immunostimulants on enhanced survival rate of shrimps when they are challenged with bacteria or viral pathogen. Itami *et al.*, (1989) observed that when kuruma shrimps treated with *V. bacterin* there was reduced mortality among the shrimps when the same were compared with untreated controls. Sung *et al.*, (1994) while working on *P. monodon* found that shrimp treated with glucan showed better survival when challenged with *V. vulnificus* cells and protective effect lasted for 18 days following the treatment. Shrimp *P. japonicus* treated orally with 1,3 glucan or fed with peptidoglycan survived in a better way after challenging by Vibrio or WSSV and when compared to untreated control (Itami *et al.*, 1994, 1998a; 1998b; Chang *et al.*, 1999). Similarly, treatment with fucoidan, also improved survival rate after challenge with WSSV in the same species of shrimp (Takahashi *et al.*, 1998). Chang *et al.*, (2000) while working on *P. monodon* noticed that the survival in the shrimp is always better when they were fed with 1, 3 glucan which is derived from *Schizophyllum commune* a causative agent of fungal sinusitis. Number of workers have suggested that to control taura virus syndrome and WSSV infection in the shrimps the application of dietary immunostimulants are effective (Brock, *et al.*, 1997; Lightner and Redman, 1998, Karunasagar and Karunasagar, 1999). Table 1 summarizes the effect of different immunostimulants on the survival rate of the shrimps.

Table1. Effect of different immunostimulants on the survival rate of the shrimps

Immunostimulants	Method of application	Effects	Reference
Vibrio bacterin	Injection, immersion spray	Improved survival on challenge on day 30	Reference
Vibrio alginolyticus bacterin	Injection	Induction of bactericidines lasting till day 5	Adams, 1991
Lactobacillus pantarum	Oral	Increase in PO, Superoxide dismutase activity, survival after challenge with *V. alginolyticus*	Chiu *et al.*, 2007
Bacillus subtilis	Oral	Increase in PO, phagocytic activity, efficiency of clearance of *V. alginolyticus*	Tseng *et al.*, 2009
1,3 glucan	Immersion	Improved survival on challenge with *Vibrio valnificus* up to day 18.	Sung *et al.*, 1994
1,3 glucan	Oral	Increased phagocytic index, improved survival on challenge with *Vibrio*, WSSV	Itami *et al.*, 1994, Chang *et al.*, 1999
1,3 glucan, zymosan, Vibrio activity	Immersion	Enhanced PO activity, oxygen burst activity	Sung *et al.*, 1996
Peptidoglycan	Oral	Enhanced phagocytic index, improved survival on challenge with WSSV	Itami *et al.*, 1998a, 1998b
1,3 glucan, Vibrio bacterin	Oral	Enhanced bactericidal, oxygen burst, PO activity, improved survival on challenge with WSSV	Devaraja *et al.*, 1998
Fucoidan	Oral	Improved survival on challenge with WSSV	Takahashi *et al.*, 1998

(**Source:** Karunasagar *et al.*, 2010 The Shrimp Book, by Victoria Alday-Sanz (Editor), Publ: Nottinghamm University Press, Nottingham, UK, 2010)

12.7 PROMISING NEW DIRECTIONS

Flegel *et al.*, (2008) reported the concept of quorum sensing control of bacterial virulence in the shrimps. Further they explained the concept mentioning that when small quantity of probiotic cells added to the shrimp pond this can control the water quality and also prevent bacterial pathogen from causing diseases without actually killing them. This they say it is possible because it has been 0known for many years that cross-takes place amongst the microbes via minute quantities of natural chemical messengers that release by the probiotic cells. This mechanism it has been termed as quorum sensing (Hardman *et al.*, 1998). Flegel *et al.*, (2002, 2008) and Defoirdt *et al.*, (2004) emphasized that this an intense area of research in many fields particularly in aquaculture. The advantage of using this kind of technology for controlling the diseases the chemical messengers which are released in to the water, mostly they are natural substances that do not kill target cells unlike when antibiotics are used for similar purpose (Flegel *et al.*, 2008)

Though number of workers have reported on the of interaction of shrimps with the pathogenic bacteria and fungi, there is very little information available on shrimps interacting with viral pathogens (Soderhall *et al.*, 1994, Soderhall, 1999, Bachere *et al.*, 2000, Lee and soderhall 2002, Stet and Arts, 2005, Flegel *et al.*, 2008). The concept of programmed cell death or cell suicide which has been discovered by some workers and this kind of cell death or cell suicide in the shrimps is also triggered by the viruses which may lead to mortality among shrimps (Sahtout *et al.*, 2001, Khanobdee *et al.*, 2002, Wongprasert *et al.*, 2003). After this investigation number of workers have pointed out to find out the key to the trigger, then perhaps a novel method of blocking this trigger can be developed to prevent shrimp mortality due to the viral infection (Flegel *et al.*, 2008). A few genes in this complex pathway have been discovered in the shrimp (Phongdara *et al.*, 2006, Bangrak *et al.*, 2002, 2004, Tonganunt *et al.*, 2005). Therefore, it is necessary to understand the molecular mechanisms thoroughly behind shrimp tolerance to viral infections so that this may lead to the development of novel methods of disease control.

12.8 SHRIMP VACCINES

About research efforts that have been made so far regarding possibilities of development and use of shrimp vaccines for effective control of viral disease, Flegel *et al.*, (2008) and Escobedo-Bonilla (2011) summarized in brief in their review papers. In this direction, earlier Tseng *et al.*, (2000) made an effort in the shrimp *P. monodon* by introducing foreign DNA into eggs and embryos through electroporation and found that the rate of success was between 37 - 19% and survival of transgenic eggs into juvenile shrimp was 0.6%. In another attempt Preston *et al.*, (2000) carried out studies in *M. japonicus* by delivering DNA to embryos using microinjection, electroporation and particle bombardment. Of these, microinjection was the most effective as high amounts of foreign DNA was delivered with this method. The protective efficacy against WSSV using plasmids encoding WSSV envelope proteins was evaluated in recent studies and delivery routes were intramuscular and oral (Kumar *et al.*, 2008, Rout *et al.*, 2007). It was observed that initial administration of foreign DNA protected the shrimps against WSSV infection resulting in low mortality for first few days but later mortality rate enhanced.

It is well known that invertebrates do not possess an adaptive defence system like vertebrates. Nonetheless some studies have indicated the presence of a specific defence response against viral infections. One work showed that some shrimp that survived a natural or experimental WSSV infection were more tolerant to a subsequent infection with WSSV (Escobedo-Bonilla 2011)

Venegas *et al.*, (2000) while studying immune response in *P. japonicus* suggested the existence of a "viral neutralizing factor" which recognizes viral molecules and hence the shrimp develops the ability to counteract infection. This they also called quasi-immune response. Further it is mentioned that those shrimps which survive after disease out breaks, they were not protected from infection (WSSV), but protected from the disease. Flegel (2007) has suggested another hypothesis called the "viral accommodation" which means that the shrimps adapt to new viral pathogens to become asymptomatic carriers without displaying disease initially. The reason why viruses remain infectious but does not cause disease to the shrimp is unknown. It is postulated that such a tolerance involves some sort of specific memory to prevent viral triggered apoptosis. These findings have paved the way to evaluate different strategies to protect the shrimps from viral pathogens, and understand in a better way shrimp viral-interaction and further develop even better methods of viral control probably in the form of shrimp vaccines. For controlling viral infection in shrimps Flegel *et al.*, (2008) mentioned the use of some new reagents to which many workers have suggested to use the term as shrimp vaccines. Further it is described by Flegel *et al.*, (2008) that vaccine is not the proper term for such reagents, however, they suggested to call these reagents as "tolerines". This kind of test for controlling viral pathogens may throw light in better understanding of shrimp-viral interaction and this may even lead to develop good protocols of viral disease control.

Considerable findings are available indicating how shrimp interact with bacterial and fungal pathogens but information on shrimp-viral interaction is scanty (Soderhall *et al.*, 1994, Soderhall, 1999, Bachere *et al.*, 2000, Lee and Soderhall, 2002, Stet and Arts, 2005). In recent years some work has been done on shrimp-viral interaction at the molecular and genetic levels (Flegel *et al.*, 2008).

In recent years several efforts are being made to design and construct shrimp antiviral DNA vaccine for control of microbial diseases in shrimps. DNA vaccines are essentially recombinant plasmid constricts capable of expressing pathogen derived antigenic proteins which protects the animals against infections when administered intramuscularly or subcutaneously (Robinson and Pertmer 2000). DNA vaccines are considered to be safer and stable than the proteins/ glycoproteins subunits of vaccines. Use of DNA vaccines in control of infectious diseases in fish has become a very popular technique. However, in shrimps the specific immune system is very rudimentary therefore application of DNA vaccines for controlling the microbial diseases has to be evaluated carefully. Efforts have been already made for the usage of RNA interference technology for controlling infectious microbial diseases in the shrimp. In this context the possibility of using DNA vaccines can be extended to construct plasmid containing short or long double stranded RNA which may inhibit the pathogen proliferation when administered into the host animals. Several workers have reported the presence of Dicer gene in number of decapods including shrimps namely *P. monodon,* (Su 2008, Li 2013), *L. vennamei* (Yao 2010), *F. chinensis* (Li, 2013) and *M. japonicus* (Huang 2012, Nishi 2014), confirming the presence of a functional RNAi pathway. In tiger shrimp *P. monodon* attempts have made to study the fate of plasmid DNA after administration into the animal either through intramuscularly or through intra peritoneal injection (Rout 2007, Krishnan 2009). Recently Chaudhari *et al.,* (2016) made an attempt to design and construct shrimp antiviral DNA vaccine expressing long and short hairpins for protection by RNA interference. In their observation it has been concluded that production of long hairpin RNA /short hairpin RNA in vivo using host machinery cuts down the cost of production compared to chemical synthesis and in vitro transcription methods.

12.9 RNA INTERFERENCE

The application RNA interference against viral infection in shrimps has been reviewed and discussed in detail by Krishnan *et al.,* (2009) and Escobedo-Bonilla (2011). RNA interference (RNAi) is one of the most advanced technologies for suppression gene expressions, which can mediate gene silencing of the virus and thus generate an antiviral response and protect shrimps from viral diseases (Giladi *et al.,* 2003; Gitlin *et al.,* 2002; Voorhoeve & Agami 2003, Lu *et al.,* 2005, Robalino *et al.,* 2005, Tirasophon *et al.,* 2005, 2007, La Fauce and Owens, 2009, Xu *et al.,* 2007). The subject has been reviewed in recent past by Robalino *et al.,* (2007). The rapid production of knockdown animals where genes of interest have been selectively silenced is just one of many possible uses for RNAi (Voorhoeve & Agami 2003, Su *et al.,* 2008). Shuey et al. (2002) suggested another reason for the use of RNAi, is the selective silencing of viral genes essential for virulence. The process consists of making small fragments of double-stranded RNA with sequences that match those of viral genes. Upon entry of RNAi, an enzyme called Dicer cleaves long dsRNA into double-stranded short interfering RNAs (siRNAs) (Dykxhoorn *et al.*, 2003). The siRNA molecules are taken up by the RNA-inducing silencing complex (RISC) comprised by various proteins which unwinds siRNAs into single stranded molecules. The antisense strand remains attached to RISC and it is coupled to its homologous target mRNA to induce endonucleolytic cleavage. Long dsRNA molecules make it possible to produce various siRNA molecules targeting a single mRNA thus increasing effective gene silencing (Dykxhoorn *et al.*, 2003). This process is evolutionarily conserved and has been found in a wide range of eukaryotic organisms (Hammond *et al.*, 2001; McManus and Sharp, 2002). When RNAi genes are injected in to shrimp or exposed to shrimp cells in culture, disease protection takes place (Flagel *et al.*, 2008). RNAi starts with the presence of RNA molecules such as double stranded RNA (Li F and Ding SW, 2006).

Number of workers have studied the RNA interference technique to prevent or control viral diseases in shrimp. The findings of such investigations have indicated that there are two pathways for controlling this kind of viral infections in shrimp: one is sequence-independent and another is sequence-specific (RNAi-mediated) (Yodmuang *et al.*, 2006) It has been reported that long sequence-independent dsRNA molecules activate the mRNA expression of the RNAi molecules. Labreuche *et al.*, (2010) mentioned that in shrimps the mechanisms involved in both the innate antiviral defence and RNAi activity are activated by the same molecular pathway to produce an efficient antiviral response like other higher organisms

Escobedo-Bonilla (2011) in order to find out specific RNAi antiviral immunity in *L. vannamei* he used non-specific dsRNA to inhibit TSV or WSSV infections. A sequence-independent, and dose dependent antiviral state was induced against TSV or WSSV using dsRNA from immunoglobulin [Ig] heavy chain from duck or pig. He found that shrimps treated with unrelated dsRNA sharply reduced mortality (50-75%) when compared to untreated controls (Robalino *et al.*, 2004). This result indicated that shrimp possesses an innate antiviral immunity. Other studies showed that using sequence-independent dsRNA induced a non-specific antiviral effect. Specific RNAi antiviral immunity by siRNA silencing of shrimp endogenous genes involved in virus infection. Further, he used sequence-specific dsRNA to inhibit virus replication in shrimp against TSV, IHHNV, YHV and WSSV. Studies done with dsRNA against a putative protease from TSV showed that sequence-specific dsRNA strongly inhibited TSV replication (11% mortality) in shrimp infected, while controls showed 100% mortality. Replication of IHHNV or HPV has successfully been inhibited with specific dsRNA (Escobedo-Bonilla, 2011). Several studies on dsRNA have been done against WSSV since this is the most lethal pathogen in shrimp aquaculture. Different efficacies in inhibition of WSSV replication have been achieved using sequence specific dsRNA against various genes encoding structural and non-structural proteins (Escobedo-Bonilla, 2011) (Table 2). Number of workers have carried out the work on triggering of RNAi with introduction of synthetic nucleotide duplexes of RNA and then study the sequence-specific inhibition of cellular mRNA, which helps in opening up possibilities for controlling replicative processes of pathogenic organisms (Elbashir *et al.*, 2001; Zamore *et al.*, 2000). RNAi has been used to specifically inhibit gene expression and replication of infectious viruses. The replication of a growing number of human pathogenic viruses has been shown to be inhibited by RNAi, including poliovirus, HIV-1, HCV, influenza virus and hepatitis B virus (Randall *et al.*, 2003). RNAi has been revealed to function as an adaptive antiviral immune mechanism (Lu *et al.*, 2005; Wilkins *et al.*, 2005). Recently the *in vivo* roles of double stranded RNA (dsRNA) and RNAi in shrimp antiviral immunity were demonstrated (Robalino *et al.*, 2004, 2005; Westenberg *et al.*, 2005, Vega *et al.*, 2008 Liu *et al.*, 2007, Maningas *et al.*, 2008). In the marine shrimp *L. vannamei*, the antiviral response can be induced by sequence-independent or sequence-specific dsRNA which may activate RNAi-like mechanisms (Robalino *et al.*, 2004, 2005). It has also been reported that siRNA could suppress the gene expression and replication of WSSV in a sequence-independent manner (Westenberg *et al.*, 2005).

In an attempt to characterize the antiviral response in shrimp, the RNAi strategy was followed by using a specific short interfering RNA (vp28-siRNA) targeting the vp28 gene of WSSV. The VP28 protein, a major envelop protein of WSSV, is involved in the attachment and penetration into shrimp cells (Wu *et al.*, 2005; Yi *et al.*, 2004). The results showed that the vp28-siRNA was capable of silencing the vp28 gene. When treated with vp28-siRNA, the expression of vp28 gene and the replication of viral DNA were significantly delayed or inhibited by siRNA, resulting in low mortality of WSSV-infected shrimp. These findings have demonstrated the potentials for the gene function research and therapeutic treatment of WSSV by RNAi.

Boonanuntanasarn (2008) has used RNAi tool to determine the function of various genes from shrimp that are involved in virus infection. This is done by silencing genes encoding certain proteins or enzymes of interest. In shrimps number of proteins involved in antiviral immunity have been studied by RNAi silencing technique, including a toll-like receptor (Yang *et al.*, 2007, Labreuche *et al.*, 2009). The innate system is the first defence line against pathogens which

Table 2. *In vivo* evaluation of dsRNA efficacy against different WSSV genes encoding structural and non-structural proteins.

Type	Gene	Administration route	Concentration (μg)	Virus dose	Mortality (%)	Duration (days)	Reference
	vp28	intramuscular injection	4	2500 SID_{50}	13	10	[108]
	vp28	intramuscular injection	5	4x10-8	15	10	[111]
	vp28	intramuscular injection	6	1-2 LD_{50}	0	7	[110]
	vp28	oral (coated in feed)	n.d	n.d	63%	15	[139]
	vp28	oral (chitosan nanoparticles)	n.d	n.d	32%	15	[139]
Structural	vp281	intramuscular injection	6	1-2 LD_{50}	20 - 47%	7	[110]
	vp26	intramuscular injection	4	2500 SID_{50}	21%	10	[108]
	vp26	intramuscular injection	25	1x10-5	100%	30	[140]
	vp24	intramuscular injection	25	1x10-5	37%	30	[140]
	vp19	intramuscular injection	25	1x10-5	66%	30	[140]
	vp15	intramuscular injection	25	1x10-5	37%	30	[140]
	RR2	intramuscular injection	5	4x10-8	22%	10	[111]
Non-structural	DNA pol	intramuscular injection	5	4x10-8	56%	10	[111]
	PK	intramuscular injection	6	1-2 LD_{50}	7%	7	[110]

SID50 - shrimp infctious dose 50% endpoint, LD50 - lethal dose 50% endpoint. N.d. not determined. RR2 -ribonucleotide reductase small subunit. DNA pol – DNA polymerase. PK - protein kinase. (**Source:** Escobedo-Bonilla, 2011) is activated by a number of molecules that recognize different pathogen-associated molecular patterns (PAMPs) (Schroder and Bowie, 2005) which include peptidoglycans, lipopolysaccharides, beta-glucans and foreign dsRNA. These in turn activate different defence responses (Yang *et al.*, 2008). In vertebrates, toll-like receptors (TLR) are involved in recognition of PAMPs and activation of defence responses against pathogens. Foreign dsRNA is recognized by TLRs involved in the activation of the RNAi antiviral response (Schroder and Bowie, 2005). TLR has also been recognised in the shrimps *P. monodon* (Arts *et al.*, 2007), *L. vannamei* (Yang *et al.*, 2007) and *F. chinensis* (Yang *et al.*, 2008).

12.10 ANTIVIRAL AND ANTIBACTERIAL SUBSTANCES IN SHRIMP

Number of workers have indicated the presence of antibacterial peptides in the shrimps to which they identified the name as penaeidins (Destoumieux *et al.*, 1999, Bachere *et al.*, 2003, 2004, Destoumieux *et al.*, 2000a, 2000b, Munoz *et al.*, 2003, Chen *et al.*, 2005, Gueguen *et al.*, 2006, Kang *et al.*, 2004, Supungul *et al.*, 2004, Flegel *et al.*, 2008). These antibacterial proteins are produced in the shrimps when they become infected with microbes. Besides penaeidins, Flegel *et al.*, (2008) mentioned that the shrimps also produce other types of proteins in response to number viral pathogens. He *et al.*, (2004) and Pan *et al.*, (2000, 2005) reported the production of 60 proteins in hepatopancreas and hemocytes of WSSV- resistant shrimp when compared to normal shrimp. He *et al.*, (2004) identified these proteins as C-type lectin and an interferon-like protein. Antibacterial and antiviral substances when they produced against bacterial and viral infection, it becomes important to understand the nature of these molecules and their specific anti nature activities against microbes. Once this known, the findings may help to evolve much better therapeutic methods for controlling the diseases.

12.11 GENOME EDITING (CRISPR-CAS TECHNIQUE)

In recent years with the introduction of new genome editing tools such as Zinc-finger nucleases (ZFNs), transcription activator like effector nucleases (TALENs) and more recently the clustered regularly inter-spaced short palindromic repeats (CRISPR)/ CRISPR- accosiated nuclease 9 (Cas 9) systems, it has become possible to knock out undesired part of the gene in different animal models. The CRISPR/ Cas 9 system has been introduced as a new class genome engineering tool even for those organisms where genome editing will be difficult (Seruggia and Montoliu 2014, Hwang *et al.*, 2013). This particular genomic pattern is naturally present in eubacteria and archaea, enabling it to respond and eliminate invading genetic material (Horvath and Barrangou 2010). These repeats were initially discovered in 1980's in *E. coli* but their function was not confirmed until 2007. Later Barrangou and colleagues demonstrated that *Streptococcus thermophilus* can acquire resistance against a bacteriophage by integrating a genome fragment of an infectious virus into its CRISPR locus (Talbot and Amacher, 2014). Three types of CRISPR mechanisms have been reported i.e type I, type II and type III, each of which is characterized by signature proteins (Cas 3, Cas 9 or Cas 10) (Makarova *et al.*, 2011). Out of three types, type II is the most studied one and in this case invading DNA from viruses or pasmids is cut into small fragments and incorporated into CRISPR locus admist a series of short repeats. The loci are transcribed and then the transcripts are processed to generate small RNAs (crRNA- CRISPR RNA) which are used to guide effector endonucleases that target invading DNA based on sequence complimentarity. To achieve site specific DNA recognition, Cas 9 must be complexed with both a crRNA and a separate trans activating crRNA that is partially complimentary to crRNA (Jinek *et al.*, 2012) (Fig 2).

In Penaeid shrimp the immune system is fragile and temporary and because of this fragile immune system they are susceptible to the viral and bacterial pathogens very frequently leading to mass mortality. Therefore, a modern genome editing CRISPR- cas 9 technology can prove as a powerful tool in order prevent the infection of viral and bacterial pathogens (Diwan *et al.*, 2017). CRISPR has potential of altering germ line of humans, animals, and other organisms. This approach can be achieved with the help of supporting tools like bioinformatics, primer design and microinjection to identify and insert the CRISPR cas 9 into the targeted genome.

12.12 CONCLUSIONS

In order flourish shrimp industry and produce quality shrimps without any disease/ infection, number of suggestions have been made by several workers. Most of the times in future the shrimp industry will have to depend on the farms to get their major source of shrimps rather than depending on capture source. Therefore, the shrimp farms will have to produce the shrimps in large quantities in captive conditions by following new guide lines stipulated by the FAO and newer technologies to produce disease free shrimps. In future many farms will be dominated by cultivation of domesticated lines of shrimp that are free of most of the significant shrimp disease. Most of the stocks used may be of improved varieties through genetic selection with desirable characteristics like faster growth and good disease tolerance capacity. Good management practices (GMP) recommended by the Global Aquaculture Alliance will be order of the shrimp farms and those have to be strictly observed.

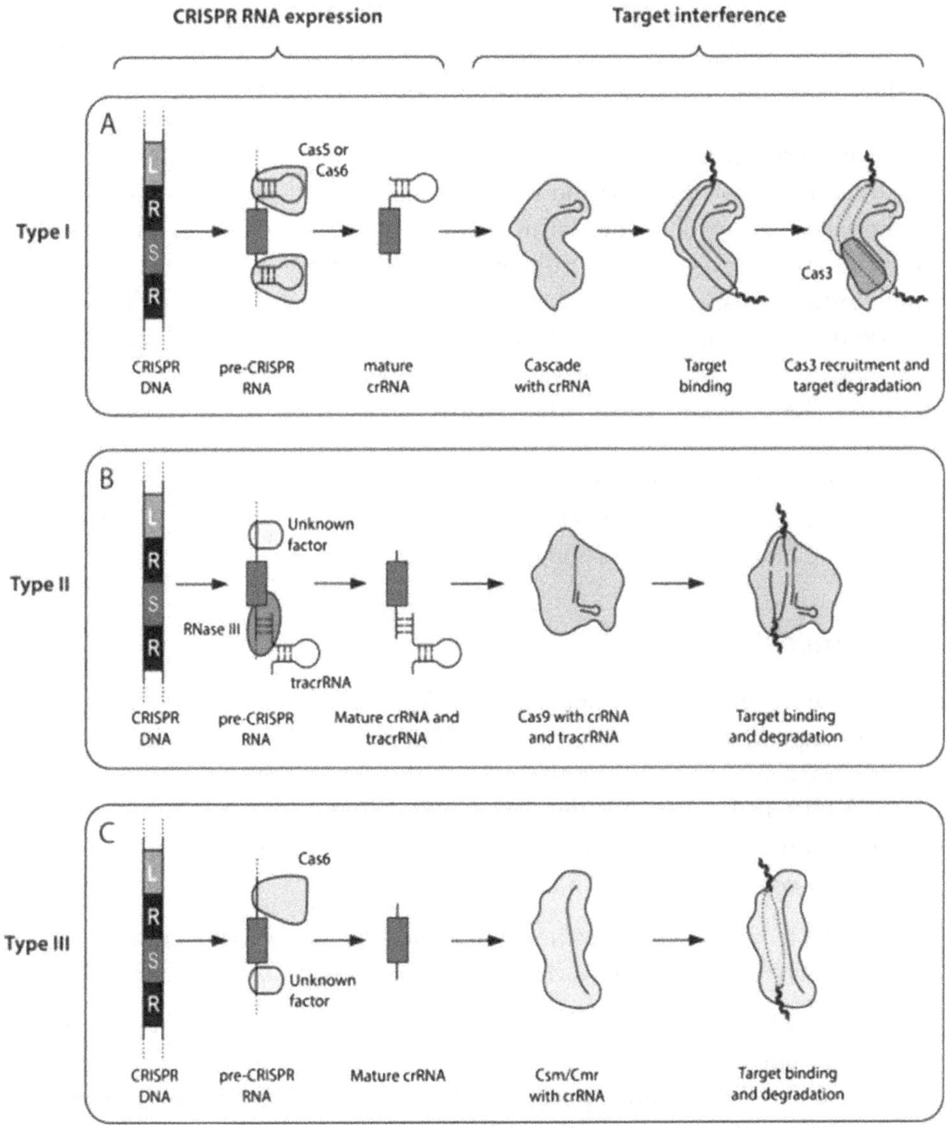

Fig. 2. Model of crRNA processing and interference. (A) In Type I systems, the pre-crRNA is processed by Cas5 or Cas6. DNA target interference requires Cas3 in addition to Cascade and crRNA. (B) Type II systems use RNase III and tracrRNA for crRNA processing together with an unknown additional factor that perform 50 end trimming. Cas9 targets DNA in a crRNA-guided manner. (C) The Type III systems also use Cas6 for crRNA processing, but in addition an unknown factor perform 30 end trimming. Here, the Type III Csm/Cmr complex is drawn as targeting DNA, but RNA may also be targeted. (**Source:** Rath *et al.,* 2015)

Shrimp genome project need to be undertaken in collaboration with the international agencies so that genetic map and complete shrimp genome sequences of individual shrimp species are made available for public use. This will constitute an important source for research on all aspects of shrimp biology including health and diseases. This will pave the way to develop molecular methods which will be promising protocols for control of various viral and bacterial diseases. Many advanced diagnostic methods and kits are available for disease identification both in the laboratory and on the farm and farmers they themselves carry out the diagnosis. In addition to this, number of tools like use of tolerines, probiotics, immunostimulants, quorum sensing modulators, shrimp vaccines, RNAi and other disease prevention methods are now being developed. Number of workers have also suggested biosecurity methods and controlled environment which will definitely help in preventing the occurrence and spread of diseases. Strict observation have to be made on new pathogens emerging, especially when living shrimps and other crustaceans are translocated without sufficient precautions. Recently developed gene editing technology can prove as a powerful tool to prevent the infections of viral and bacterial pathogens in shrimp.

References

Amalea Dulcene D, Nicolasora Benedict A, Maralit, Christopher Marlowe A. Caipang, Mudjekeewis D. Santos, Adelaida Calpe, and Mary Beth B. Maningas, 2014. Utilization of loop-mediated isothermal amplification (LAMP) technology for detecting white Spot Syndrome Virus (WSSV) and Vibrio spp. in *Litopenaeus vannamei* in selected sites in the Philippines. *Philippine Science Letters.* 7,2: 309-316.

Anderson DP, 1992. Immunostimulants, adjuvants and vaccine carriers in fish: applications to aquaculture *Annual Review of Fish Diseases,* 2:281-307.

Alday-Sanz, Stentiford GD, Bonami JR, V, 2009. A critical review of the susceptibility of crustaceans to Taura syndrome, Yellow head disease and White spot disease and implications of inclusion of these diseases in European legislation. *Aquaculture.* 291:1–17

Arts JA, Cornelissen FH, Cijsouw T, Hermsen T, Savelkoul HF, Stet R J, 2007. Molecular cloning and expression of a Toll receptor in the giant tiger shrimp, *Penaeus monodon. Fish & shellfish immunology* 23: 504-513.

Arunrut N, Kampeera J, Suebsing R, Kiatpathomchai W. 2013 Rapid and sensitive detection of shrimp infectious myonecrosis virus using a reverse transcription loop-mediated isothermal amplification and visual colorogenic nano-gold hybridization probe assay. *J Virol Methods.*193,2: 542-7.

Bachère E, Fuchs R and Söderhäll K, 2000. Special issue on shrimp response to pathogens. *Aquaculture* 191.

Bachère, E. 2003. Anti-infectious immune effectors in marine invertebrates: Potential tools for disease control in larviculture. *Aquaculture* 227:427-438.

Bangrak, P., Graidist, P., Chotigeat, W., Supamattaya, K. and Phongdara, A. 2002. A syntenin-like protein with postsynaptic density protein (PDZ) domains produced by black tiger shrimp *Penaeus monodon* in response to white spot syndrome virus infection. *Dis. Aquat. Org.* 49:19-25.

Barrangou R, Fremaux C, Deveau H, Richards M, Boyaval P, Moineau S, Romero DA, Horvath P. CRISPR provides acquired resistance against viruses in prokaryotes. *Science.* 2007;315: 1709–1712.

Boonanuntanasarn S, 2008. Gene knockdown: a powerful tool for gene function study in fish. *Journal of the World Aquaculture Society* 39: 31-323.

Boyd C and Gross A, 1998. Use of probiotics for improving soil and water quality in aquaculture ponds. In: Flegel, T. (Ed.), *Advances in shrimp biotechnology.* pp 101-106. National Center for Genetic Engineering and Biotechnology, Bangkok.

Brock JA, Gose RB, Lightner DV and Hasson KW, 1997. Recent developments and an overview of Taura syndrome of farmed shrimp in the Americas. In *Diseases in Asian Aquaculture III* pp 275-283 Eds Flegel TW and MacRae JH *Asian Fisheries Society*, Manila

Chang C, Su M, Chen H, Lo C, Kou G and Liao I, 1999. Effect of dietary 1-3 glucan on resistance to white spot syndrome virus (WSSV) in post larval and juvenile *Penaeus monodon Diseases of Aquatic Organisms,* 36 :163-168.

Chang, C. F., Chen, H. Y., Su, M. S. and Liao, I. C. 2000. Immunomodulation by dietary α 1-3 glucan in the brooders of the black tiger shrimp *Penaeus monodon. Fish Shellfish Immunol.*, 10: 505-514.

Charoensapsri W, Amparyup P, Hirono I, Aoki T, Tassanakajon A, 2009. Gene silencing of a prophenoloxidase activating enzyme in the shrimp, *Penaeus monodon*, increases susceptibility to *Vibrio harveyi* infection. *Developmental and comparative immunology* 33: 811-820.

Chaudhari A, Pathakota G B, Annam P K, 2016. Design and Construction of Shrimp Antiviral DNA Vaccines Expressing Long and Short Hairpins for Protection by RNA Interference. Methods Mol Biol. 1404:225-40.

Chen D and Ainsworth A.J ,1992. Glucan administration poteniates immune defence mechanisms of channel catfish, *lctalurus punctatus* rafineque. *Journal of Fish Diseases* 15: 295-304.

Chen, J.-Y., Chuang, H., Pan, C.-Y., Kuo, C.-M. 2005. cDNA sequence encoding an antimicrobial peptide of chelonianin from the tiger shrimp *Penaeus monodon. Fish Shellfish Immunol.* 18:179-183.

Chythanya R, Karunasagar I and Karunasagar I, 2002. Inhibition of shrimp pathogenic *Vibrio* spp by marine *Pseudomonas* I-2 strain *Aquaculture* 208: 1-10

Cowley JA, Walker PJ, 2002. The complete genome sequence of gillassociated virus of *Penaeus monodon* prawns indicates a gene organization unique among nidoviruses. *Archives of virololgy* 147: 1977-1987.

Dalmo R A and Bogwald J, 2008. Glucans as conductors of immune symphonies *Fish and Shellfish Immunology.* 25 :384-396.

De la Vega E, O'Leary NA, Shockey JE, Robalino J, Payne C, Browdy CL, Warr GW and Gross PS, 2008. Anti-lipopolysaccharide factor in Litopenaeus vannamei (LvALF): A broad spectrum antimicrobial peptide essential for shrimp immunity against bacterial and fungal infection. *Mol Immunol* .45 :1916-1925.

Decamp O, Moriarty DJW, Lavens P, 2008. Probiotics for shrimp larviculture: review of feld data from Asia and Latin America *Aquaculture Research* 39: 334-338.

Defoirdt, T., Boona, N., Bossierb, P. and Verstraete, W. 2004. Disruption of bacterial quorum sensing: an unexplored strategy to fight infections in aquaculture. *Aquaculture* 240:69-88.

Devaraja TN. Otta SK, Shubha G, Karunasagar I. Tauro P and Karunasagar I, 1998 Immunistimulation of Shrimp through oral administration of Vibrio bacteria and yeast glucans In *Advances in shrimp biotechnology* pp167-170 Ed Flegal TW. National Centre for Genetic Engineering and Biotechnology, Bangkok.

Destoumieux, D., Bulet, P., Strub, J.M., Van Dorsselaer, A. and Bachère, E. 1999. Recombinant expression and range of activity of penaeidins, antimicrobial peptides from penaeid shrimp. *Eur. J. Biochem.*/FEBS266:335-346.

Destoumieux, D., Munoz, M., Bulet, P. and Bachère, E. 2000a. Penaeidins, a family of antimicrobial peptides from penaeid shrimp (Crustacea, Decapoda). *Cell. Mol. Life Sci.* 57:1260-1271.

Destoumieux, D., Munoz, M., Cosseau, C., Rodriguez, J., Bulet, P., Comps, M. and Bachere, E. 2000b. Penaeidins, antimicrobial peptides with chitin-binding activity, are produced and stored in shrimp granulocytes and released after microbial challenge. *J. Cell Sci.* 113:461-469.

Diwan, A.D., Ninawe, A.S. and Harke, S.N. 2017. Gene editing (CRISPR-Cas) technology and fisheries sector. *Canadian Journal of Biotechnology*, 1(2): 65-72.

Dykxhoorn, DM, Novina CD, Sharp PA, 2003. Killing the messenger: short RNAs that silence gene expression. *Nature reviews* 4: 457-466.

Elbashir, S. M., J. Harborth, W. Lendeckel, A. Yalcin, K. Weber and T. Tuschl. 2001. Duplexes of 21-nucleotide RNAs mediate RNA interference in cultured mammalian cells. *Nature.* 411,6836: 494-8.

Erickson HS, Poulos BT, Tang KFJ, Bradley-Dunlop D and Lightner DV, 2005. Taura syndrome virus from Belize represents a unique variant. *Dis Aquat Org*.64: 91-98.

Escobedo-Bonilla CM, 2011. Application of RNA Interference (RNAi) against Viral Infections in Shrimp: A Review. *J Antivir Antiretrovir.* S9-001.

Flegel, T.W., 1997. Major viral diseases of the black tiger prawn (*Penaeus monodon*). *World J. Microbiol. Biotechnol.*13: 433–442

Flegel TW, 2006. Disease testing and treatment. In: Boyd, C.E., Jory, D., Chamberlain, G.W. (Eds.), Operating procedures for shrimp farming. Global shrimp OP survey results and recommendations. Global Aquaculture Alliance, St. Louis. 98-103.

Flegel TW, Lightner DV, Lo CF and Owens L, 2008. Shrimp disease control: past, present and future. In: Bondad-Reantaso, M.G., Mohan, C.V., Crumlish, M., Subasinghe, R.P. (Eds.), *Diseases in Asian Aquaculture* VI. 355-378 Fish Health Section, Asian Fisheries Society, Manila, Philippines.

Gatesoupe FJ, 1999. The use of probiotics in aquaculture *Aquaculture* 180: 147-165.

Garriques D and Arevalo G, 1995. An evaluation of the production and use of live bacterial isolate to manipulate the microflora in the commercial production of *Litopenaeus vannamei* postlarvae in Ecuador In *Swimming through the troubled waters: Proceedings of the Special Session on Shrimp farming, Aquaculture'95* : 53-59 Ed Browdy CL and Hopkins JS World Aquaculture Society Baton Rouge La.

Giladi, H., Ketzinel-Gilad, M., Rivkin, L., Felig, Y., Nussbaum, O., Galun, E., 2003. Small interfering RNA inhibits hepatitis B virus replication in mice. *Mol. Ther.* 8, 769–776.

Gitlin, L., Karelsky, S., Andino, R., 2002. Short interfering RNA confers intra-cellular antiviral immunity in human cells. *Nature* 418, 430–434.

Guan Y, Yu Z, Li C, 2003. The effects of temperature on white spot syndrome infections in *Marsupenaeus japonicus*. *Journal of invertebrate pathology* 83:257-260.

Gueguen, Y., Garnier, J., Robert, L., De Lorgeril, J., Janech, M., Bachère, E., Lefranc, M.-P., Mougenot, I., Gross, P.S., Warr, G.W., Cuthbertson, B., Barracco, M.A., Bulet, P., Aumelas, A., Yang, Y., Bo, D., Xiang, J., Tassanakajon, A. and Piquemal, D. 2006. PenBase, the shrimp antimicrobial peptide penaeidin database: Sequence-based classification and recommended nomenclature. *Developmental and comparative immunology*. 30:283-288.

Hardman, A.M., Stewart, G.S. and Williams, P. 1998. Quorum sensing and the cell-cell communication dependent regulation of gene expression in pathogenic and non-pathogenic bacteria. *Antonie Van Leeu* 74:199-210.

Hammond, S.M., Caudy, A.A., Hannon, G.J., 2001. Post-transcriptional gene silencing by double-stranded RNA. *Nat. Rev. Genet.* 2, 110–119.

Horvath P and Barrangou R, 2010. CRISPR/Cas, the immune system of bacteria and archaea. *Science*. 327:167–170.

He, N., Liu, H. and Xu, X. 2004. Identification of genes involved in the response of haemocytes of *Penaeus japonicus* by suppression subtractive hybridization (SSH) following microbial challenge. *Fish Shellfish Immunol.*17:121-12.

Huang TZ, Zhang XB, 2012. Characterization of two members of the dicer family in shrimp *Marsupenaeus japonicus*. Unpublished Source: GenBank Accession number: JQ349041

Hwang WY, Fu Y, Reyon D, Maeder ML, Tsai SQ, Sander JD, Peterson RT, Yeh JR, Joung JK, 2013. Efficient genome editing in zebrafish using a CRISPR-Cas system. *Nature Biotechnology*. 31: 227–229.

Inouye, K., Miwa, S., Oseko, N., Nakono, H., Kimura, T., Momoyama, K., Hiraoka, M., 1994. Mass mortalities of cultured kuruma shrimp, *Penaeus japonicus*, in Japan in 1993: electron microscopic evidence of the causative virus. *Fish Pathol.* 29, 149–158.

Inouye, K., Yamano, K., Ikeda, N., Kimura, T., Nakano, H., Momoyama, K., Kobayashi, J., Miyajima, S., 1996. The penaeid rod shaped DNA virus (PRDV), which causes penaeid acute viremia (PAV). *Fish Pathol.* 31, 39–45

Itami T, Takahashi Y, Tsuchiharo E. and Lgusa H, 1994. Enhancement of disease resistance of kuruma prawn *Penaeus japonicus* and increase in phagocytic activity of prawn haemocytes ctiv oral administration of ? 1-3 glucan (Schizophyllan) In *Proceedings of The Third Asian Fisheries Forum* pp 375-

378 Eds Chou LM, Munro AD, Lam TJ, Chen TW, Cheong LKK, Ding JK, Hooi KK, Khoo HW, Phang VPE, Shim KF and Tan CH Asian Fisheries Society. Manila.

Itami T, Kubono K, Asano M, Tokushige K, Takeno N,.Nishimura H, Kendo M and Takahashi Y, 1998a. Enhanced disease resistances of Kuruma shrimp *Penaeus japonicus* after oral administration of peptiodoglycan derived from *Bifidobacterium thermophilum Aquaculture,* 164: 277-288.

Itami T, Maeda M. Suzuki N, Tokushige K, Nagagawa HO, Kondo M., Kosornchandra J, Hirono I, Aoki T, Kusuda R. and Takahashi T, 1998b. Possible prevention of white spot syndrome (WSS) in kuruma shrimp, *Penaeus japonicus* in Japan. In *Advances in shrimp biotechnology* pp 291-295 Ed Flegel TW National Centre for Genetic Engineering and Biotechnology. Bangkok.

Jinek, M., Krzysztof Chylinski, Ines Fonfara, Michael Hauer, Jennifer A. Doudna, Emmanuelle Charpentier, 2012. A Programmable Dual-RNA–Guided DNA Endonuclease in Adaptive Bacterial Immunity,. *Science,* 337:816–821.

Kang, C.-J., Wang, J.-X., Zhao, X.-F., Yang, X.-M., Shao, H.-L. and Xiang, J.-H. 2004. Molecular cloning and expression analysis of Ch-penaeidin, an antimicrobial peptide from Chinese shrimp, Fenneropenaeus chinensis. *Fish Shellfish Immunol.*16:513-52.

Karunasagar I and Karunasagar I, 1999. Diagnosis, treatment and prevention of microbial diseases of fish and shellfish. *Current Science,* 76 :387-399.

Karunasagar I, Vinod MG, Bob Kennedy MD, Atnur V, Deepanjali A, Umesha KR and Karunasagar I, 2005. Biocontrol of bacterial pathogens in aquaculture with emphasis on phage therapy. In *Diseases in Asian Aquaculture V*: 535-542 Eds Walker P, Lester R and Bondad-Reantaso MG, Fish Health Section, Asian Fisheries Society, Manila

Karunasagar I, Shivu MM, Girisha SK, Krohne G and Karunasagar I, 2007. Biocontrol of pathogens in shrimp hatcheries using bacteriophages *Aquaculture* 268: 288-292.

Karunasagar I, Iddya Karunasagar and Victoria Alday-Sanz, 2010 The Shrimp Book, by Victoria Alday-Sanz (Editor), Publ: Nottingham University Press, Nottingham, UK, 2010.

Kajita Y, Sakai M., Atsuta S. and Kobayashi M, 1990. The immunomodulatory effects of levamisole on rainbow trout. *Oncorhynchus mykiss. Fish Pathology.* 25 :93-98.

Khanobdee K, Soowannayan C, Flegel TW, Ubol S and Withyachumnarnkul B, 2002. Evidence for apoptosis correlated with mortality in the giant black tiger shrimp *Penaeus monodo*n infected with yellow head virus. *Dis Aquat Org.* 48: 79-90.

Krishnan P, Gireesh BP, Saravanan S, Rajendran KV, Chaud-hari A, 2009. DNA constructs expressing long-hairpin RNA (lhRNA) protect *Penaeus monodon* against white spot syndrome virus. *Vaccine* 27:3849 –3855.

Kumar RS, Ishaq Ahamed VP, Sarathi M, Nazeer Basha A, Sahul Hameed AS, 2008. Immunological responses of *Penaeus monodon* to DNA vaccine and its efficacy to protect shrimp against white spot syndrome virus (WSSV). *Fish & shellfish immunology* 24: 467-478.

Labreuche Y, O´Leary NA, de la Vega E, veloso A, Gross PS, Chapman RW, Browdy CL, Warr GW, 2009. Lack of evidence for *Litopenaeus vannamei* Toll receptor (lToll) involvement in activation of sequence-independent antiviral immunity in shrimp. *Developmental and comparative immunology* 33: 806-810.

Labreuche Y, Veloso A, de la Vega E, Gross PS, Chapman RW, Browdy CL, Warr GW, 2010. Non-specific activation of antiviral immunity and induction of RNA interference may engage the same pathway in the Pacific white leg shrimp *Litopenaeus vannamei. Developmental and Comparative Immunology,* 34: 1209-1218.

La Fauce KA and Owens L, 2009. RNA interference reduces PmergDNV expression and replication in an in vivo cricket model. *J Invertebr Pathol* In Press, Accepted Manuscript.

Lee SY, Soderhall K, 2002: Early events in crustacean innate immunity. *Fish and Shell fish Immunology* 12, 421–437.

Li F, Ding SW, 2006. Virus counter defence: diverse strategies for evading the RNA-silencing immunity. *Annual Reviews in Microbiology* 60: 503-531.

Li X, Yang L, Jiang S, Fu M, Huang J, Jiang S, 2013. Identification and expression analysis of Dicer2 in black tiger shrimp (*Penaeus monodon*) responses to immune challenges. *Fish Shellfish Immunol* 35:1–8.

Li S, Zhang X, Sun Z, Li F, Xiang J, 2013. Transcriptome analysis on Chinese shrimp *Fenneropenaeus chinensis* during WSSV acute infection. *PLoS On.e* 8,3:58627.

Lightner DV, Redman RM, 1998. Shrimp diseases and current diagnostic methods. *Aquaculture* 164: 201-220.

Liu B, Yu Z, Song X, Guan Y, Jian X and He J, 2006. The effect of acute salinity change on white spot syndrome (WSS) outbreaks in Fenneropenaeus chinensis. *Aquaculture* 253 163-170.

Liu YC, Li FH, Dong B, Wang B, Luan W, Zhang XJ, Zhang LS and Xiang JH, 2007. Molecular cloning, characterization and expression analysis of a putative C-type lectin (Fclectin) gene in Chinese shrimp Fenneropenaeus chinensis. Mol Immunol 44 598-607.

Liu CH, Chiu CS, Ho PL, Wang WW, 2009. Improvement of growth performance of white shrimp *Litopenaeus vannamei* by a protease producing probiotic, *Bacillus subtilis* E20, from natto *Journal of Applied Microbiology* 107: 1031-1041.

Loessner M J, 2005. Bacteriophage endolysins- current state of research and applications *Current Opinions in Microbiology* 8: 480-487.

Lu Y and Sun PS, 2005. Viral resistance in shrimp that express an antisense Taura syndrome virus coat protein gene. *Antiviral Research,* 67 :141-146.

McManus, M.T., Sharp, P.A., 2002. Gene silencing in mammals by small interfering RNAs. *Nat. Rev. Genet.* 3, 737–747.

Makarova KS, Haft DH, Barrangou R, Brouns SJ, Charpentier E, Horvath P, Moineau S, Mojica FJ, Wolf YI, Yakunin AF, et al. Evolution and classification of the CRISPR-Cas systems. *Nat. Rev. Microbiol.* 2011, 9:467–477

Maningas MB, Kondo H, Hirono I, Saito-Taki T and Aoki T, 2008. Essential function of transglutaminase and clotting protein in shrimp immunity. *Mol Immunol* 45 1269-1275.

Moriarty D, 1998. Control of luminous *Vibrio* species in aquaculture ponds *Aquaculture* 164: 351-358.

Munoz, M., Vandenbulcke, F., Gueguen, Y. and Bachere, E. 2003. Expression of penaeidin antimicrobial peptides in early larval stages of the shrimp *Penaeus vannamei. Dev. Comp. Immunol.* 27:283-289.

Nakai T and Park SC, 2002. Bacteriophage therapy of infectious diseases in aquaculture *Research in Microbiology* 153: 13-18

Nakai T, Sugimoto R, Park KH, Matsuoaka S, Mori K, Nishioka T, and Maruyama K, 1999. Protective effects of bacteriophage on experimental *Lactococcus garviae* infection in yellow tail. *Dis Aqu Org,* 37: 33-41.

Nishi J, Mekata T, Okugawa S, Yoshimine M, Sakai M, Itami T (2014) Identification of dicer-2 from *Marsupenaeus.* Unpublished (source: GenBank accession number: AB618794).

Park SC, Shimamura I, Fukunaga M, Mori K, Nakai T, 2000. Isolation of bacteriophages specific to a fish pathogen *Pseudomonas plecoglossicida* as a candidate for disease control Applied and Environmental Microbiology 66: 1416-1422.

Pan, D., He, N., Yang, Z., Liu, H. and Xu, X. 2005. Differential gene expression profile in hepatopancreas of WSSV-resistant shrimp (*Penaeus japonicus*) by suppression subtractive hybridization. *Dev. Comp. Immunol.* 29:103-112.

Pan, J., Kurosky, A., Xu, B., Chopra, A.K., Coppenhaver, D.H., Singh, I.P. and Baron, S. 2000. Broad antiviral activity in tissues of crustaceans. *Antiviral Res.* 48:39-47.

Poulos BT, Tang KFJ, Pantoja CR, Bonami JR and Lightner DV, 2006. Purification and characterization of infectious myonecrosis virus of penaeid shrimp. *J Gen Virol* 87 987-996.

Phromjai J, Boonsaeng V, Withyachumnarnkul B and Flegel TW, 2002. Detection of hepatopancreatic parvovirus in Thai shrimp *Penaeus monodon* by in situ hybridization, dot blot hybridization and PCR amplification. *Dis Aquat Org.* 51: 227-232.

Phongdara, A., Wanna, W. and Chotigeat, W. 2006. Molecular cloning and expression of caspase from white shrimp Penaeus merguiensis. *Aquaculture* 252:114-120.

Preston, N. P., V. J. Baule, R. Leopold, J. Henderling, P. W. Atkinson, and S. Whyard, 2000. "Delivery of DNA to early embryos of the Kuruma prawn, *Penaeus japonicus.*" *Aquaculture* 181, 3: 225-234.

Randall, G., Grakoui, A., Rice, C.M., 2003. Clearance of replicating hepatitis C virus replicon RNAs in cell culture by small interfering RNAs. Proc. Natl. Acad. Sci. USA 100, 235–240.

Rath D, Lina Amlinger, Archana Rath, Magnus Lundgren, 2015. The CRISPR-Cas immune system: Biology, mechanisms and applications. *Biochimie* 117:119-128.

Ravi M, Nazeer Basha A, Sarathi M, Rosa Idalia HH, Sri Widada J, Bonami JR, Sahul Hameed AS, 2009. Studies on the occurrence of white tail disease (WTD) caused by MrNV and XSV in hatchery-reared post-larvae of *Penaeus indicus* and *P. monodon. Aquaculture* 292 117-120.

Rengpipat S, Phianphak W, Piyatiratitivorakul S, Menasveta P, 1998. Effect of probiotic bacterium on black tiger shrimp survival and growth *Aquaculture* 167: 301-313.

Rengpipat S, Rukpratanporn S, Piyatiratitivorakul S, Menasveta P, 2000. Immunity enhancement in black tiger shrimp (*Penaeus monodon*) by a probiont bacterium (*Bacillus* S11). *Aquaculture* 191: 271-288.

Robalino J, Browdy CL, Prior S, Metz A, Parnell P, Gross P, Warr G, 2004. Induction of antiviral immunity of double-stranded RNA in a marine invertebrate. *Journal of Virology* 78: 10442-10448.

Robalino J, Bartlett T, Shepard EF, Prior S, Jaramillo G, Scura E, Chapman RW, Gross PS, Browdy CL, Warr GW, 2005. Double stranded RNA induces sequence-specific antiviral silencing in addition to nonspecific immunity in marine shrimp: convergence of RNA interference and innate immunity in the invertebrate antiviral response. *Journal of Virology* 79:13561-13571.

Robinson HL, Pertmer TM, 2000. DNA vaccines for viral infections: basic studies and applications. Adv Virus Res. 2000; 55:1-74.

Rout N, Kumar S, Jaganmohan S, Murugan V, 2007. DNA vaccines encoding viral envelope proteins confer protective immunity against WSSV in black tiger shrimp. *Vaccine* 25: 2778-2786.

Roux MM, Pain A, Klimpel KR and Dhar AK, 2002. The lipopolysaccharide and 1,3 glucan binding protein gene is upregulated in white spot virus infected shrimp (*Penaeus styloristris*) *Journal of Virology* 76: 7140-7149.

Sakai M, Otubo T., Atsuta S. and Kobayashi M, 1993. Enhancement of resistance to bacterial infection in rainbow trout, *Oncorhynchus mykiss* (Walbaum) by oral administration of bovine lactoferrin. *Journal of Fish. Diseases.* 16 :239-247.

Sahtout AH, Hassan MD and Shariff M, 2001. DNA fragmentation, an indicator of apoptosis, in cultured black tiger shrimp *Penaeus monodon* infected with white spot syndrome virus (WSSV). *Dis Aquat Org* 44 155-159.

Schröder M, Bowie AG, 2005. TLR3 in antiviral immunity: key player or by stander. *TRENDS in Immunology* 26: 462-468.

Seruggia, D. and L. Montoliu, 2014. The new CRISPR–Cas system: RNA-guided genome engineering to efficiently produce any desired genetic alteration in animals. Transgenic research. 23,5: 707-716.

Skurnik M and Strauch E, 2006. Phage therapy: facts and fiction *International Journal of Medical Microbiology* 296: 5-14.

Smith VJ and Chisholm JRS, 1992. Non-cellular immunity in crustaceans. *Fish and Shellfish Immunology.* 2: 1-31.

Shivu MM, Rajeeva BC, Girisha SK, Karunasagar I, Krohne G and Karunasagar I, 2007. Molecular characterization of *Vibrio harveyi* bacteriophage isolated from aquaculture environments along the coast of India *Environmental Microbiology* 9: 322-331.

Shuey D J, McCallus DE, Giordano T, 2002. RNAi: gene- silencing in therapeutic intervention. *Drug DiscovToday* 7:1040–1046.

Sittidilokratna N, Hodgson RA, Cowley JA, Jitrapakdee S, Boonsaeng V, Panyim S and Walker PJ, 2002. Complete ORF1b-gene sequence indicates yellow head virus is an invertebrate nidovirus. *Dis Aquat Org*.50: 87-93.

Sittidilokratna N, Dangtip S, Cowley JA and Walker PJ, 2008. RNA transcription analysis and completion of the genome sequence of yellow head nidovirus. *Virus Research* 136 157-165.

Söderhäll K, 1999. Invertebrate immunity. *Dev Comp Immunol.* 23:263-442.

Soderhall, K., Cerenius, L. and Johansson, M.W. 1994. The prophenoloxidase activating system and its role in invertebrate defence. Ann. N. Y. *Acad. Sci*.712:155-161

Sritunyalucksana K, Sithisarin, P, Whitayachumnarnkul B and Flegel TW, 1999. Activation of prophenol oxidase, agglutinin and antibacterial proteins in black tiger prawn, *Penaeus monodon* by immunostimulants *Fish Shellfish Immunology*.9: 21-30.

Stet, R.J. and Arts, J.A. 2005. Immune functions in crustaceans: lessons from flies. *Devel. Biologicals* 121:33-43.

Su J, Oanh DTH, Lyons RE, Leeton L, van Hulten MCW, Tan S-H, Song L, Rajendran KV and Walker PJ, 2008. A key gene of the RNA interference pathway in the black tiger shrimp, *Penaeus monodon*: Identification and functional characterisation of Dicer-1. *Fish shellfish Immunol.* 24 :223-233.

Supungul, P., Klinbunga, S., Pichyangkura, R., Hirono, I., Aoki, T. and Tassanakajon, A. 2004 Antimicrobial peptides discovered in the black tiger shrimp *Penaeus monodon* using the ESTapproach. *Dis. Aquat. Org*.61:123-135.

Sung HH, Kou GH and Song YL, 1994. Vibriosis resistance induced by glucans treatment in Tiger shrimp (*Penaeus monodon) Fish Pathology* 29 11-17.

Sung H.H, Yang YL and Song YL, 1996. Enhancement of microbial activity in the tiger shrimp *Penaeus monodon* via immunostimulation *Journal of Crustacean Biology* 16 278-28.

Takahashi Y, Uehara K, Watanabe R, Okumura T, Yamashita Omura H, Yomo T, Kawano T, Kanemitsu A, Narasaka H, Suzuki N. and Itami T, 1998. Efficacy of oral administration of fucoidan, a sulphated polysaccharide in controlling white spot syndrome in kuruma shrimp in Japan. In: *Advances in Shrimp Biotechnology* pp 171-173 Ed Flegel TW National Centre for Genetic Engineering and Biotechnology Bangkok.

Takahashi Y, Kondo M, Itami T, Honda T, Inagawa H, Nishizawa T, Soma GI, Yokomizo Y, 2000. Enhancement of disease resistance against penaeid acute viraemia and induction of virus inactivating activity in haemolymph of kuruma shrimp, *Penaeus japonicus*, by oral administration of *Pantoea agglomerans* lipopolysaccharide (LPS). *Fish & shellfish immunology* 10: 555-558.

Talbot JC and Amacher SL, 2014. A streamlined CRISPR pipeline to reliably generate zebrafish frameshifting alleles. *Zebrafish*.11:583–585.

Tang KFJ, Pantoja CR, Redman RM and Lightner DV, 2007. Development of in situ hybridization and RT-PCR assay for the detection of a nodavirus (PvNV) that causes muscle necrosis in *Penaeus vannamei*. *Dis Aquat Org* ,75 :183-190.

Thakur PC, Corsin F, Turnbull JF, Shankar KM, Hao NV, Padiyar PA, Madhusudhan M, Morgan KL and Mohan CV, 2002. Estimation of prevalence of white spot syndrome virus (WSSV) by polymerase chain reaction in *Penaeus monodon* postlarvae at time of stocking in shrimp farms of Karnataka, India: a population-based study. *Dis Aquat Org* 49 235-243.

Tseng FS, Tsai HJ, Liao IC, Song YL, 2000. Introducing foreign DNA into tiger shrimp *Penaeus monodon* by electroporation. *Theriogenology* 54: 1421-1432.

Tseng DY, Ho PL, Huang, SY, Cheng, SC, Shiu YL, Chiu CS, Liu CH, 2009. Enhancement of immunity and disease resistance in the white shrimp *Litopenaeus vannamei*, by the probiotic *Bacillus subtlis* E20 *Fish and shellfish Immunology*, 26: 339-344.

Tirasophon W, Roshorm Y, Panyim S, 2005. Silencing of yellow head virus replication in penaeid shrimp cells by dsRNA. *Biochemical and biophysical research communications* 334: 102-107.

Tonganunt, M., Phongdara, A., Chotigeat, W. and Fujise, K. 2005. Identification and characterization of syntenin binding protein in the black tiger shrimp *Penaeus monodon. J. Biotechnol.*120:135-145.

Venegas CA, Nonaka L, Mushiake K, Nishizawa T, Muroga K, 2000. Quasiimmune response of *Penaeus japonicus* to penaeid rod-shaped DNA virus (PRDV). *Dis aquat org* 42: 83-89.

Verschuere L, Rombaut G, Sorgeloos P, Verstraete W, 2000. Probiotic bacteria as biological control agents in aquaculture *Microbiology and Molecular Biology Reviews* 64: 655-671.

Vinod MG, Shivu, MM, Umesha KR, Rajeeva BC, Krohne G, Karunasagar I and Karunasagar I, 2006. Isolation of Vibrio harveyi bacteriophage with a potential for biocontrol of luminous vibriosis in hatchery environments *Aquaculture* 255: 117-124.

Voorhoeve PM and Agami R, 2003. Knockdown stands up. *Trends Biotechnol* 21:2– 4.

Westenberg M, Heinhuis B, Zuidema D, Vlak JM, 2005. siRNA injection induces sequence-independent protection in *Penaeus monodon* against white spot syndrome virus. *Virus research* 114: 133-139.

Wongprasert K, Khanobdee K, Glunukarn S, Meeratana P and Withyachumnarnkul B, 2003. Time-course and levels of apoptosis in various tissues of black tiger shrimp Penaeus monodon infected with white spot syndrome virus. *Dis Aquat Org* 55 3-10.

Wilkins, C., Dishongh, R., Moore, S.C., Whitt, M.A., Chow, M., Machaca, K, 2005. RNA interference is an antiviral defence mechanism in *Caenorhabditis elegans. Nature* 436:1044–1047.

Wu, W., Wang, L., Zhang, X, 2005. Identification of white spot syndrome virus (WSSV) envelope proteins involved in shrimp infection. *Virology* 332: 578–583.

Wyban J, 2007. Domestication of pacific white shrimp revolutionizes aquaculture. *Global Aquaculture* Advocate, 42-44.

Xu J, Han F, Zhang X, 2007. Silencing shrimp white spot syndrome virus (WSSV) genes by siRNA. *Antiviral research* 73: 126-131.

Yao X, Wang L, Song L, Zhang H, Dong C, Zhang Y, Qiu L, Shi Y, Zhao J, Bi Y, 2010. A Dicer-1 gene from white shrimp *Litopenaeus vannamei*: expression pattern in the processes of immune response and larval development. *Fish Shellfish Immunol* 29:565–570.

Yang LS, Yin ZX, Liao JX, Huang XD, Guo CJ, Weng SP, Chan SM, Yu XQ, He JG, 2007. A Toll receptor in shrimp. *Molecular immunology* 44: 1999-2008.

Yang C, Zhang J, Li F, Ma H, Zhang Q, Jose Priya TA, Zhang X, Xiang J, 2008. A Toll receptor from Chinese shrimp *Fenneropenaeus chinensis* is responsive to *Vibrio anguillarum* infection. Fish & shellfish immunology 24: 564-574.

Yi, G., Wang, Z., Qi, Y., Yao, L., Qian, J., Hu, L., 2004. VP28 of shrimp white spot syndrome virus is involved in the attachment and penetration into shrimp cells. *J. Biochem. Mol. Biol.* 37, 726–734.

Yodmuang S, Tirasophon W, Roshorm Y, Chinnirunvong W, Panyim S, 2006. YHV-protease dsRNA inhibits YHV replication in *Penaeus monodon* and prevents mortality. *Biochemical and biophysical research communications* 341: 351-356.

Yoshida T, Sakai MT, Khlil SM, Araki S, Saitoh R, Ineno T and Inglis V, 1993. Immunodulatory effects of the fermented products of chicken egg. EF203, on rainbow trout, *Oncorhynchus mykiss. Aquaculture.* 109: 207-214.

Zamore, P.D., Tuschl, T., Sharp, P.A., Bartel, D.P., 2000. RNAi: double-stranded RNA directs the ATP-dependent cleavage of mRNA at 21 to 23 nucleotide intervals. Cell 101, 25–33.

SUBJECT INDEX

A

B

C

D

E

I

L

M

N

O

P

R

S

T

U

V

W

X

Y

Z